METHODS
OF
ALGEBRAIC GEOMETRY

METHODS
OF
ALGEBRAIC GEOMETRY

by

W. V. D. HODGE, Sc.D., F.R.S.

Formerly Lowndean Professor of Astronomy and Geometry, and Fellow of Pembroke College, Cambridge

and

D. PEDOE, Ph.D.

Emeritus Professor of Mathematics University of Minnesota

VOLUME III

BOOK V: BIRATIONAL GEOMETRY

CAMBRIDGE
UNIVERSITY PRESS

Published by the Press Syndicate of the University of Cambridge
The Pitt Building, Trumpington Street, Cambridge CB2 1RP
40 West 20th Street, New York, NY 10011–4211, USA
10 Stamford Road, Oakleigh, Melbourne 3166, Australia

First published 1954
Reissued in the Cambridge Mathematical Library 1994

ISBN 0 521 46775 6 paperback

Transferred to digital printing 2003

CONTENTS

BOOK V

BIRATIONAL GEOMETRY

CHAPTER XV: IDEAL THEORY OF COMMUTATIVE RINGS

CHAPTER XVI: THE ARITHMETIC THEORY OF VARIETIES

CHAPTER XVII: VALUATION THEORY

CHAPTER XVIII: BIRATIONAL TRANSFORMATIONS

PREFACE

THE PURPOSE of this volume is to provide an account of the modern algebraic methods available for the investigation of the birational geometry of algebraic varieties. An account of these methods has already been published by Professor André Weil in his *Foundations of Algebraic Geometry* (New York, 1946), and when Professor Zariski's Colloquium Lectures, delivered in 1947 to the American Mathematical Society, are published, another full account of this branch of geometry will be available. The excuse for a third work dealing with this subject is that the present volume is designed to appeal to a different class of reader. It is written to meet the needs of those geometers trained in the classical methods of algebraic geometry who are anxious to acquire the new and powerful tools provided by modern algebra, and who also want to see what they mean in terms of ideas familiar to them. Thus in this volume we are primarily concerned with methods, and not with the statement of original results or with a unified theory of varieties.

Such a purpose in writing this volume has had several effects on the plan of the work. In the first place, we have confined our attention to varieties defined over a ground field without characteristic. This is partly because the geometrical significance of the algebraic methods and results is more easily comprehended by a classical geometer in this case; also, though others have shown that modern algebraic methods have enabled us to make great strides in the theory of algebraic varieties over a field of finite characteristic, many of the theorems which the classical geometer regards as fundamental have only been proved, as yet, in the restricted case.

Secondly, a chapter and a half have had to be devoted entirely to algebraic theory. The ideal theory of rings and the theory of valuations are central topics in modern algebra, and it might have seemed reasonable to refer the reader to other text-books for this. But our purpose is to bring together the materials the geometer needs, and for this reason we felt justified in including an account of the algebraic theories. The algebraic parts of the book have been written with the needs of geometry in mind, and this has often influenced the emphasis given to different concepts. Nevertheless, at the earliest stages some attempt has been made to treat the subject generally, with a view to giving the reader who is fresh to the subject some slight idea of its general significance.

Thirdly, in order to provide the reader with as many tools as possible, we have not restricted ourselves to a single method, and have, from time to time, used devices which may appear inelegant to a master of the subject. For instance, much of the earlier work on birational transformations might have been done without reference to valuation theory, as Professor Weil has shown, and it might have been more aesthetically pleasing to base the theory of birational transformations on the theory of specialisations. But it is extremely unlikely that a complete theory of birational transformations can be built up without valuation theory, and it seems necessary that a geometer should have this tool at his disposal. In order to illustrate the methods described we have therefore used valuation theory where it is not strictly necessary. Sometimes, too, a result has been proved twice, to illustrate different methods, or to save the reader from having to refer back too often.

Fourthly, it has not been our object to push the subject as far as it can go. Many important results in the algebraic theory of varieties have been omitted altogether, even from within the range to which we have confined ourselves. It is our hope that the reader may find this volume sufficient to enable him to familiarise himself with the modern methods described, and that on completing the book he will feel more ready to read works which go further in the subject. An apparent exception to our rule should be mentioned. While most of the volume is concerned with basic principles, the second part of Chapter XVIII deals with the application of the methods to certain fundamental problems of considerable difficulty, namely, the Local Uniformisation Theorem and the reduction of singularities of an algebraic surface. The reason is that to a classical geometer these problems concern the foundations of his subject, and as the arithmetic theory of varieties described has reached a state of definiteness with regard to these problems, and has achieved most striking results, it seems desirable to give an account of these topics as an illustration of the methods previously described.

It may seem strange that a volume dealing with the birational theory of varieties makes no reference to the work of the great Italian school of geometers. The reason for this is that we are concerned only with certain methods, and these have been developed elsewhere, while the properties of varieties, the discovery of which we owe to such mathematicians as Segre, Castelnuovo, Enriques and Severi, are beyond the scope of this work. When the methods described in this volume are more fully employed in the theory of the birational invariants of algebraic varieties, the work of these geometers will play an outstanding role in any account of the subject.

We should like to take this opportunity of removing a misunderstanding which has apparently arisen in the minds of some of our friends concerning the Bibliographical Notes given at the end of each of the three volumes. The purpose of the Notes has been merely to acknowledge the principal sources on which we have drawn for the methods and results described in the volumes; they do not even record the full extent of our obligations, though it is hoped that they will enable a reader to trace to their source the various ideas developed. They are in no sense intended to be an account of the literature on the foundations of algebraic geometry, or even of the literature dealing with the methods described. It has been suggested to us that we ought to have given full references to the various accounts of the foundations of algebraic geometry, and that we have been discourteous in not referring to all the geometers who have written on this subject. It is hardly necessary to say that no discourtesy is intended. At the outset of our work we indeed gave careful thought to the possibility of full documentation, but soon came to the conclusion that it was impracticable. So much has been written on the subjects treated in these volumes that it would be a major work of research to analyse and record the contributions of every mathematician to the topics covered, and indeed this could only be done in a history of a major portion of mathematics. Any intermediate course would, on the one hand, have been open to the charge of important omissions, and on the other, would have delayed the completion of our task by many months, and would, at the same time, have added considerably to the length of a work which is already too long for its purpose.

Professor T. A. A. Broadbent has again rendered us valuable assistance with the manuscript of this volume, and with the proofs. We wish to put on record our gratitude to him for the immense amount of time and trouble he has expended on our account during the years in which we have been engaged in writing this work. We are also in the debt of the Cambridge University Press for much patience and ingenuity shown in meeting our requirements.

Note. This volume contains Chapters XV–XVIII. Vol. I contains Chapters I–IX, Vol. II, Chapters X–XIV. The reference [II, §4, Th. II] is to Theorem II in §4 of Chapter II. If a reference is to the same chapter or section, the corresponding numeral or numerals are omitted.

W. V. D. H.
D. P.

CAMBRIDGE
May 1953

BOOK V

BIRATIONAL GEOMETRY

CHAPTER XV

IDEAL THEORY OF COMMUTATIVE RINGS

IN Volume II we were concerned mainly with the geometry of varieties in projective space, regarded as subvarieties of the space. We had, however, occasion to consider relations between different varieties of the same space, or of different spaces; for this we used the correspondence theory of Chapter XI. In particular, use was made from time to time of birational correspondences between irreducible varieties; if U and V are irreducible varieties in spaces with coordinate systems $(x_0, \ldots, x_n)$ and $(y_0, \ldots, y_m)$ respectively, they are in birational correspondence if there exists a correspondence between them whose equations include equations of the form

$$x_i f_j(y_0, \ldots, y_m) - x_j f_i(y_0, \ldots, y_m) = 0 \quad (i,j = 0, \ldots, n),$$
$$y_i g_j(x_0, \ldots, x_n) - y_j g_i(x_0, \ldots, x_n) = 0 \quad (i,j = 0, \ldots, m),$$

where not all forms $f_i(y)$ vanish on V, and not all forms $g_i(x)$ vanish on U. An important branch of the theory of algebraic varieties deals with the investigation of properties common to birationally equivalent varieties. In this theory we are concerned, not so much with the properties of individual varieties as varieties in projective space, as with properties of sets of algebraic varieties which are birationally equivalent to one another. This branch of the theory of varieties is called *birational geometry.*

The purpose of this volume is to introduce the reader to the algebraic methods which have proved most useful in birational geometry, and to establish certain basic results with which a geometer must be familiar before he embarks on a systematic study of birational geometry. For this purpose, it is necessary to develop more fully certain algebraic concepts introduced in Volume I, and to introduce new ones. In Chapter I the notion of a ring was introduced, and mention has also been made in earlier chapters of ideals in a ring, but only the most elementary properties of these have been used. It is now necessary to study commutative rings, and ideals in them, more systematically, and the present chapter is devoted to this end.

1. Ideals in a commutative ring. Let $\mathfrak{R}$ be any commutative ring. A non-empty set $\mathfrak{i}$ of elements of $\mathfrak{R}$ is said to form an *ideal* in $\mathfrak{R}$ if it has the two properties: (i) if α and β are any elements belonging to $\mathfrak{i}$, then $\alpha-\beta$ belongs to $\mathfrak{i}$, (ii) if α belongs to $\mathfrak{i}$ and ρ is any element of $\mathfrak{R}$, then $\rho\alpha$ belongs to $\mathfrak{i}$.

The ring $\mathfrak{R}$ contains a zero element; if we take this to be ρ in (ii), then $0.\alpha = 0$ is in $\mathfrak{i}$. Thus every ideal in $\mathfrak{R}$ contains the zero of $\mathfrak{R}$. On the other hand, if a set $\mathfrak{i}$ consists solely of the zero of $\mathfrak{R}$, it satisfies the conditions (i) and (ii); hence it is an ideal. We call this the *zero ideal* of $\mathfrak{R}$. Again, if $\mathfrak{i}$ includes every element of $\mathfrak{R}$, it also satisfies conditions (i) and (ii); hence it forms an ideal, which we call the *unit ideal* of $\mathfrak{R}$. Thus every commutative ring contains at least two ideals, the zero ideal and the unit ideal. These two ideals are sometimes called *improper ideals*, and any ideal distinct from the zero ideal and the unit ideal is called a *proper ideal*.

If $\mathfrak{R}$ has unity e and $\mathfrak{i}$ contains e, $\mathfrak{i}$ contains $\rho e = \rho$, where ρ is any element of $\mathfrak{R}$. Hence $\mathfrak{i}$ is the unit ideal if it contains e.

It is convenient to determine at the outset which commutative rings possess no ideals other than the two improper ideals. We consider two cases.

Case I. Suppose that $\mathfrak{R}$ contains two elements ρ, ω such that $\rho\omega \neq 0$. (This is certainly the case when the ring has unity.) For such an element ω we consider the set $\mathfrak{i}$ of elements $\sigma\omega$, where σ can be any element in $\mathfrak{R}$. It is clear that $\mathfrak{i}$ satisfies the conditions (i) and (ii), and hence it is an ideal, and since $\mathfrak{i}$ contains $\rho\omega \neq 0$ it is not the zero ideal. If $\mathfrak{R}$ has only improper ideals, it follows that $\mathfrak{i} = \mathfrak{R}$, and hence if ν is any element of $\mathfrak{R}$, there exists an element x in $\mathfrak{R}$ such that

$$x\omega = \nu.$$

Let e be a solution of this equation when $\nu = \omega$, and let x be a solution for an arbitrarily chosen element ν of $\mathfrak{R}$. Then

$$e\nu = ex\omega = x(e\omega) = x\omega = \nu.$$

It follows that $\mathfrak{R}$ has unity, namely, e. Now let ν be any non-zero element of $\mathfrak{R}$. The set of elements of the form $\sigma\nu$, where σ is any element of $\mathfrak{R}$, forms an ideal $\mathfrak{j}$ which contains the element $e\nu = \nu$, and $\mathfrak{j}$ is therefore the unit ideal. Then there exists an element ν' in $\mathfrak{R}$ such that $\nu'\nu = e$. Hence every non-zero element in $\mathfrak{R}$ has an inverse. $\mathfrak{R}$ is therefore a field.

Conversely, suppose that $\mathfrak{R}$ is a field, and let $\mathfrak{i}$ be any ideal in $\mathfrak{R}$ which is not the zero ideal. $\mathfrak{i}$ contains a non-zero element α. Let β be any element of $\mathfrak{R}$. By property (ii), $\mathfrak{i}$ contains $\beta\alpha^{-1}.\alpha = \beta$. Hence $\mathfrak{i}$ contains every element of $\mathfrak{R}$ and is therefore the unit ideal.

Case II. Suppose that $\mathfrak{R}$ is a commutative ring, not consisting solely of the zero element, such that, if α, β are any two elements of $\mathfrak{R}$, $\alpha\beta = 0$, and let ω be any non-zero element of $\mathfrak{R}$. The set of elements $0, \pm\omega, \pm 2\omega, \pm 3\omega, \ldots$ clearly forms an ideal $\mathfrak{i}$ in $\mathfrak{R}$, different from the zero ideal. Hence, if the only ideals in $\mathfrak{R}$ are improper, $\mathfrak{i} = \mathfrak{R}$. The elements $0, \pm\omega, \pm 2\omega, \pm 3\omega, \ldots$ cannot all be distinct, for, if they were, the elements $0, \pm 2\omega, \pm 4\omega, \ldots$ would constitute a proper ideal in $\mathfrak{R}$, and we are assuming that every ideal in $\mathfrak{R}$ is improper. Let m be the smallest positive integer such that $m\omega = 0$. Then $0, \omega, 2\omega, \ldots, (m-1)\omega$ are the elements of $\mathfrak{R}$. If m is composite, say $m = ab$ where $a > 1, b > 1$, the set $0, a\omega, 2a\omega, \ldots, (b-1)a\omega$ would, clearly, constitute a proper ideal in $\mathfrak{R}$, contrary to our hypothesis. Hence m is a prime number p. From this it follows that if Ω is the 2×2 matrix over the ring of integers modulo p,

$$\Omega = \begin{pmatrix} 0 & 1 \\ 0 & 0 \end{pmatrix},$$

$\mathfrak{R}$ is isomorphic with the ring consisting of $0, \Omega, 2\Omega, \ldots, (p-1)\Omega$, and we can verify at once that if $\mathfrak{R}$ is isomorphic with this ring, the only ideals in it are improper. Hence we have

THEOREM I. *A ring $\mathfrak{R}$ which is a field contains only improper ideals. Conversely, any ring whose only ideals are improper ideals is either a field or is isomorphic with the ring of matrices of the form*

$$\begin{pmatrix} 0 & k \\ 0 & 0 \end{pmatrix}$$

over the ring of integers reduced modulo p, for some prime number p.

We now return to the general theory of ideals in a commutative ring $\mathfrak{R}$. We can construct an ideal in $\mathfrak{R}$ as follows. Let $\omega_1, \ldots, \omega_r$ be a finite set of elements in $\mathfrak{R}$, and consider the set $\mathfrak{i}$ of elements in $\mathfrak{R}$ which can be written in the form

$$\alpha_1\omega_1 + \ldots + \alpha_r\omega_r + n_1\omega_1 + \ldots + n_r\omega_r, \tag{1}$$

where $\alpha_1, \ldots, \alpha_r$ are any elements of $\mathfrak{R}$, and $n_i \omega_i$ stands for the sum of n_i elements each equal to ω_i. If $\mathfrak{R}$ has unity e, we can write the element as

$$(\alpha_1 + n_1 e)\,\omega_1 + \ldots + (\alpha_r + n_r e)\,\omega_r = \alpha_1' \omega_1 + \ldots + \alpha_r' \omega_r,$$

where $\alpha_1', \ldots, \alpha_r'$ are in $\mathfrak{R}$, so that it is unnecessary to include the terms $n_1 \omega_1, \ldots, n_r \omega_r$ in (1). It can be verified at once that the set $\mathfrak{i}$ satisfies the conditions (i) and (ii), and hence forms an ideal. The ideal is usually described as the ideal *generated by* $\omega_1, \ldots, \omega_r$, or *having the basis* $\omega_1, \ldots, \omega_r$, and is denoted by $\mathfrak{R}.(\omega_1, \ldots, \omega_r)$. From the definition of the basis, it is clear that the elements of a basis for an ideal $\mathfrak{i}$ belong to $\mathfrak{i}$. It is not, however, clear that every ideal in $\mathfrak{R}$ has a finite basis, that is that, given an ideal $\mathfrak{i}$ in $\mathfrak{R}$, there exists a finite set of elements of $\mathfrak{i}$, say, $\omega_1, \ldots, \omega_r$, such that $\mathfrak{i} = \mathfrak{R}.(\omega_1, \ldots, \omega_r)$. If the ring $\mathfrak{R}$ has the property that every ideal in $\mathfrak{R}$ possesses a finite basis, we say that *the Basis Theorem holds in* $\mathfrak{R}$. Many results can be proved for rings in which the Basis Theorem holds which are not true for more general rings. An example of a ring in which the Basis Theorem holds is provided by the ring of polynomials in r indeterminates over a commutative field [IV, § 2, Th. I]. For certain special rings, such as the ring of natural integers, or the ring of polynomials in one indeterminate over a commutative field, it can be proved that every ideal has a basis consisting of a single element. An ideal in a ring $\mathfrak{R}$ generated by a single element is called a *principal ideal*, and a ring in which every ideal is a principal ideal is called a *principal ideal ring*.

We now introduce certain notational conventions. If α is any element of the ideal $\mathfrak{i}$ we shall write $\alpha \in \mathfrak{i}$, or, more usually,

$$\alpha = 0 \ \ (\mathfrak{i}),$$

and when $\mathfrak{i} = \mathfrak{R}.(\omega_1, \ldots, \omega_r)$, we shall also write this in the form

$$\alpha = 0 \quad (\mathrm{mod}\, \omega_1, \ldots, \omega_r).$$

If $\mathfrak{j}$ is an ideal every element of which is in $\mathfrak{i}$, we shall write $\mathfrak{j} \subseteq \mathfrak{i}$, or, more usually,

$$\mathfrak{j} = 0 \ \ (\mathfrak{i}).$$

It is clear that if we have, simultaneously,

$$\mathfrak{j} = 0 \ \ (\mathfrak{i}) \quad \text{and} \quad \mathfrak{i} = 0 \ \ (\mathfrak{j}),$$

then $\mathfrak{i} = \mathfrak{j}$.

We now define the elementary operations which can be performed on ideals.

I. Let $\mathfrak{i}$, $\mathfrak{j}$ be ideals in the commutative ring $\mathfrak{R}$. Consider the set $\mathfrak{k}$ of elements of $\mathfrak{R}$ which can be written in the form $\alpha+\beta$, where

$$\alpha = 0 \ (\mathfrak{i}) \quad \text{and} \quad \beta = 0 \ (\mathfrak{j}).$$

If $\alpha+\beta$ and $\alpha'+\beta'$ are two such elements,

$$(\alpha+\beta)-(\alpha'+\beta') = (\alpha-\alpha')+(\beta-\beta'),$$

where $$\alpha-\alpha' = 0 \ (\mathfrak{i}) \quad \text{and} \quad \beta-\beta' = 0 \ (\mathfrak{j}),$$

and if ρ is any element of $\mathfrak{R}$,

$$\rho(\alpha+\beta) = \rho\alpha+\rho\beta,$$

where $$\rho\alpha = 0 \ (\mathfrak{i}), \quad \rho\beta = 0 \ (\mathfrak{j}).$$

Hence $\mathfrak{k}$ is an ideal, which we denote by $(\mathfrak{i}, \mathfrak{j})$; it is called the *join* of $\mathfrak{i}$ and $\mathfrak{j}$.

The join $(\mathfrak{i}, \mathfrak{j})$ has the following properties, the proofs of which are immediate:

(i) $$\mathfrak{i} \subseteq (\mathfrak{i}, \mathfrak{j}), \quad \mathfrak{j} \subseteq (\mathfrak{i}, \mathfrak{j});$$

(ii) $$(\mathfrak{i}, \mathfrak{j}) = (\mathfrak{j}, \mathfrak{i});$$

(iii) $$(\mathfrak{i}, \mathfrak{i}) = \mathfrak{i};$$

(iv) if $\mathfrak{a} = (\mathfrak{i}, \mathfrak{j})$, $\mathfrak{b} = (\mathfrak{j}, \mathfrak{k})$, where $\mathfrak{k}$ is any other ideal of $\mathfrak{R}$,

$$(\mathfrak{a}, \mathfrak{k}) = (\mathfrak{i}, \mathfrak{b}),$$

since these consist of the elements of $\mathfrak{R}$ which can be written in the form $\alpha+\beta+\gamma$, where

$$\alpha = 0 \ (\mathfrak{i}), \quad \beta = 0 \ (\mathfrak{j}), \quad \gamma = 0 \ (\mathfrak{k}).$$

Without ambiguity we can write $(\mathfrak{a}, \mathfrak{k}) = (\mathfrak{i}, \mathfrak{b}) = (\mathfrak{i}, \mathfrak{j}, \mathfrak{k})$. By extension, we can define the join of any finite number of ideals.

(v) If $\mathfrak{i}$ and $\mathfrak{j}$ both possess finite bases:

$$\mathfrak{i} = \mathfrak{R}.(\omega_1, \ldots, \omega_r),$$
$$\mathfrak{j} = \mathfrak{R}.(\nu_1, \ldots, \nu_s),$$

then $$(\mathfrak{i}, \mathfrak{j}) = \mathfrak{R}.(\omega_1, \ldots, \omega_r, \nu_1, \ldots, \nu_s).$$

II. If $\mathfrak{i}$, $\mathfrak{j}$ are ideals in $\mathfrak{R}$, the set of elements α of $\mathfrak{R}$ such that

$$\alpha = 0 \ (\mathfrak{i}) \quad \text{and} \quad \alpha = 0 \ (\mathfrak{j})$$

clearly form an ideal. We denote this by $[\mathfrak{i}, \mathfrak{j}]$, and call it the *intersection* of $\mathfrak{i}$ and $\mathfrak{j}$. The following properties of the intersection are immediate:

(i) $$[\mathfrak{i}, \mathfrak{j}] \subseteq \mathfrak{i}, \quad [\mathfrak{i}, \mathfrak{j}] \subseteq \mathfrak{j};$$

(ii) $$[\mathfrak{i}, \mathfrak{j}] = [\mathfrak{j}, \mathfrak{i}];$$

(iii) $$[\mathfrak{i}, \mathfrak{i}] = \mathfrak{i};$$

(iv) $$[\mathfrak{i}, [\mathfrak{j}, \mathfrak{k}]] = [[\mathfrak{i}, \mathfrak{j}], \mathfrak{k}],$$

where $\mathfrak{k}$ is any other ideal in $\mathfrak{R}$. We can, without ambiguity, write $[\mathfrak{i}, [\mathfrak{j}, \mathfrak{k}]] = [\mathfrak{i}, \mathfrak{j}, \mathfrak{k}]$. We can, similarly, define the intersection of any number of ideals in $\mathfrak{R}$.

III. If $\mathfrak{i}$ and $\mathfrak{j}$ are ideals in $\mathfrak{R}$, the set of elements of $\mathfrak{R}$ of the form $\alpha\beta$, where

$$\alpha = 0 \ (\mathfrak{i}) \quad \text{and} \quad \beta = 0 \ (\mathfrak{j}),$$

do not form an ideal in $\mathfrak{R}$. But the elements of $\mathfrak{R}$ which can be written as finite sums $\sum_{1}^{s} \alpha_k \beta_k$, where

$$\alpha_k = 0 \ (\mathfrak{i}) \quad \text{and} \quad \beta_k = 0 \ (\mathfrak{j}),$$

for $k = 1, ..., s$, do form an ideal, as can be verified at once. We call this the *product* of $\mathfrak{i}$ and $\mathfrak{j}$, and denote it by $\mathfrak{ij}$. It can be verified at once that the multiplication of ideals in a commutative ring is commutative and associative:

$$\mathfrak{ij} = \mathfrak{ji}, \quad \mathfrak{i}(\mathfrak{jk}) = (\mathfrak{ij})\mathfrak{k} = \mathfrak{ijk}.$$

Moreover, we have $$\mathfrak{ij} = 0 \ (\mathfrak{i}),$$

$$\mathfrak{ij} = 0 \ (\mathfrak{j}),$$

hence $$\mathfrak{ij} \subseteq [\mathfrak{i}, \mathfrak{j}].$$

If we take $\mathfrak{j} = \mathfrak{i}$, we can define the powers $\mathfrak{i}^2, \mathfrak{i}^3, ..., \mathfrak{i}^\rho, ...$ of $\mathfrak{i}$ for any integer ρ ($\rho \geqslant 1$). If $\mathfrak{R}$ has unity, and (1) is the unit ideal, any element α of $\mathfrak{i}$ can be written as $e\alpha$, and is therefore in $\mathfrak{i}(1)$. Hence $\mathfrak{i} = 0 \ (\mathfrak{i}(1))$.

But $$\mathfrak{i}(1) = 0 \ (\mathfrak{i}).$$

Therefore $$\mathfrak{i}(1) = \mathfrak{i}.$$

IV. If $\mathfrak{i}$ and $\mathfrak{j}$ are two ideals of $\mathfrak{R}$, let us consider the elements γ of $\mathfrak{R}$ such that

$$\gamma\alpha = 0 \ \ (\mathfrak{i}),$$

for every element α of $\mathfrak{j}$. If γ and γ' are two such elements, and ρ is any element of $\mathfrak{R}$, we see that

$$(\gamma - \gamma')\,\alpha = 0 \ \ (\mathfrak{i}) \quad \text{and} \quad \rho\gamma\alpha = 0 \ \ (\mathfrak{i})$$

for every element α of $\mathfrak{j}$. Hence the elements γ with the stated property form an ideal in $\mathfrak{R}$. We call this ideal the *quotient of* $\mathfrak{i}$ *by* $\mathfrak{j}$, and denote it by $\mathfrak{i}:\mathfrak{j}$. We have, at once,

$$\mathfrak{i} = 0 \ \ (\mathfrak{i}:\mathfrak{j}),$$

and

$$\mathfrak{i}:\mathfrak{i} = \mathfrak{R}.$$

We now prove certain properties of the four operations which we have defined.

THEOREM II. $\qquad \mathfrak{i}(\mathfrak{j}, \mathfrak{k}) = (\mathfrak{i}\mathfrak{j}, \mathfrak{i}\mathfrak{k}).$

Let $\alpha_1, \alpha_2, \ldots$ be elements of $\mathfrak{i}$, $\beta_1, \beta_2, \ldots$ elements of $\mathfrak{j}$, and $\gamma_1, \gamma_2, \ldots$ be elements of $\mathfrak{k}$. Any element of $\mathfrak{i}(\mathfrak{j}, \mathfrak{k})$ is of the form

$$\sum_1^s \alpha_a(\beta_a + \gamma_a) = \sum_1^s \alpha_a\beta_a + \sum_1^s \alpha_a\gamma_a \in (\mathfrak{i}\mathfrak{j}, \mathfrak{i}\mathfrak{k});$$

hence

$$\mathfrak{i}(\mathfrak{j}, \mathfrak{k}) \subseteq (\mathfrak{i}\mathfrak{j}, \mathfrak{i}\mathfrak{k}). \tag{2}$$

Conversely, any element of $(\mathfrak{i}\mathfrak{j}, \mathfrak{i}\mathfrak{k})$ is of the form

$$\sum_1^s \alpha_a\beta_a + \sum_1^t \alpha_{s+b}\gamma_{s+b} = \sum_1^{s+t} \alpha_a(\beta_a + \gamma_a) \in \mathfrak{i}(\mathfrak{j}, \mathfrak{k}),$$

where $\beta_{s+1} = \ldots = \beta_{s+t} = \gamma_1 = \ldots = \gamma_s = 0$. Hence

$$(\mathfrak{i}\mathfrak{j}, \mathfrak{i}\mathfrak{k}) \subseteq \mathfrak{i}(\mathfrak{j}, \mathfrak{k}). \tag{3}$$

The result follows from (2) and (3).

THEOREM III.

$$[\mathfrak{i}_1, \mathfrak{i}_2, \ldots, \mathfrak{i}_r]:\mathfrak{j} = [\mathfrak{i}_1:\mathfrak{j}, \mathfrak{i}_2:\mathfrak{j}, \ldots, \mathfrak{i}_r:\mathfrak{j}].$$

If

$$\gamma\mathfrak{j} = 0 \ \ ([\mathfrak{i}_1, \mathfrak{i}_2, \ldots, \mathfrak{i}_r]),$$

then

$$\gamma\mathfrak{j} = 0 \ \ (\mathfrak{i}_a) \quad (a = 1, \ldots, r).$$

Hence

$$\gamma = 0 \ \ (\mathfrak{i}_a:\mathfrak{j}) \quad (a = 1, \ldots, r),$$

and therefore

$$\gamma = 0 \ \ ([\mathfrak{i}_1:\mathfrak{j}, \mathfrak{i}_2:\mathfrak{j}, \ldots, \mathfrak{i}_r:\mathfrak{j}]).$$

Therefore $$[\mathfrak{i}_1, \mathfrak{i}_2, \ldots, \mathfrak{i}_r] : \mathfrak{j} \subseteq [\mathfrak{i}_1 : \mathfrak{j}, \mathfrak{i}_2 : \mathfrak{j}, \ldots, \mathfrak{i}_r : \mathfrak{j}]. \tag{4}$$

Conversely, if $$\gamma \in [\mathfrak{i}_1 : \mathfrak{j}, \mathfrak{i}_2 : \mathfrak{j}, \ldots, \mathfrak{i}_r : \mathfrak{j}],$$

$$\gamma\mathfrak{j} = 0 \ \ (\mathfrak{i}_a) \quad (a = 1, \ldots, r),$$

and hence $$\gamma\mathfrak{j} = 0 \ \ ([\mathfrak{i}_1, \mathfrak{i}_2, \ldots, \mathfrak{i}_r]),$$

that is, $$\gamma \in [\mathfrak{i}_1, \mathfrak{i}_2, \ldots, \mathfrak{i}_r] : \mathfrak{j}.$$

Hence $$[\mathfrak{i}_1 : \mathfrak{j}, \mathfrak{i}_2 : \mathfrak{j}, \ldots, \mathfrak{i}_r : \mathfrak{j}] \subseteq [\mathfrak{i}_1, \mathfrak{i}_2, \ldots, \mathfrak{i}_r] : \mathfrak{j}, \tag{5}$$

and the theorem follows from (4) and (5).

THEOREM IV. $$[\mathfrak{i}, \mathfrak{j}](\mathfrak{i}, \mathfrak{j}) = 0 \ \ (\mathfrak{i}\mathfrak{j}).$$

By Theorem II, $$[\mathfrak{i}, \mathfrak{j}](\mathfrak{i}, \mathfrak{j}) = ([\mathfrak{i}, \mathfrak{j}]\mathfrak{i}, [\mathfrak{i}, \mathfrak{j}]\mathfrak{j}).$$

Since $$[\mathfrak{i}, \mathfrak{j}] = 0 \ \ (\mathfrak{i}) \quad \text{and} \quad [\mathfrak{i}, \mathfrak{j}] = 0 \ \ (\mathfrak{j}),$$

$$[\mathfrak{i}, \mathfrak{j}]\mathfrak{i} = 0 \ \ (\mathfrak{i}\mathfrak{j}), \quad [\mathfrak{i}, \mathfrak{j}]\mathfrak{j} = 0 \ \ (\mathfrak{i}\mathfrak{j}),$$

and the result follows.

Corollary I. If $(\mathfrak{i}, \mathfrak{j}) = \mathfrak{R}$, *and* $\mathfrak{R}$ *has unity, then* $[\mathfrak{i}, \mathfrak{j}] = \mathfrak{i}\mathfrak{j}$. For

$$[\mathfrak{i}, \mathfrak{j}](\mathfrak{i}, \mathfrak{j}) = [\mathfrak{i}, \mathfrak{j}]\mathfrak{R} = [\mathfrak{i}, \mathfrak{j}].$$

Hence $$[\mathfrak{i}, \mathfrak{j}] \subseteq \mathfrak{i}\mathfrak{j} \subseteq [\mathfrak{i}, \mathfrak{j}].$$

A *maximal* ideal in $\mathfrak{R}$ is defined as an ideal $\mathfrak{i} \neq \mathfrak{R}$ such that if $\mathfrak{j}$ is any ideal with the properties

$$\mathfrak{i} = 0 \ \ (\mathfrak{j}), \quad \mathfrak{j} \neq 0 \ \ (\mathfrak{i}),$$

then $\mathfrak{j} = \mathfrak{R}$. From Theorem IV we deduce

Corollary II. If R has unity and $\mathfrak{i}$ *is a maximal ideal in* $\mathfrak{R}$, *and* $\mathfrak{j} \neq 0 \ (\mathfrak{i})$, *then* $[\mathfrak{i}, \mathfrak{j}] = \mathfrak{i}\mathfrak{j}$. Since $\mathfrak{j} \neq 0 \ (\mathfrak{i})$, $(\mathfrak{i}, \mathfrak{j})$ contains $\mathfrak{i}$, and at least one element not in $\mathfrak{i}$. Since $\mathfrak{i}$ is maximal, we must have

$$(\mathfrak{i}, \mathfrak{j}) = \mathfrak{R},$$

the unit ideal, and the result follows from Corollary I.

THEOREM V. $$\mathfrak{i} : (\mathfrak{j}_1, \ldots, \mathfrak{j}_r) = [\mathfrak{i} : \mathfrak{j}_1, \ldots, \mathfrak{i} : \mathfrak{j}_r].$$

If $\gamma \in \mathfrak{i} : (\mathfrak{j}_1, \ldots, \mathfrak{j}_r)$,

$$\gamma\mathfrak{j}_a = 0 \ \ (\mathfrak{i}) \quad (a = 1, \ldots, r).$$

Hence $$\gamma \in [\mathfrak{i} : \mathfrak{j}_1, \ldots, \mathfrak{i} : \mathfrak{j}_r]. \tag{6}$$

On the other hand, if $\gamma \in [\mathfrak{i}:\mathfrak{j}_1, \ldots, \mathfrak{i}:\mathfrak{j}_r]$,

$$\gamma \mathfrak{j}_a = 0 \ (\mathfrak{i}) \quad (a = 1, \ldots, r),$$

and hence $$\gamma \in \mathfrak{i}:(\mathfrak{j}_1, \ldots, \mathfrak{j}_r). \tag{7}$$

The result follows from (6) and (7).

We conclude this section with a discussion of some consequences of the assumption that the Basis Theorem holds in $\mathfrak{R}$. We take $\mathfrak{R}$ to be any commutative ring for which the Basis Theorem holds. Suppose that $\mathfrak{i}_1, \mathfrak{i}_2, \mathfrak{i}_3, \ldots$ is a sequence of ideals in $\mathfrak{R}$ such that

$$\mathfrak{i}_1 = 0 \ (\mathfrak{i}_2), \quad \mathfrak{i}_2 = 0 \ (\mathfrak{i}_3), \ldots, \quad \mathfrak{i}_r = 0 \ (\mathfrak{i}_{r+1}), \ldots.$$

Let us consider the set $\mathfrak{i}$ of elements of $\mathfrak{R}$ which can be written as finite sums $\sum_{j=1}^{k} \alpha_j$, where each α_j belongs to some ideal $\mathfrak{i}_j$ of the sequence. It is clear that $\mathfrak{i}$ satisfies the two properties which define an ideal. Since the Basis Theorem holds in $\mathfrak{R}$, there exists in $\mathfrak{i}$ a finite set of elements $\omega_1, \ldots, \omega_r$ forming a basis for $\mathfrak{i}$. Since ω_a is in $\mathfrak{i}$,

$$\omega_a = \sum_{j=1}^{s_a} \alpha_{aj},$$

where $\alpha_{aj} \in \mathfrak{i}_j$. Let $t = \max[s_1, \ldots, s_r]$. Since

$$\alpha_{aj} = 0 \ [\mathfrak{i}_j], \quad \mathfrak{i}_j = 0 \ (\mathfrak{i}_t),$$

for all relevant values of j, $\alpha_{aj} \in \mathfrak{i}_t$, and hence

$$\omega_a = 0 \ (\mathfrak{i}_t) \quad (a = 1, \ldots, r),$$

and therefore $$\mathfrak{i} = 0 \ (\mathfrak{i}_t). \tag{8}$$

Let k be any integer greater than t. If $\alpha \in \mathfrak{i}_k$, then $\alpha \in \mathfrak{i}$, from the definition of $\mathfrak{i}$, and hence

$$\mathfrak{i}_k = 0 \ (\mathfrak{i}).$$

Therefore, by (8), $$\mathfrak{i}_k = 0 \ (\mathfrak{i}_t).$$

But since $k > t$, $$\mathfrak{i}_t = 0 \ (\mathfrak{i}_k),$$

and therefore $$\mathfrak{i}_k = \mathfrak{i}_t.$$

Thus for the sequence $\mathfrak{i}_1, \mathfrak{i}_2, \ldots$ there exists a finite integer t such that

$$\mathfrak{i}_t = \mathfrak{i}_{t+1} = \mathfrak{i}_{t+2} = \ldots.$$

Conversely, suppose that the ring $\mathfrak{R}$ has the property that if $\mathfrak{i}_1, \mathfrak{i}_2, \ldots$ is any sequence of ideals in $\mathfrak{R}$ such that

$$\mathfrak{i}_1 = 0 \ (\mathfrak{i}_2), \quad \mathfrak{i}_2 = 0 \ (\mathfrak{i}_3), \ldots,$$

then there necessarily exists a finite integer t such that

$$\mathfrak{i}_t = \mathfrak{i}_{t+1} = \mathfrak{i}_{t+2} = \ldots.$$

We show that this implies that the Basis Theorem holds in $\mathfrak{R}$. Let $\mathfrak{j}$ be any ideal in $\mathfrak{R}$, and let ω_1 be any element of $\mathfrak{j}$. Then, clearly,

$$\mathfrak{i}_1 = \mathfrak{R}.(\omega_1) = 0 \ (\mathfrak{j}).$$

If $\mathfrak{i}_1$ is not equal to $\mathfrak{j}$, $\mathfrak{j}$ contains an element ω_2 not in $\mathfrak{i}_1$, and we have

$$\mathfrak{i}_2 = \mathfrak{R}.(\omega_1, \omega_2) = 0 \ (\mathfrak{j}),$$
$$\mathfrak{i}_1 = 0 \ (\mathfrak{i}_2).$$

If $\mathfrak{i}_2$ is not equal to $\mathfrak{j}$, we select an element ω_3 in $\mathfrak{j}$ but not in $\mathfrak{i}_2$, and proceed as before. We apply the hypothesis on $\mathfrak{R}$ to the sequence $\mathfrak{i}_1, \mathfrak{i}_2, \ldots$. Then there exists an integer t such that

$$\mathfrak{i}_t = \mathfrak{i}_{t+1} = \ldots$$

(unless the sequence ends at $\mathfrak{i}_t$). From the equation $\mathfrak{i}_t = \mathfrak{i}_{t+1}$, it follows that $\omega_{t+1} \in \mathfrak{i}_t$, which conflicts with the method of constructing ω_{t+1}. Hence the sequence ends at $\mathfrak{i}_t$. But the sequence can only end there if $\mathfrak{i}_t = \mathfrak{j}$. It follows that $\mathfrak{j} = \mathfrak{i}_t = \mathfrak{R}.(\omega_1, \ldots, \omega_t)$. Hence $\mathfrak{j}$ has a finite basis.

If $\mathfrak{i}$ and $\mathfrak{j}$ are two ideals of $\mathfrak{R}$ such that

$$\mathfrak{i} = 0 \ (\mathfrak{j}),$$

$\mathfrak{i}$ is said to be a *multiple* of $\mathfrak{j}$, and $\mathfrak{j}$ is said to be a *factor* of $\mathfrak{i}$. If

$$\mathfrak{j} \neq 0 \ (\mathfrak{i}),$$

$\mathfrak{i}$ is a *proper* multiple of $\mathfrak{j}$, and $\mathfrak{j}$ is a *proper* factor of $\mathfrak{i}$. In this terminology the result just proved is equivalent to

THEOREM VI. *A necessary and sufficient condition that the Basis Theorem hold in $\mathfrak{R}$ is that any sequence of ideals in $\mathfrak{R}$ with the property that each ideal of the sequence is a proper multiple of its successor is a finite sequence.*

Sometimes it is convenient to take as the fundamental property of $\mathfrak{R}$ the property that any sequence of ideals such that each is a proper multiple of its successor is finite. We then say that the 'ascending chain-condition' holds in $\mathfrak{R}$.

Two corollaries of Theorem VI may be noted.

Corollary I. If the Basis Theorem holds in $\mathfrak{R}$, any non-vacuous set of ideals contains at least one ideal which is not a proper multiple of any ideal of the set.

The proof is obvious.

Corollary II. (Principle of Induction for Ideals.) If the Basis Theorem holds in $\mathfrak{R}$, and E is any property which (i) *holds for the unit ideal,* (ii) *holds for an ideal $\mathfrak{i}$ when it holds for all proper factors of $\mathfrak{i}$, then E holds for all ideals in $\mathfrak{R}$.*

Suppose that the corollary is false, and consider the set S of ideals for which E does not hold. By Corollary I, there exists in the set an ideal $\mathfrak{i}$ which is not a proper multiple of any ideal of the set. $\mathfrak{i}$ is not the unit ideal since, by hypothesis, E holds for the unit ideal. $\mathfrak{i}$ has proper factors (for instance, the unit ideal, which satisfies our definition of a proper factor), and since none of these proper factors can belong to S, $\mathfrak{i}$ being maximal in S, it follows from condition (ii) that E holds for $\mathfrak{i}$. But $\mathfrak{i}$ is in S, and hence E does not hold for it. We thus have a contradiction. It follows that S must be vacuous, and our assumption that the corollary is not true is invalid.

2. Prime ideals and primary ideals. An ideal $\mathfrak{p}$ in the commutative ring $\mathfrak{R}$ is said to be a *prime ideal* if the equation

$$\alpha\beta = 0 \ (\mathfrak{p})$$

implies either

$$\alpha = 0 \ (\mathfrak{p})$$

or

$$\beta = 0 \ (\mathfrak{p}).$$

It is clear that the unit ideal is always prime, and that the zero ideal is prime if and only if $\mathfrak{R}$ contains no divisors of zero. Again, if $\mathfrak{p}$ is prime, and $\mathfrak{i}$ and $\mathfrak{j}$ are two ideals of $\mathfrak{R}$ such that

$$\mathfrak{i}\mathfrak{j} = 0 \ (\mathfrak{p}),$$

then either

$$\mathfrak{i} = 0 \ (\mathfrak{p})$$

or

$$\mathfrak{j} = 0 \ (\mathfrak{p}).$$

Indeed, suppose that $\quad \mathfrak{i} \neq 0 \ (\mathfrak{p}).$

Then there exists an element α in $\mathfrak{i}$ which is not in $\mathfrak{p}$. If β is any element of $\mathfrak{j}$, $\alpha\beta \in \mathfrak{ij}$, and hence

$$\alpha\beta = 0 \quad (\mathfrak{p}).$$

Since α is not in $\mathfrak{p}$, and $\mathfrak{p}$ is prime, it follows that β is in $\mathfrak{p}$. Hence

$$\mathfrak{j} = 0 \quad (\mathfrak{p}).$$

An ideal $\mathfrak{q}$ in $\mathfrak{R}$ is said to be *primary* if the equation

$$\alpha\beta = 0 \quad (\mathfrak{q}),$$

together with the inequality

$$\alpha \neq 0 \quad (\mathfrak{q}),$$

implies

$$\beta^\rho = 0 \quad (\mathfrak{q}),$$

for a suitable integer ρ. A prime ideal is, of course, primary.

If $\mathfrak{R}$ is the principal ideal ring formed by the natural integers, the ideal $\mathfrak{R}.(d)$ is prime if and only if d is a prime number, and is primary if and only if d is a power of a prime number. For our purposes, however, a more suggestive example is provided by the ring $\mathfrak{R} = K[x, y]$ of polynomials in two independent indeterminates over the ground field K. Let $\mathfrak{p} = \mathfrak{R}.(x, y)$. The elements of $\mathfrak{p}$ are just those polynomials whose constant term is zero. If α and β are two polynomials whose product has constant term zero, then either α or β must have constant term zero and hence belongs to $\mathfrak{p}$. Thus $\mathfrak{p}$ is prime. Next, let $\mathfrak{q} = \mathfrak{R}.(x, y^2)$. Any element of $\mathfrak{q}$ is of the form

$$ax + (cx^2 + 2dxy + ey^2) + (lx^3 + 3mx^2y + 3nxy^2 + py^3) + \dots.$$

Let

$$\alpha = a_1 + (b_1 x + c_1 y) + \dots$$

and

$$\beta = a_2 + (b_2 x + c_2 y) + \dots.$$

If $\alpha\beta \in \mathfrak{q}$, we must have

$$a_1 a_2 = 0, \quad a_1 c_2 + a_2 c_1 = 0.$$

If α does not belong to $\mathfrak{q}$, a_1 and c_1 cannot both be zero. It follows at once that we must have $a_2 = 0$. But if a_2 is zero,

$$\beta^2 = (b_2^2 x^2 + 2b_2 c_2 xy + c_2^2 y^2) + \dots$$

has the form of an element of $\mathfrak{q}$. Hence $\mathfrak{q}$ is primary. If $c_2 \neq 0$, β does not belong to $\mathfrak{q}$, hence $\mathfrak{q}$ is not a prime ideal.

Now let $\mathfrak{q}$ be any primary ideal in a commutative ring $\mathfrak{R}$, and consider the set of elements $\alpha, \beta, \ldots$ of $\mathfrak{R}$ which have the property that some power of them belongs to $\mathfrak{q}$. If

$$\alpha^\rho = 0 \ (\mathfrak{q}) \quad \text{and} \quad \beta^\sigma = 0 \ (\mathfrak{q}),$$

then $(\alpha-\beta)^{\rho+\sigma-1}$ is equal to a sum of terms each of which is either the product of α^ρ by an element of $\mathfrak{R}$ or the product of β^σ by an element of $\mathfrak{R}$, and is therefore in $\mathfrak{q}$. Hence $(\alpha-\beta)^{\rho+\sigma-1} \in \mathfrak{q}$, and so $\alpha-\beta$ is in the set. Similarly, if ξ is any element of $\mathfrak{R}$,

$$(\xi\alpha)^\rho = \xi^\rho\alpha^\rho = 0 \ (\mathfrak{q}).$$

Hence $\xi\alpha$ is in the set, which is therefore an ideal, which we denote by $\mathfrak{p}$. We now show that $\mathfrak{p}$ is a prime ideal. Suppose that α and β are such that

$$\alpha\beta = 0 \ (\mathfrak{p}), \quad \alpha \neq 0 \ (\mathfrak{p}).$$

Since $\alpha\beta \in \mathfrak{p}$, there exists an integer ρ such that

$$\alpha^\rho\beta^\rho = 0 \ (\mathfrak{q}).$$

Since α is not in $\mathfrak{p}$, α^ρ is not in $\mathfrak{q}$, and, since $\mathfrak{q}$ is primary, there exists an integer τ such that $\beta^{\rho\tau} \in \mathfrak{q}$. Hence β is in $\mathfrak{p}$. This proves that $\mathfrak{p}$ is prime. $\mathfrak{p}$ is called the prime ideal *belonging to* $\mathfrak{q}$, or the *radical* of $\mathfrak{q}$. It is clear that

$$\mathfrak{q} = 0 \ (\mathfrak{p}),$$

and that $\mathfrak{q} = \mathfrak{p}$ if and only if $\mathfrak{q}$ is prime.

From the definition of a primary ideal and its radical, we see that if

$$\alpha\beta = 0 \ (\mathfrak{q}) \quad \text{and} \quad \alpha \neq 0 \ (\mathfrak{q}),$$

then

$$\beta = 0 \ (\mathfrak{p}).$$

Hence $\mathfrak{q}$ and its radical $\mathfrak{p}$ are related by the following properties:

(i) if $\alpha\beta = 0 \ (\mathfrak{q})$ and $\alpha \neq 0 \ (\mathfrak{q})$, then $\beta = 0 \ (\mathfrak{p})$;

(ii) $\mathfrak{q} = 0 \ (\mathfrak{p})$,

(iii) $\beta = 0 \ (\mathfrak{p})$ implies $\beta^\rho = 0 \ (\mathfrak{q})$,

for a suitable integer ρ.

We may note that if $\mathfrak{R}$ has unity and $\mathfrak{q}$ is not the unit ideal, then (i) implies (ii). (If $\mathfrak{q}$ is the unit ideal, (i) does not define $\mathfrak{p}$.) If β is any element of $\mathfrak{q}$, we have

$$1.\beta = 0 \ (\mathfrak{q}),$$

and since $\mathfrak{q}$ is not the unit ideal it does not contain 1. Hence, by (i),

$$\beta = 0 \ (\mathfrak{p}).$$

Therefore
$$\mathfrak{q} = 0 \ (\mathfrak{p}).$$

We now show that if $\mathfrak{q}$ and $\mathfrak{p}$ are two ideals in $\mathfrak{R}$ satisfying the properties (i), (ii), (iii), then $\mathfrak{q}$ is primary, and $\mathfrak{p}$ is its radical. More briefly, we shall say that $\mathfrak{q}$ is $\mathfrak{p}$-*primary*. From (i) and (iii) it follows that $\mathfrak{q}$ is primary. To show that $\mathfrak{p}$ is its radical, let β be any element of $\mathfrak{R}$ such that $\beta^\rho \in \mathfrak{q}$ for some integer ρ. If $\beta \in \mathfrak{q}$, then $\beta \in \mathfrak{p}$, by (ii). If β does not belong to $\mathfrak{q}$, let ρ $(\rho > 1)$ be the smallest integer such that $\beta^\rho \in \mathfrak{q}$. Then

$$\beta . \beta^{\rho-1} = 0 \ (\mathfrak{q}),$$

and
$$\beta^{\rho-1} \neq 0 \ (\mathfrak{q}),$$

hence, by (i),
$$\beta = 0 \ (\mathfrak{p}).$$

This, taken with condition (iii), gives

THEOREM I. *A primary ideal* $\mathfrak{q}$ *and its radical* $\mathfrak{p}$ *are characterised by the properties* (i), (ii), (iii).

THEOREM II. *If* $\mathfrak{q}$ *is* $\mathfrak{p}$-*primary, and* $\mathfrak{i}$ *and* $\mathfrak{j}$ *are ideals of* $\mathfrak{R}$ *such that*

$$\mathfrak{i}\mathfrak{j} = 0 \ (\mathfrak{q}), \quad \mathfrak{i} \neq 0 \ (\mathfrak{q}),$$

then
$$\mathfrak{j} = 0 \ (\mathfrak{p}).$$

Since $\mathfrak{i}$ is not contained in $\mathfrak{q}$, there is an element α of $\mathfrak{i}$ not contained in $\mathfrak{q}$. If β is any element of $\mathfrak{j}$, $\alpha\beta \in \mathfrak{q}$, and α is not in $\mathfrak{q}$. Hence, by property (i) above, $\beta \in \mathfrak{p}$, that is

$$\mathfrak{j} = 0 \ (\mathfrak{p}).$$

Corollary I. If

$$\mathfrak{i}\mathfrak{j} = 0 \ (\mathfrak{q}) \quad \textit{and} \quad \mathfrak{j} \neq 0 \ (\mathfrak{p}), \quad \textit{then} \quad \mathfrak{i} = 0 \ (\mathfrak{q}).$$

Corollary II. If

$$\mathfrak{j} \neq 0 \ (\mathfrak{p}), \quad \textit{then} \quad \mathfrak{q} : \mathfrak{j} = \mathfrak{q}.$$

THEOREM III. *If* $\mathfrak{q}$ *and* $\mathfrak{q}'$ *are* $\mathfrak{p}$-*primary, then* $[\mathfrak{q}, \mathfrak{q}']$ *is* $\mathfrak{p}$-*primary.*

(i) If
$$\alpha\beta = 0 \ ([\mathfrak{q}, \mathfrak{q}']) \quad \text{and} \quad \alpha \neq 0 \ ([\mathfrak{q}, \mathfrak{q}']),$$

we have
$$\alpha\beta = 0 \ (\mathfrak{q}) \quad \text{and} \quad \alpha\beta = 0 \ (\mathfrak{q}'),$$

and either
$$\alpha \neq 0 \ (\mathfrak{q}) \quad \text{or} \quad \alpha \neq 0 \ (\mathfrak{q}').$$

Suppose, for instance, that

$$\alpha\beta = 0 \ (\mathfrak{q}) \quad \text{and} \quad \alpha \neq 0 \ (\mathfrak{q}).$$

Then since $\mathfrak{q}$ is $\mathfrak{p}$-primary,

$$\beta = 0 \ (\mathfrak{p}).$$

If

$$\alpha \neq 0 \ (\mathfrak{q}'),$$

a similar argument gives $\beta = 0 \ (\mathfrak{p})$.

(ii)

$$\mathfrak{q} = 0 \ (\mathfrak{p}) \quad \text{and} \quad \mathfrak{q}' = 0 \ (\mathfrak{p});$$

hence

$$[\mathfrak{q}, \mathfrak{q}'] = 0 \ (\mathfrak{p}).$$

(iii) If $\beta \in \mathfrak{p}$, there exist integers ρ, σ such that

$$\beta^\rho = 0 \ (\mathfrak{q}) \quad \text{and} \quad \beta^\sigma = 0 \ (\mathfrak{q}').$$

If $\tau = \max[\rho, \sigma]$,

$$\beta^\tau = 0 \ ([\mathfrak{q}, \mathfrak{q}']).$$

It follows from Theorem I that $[\mathfrak{q}, \mathfrak{q}']$ is $\mathfrak{p}$-primary.

THEOREM IV. *If* $\mathfrak{q}$ *is* $\mathfrak{p}$*-primary and* $\mathfrak{q}'$ *is* $\mathfrak{p}'$*-primary, where* $\mathfrak{p} \neq \mathfrak{p}'$, *and if*

$$[\mathfrak{q}, \mathfrak{q}'] \neq \mathfrak{q} \quad \text{and} \quad [\mathfrak{q}, \mathfrak{q}'] \neq \mathfrak{q}',$$

then $[\mathfrak{q}, \mathfrak{q}']$ *is not primary.*

Since $\mathfrak{p} \neq \mathfrak{p}'$, there exists an element α which is in one of these ideals, but not in the other. Suppose that

$$\alpha = 0 \ (\mathfrak{p}), \quad \alpha \neq 0 \ (\mathfrak{p}').$$

Then no power of α belongs to $\mathfrak{q}'$, while there exists an integer ρ such that

$$\alpha^\rho = 0 \ (\mathfrak{q}).$$

Then

$$\alpha^\rho \neq 0 \ ([\mathfrak{q}, \mathfrak{q}']).$$

Since $[\mathfrak{q}, \mathfrak{q}'] \neq \mathfrak{q}'$, there exists an element β which is in $\mathfrak{q}'$ but not in $[\mathfrak{q}, \mathfrak{q}']$, and hence not in $\mathfrak{q}$. Then

$$\alpha^\rho \beta \in \mathfrak{q}\mathfrak{q}' \subseteq [\mathfrak{q}, \mathfrak{q}'],$$

[§ 1, p. 6]. Now

$$\beta \neq 0 \ ([\mathfrak{q}, \mathfrak{q}']),$$

and it would follow, if $[\mathfrak{q}, \mathfrak{q}']$ were primary, that some power of α^ρ lay in $[\mathfrak{q}, \mathfrak{q}']$, and hence that some power of α was in $\mathfrak{q}'$. Since α is not in $\mathfrak{p}'$, this contradicts the fact that $\mathfrak{q}'$ is $\mathfrak{p}'$-primary. Hence we conclude that $[\mathfrak{q}, \mathfrak{q}']$ is not primary.

The theorems so far proved in this section do not require the assumption that the Basis Theorem holds in $\mathfrak{R}$. We now go on to prove some results which depend on the assumption that any ideal in $\mathfrak{R}$ has a finite basis. *For the remainder of this section we assume that the Basis Theorem holds in* $\mathfrak{R}$.

Let $\mathfrak{q}$ be a $\mathfrak{p}$-primary ideal in $\mathfrak{R}$, and let $\omega_1, \ldots, \omega_r$ be a basis for $\mathfrak{p}$. Corresponding to ω_i there exists an integer σ_i such that

$$\omega_i^{\sigma_i} = 0 \ \ (\mathfrak{q}).$$

Let
$$\rho = \sum_{i=1}^{r} (\sigma_i - 1) + 1.$$

Any element of $\mathfrak{p}^\rho$ is of the form

$$\sum_{i=1}^{k} \alpha_i \Omega_i + \sum_{i=1}^{k} n_i \Omega_i,$$

where $\alpha_1, \ldots, \alpha_k$ are in $\mathfrak{R}$, and $n_1, \ldots, n_k$ are integers, and $\Omega_1, \ldots, \Omega_k$ are products of degree ρ in $\omega_1, \ldots, \omega_r$. If, for instance,

$$\Omega_i = \omega_1^{\rho_1} \ldots \omega_r^{\rho_r},$$

then
$$\sum_{i=1}^{r} \rho_i = \rho = \sum_{i=1}^{r} (\sigma_i - 1) + 1,$$

and hence, for some value of j, $\rho_j \geqslant \sigma_j$. Thus $\Omega_i \in \mathfrak{q}$, and hence

$$\mathfrak{p}^\rho = 0 \ \ (\mathfrak{q}).$$

There may, of course, be an integer σ $(\sigma < \rho)$ such that

$$\mathfrak{p}^\sigma = 0 \ \ (\mathfrak{q});$$

the smallest integer σ with this property is called the *index* of $\mathfrak{q}$. A primary ideal is prime if and only if its index is 1.

THEOREM V. *Let* $\mathfrak{q}$ *be a primary ideal and* $\mathfrak{p}$ *a prime ideal. If*

$$\mathfrak{q} = 0 \ \ (\mathfrak{p}), \quad \mathfrak{p}^\sigma = 0 \ \ (\mathfrak{q}),$$

for some integer σ, *then* $\mathfrak{q}$ *is* $\mathfrak{p}$*-primary.*

Let $\mathfrak{p}'$ be the radical of $\mathfrak{q}$; there exists an integer ρ such that

$$\mathfrak{p}'^\rho = 0 \ \ (\mathfrak{q}).$$

Hence
$$\mathfrak{p}'^\rho \subseteq \mathfrak{q} \subseteq \mathfrak{p},$$

that is,
$$\mathfrak{p}'^\rho = 0 \ \ (\mathfrak{p}),$$

and, since $\mathfrak{p}$ is prime, we deduce that

$$\mathfrak{p}' = 0 \ (\mathfrak{p}). \tag{1}$$

On the other hand, we have

$$\mathfrak{q} = 0 \ (\mathfrak{p}'),$$

and hence $$\mathfrak{p}^{\sigma} \subseteq \mathfrak{q} \subseteq \mathfrak{p}'.$$

Therefore, since $\mathfrak{p}'$ is prime,

$$\mathfrak{p} = 0 \ (\mathfrak{p}'). \tag{2}$$

From (1) and (2) we deduce that $\mathfrak{p} = \mathfrak{p}'$.

An ideal $\mathfrak{i}$ in $\mathfrak{R}$ is said to be *reducible* if it can be written in the form

$$\mathfrak{i} = [\mathfrak{j}, \mathfrak{k}],$$

where $\mathfrak{j}$ and $\mathfrak{k}$ are *proper* factors of $\mathfrak{i}$; otherwise it is *irreducible*. A prime ideal (and, in particular, the unit ideal) is irreducible. For if $\mathfrak{i}$ is prime and

$$\mathfrak{i} = [\mathfrak{j}, \mathfrak{k}],$$

we have $$\mathfrak{j}\mathfrak{k} \subseteq [\mathfrak{j}, \mathfrak{k}] = \mathfrak{i},$$

and hence $$\mathfrak{j} = 0 \ (\mathfrak{i}) \quad \text{or} \quad \mathfrak{k} = 0 \ (\mathfrak{i}),$$

and therefore $\mathfrak{j}$ and $\mathfrak{k}$ cannot both be proper factors of $\mathfrak{i}$.

It is easy to show, by means of the Principle of Induction for ideals [§ 1, Th. VI, Cor. II], that any ideal $\mathfrak{i}$ is the intersection of a finite number of irreducible ideals. If $\mathfrak{i}$ is irreducible, there is nothing to prove. Suppose that $\mathfrak{i}$ is reducible:

$$\mathfrak{i} = [\mathfrak{j}, \mathfrak{k}],$$

where $\mathfrak{j}$ and $\mathfrak{k}$ are proper factors of $\mathfrak{i}$. If

$$\mathfrak{j} = [\mathfrak{j}_1, \mathfrak{j}_2, \ldots, \mathfrak{j}_r],$$

where $\mathfrak{j}_1, \ldots, \mathfrak{j}_r$ are irreducible ideals, and if

$$\mathfrak{k} = [\mathfrak{k}_1, \mathfrak{k}_2, \ldots, \mathfrak{k}_s],$$

where $\mathfrak{k}_1, \ldots, \mathfrak{k}_s$ are irreducible ideals, then

$$\mathfrak{i} = [\mathfrak{j}_1, \ldots, \mathfrak{j}_r, \mathfrak{k}_1, \ldots, \mathfrak{k}_s]$$

is the intersection of a finite number of irreducible ideals. The property of being the intersection of a finite number of irreducible ideals is true for the unit ideal, and also for any ideal when it is true for its proper factors. Hence it is true for all ideals in $\mathfrak{R}$. Thus we have

THEOREM VI. *Every ideal in a ring for which the Basis Theorem holds is the intersection of a finite number of irreducible ideals.*

THEOREM VII. *Every irreducible ideal is primary.*

Suppose that the ideal $\mathfrak{i}$ is not primary. We can then find two elements α and β in $\mathfrak{R}$ such that

$$\alpha\beta = 0 \ (\mathfrak{i}), \quad \alpha \neq 0 \ (\mathfrak{i}), \quad \beta^\rho \neq 0 \ (\mathfrak{i}),$$

for every positive integer ρ. The elements of $\mathfrak{R}$ of the form $\xi\beta^a$, where ξ is any element of $\mathfrak{R}$ and a is a fixed integer, clearly form an ideal, which we denote by $\mathfrak{j}_a$; if R has unity, $\mathfrak{j}_a = \mathfrak{R}.(\beta^a)$, but in general if $\mathfrak{R}$ has not unity $\mathfrak{j}_a$ is a proper multiple of $\mathfrak{R}.(\beta^a)$. We see immediately that

$$\mathfrak{i}:\mathfrak{j}_1 \subseteq \mathfrak{i}:\mathfrak{j}_2 \subseteq \mathfrak{i}:\mathfrak{j}_3 \subseteq \ldots.$$

By § 1, Th. VI, there exists an integer r such that

$$\mathfrak{i}:\mathfrak{j}_r = \mathfrak{i}:\mathfrak{j}_{r+1} = \ldots.$$

Let $\mathfrak{k} = \mathfrak{R}.(\alpha)$. The ideal $(\mathfrak{i}, \mathfrak{k})$ is a proper factor of $\mathfrak{i}$, since it contains $\mathfrak{i}$ and also α, which is not in $\mathfrak{i}$. Similarly, the ideal $(\mathfrak{i}, \mathfrak{j}_{r+1})$ is a proper factor of $\mathfrak{i}$, since it contains $\mathfrak{i}$ and $\beta^{r+2} = \beta.\beta^{r+1}$, which is not in $\mathfrak{i}$. The theorem will be proved if we can show that

$$\mathfrak{i} = [(\mathfrak{i}, \mathfrak{k}), (\mathfrak{i}, \mathfrak{j}_{r+1})].$$

Clearly, we have $\mathfrak{i} \subseteq [(\mathfrak{i}, \mathfrak{k}), (\mathfrak{i}, \mathfrak{j}_{r+1})]$. To prove that $[(\mathfrak{i}, \mathfrak{k}), (\mathfrak{i}, \mathfrak{j}_{r+1})] \subseteq \mathfrak{i}$, we consider any element η common to $(\mathfrak{i}, \mathfrak{k})$ and $(\mathfrak{i}, \mathfrak{j}_{r+1})$, and show that it lies in $\mathfrak{i}$. Since $\eta \in (\mathfrak{i}, \mathfrak{k})$,

$$\eta = \gamma + \rho\alpha + n\alpha,$$

where $\gamma \in \mathfrak{i}$, $\rho \in \mathfrak{R}$, and n is an integer. Since $\eta \in (\mathfrak{i}, \mathfrak{j}_{r+1})$,

$$\eta = \delta + \sigma\beta^{r+1},$$

where $\delta \in \mathfrak{i}$ and $\sigma \in \mathfrak{R}$. We therefore have

$$\beta\delta + \sigma\beta^{r+2} = \beta\eta = \beta\gamma + \rho\alpha\beta + n\alpha\beta = 0 \ (\mathfrak{i}),$$

since γ and $\alpha\beta$ belong to $\mathfrak{i}$. Since δ is also in $\mathfrak{i}$, we have

$$\sigma\beta^{r+2} = 0 \ (\mathfrak{i}),$$

and hence $\sigma \in \mathfrak{i} : \mathfrak{j}_{r+2}$. But $\mathfrak{i} : \mathfrak{j}_{r+2} = \mathfrak{i} : \mathfrak{j}_r$. Hence σ belongs to $\mathfrak{i} : \mathfrak{j}_r$. But $\sigma\beta^{r+1} = \sigma(\beta . \beta^r)$, and hence $\sigma\beta^{r+1}$ is contained in $\mathfrak{i}$. It follows that η is in $\mathfrak{i}$. Theorem VII is therefore proved.

Theorems VI and VII, taken together, tell us that every ideal $\mathfrak{i}$ in $\mathfrak{R}$ can be expressed as the intersection of a finite number of primary ideals. Let

$$\mathfrak{i} = [\mathfrak{q}_1, \ldots, \mathfrak{q}_r],$$

where $\mathfrak{q}_i$ is $\mathfrak{p}_i$-primary. If $\mathfrak{p}_i = \mathfrak{p}_j$, it follows from Theorem III that $\mathfrak{q}_{ij} = [\mathfrak{q}_i, \mathfrak{q}_j]$ is also $\mathfrak{p}_i$-primary, and in the above expression for $\mathfrak{i}$ as an intersection of primary ideals we can omit $\mathfrak{q}_i$ and $\mathfrak{q}_j$, and put $\mathfrak{q}_{ij}$ in their place. Proceeding in this way, we can combine all the primary ideals $\mathfrak{q}_i$ which have the same radical into a single primary ideal, and so obtain a representation of $\mathfrak{i}$,

$$\mathfrak{i} = [\mathfrak{Q}_1, \ldots, \mathfrak{Q}_s],$$

as the intersection of primary ideals, where $\mathfrak{Q}_i$ is $\mathfrak{P}_i$-primary, and $\mathfrak{P}_1, \ldots, \mathfrak{P}_s$ are all different. $\mathfrak{Q}_1, \ldots, \mathfrak{Q}_s$ are called *primary components* of $\mathfrak{i}$.

Suppose that $\mathfrak{Q}_i$ is such that

$$\mathfrak{Q}_i \supseteq \mathfrak{J}_i = [\mathfrak{Q}_1, \ldots, \mathfrak{Q}_{i-1}, \mathfrak{Q}_{i+1}, \ldots, \mathfrak{Q}_s].$$

Then $\mathfrak{Q}_i$ is said to be an *irrelevant* primary component of $\mathfrak{i}$. In this case

$$\mathfrak{i} = [\mathfrak{Q}_i, \mathfrak{J}_i] = \mathfrak{J}_i,$$

and we can omit the component $\mathfrak{Q}_i$ in the representation of $\mathfrak{i}$. Proceeding, we eventually obtain a representation, say

$$\mathfrak{i} = [\mathfrak{Q}_1, \ldots, \mathfrak{Q}_k],$$

of $\mathfrak{i}$ as the intersection of primary ideals, the radicals of which are all different, and none of which is irrelevant. If we combine two of the components in this representation into one, we get a component which is not primary [Th. IV], and if we omit one of the components we clearly alter the intersection. For this reason the representations $[\mathfrak{Q}_1, \ldots, \mathfrak{Q}_k]$ of $\mathfrak{i}$ is said to be *uncontractible*. For practical purposes it is the uncontractible representations of the ideal, rather than its representations as the intersection of irreducible ideals, which are important.

Examples can be given to show that an ideal $\mathfrak{i}$ may have two distinct uncontractible representations. But there are certain uniqueness theorems which are important, and these we now prove.

THEOREM VIII. *Let* $[\mathfrak{q}_1, \ldots, \mathfrak{q}_k]$ *and* $[\mathfrak{q}'_1, \ldots, \mathfrak{q}'_l]$ *be two uncontractible representations of an ideal* $\mathfrak{i}$ *of* $\mathfrak{R}$, *where* $\mathfrak{q}_j$ *is* $\mathfrak{p}_j$*-primary and* $\mathfrak{q}'_j$ *is* $\mathfrak{p}'_j$*-primary. Then* $k = l$, *and, if the components* $\mathfrak{q}'_j$ *are suitably arranged*, $\mathfrak{p}'_j = \mathfrak{p}_j$ $(j = 1, \ldots, k)$.

By a characteristic property of an uncontractible representation, $\mathfrak{p}_1, \ldots, \mathfrak{p}_k$ are all distinct, and $\mathfrak{p}'_1, \ldots, \mathfrak{p}'_l$ are all distinct. Amongst these $k+l$ prime ideals we can find at least one which is not a proper multiple of any other, and without loss of generality we may assume that $\mathfrak{p}_1$ has this property; $\mathfrak{p}_1$ is not equal to any $\mathfrak{p}_j$ $(j > 1)$, and is at most equal to one $\mathfrak{p}'_j$.

Suppose that $\mathfrak{p}_1$ is not equal to any $\mathfrak{p}'_j$. For any given value of j $(j > 1)$ there exists an element α which is in $\mathfrak{p}_1$ but not in $\mathfrak{p}_j$, since

$$\mathfrak{p}_1 \not\equiv 0 \quad (\mathfrak{p}_j).$$

Let ρ be an integer such that $\alpha^\rho \in \mathfrak{q}_1$. If α is not in $\mathfrak{p}_j$, α^ρ is not in $\mathfrak{p}_j$, and hence

$$\mathfrak{q}_1 \not\equiv 0 \quad (\mathfrak{p}_j).$$

This holds for $j = 2, \ldots, k$. Again, since

$$\mathfrak{p}_1 \not\equiv 0 \quad (\mathfrak{p}'_j),$$

a similar argument shows that

$$\mathfrak{q}_1 \not\equiv 0 \quad (\mathfrak{p}'_j),$$

for $j = 1, \ldots, l$. Now [§ 1, Th. III]

$$[\mathfrak{q}_1 : \mathfrak{q}_1, \mathfrak{q}_2 : \mathfrak{q}_1, \ldots, \mathfrak{q}_k : \mathfrak{q}_1] = \mathfrak{i} : \mathfrak{q}_1 = [\mathfrak{q}'_1 : \mathfrak{q}_1, \mathfrak{q}'_2 : \mathfrak{q}_1, \ldots, \mathfrak{q}'_l : \mathfrak{q}_1].$$

Clearly
$$\mathfrak{q}_1 : \mathfrak{q}_1 = \mathfrak{R},$$

and from Theorem II, Corollary II,

$$\mathfrak{q}_j : \mathfrak{q}_1 = \mathfrak{q}_j \quad (j > 1),$$

and
$$\mathfrak{q}'_j : \mathfrak{q}_1 = \mathfrak{q}'_j \quad (j \geqslant 1).$$

Hence
$$[\mathfrak{q}_2, \ldots, \mathfrak{q}_k] = [\mathfrak{q}'_1, \ldots, \mathfrak{q}'_l] = \mathfrak{i},$$

contradicting the assumption that $[\mathfrak{q}_1, \ldots, \mathfrak{q}_k]$ is an uncontractible representation of $\mathfrak{i}$. It follows that $\mathfrak{p}_1$ must be equal to one $\mathfrak{p}'_j$, and by arranging the components $\mathfrak{q}'_1, \ldots, \mathfrak{q}'_l$ suitably we may suppose that $\mathfrak{p}_1 = \mathfrak{p}'_1$.

By Theorem III, $\mathfrak{k} = [\mathfrak{q}_1, \mathfrak{q}'_1]$ is $\mathfrak{p}_1$-primary. Just as above, we can prove that

$$\mathfrak{q}_j : \mathfrak{k} = \mathfrak{q}_j \quad \text{and} \quad \mathfrak{q}'_j : \mathfrak{k} = \mathfrak{q}'_j,$$

provided that $j > 1$ in both cases; and since $\mathfrak{k} \subseteq \mathfrak{q}_1$, $\mathfrak{k} \subseteq \mathfrak{q}'_1$,

$$\mathfrak{q}_1 : \mathfrak{k} = \mathfrak{R} = \mathfrak{q}'_1 : \mathfrak{k}.$$

Hence, from the equation

$$[\mathfrak{q}_1, \ldots, \mathfrak{q}_k] : \mathfrak{k} = [\mathfrak{q}'_1, \ldots, \mathfrak{q}'_l] : \mathfrak{k},$$

we obtain the equation

$$[\mathfrak{q}_2, \ldots, \mathfrak{q}_k] = [\mathfrak{q}'_2, \ldots, \mathfrak{q}'_l].$$

We show that $[\mathfrak{q}_2, \ldots, \mathfrak{q}_k]$ is an uncontractible representation of this ideal. Indeed, if

$$[\mathfrak{q}_2, \ldots, \mathfrak{q}_{i-1}, \mathfrak{q}_{i+1}, \ldots, \mathfrak{q}_k] = 0 \quad (\mathfrak{q}_i),$$

then
$$[\mathfrak{q}_1, \mathfrak{q}_2, \ldots, \mathfrak{q}_{i-1}, \mathfrak{q}_{i+1}, \ldots, \mathfrak{q}_k] = 0 \quad (\mathfrak{q}_i),$$

contradicting the hypothesis that $[\mathfrak{q}_1, \ldots, \mathfrak{q}_k]$ is an uncontractible representation of $\mathfrak{i}$.

Theorem VIII now follows by a simple argument, using induction on the minimum number of components required to represent an ideal $\mathfrak{i}$ as an uncontractible intersection of primaries. Let k be this number. If $k = 1$, $\mathfrak{i}$ is primary, and $\mathfrak{q}_1 = \mathfrak{i}$. If $[\mathfrak{q}'_1, \ldots, \mathfrak{q}'_l]$ is another uncontractible representation of $\mathfrak{i}$, the proof given above shows that

$$[\mathfrak{q}'_2, \ldots, \mathfrak{q}'_l] = \mathfrak{R},$$

and hence that
$$\mathfrak{q}'_i = \mathfrak{R} \quad (i = 2, \ldots, l).$$

Hence $[\mathfrak{q}'_1, \ldots, \mathfrak{q}'_l]$ is only uncontractible if $l = 1$, and then $\mathfrak{q}'_1 = \mathfrak{q}_1$ and therefore $\mathfrak{p}'_1 = \mathfrak{p}_1$. If we now assume the truth of the theorem for ideals which can be represented as an uncontractible intersection of $k-1$ primaries, and consider an ideal $\mathfrak{i}$ which requires k components:

$$\mathfrak{i} = [\mathfrak{q}_1, \ldots, \mathfrak{q}_k] = [\mathfrak{q}'_1, \ldots, \mathfrak{q}'_l],$$

the reasoning above shows that:

(i) for a suitable arrangement of the components,

$$\mathfrak{p}_1 = \mathfrak{p}'_1;$$

(ii)
$$[\mathfrak{q}_2, \ldots, \mathfrak{q}_k] = [\mathfrak{q}'_2, \ldots, \mathfrak{q}'_l];$$

and the hypothesis of induction tells us that $k-1 = l-1$, that is, $k = l$, and the components can be arranged so that

$$\mathfrak{p}_i = \mathfrak{p}'_i \quad (i = 2, \ldots, k).$$

The proof is complete.

To prove our second uniqueness theorem we must first define the isolated and embedded components of an ideal $\mathfrak{i}$. Let $[\mathfrak{q}_1, \ldots, \mathfrak{q}_k]$ be an uncontractible representation of $\mathfrak{i}$, and let $\mathfrak{p}_j$ be the prime ideal belonging to $\mathfrak{q}_j$. By Theorem VIII, $\mathfrak{p}_1, \ldots, \mathfrak{p}_k$ are prime ideals associated with $\mathfrak{i}$ in a unique way, independent of the uncontractible representation chosen. If $\mathfrak{p}_j$ is not a proper factor of any other prime ideal $\mathfrak{p}_i$ of the set, $\mathfrak{q}_j$ is called an *isolated component* of $\mathfrak{i}$, and if $\mathfrak{p}_j$ is a proper factor of some $\mathfrak{p}_i$, $\mathfrak{q}_j$ is said to be *embedded*. While the embedded components of $\mathfrak{i}$ need not be uniquely determined we can prove

THEOREM IX. *The isolated components of an ideal* $\mathfrak{i}$ *are uniquely determined.*

Let $[\mathfrak{q}_1, \ldots, \mathfrak{q}_k]$ and $[\mathfrak{q}'_1, \ldots, \mathfrak{q}'_k]$ be two uncontractible representations of $\mathfrak{i}$, and suppose that the components are arranged so that $\mathfrak{q}_j$ and $\mathfrak{q}'_j$ are both $\mathfrak{p}_j$-primary. Let $\mathfrak{q}_1$ be an isolated component of $\mathfrak{i}$. Then $\mathfrak{p}_1$ is not a proper factor of any $\mathfrak{p}_j$ $(j > 1)$. Hence $\mathfrak{q}'_1$ is also an isolated component. For convenience, we write

$$\mathfrak{j} = [\mathfrak{q}_2, \ldots, \mathfrak{q}_k], \quad \mathfrak{j}' = [\mathfrak{q}'_2, \ldots, \mathfrak{q}'_k],$$

so that

$$[\mathfrak{q}_1, \mathfrak{j}] = \mathfrak{i} = [\mathfrak{q}'_1, \mathfrak{j}'].$$

Let $\mathfrak{q}_j$ be of index ρ_j. Then

$$\mathfrak{p}_2^{\rho_2} \ldots \mathfrak{p}_k^{\rho_k} = 0 \ \ (\mathfrak{j}).$$

If $\mathfrak{j} \subseteq \mathfrak{p}_1$, we have

$$\mathfrak{p}_2^{\rho_2} \ldots \mathfrak{p}_k^{\rho_k} = 0 \ \ (\mathfrak{p}_1),$$

and hence

$$\mathfrak{p}_i = 0 \ \ (\mathfrak{p}_1)$$

for some value of i greater than 1, contrary to the hypothesis that $\mathfrak{q}_1$ is isolated. Hence

$$\mathfrak{j} \neq 0 \ \ (\mathfrak{p}_1).$$

Therefore [Th. II, Cor. II],

$$\mathfrak{q}_1 = [\mathfrak{q}_1, \mathfrak{j}] : \mathfrak{j} = \mathfrak{i} : \mathfrak{j} = [\mathfrak{q}'_1, \mathfrak{j}'] : \mathfrak{j},$$

and hence

$$\mathfrak{q}_1 = 0 \ \ (\mathfrak{q}'_1).$$

Similarly, we show that

$$\mathfrak{j}' \neq 0 \ \ (\mathfrak{p}_1),$$

and deduce that

$$\mathfrak{q}'_1 = 0 \ \ (\mathfrak{q}_1).$$

Hence $\mathfrak{q}_1 = \mathfrak{q}'_1$, and the theorem follows.

A case of particular interest is that in which $\mathfrak{i} = \mathfrak{p}^\rho$, a power of a prime ideal $\mathfrak{p}$. Let

$$\mathfrak{p}^\rho = [\mathfrak{q}_1, \ldots, \mathfrak{q}_k]$$

be an uncontractible representation of $\mathfrak{p}^\rho$, and suppose that $\mathfrak{q}_i$ is $\mathfrak{p}_i$-primary. Then

$$\mathfrak{p}^\rho = 0 \ \ (\mathfrak{q}_j),$$

and hence

$$\mathfrak{p}^\rho = 0 \ \ (\mathfrak{p}_j),$$

from which it follows that

$$\mathfrak{p} = 0 \ \ (\mathfrak{p}_j).$$

Thus $\mathfrak{p}$ is a multiple of every $\mathfrak{p}_j$. Since $\mathfrak{p}_1, \ldots, \mathfrak{p}_k$ are all different, $\mathfrak{p}$ can be equal to at most one $\mathfrak{p}_j$. We show that it is equal to exactly one of them. For, if $\mathfrak{p}$ is a proper multiple of each $\mathfrak{p}_j$, we can find in $\mathfrak{p}_j$ an element α_j not in $\mathfrak{p}$ $(j = 1, \ldots, k)$. Let σ_j be an integer such that

$$\alpha_j^{\sigma_j} = 0 \ \ (\mathfrak{q}_j).$$

Then, if $\sigma = \max[\sigma_1, \ldots, \sigma_k]$,

$$\begin{aligned}(\alpha_1 \ldots \alpha_k)^\sigma &= 0 \ \ ([\mathfrak{q}_1, \ldots, \mathfrak{q}_k]) \\ &= 0 \ \ (\mathfrak{p}),\end{aligned}$$

and since $\mathfrak{p}$ is prime it follows that some α_j lies in $\mathfrak{p}$, and we have a contradiction. It follows that there is one $\mathfrak{p}_j$ which is equal to $\mathfrak{p}$, and we may suppose the components arranged so that this is $\mathfrak{p}_1$. Since

$$\mathfrak{p}_1 = 0 \ \ (\mathfrak{p}_j) \quad (j = 2, \ldots, k),$$

$\mathfrak{q}_1$ is an isolated component of $\mathfrak{p}^\rho$, and $\mathfrak{q}_2, \ldots, \mathfrak{q}_k$ are embedded. By Theorem IX, $\mathfrak{q}_1$ does not depend on the uncontractible representation of $\mathfrak{p}^\rho$ chosen. We call it the *symbolic ρth power* of $\mathfrak{p}$, and denote it by $\mathfrak{p}^{(\rho)}$.

We conclude this section with some further results, which we shall need later, which concern integral domains in which the Basis Theorem holds.

Let $\mathfrak{R}$ be an integral domain in which the Basis Theorem holds, and let $\mathfrak{i}$ and $\mathfrak{j}$ be two ideals in it. We shall assume that $\mathfrak{i}$ is a proper ideal, but $\mathfrak{j}$ may be the zero ideal. The case in which $\mathfrak{j}$ is the unit ideal is of no interest. We consider the set of elements of $\mathfrak{R}$ which belong to $(\mathfrak{i}^n, \mathfrak{j})$ for all positive integral values of n. Let α and β be two such elements of $\mathfrak{R}$, and let ρ denote any element of $\mathfrak{R}$. Since

$$\alpha = 0 \ \ ((\mathfrak{i}^n, \mathfrak{j})) \quad \text{and} \quad \beta = 0 \ \ ((\mathfrak{i}^n, \mathfrak{j}))$$

for all values of n, we have

$$\alpha-\beta=0\ ((\mathfrak{i}^n,\mathfrak{j}))\quad\text{and}\quad\rho\alpha=0\ ((\mathfrak{i}^n,\mathfrak{j}))$$

for all values of n. Hence the elements of $\mathfrak{R}$ which belong to $(\mathfrak{i}^n,\mathfrak{j})$ for all values of n form an ideal, which we denote by $\mathfrak{k}$. We first prove the result:

$$(\mathfrak{i}\mathfrak{k},\mathfrak{j})=\mathfrak{k}.$$

Since $$\mathfrak{i}\mathfrak{k}\subseteq\mathfrak{k}\quad\text{and}\quad\mathfrak{j}\subseteq\mathfrak{k},$$

we have $$(\mathfrak{i}\mathfrak{k},\mathfrak{j})\subseteq\mathfrak{k}. \tag{3}$$

Let $[\mathfrak{q}_1,\ldots,\mathfrak{q}_r]$ be a representation of $(\mathfrak{i}\mathfrak{k},\mathfrak{j})$ as an uncontractible intersection of primaries, and let $\mathfrak{p}_h$ be the radical of $\mathfrak{q}_h$ $(h=1,\ldots,r)$. (If $(\mathfrak{i}\mathfrak{k},\mathfrak{j})$ is the zero ideal, $r=1$, and $\mathfrak{p}_1=\mathfrak{q}_1$, since $\mathfrak{R}$ is an integral domain.)

Case I. If $$\mathfrak{i}\neq 0\ (\mathfrak{p}_h),$$

we have $$\mathfrak{k}\subseteq\mathfrak{i}\mathfrak{k}:\mathfrak{i}\subseteq(\mathfrak{i}\mathfrak{k},\mathfrak{j}):\mathfrak{i}\subseteq\mathfrak{q}_h:\mathfrak{i}=\mathfrak{q}_h$$
[Th. II, Cor. II].

Case II. If $$\mathfrak{i}=0\ (\mathfrak{p}_h),$$

we have $$\mathfrak{i}^n=0\ (\mathfrak{q}_h),$$

for all n not less than the index of $\mathfrak{q}_h$. Also

$$\mathfrak{j}\subseteq(\mathfrak{i}\mathfrak{k},\mathfrak{j})\subseteq\mathfrak{q}_h,$$

so that $$\mathfrak{k}\subseteq(\mathfrak{i}^n,\mathfrak{j})\subseteq\mathfrak{q}_h.$$

Hence in either case $$\mathfrak{k}\subseteq\mathfrak{q}_h,$$

and since this holds for all values of h,

$$\mathfrak{k}\subseteq(\mathfrak{i}\mathfrak{k},\mathfrak{j}). \tag{4}$$

From (3) and (4) we conclude that

$$(\mathfrak{i}\mathfrak{k},\mathfrak{j})=\mathfrak{k}.$$

Now let $\gamma_1,\ldots,\gamma_s$ be a basis for $\mathfrak{k}$. It follows that

$$\gamma_i=\sum_{j=1}^{s}\alpha_{ij}\gamma_j+\beta_i\quad(i=1,\ldots,s),$$

where $\alpha_{ij}\in\mathfrak{i}$ and $\beta_i\in\mathfrak{j}$. If $\Delta=|\delta_{ij}-\alpha_{ij}|$, we obtain from these equations the equation

$$\Delta\gamma_i=0\ (\mathfrak{j}), \tag{5}$$

and from the definition of Δ,

$$\Delta - 1 = 0 \ (\mathfrak{i}).$$

Hence $$\Delta \neq 0 \ (\mathfrak{i}),$$
since $\mathfrak{i}$ is not the unit ideal.

We now consider some special cases.

A. Suppose that $\mathfrak{j}$ is the zero ideal. From (5) we have $\gamma_i = 0$ $(i = 1, \ldots, s)$, since Δ is not zero, and $\mathfrak{R}$ is an integral domain. Hence $\mathfrak{k}$ is the zero ideal.

B. Suppose that $\mathfrak{j}$ is $\mathfrak{p}$-primary, and that $\mathfrak{p} \subseteq \mathfrak{i}$. Then Δ is not contained in $\mathfrak{p}$, so that [Th. II, Cor. I] $\gamma_i \in \mathfrak{j}$ $(i = 1, \ldots, s)$, and

$$\mathfrak{k} \subseteq \mathfrak{j}.$$

But $\mathfrak{j} \subseteq \mathfrak{k}$. Hence $$\mathfrak{k} = \mathfrak{j}.$$

C. Suppose that $(\mathfrak{i}, \mathfrak{j})$ is the unit ideal. Then $(\mathfrak{i}^n, \mathfrak{i}^{n-1}\mathfrak{j}, \ldots, \mathfrak{i}\mathfrak{j}^{n-1}, \mathfrak{j}^n)$ is the unit ideal, and since this is contained in $(\mathfrak{i}^n, \mathfrak{j})$, this last ideal is the unit ideal, for all values of n. In this case it is clear that $\mathfrak{k}$ is the unit ideal.

D. Suppose that $\mathfrak{i} = \mathfrak{p}$ is a maximal ideal of $\mathfrak{R}$.† Let

$$\mathfrak{j} = [\mathfrak{q}'_1, \ldots, \mathfrak{q}'_l, \mathfrak{q}''_1, \ldots, \mathfrak{q}''_m],$$

where $\mathfrak{q}'_h$ is $\mathfrak{p}'_h$-primary, and $\mathfrak{q}''_h$ is $\mathfrak{p}''_h$-primary. Suppose that

$$\mathfrak{p}'_h \subseteq \mathfrak{p} \quad (h = 1, \ldots, l) \quad \text{and} \quad \mathfrak{p}''_h \nsubseteq \mathfrak{p} \quad (h = 1, \ldots, m).$$

If $\mathfrak{k}'_h$ is the intersection of the ideals $(\mathfrak{p}^n, \mathfrak{q}'_h)$ for all n, and $\mathfrak{k}''_h$ is the intersection of ideals $(\mathfrak{p}^n, \mathfrak{q}''_h)$ for all n, it follows from B that $\mathfrak{k}'_h = \mathfrak{q}'_h$, and it is easily proved that $\mathfrak{k}''_h$ is the unit ideal. Indeed, since $\mathfrak{p}''_h$ is not contained in $\mathfrak{p}$, $(\mathfrak{p}''_h, \mathfrak{p})$ is a factor of $\mathfrak{p}$ not coincident with $\mathfrak{p}$, so that since $\mathfrak{p}$ is a maximal ideal, $(\mathfrak{p}''_h, \mathfrak{p})$ is the unit ideal. If ρ is the index of $\mathfrak{q}''_h$,

$$(\mathfrak{q}''_h, \mathfrak{p}) \supseteq (\mathfrak{p}''_h, \mathfrak{p})^\rho,$$

and hence it is the unit ideal. We therefore deduce the result from C.

Since $$(\mathfrak{p}^n, \mathfrak{j}) \subseteq (\mathfrak{p}^n, \mathfrak{q}'_h),$$
and $$(\mathfrak{p}^n, \mathfrak{j}) \subseteq (\mathfrak{p}^n, \mathfrak{q}''_h),$$
for all values of n, we deduce that

$$\mathfrak{k} \subseteq [\mathfrak{k}'_1, \ldots, \mathfrak{k}'_l, \mathfrak{k}''_1, \ldots, \mathfrak{k}''_m]$$
$$= [\mathfrak{q}'_1, \ldots, \mathfrak{q}'_l].$$

† It will be proved in §3, Theorem III, that every maximal ideal in an integral domain $\mathfrak{R}$ is prime. This result is not, however, assumed here.

Let
$$\mathfrak{j}_1 = [\mathfrak{q}'_1, \ldots, \mathfrak{q}'_l] \quad \text{and} \quad \mathfrak{j}_2 = [\mathfrak{q}''_1, \ldots, \mathfrak{q}''_m].$$
Then $\mathfrak{j}_2$ is not contained in $\mathfrak{p}$, and, since $\mathfrak{p}$ is maximal, we have
$$(\mathfrak{j}_2, \mathfrak{p}^n) \supseteq (\mathfrak{j}_2, \mathfrak{p})^n = \mathfrak{R},$$
so that
$$(\mathfrak{j}_2, \mathfrak{p}^n) = \mathfrak{R}.$$
But
$$\mathfrak{j}_1 = \mathfrak{j}_1(\mathfrak{j}_2, \mathfrak{p}^n) = (\mathfrak{j}_1\mathfrak{j}_2, \mathfrak{j}_1\mathfrak{p}^n) \subseteq (\mathfrak{j}, \mathfrak{p}^n)$$
for all values of n. Hence $\mathfrak{j}_1 \subseteq \mathfrak{k}$. But we have already seen that $\mathfrak{k} \subseteq \mathfrak{j}_1$. Hence
$$\mathfrak{k} = [\mathfrak{q}'_1, \ldots, \mathfrak{q}'_l].$$

We sum up the results which will be required in the sequel in

THEOREM X. *If $\mathfrak{i}$ and $\mathfrak{j}$ are ideals in an integral domain $\mathfrak{R}$ in which the Basis Theorem holds, and $\mathfrak{i}$ is a proper ideal, then:*

(1) *the only element of $\mathfrak{R}$ which lies in $\mathfrak{i}^n$ for all values of n is the zero of $\mathfrak{R}$;*

(2) *if $\mathfrak{i} = \mathfrak{p}$ is a maximal ideal of $\mathfrak{R}$ and $\mathfrak{j} = [\mathfrak{q}'_1, \ldots, \mathfrak{q}'_l, \mathfrak{q}''_1, \ldots, \mathfrak{q}''_m]$, where $\mathfrak{q}'_1, \ldots, \mathfrak{q}'_l$ are primary components of $\mathfrak{j}$ which are multiples of $\mathfrak{p}$, and $\mathfrak{q}''_1, \ldots, \mathfrak{q}''_m$ are primary components of $\mathfrak{j}$ which are not multiples of $\mathfrak{p}$, then the intersection of the ideals $(\mathfrak{p}^n, \mathfrak{j})$ for $n = 1, 2, \ldots$ is the ideal $[\mathfrak{q}'_1, \ldots, \mathfrak{q}'_l]$.*

It is convenient to prove here a lemma which follows from this result.

Let $\mathfrak{p}$ be a maximal ideal in $\mathfrak{R}$, and let $\mathfrak{q}'$ be a primary ideal in $\mathfrak{R}$, with radical $\mathfrak{p}'$. We have already seen that if $\mathfrak{p}'$ is not contained in $\mathfrak{p}$,
$$(\mathfrak{q}', \mathfrak{p}) = (\mathfrak{q}', \mathfrak{p}^2) = \ldots = \mathfrak{R}.$$
Suppose, now, that $\mathfrak{p}' \subseteq \mathfrak{p}$. Then
$$\mathfrak{p} = (\mathfrak{q}', \mathfrak{p}) \supseteq (\mathfrak{q}', \mathfrak{p}^2) \supseteq (\mathfrak{q}', \mathfrak{p}^3) \supseteq \ldots.$$
We can show that $(\mathfrak{q}', \mathfrak{p}^n)$ is $\mathfrak{p}$-primary. Let $[\mathfrak{q}_1, \ldots, \mathfrak{q}_k]$ be an uncontractible representation of $(\mathfrak{q}', \mathfrak{p}^n)$, and let $\mathfrak{p}_i$ be the radical of $\mathfrak{q}_i$. Then
$$\mathfrak{p}^n = 0 \ \ (\mathfrak{q}_i),$$
and hence
$$\mathfrak{p}^n = 0 \ \ (\mathfrak{p}_i);$$
therefore
$$\mathfrak{p} = 0 \ \ (\mathfrak{p}_i).$$
Since $\mathfrak{p}$ is maximal, $\mathfrak{p}_i = \mathfrak{p}$, and since $\mathfrak{p}_1, \ldots, \mathfrak{p}_k$ are all distinct we must have $k = 1$, that is,
$$(\mathfrak{q}', \mathfrak{p}^n) = \mathfrak{q}_1.$$

We are interested in the case in which $(\mathfrak{q}', \mathfrak{p}^2) = \mathfrak{p}$. We show that this implies $(\mathfrak{q}', \mathfrak{p}^n) = \mathfrak{p}$ for all values of n. Let $\omega_1, \ldots, \omega_r$ be a basis for $\mathfrak{p}$. Then since $\mathfrak{p} = (\mathfrak{q}', \mathfrak{p}^2)$,

$$\omega_i = \beta_i + f_i^2(\omega_1, \ldots, \omega_r) \quad (i = 1, \ldots, r),$$

where $f_i^k(x_1, \ldots, x_r)$ denotes a homogeneous polynomial in $x_1, \ldots, x_r$ of degree k, with coefficients in $\mathfrak{R}$, and β_i is in $\mathfrak{q}'$. Hence

$$\begin{aligned} \omega_i &= \beta_i + f_i^2(\beta_1 + f_1^2(\omega), \ldots, \beta_r + f_r^2(\omega)) \\ &= \beta_i' + f_i^4(\omega_1, \ldots, \omega_r) \quad (i = 1, \ldots, r), \end{aligned}$$

where β_i' is in q'. Hence $\mathfrak{p} \subseteq (\mathfrak{q}', \mathfrak{p}^4)$, and since

$$\mathfrak{p} \supseteq (\mathfrak{q}', \mathfrak{p}^3) \supseteq (\mathfrak{q}', \mathfrak{p}^4),$$

we obtain
$$\mathfrak{p} = (\mathfrak{q}', \mathfrak{p}^3) = (\mathfrak{q}', \mathfrak{p}^4).$$

Proceeding in this way, we obtain

$$(\mathfrak{q}', \mathfrak{p}^n) = \mathfrak{p}$$

for all values of n.

It follows from this that the intersection of ideals $(\mathfrak{q}', \mathfrak{p}^n)$ for $n = 1, 2, \ldots$ is the ideal $\mathfrak{p}$. But, from Theorem X, we know that it is also $\mathfrak{q}'$. Hence $\mathfrak{q}' = \mathfrak{p}$. Thus we have the

Lemma. Let $\mathfrak{p}$ be a maximal ideal in an integral domain $\mathfrak{R}$ in which the Basis Theorem holds, and let $\mathfrak{q}'$ be a $\mathfrak{p}'$-primary ideal in $\mathfrak{R}$. Then

(i) *if* $\mathfrak{p}' \neq 0 \ (\mathfrak{p})$,
$$(\mathfrak{q}', \mathfrak{p}) = (\mathfrak{q}', \mathfrak{p}^2) = \ldots = \mathfrak{R};$$

(ii) *if* $\mathfrak{p}' = 0 \ (\mathfrak{p})$, $(\mathfrak{q}', \mathfrak{p}^n)$ *is $\mathfrak{p}$-primary, and is prime if and only if* $\mathfrak{q}' = \mathfrak{p}$. *The conditions* $\mathfrak{q}' = \mathfrak{p}$ *and* $(\mathfrak{q}', \mathfrak{p}^2) = \mathfrak{p}$ *are equivalent.*

3. Remainder-class rings. Let $\mathfrak{i}$ be an ideal in the commutative ring $\mathfrak{R}$. Two elements α and β of $\mathfrak{R}$ are said to be congruent with respect to $\mathfrak{i}$ if

$$\alpha - \beta = 0 \ (\mathfrak{i}).$$

It can be seen at once that this congruence relation is reflexive, symmetric and transitive, and is therefore a true equivalence relation. The elements of $\mathfrak{R}$ can therefore be divided into mutually exclusive sets of elements congruent with respect to $\mathfrak{i}$. We denote the set to which α belongs by $\bar{\alpha}$. Thus $\bar{\alpha} = \bar{\beta}$ if and only if $\alpha - \beta \in \mathfrak{i}$. If

$$\alpha - \alpha' = 0 \ (\mathfrak{i}) \quad \text{and} \quad \beta - \beta' = 0 \ (\mathfrak{i}),$$

then
$$(\alpha+\beta)-(\alpha'+\beta') = 0 \ (\mathfrak{i}),$$
and
$$\alpha\beta-\alpha'\beta' = \alpha(\beta-\beta')+\beta'(\alpha-\alpha') = 0 \ (\mathfrak{i}).$$

Hence we can define addition and multiplication of sets, and we can verify at once that with these definitions the operations have the properties required of addition and multiplication in a ring. Thus the sets constitute a ring, which we denote by $\overline{\mathfrak{R}}$. There is a mapping of $\mathfrak{R}$ on $\overline{\mathfrak{R}}$, given by $\alpha \to \overline{\alpha}$, with the properties

$$\overline{\alpha+\beta} = \overline{\alpha}+\overline{\beta}, \quad \overline{\alpha\beta} = \overline{\alpha}\overline{\beta};$$

a mapping with these properties is called *a homomorphism*. The elements of $\mathfrak{R}$ which map on the zero of $\overline{\mathfrak{R}}$ are the elements of $\mathfrak{i}$.

Conversely, if we have any homomorphism $\alpha \to \overline{\alpha}$ of the ring $\mathfrak{R}$ on a ring $\overline{\mathfrak{R}}$, we consider the elements of $\mathfrak{R}$ which map on the zero of $\overline{\mathfrak{R}}$. If α and β are such that $\overline{\alpha} = \overline{\beta} = 0$, and if ρ is any element of $\mathfrak{R}$, we have

$$\overline{\alpha-\beta} = \overline{\alpha}-\overline{\beta} = 0 \quad \text{and} \quad \overline{\rho\alpha} = \overline{\rho}\,\overline{\alpha} = 0.$$

Hence the elements of $\mathfrak{R}$ which map on the zero of $\overline{\mathfrak{R}}$ form an ideal, $\mathfrak{i}$, say, in $\mathfrak{R}$. Two elements α and β of $\mathfrak{R}$ map on the same element of $\overline{\mathfrak{R}}$ if and only if $\alpha-\beta$ maps on the zero of $\overline{\mathfrak{R}}$, that is, if and only if

$$\alpha-\beta = 0 \ (\mathfrak{i}),$$

and hence any homomorphism of $\mathfrak{R}$ is obtained in the manner described for a suitable choice of the ideal $\mathfrak{i}$. The ring $\overline{\mathfrak{R}}$ obtained from the ideal $\mathfrak{i}$ of $\mathfrak{R}$ is called a *remainder-class ring*, and is usually denoted by $\mathfrak{R}/\mathfrak{i}$. If $\mathfrak{i}$ is the zero ideal, $\mathfrak{R}/\mathfrak{i}$ is clearly isomorphic with $\mathfrak{R}$, and if $\mathfrak{i}$ is the unit ideal, $\mathfrak{R}/\mathfrak{i}$ contains only zero.

Since a field has only two ideals, the unit ideal and the zero ideal, we deduce

THEOREM I. *A homomorphism of a field either maps every element on zero, or is an isomorphism.*

Let $\mathfrak{i}$, $\mathfrak{j}$ be two ideals of $\mathfrak{R}$, and consider the map of $\mathfrak{j}$ on $\mathfrak{R}/\mathfrak{i}$. If $\mathfrak{j} \subseteq \mathfrak{i}$, every element of $\mathfrak{j}$ maps on the zero of $\mathfrak{R}/\mathfrak{i}$, and the map of $\mathfrak{j}$ is the zero ideal of $\mathfrak{R}/\mathfrak{i}$. Now suppose that $\mathfrak{j}$ is not contained in $\mathfrak{i}$. If $\overline{\alpha}$ and $\overline{\beta}$ are elements of $\mathfrak{R}/\mathfrak{i}$ which are maps of elements α and β of $\mathfrak{j}$, and $\overline{\rho}$ is any element of $\mathfrak{R}/\mathfrak{i}$, the map of an element ρ of $\mathfrak{R}$, then $\overline{\alpha}-\overline{\beta}$ and $\overline{\rho}\overline{\alpha}$ are the maps of elements $\alpha-\beta$ and $\rho\alpha$ of $\mathfrak{j}$. Hence the elements of $\mathfrak{R}/\mathfrak{i}$ which are maps of elements of $\mathfrak{j}$ form an ideal in $\mathfrak{R}/\mathfrak{i}$, which we denote by $\mathfrak{j}/\mathfrak{i}$.

Now let $\bar{\mathfrak{k}}$ be any ideal of $\mathfrak{R}/\mathfrak{i}$, and let α and β be any elements of $\mathfrak{R}$ which map on elements $\bar{\alpha}$ and $\bar{\beta}$ of $\bar{\mathfrak{k}}$. If ρ is any element of $\mathfrak{R}$ and $\bar{\rho}$ its map on $\mathfrak{R}/\mathfrak{i}$, $\alpha-\beta$ and $\rho\alpha$ map on $\bar{\alpha}-\bar{\beta}$ and $\bar{\rho}\bar{\alpha}$, which are elements of $\bar{\mathfrak{k}}$. Hence the elements of $\mathfrak{R}$ which map on elements of $\bar{\mathfrak{k}}$ form an ideal $\mathfrak{k}$ in $\mathfrak{R}$. Since any element of $\mathfrak{i}$ maps on the zero of $\mathfrak{R}/\mathfrak{i}$, which lies in $\bar{\mathfrak{k}}$, $\mathfrak{i} \subseteq \mathfrak{k}$. If $\bar{\mathfrak{k}} = \mathfrak{j}/\mathfrak{i}$, clearly $\mathfrak{j} \subseteq \mathfrak{k}$, but we do not necessarily have $\mathfrak{j} = \mathfrak{k}$. Indeed, a necessary and sufficient condition that an element α of $\mathfrak{R}$ map on an element of $\mathfrak{j}/\mathfrak{i}$ is that there should exist an element β of $\mathfrak{j}$ such that $\alpha-\beta \in \mathfrak{i}$. Hence, if $\bar{\mathfrak{k}} = \mathfrak{j}/\mathfrak{i}$,

$$\mathfrak{k} = (\mathfrak{i}, \mathfrak{j}),$$

and a necessary and sufficient condition that $\mathfrak{k} = \mathfrak{j}$ is therefore that

$$\mathfrak{i} = 0 \quad (\mathfrak{j}).$$

From this, we have at once

THEOREM II. *Any ideal* $\mathfrak{j}$ *of* $\mathfrak{R}$ *maps on an ideal* $\mathfrak{j}/\mathfrak{i}$ *of* $\mathfrak{R}/\mathfrak{i}$, *and the largest ideal of* $\mathfrak{R}$ *which maps on* $\mathfrak{j}/\mathfrak{i}$ *is* $(\mathfrak{i}, \mathfrak{j})$, *which contains all ideals which map on* $\mathfrak{j}/\mathfrak{i}$. *There is a one-to-one correspondence between the ideals of* $\mathfrak{R}$ *which are factors of* $\mathfrak{i}$ *and the ideals of* $\mathfrak{R}/\mathfrak{i}$.

In virtue of this result, when we consider the relation between ideals in $\mathfrak{R}$ and ideals in $\mathfrak{R}/\mathfrak{i}$, we may confine ourselves to ideals in $\mathfrak{R}$ which are factors of $\mathfrak{i}$.

If the ring $\mathfrak{R}$ has unity, and $\mathfrak{i}$ is not the unit ideal, it is seen at once that the image of the unity of $\mathfrak{R}$ in $\mathfrak{R}/\mathfrak{i}$ is the unity of $\mathfrak{R}/\mathfrak{i}$.

We now use remainder-class rings to prove

THEOREM III. *A maximal ideal of any commutative ring* $\mathfrak{R}$ *is primary. If* $\mathfrak{R}$ *has unity, any maximal ideal of* $\mathfrak{R}$ *is prime.*

Let $\mathfrak{i}$ be a maximal ideal of $\mathfrak{R}$. Then $\mathfrak{R}/\mathfrak{i}$ can have no proper ideals. For, if $\bar{\mathfrak{j}}$ is any ideal of $\mathfrak{R}/\mathfrak{i}$, the largest corresponding ideal of $\mathfrak{R}$ is a factor of $\mathfrak{i}$, and hence, since $\mathfrak{i}$ is maximal, is either $\mathfrak{i}$ or the unit ideal. Hence $\bar{\mathfrak{j}}$ is either the zero ideal or the unit ideal of $\mathfrak{R}/\mathfrak{i}$.

Since $\mathfrak{R}/\mathfrak{i}$ has no proper ideals, it is either a field, or a ring consisting of a finite number of elements $\bar{\alpha}_1, \ldots, \bar{\alpha}_p$, where $\bar{\alpha}_i\bar{\alpha}_j = 0$ [§ 1, Th. I].

(i) Suppose that $\mathfrak{R}/\mathfrak{i}$ is a field. This is certainly the case if $\mathfrak{R}$ has unity, for then $\mathfrak{R}/\mathfrak{i}$ has unity. Let α and β be two elements of $\mathfrak{R}$ such that $\alpha\beta \in \mathfrak{i}$. Then, if $\bar{\alpha}$ and $\bar{\beta}$ are the corresponding elements of $\mathfrak{R}/\mathfrak{i}$,

$$\bar{\alpha}\bar{\beta} = 0.$$

Since $\mathfrak{R}/\mathfrak{i}$ is a field, either $\bar{\alpha}$ or $\bar{\beta}$ is zero, and hence either α or β is in $\mathfrak{i}$. Hence $\mathfrak{i}$ is prime.

(ii) Suppose that $\mathfrak{R}/\mathfrak{i}$ consists of a finite number of elements $\bar{\alpha}_1, \ldots, \bar{\alpha}_p$. If α is any element of $\mathfrak{R}$, the corresponding element $\bar{\alpha}$ of $\mathfrak{R}/\mathfrak{i}$ is an $\bar{\alpha}_i$, and $\bar{\alpha}_i^2 = 0$. Therefore $\alpha^2 \in \mathfrak{i}$. In particular, if α and β are elements of $\mathfrak{R}$ such that

$$\alpha\beta = 0 \ (\mathfrak{i}) \quad \text{and} \quad \alpha \neq 0 \ (\mathfrak{i}),$$

then

$$\beta^2 = 0 \ (\mathfrak{i}),$$

and hence $\mathfrak{i}$ is primary. We note that since the square of any element of $\mathfrak{R}$ is in $\mathfrak{i}$, the radical $\mathfrak{p}$ of $\mathfrak{i}$ is the unit ideal, and

$$\mathfrak{p}^2 = 0 \ (\mathfrak{i}).$$

Since $\mathfrak{p} \not\subset \mathfrak{i}$, $\mathfrak{i}$ is not prime.

A further consequence of our reasoning is that if $\mathfrak{i}$ is maximal and prime, $\mathfrak{R}/\mathfrak{i}$ is a field. We can show, conversely, that if $\mathfrak{R}/\mathfrak{i}$ is a field, $\mathfrak{i}$ is maximal and prime. Indeed, if $\mathfrak{i}$ were not maximal, it would have a proper factor $\mathfrak{j}$ which is not the unit ideal, and hence $\mathfrak{j}/\mathfrak{i}$ would be a proper ideal of $\mathfrak{R}/\mathfrak{i}$, whereas we know that a field has no proper ideals. If α and β are any two elements of $\mathfrak{R}$ such that $\alpha\beta \in \mathfrak{i}$, and $\bar{\alpha}$ and $\bar{\beta}$ are the corresponding elements of $\mathfrak{R}/\mathfrak{i}$,

$$\bar{\alpha}\bar{\beta} = 0,$$

and hence $\bar{\alpha}$ or $\bar{\beta}$ is zero; that is, α or β is in $\mathfrak{i}$. Hence $\mathfrak{i}$ is prime. Thus we have

THEOREM IV. *A necessary and sufficient condition that* $\mathfrak{R}/\mathfrak{i}$ *be a field is that* $\mathfrak{i}$ *be maximal and prime.*

Similar reasoning shows that a necessary and sufficient condition that $\mathfrak{R}/\mathfrak{i}$ be without divisors of zero is that $\mathfrak{i}$ be prime. For, if $\alpha\beta \in \mathfrak{i}$, then $\bar{\alpha}\bar{\beta}$ is zero, and if $\mathfrak{R}/\mathfrak{i}$ has no divisors of zero, then either $\bar{\alpha}$ or $\bar{\beta}$ is zero; that is, either α or β is in $\mathfrak{i}$; hence $\mathfrak{i}$ is prime. Conversely, if $\mathfrak{i}$ is prime, $\bar{\alpha}\bar{\beta} = 0$ implies $\alpha\beta \in \mathfrak{i}$, so that α or β is in $\mathfrak{i}$. Therefore $\bar{\alpha}$ or $\bar{\beta}$ is zero, and we conclude that $\mathfrak{R}/\mathfrak{i}$ has no divisors of zero. This fact is sometimes taken as the basis of another definition of a prime ideal.

THEOREM V. *If* $\mathfrak{j}$ *is a factor of* $\mathfrak{i}$, $\mathfrak{j}/\mathfrak{i}$ *is primary if and only if* $\mathfrak{j}$ *is primary. If* $\mathfrak{j}$ *is* $\mathfrak{p}$*-primary,* $\mathfrak{j}/\mathfrak{i}$ *is* $(\mathfrak{p}/\mathfrak{i})$*-primary.*

Let us suppose, first, that $\mathfrak{j}$ is primary and that $\mathfrak{p}$ is its radical. Then $\mathfrak{p} \supseteq \mathfrak{j} \supseteq \mathfrak{i}$.

(i) Let $\bar{\alpha}$, $\bar{\beta}$ be elements of $\mathfrak{R}/\mathfrak{i}$ such that

$$\bar{\alpha}\bar{\beta} = 0 \ \ (\mathfrak{j}/\mathfrak{i}), \quad \bar{\alpha} \neq 0 \ \ (\mathfrak{j}/\mathfrak{i}).$$

If α and β are any two elements of $\mathfrak{R}$ which map on $\bar{\alpha}$ and $\bar{\beta}$, respectively, we have

$$\alpha\beta = 0 \ \ (\mathfrak{j}), \quad \alpha \neq 0 \ \ (\mathfrak{j}).$$

Since $\mathfrak{j}$ is $\mathfrak{p}$-primary, it follows that $\beta \in \mathfrak{p}$. Hence

$$\bar{\beta} = 0 \ \ (\mathfrak{p}/\mathfrak{i}).$$

(ii) Since
$$\mathfrak{j} = 0 \ \ (\mathfrak{p}),$$
it follows that
$$\mathfrak{j}/\mathfrak{i} = 0 \ \ (\mathfrak{p}/\mathfrak{i}).$$

(iii) If
$$\bar{\beta} = 0 \ \ (\mathfrak{p}/\mathfrak{i}),$$
we have
$$\beta = 0 \ \ (\mathfrak{p}),$$
and hence
$$\beta^{\rho} = 0 \ \ (\mathfrak{j}),$$
for some integer ρ. Hence
$$\bar{\beta}^{\rho} = 0 \ \ (\mathfrak{j}/\mathfrak{i}).$$

It follows from § 2, Th. I, that $\mathfrak{j}/\mathfrak{i}$ is $(\mathfrak{p}/\mathfrak{i})$-primary.

Next, suppose that $\mathfrak{j}/\mathfrak{i}$ is primary. Let $\bar{\mathfrak{p}}$ be its radical, $\mathfrak{p}$ the largest ideal in $\mathfrak{R}$ which maps on $\bar{\mathfrak{p}}$. Then $\bar{\mathfrak{p}} = \mathfrak{p}/\mathfrak{i}$. We again denote elements of $\mathfrak{R}$ by $\alpha, \beta, \ldots$ and their maps on $\mathfrak{R}/\mathfrak{i}$ by $\bar{\alpha}, \bar{\beta}, \ldots$.

(i) If α and β are elements of $\mathfrak{R}$ such that

$$\alpha\beta = 0 \ \ (\mathfrak{j}), \quad \alpha \neq 0 \ \ (\mathfrak{j}),$$

then

$$\bar{\alpha}\bar{\beta} = 0 \ \ (\mathfrak{j}/\mathfrak{i}), \quad \bar{\alpha} \neq 0 \ \ (\mathfrak{j}/\mathfrak{i}).$$

Since $\mathfrak{j}/\mathfrak{i}$ is $(\mathfrak{p}/\mathfrak{i})$-primary, $\bar{\beta} \in (\mathfrak{p}/\mathfrak{i})$, and so $\beta \in \mathfrak{p}$.

(ii) Since $\mathfrak{j}/\mathfrak{i} = 0 \ \ (\mathfrak{p}/\mathfrak{i})$, we have $\mathfrak{j} = 0 \ \ (\mathfrak{p})$.

(iii) If $\beta \in \mathfrak{p}$, $\bar{\beta} \in \mathfrak{p}/\mathfrak{i}$; hence, for some integer ρ, $\bar{\beta}^{\rho} \in \mathfrak{j}/\mathfrak{i}$. Therefore $\beta^{\rho} \in \mathfrak{j}$. Hence [§ 2, Th. I], $\mathfrak{j}$ is $\mathfrak{p}$-primary.

Suppose, now, that there exists an integer σ such that

$$(\mathfrak{p}/\mathfrak{i})^{\sigma} = 0 \ \ (\mathfrak{j}/\mathfrak{i}).$$

If $\beta \in \mathfrak{p}^{\sigma}$,
$$\beta = \sum_i \alpha_{i1} \ldots \alpha_{i\sigma}, \quad \text{where} \quad \alpha_{ij} \in \mathfrak{p}.$$

Hence
$$\bar{\beta} = \sum_i \bar{\alpha}_{i1} \ldots \bar{\alpha}_{i\sigma} = 0 \ \ ((\mathfrak{p}/\mathfrak{i})^{\sigma}).$$

By hypothesis $\bar{\beta} \in \mathfrak{j}/\mathfrak{i}$, and therefore $\beta \in \mathfrak{j}$. Hence

$$\mathfrak{p}^{\sigma} = 0 \ \ (\mathfrak{j}).$$

Conversely, we show in a similar manner that if

$$\mathfrak{p}^{\sigma} = 0 \ \ (\mathfrak{j}),$$

then
$$(\mathfrak{p}/\mathfrak{i})^{\sigma} = 0 \ \ (\mathfrak{j}/\mathfrak{i}).$$

We thus have

Corollary I. If $\mathfrak{j} \supseteq \mathfrak{i}$ *is* $\mathfrak{p}$*-primary and* $\mathfrak{p}^{\sigma} = 0 \ (\mathfrak{j})$, *then*

$$(\mathfrak{p}/\mathfrak{i})^{\sigma} = 0 \ \ (\mathfrak{j}/\mathfrak{i});$$

and if there exists an integer ρ *such that*

$$(\mathfrak{p}/\mathfrak{i})^{\rho} = 0 \ \ (\mathfrak{j}/\mathfrak{i}), \quad \textit{then} \quad \mathfrak{p}^{\rho} = 0 \ \ (\mathfrak{j}).$$

Corollary II. If $\mathfrak{j} \supseteq \mathfrak{i}$, $\mathfrak{j}$ *is prime if and only if* $\mathfrak{j}/\mathfrak{i}$ *is prime.*

This follows from Corollary I by taking $\rho = \sigma = 1$.

Corollary I implies that when the Basis Theorem holds in $\mathfrak{R}$ and in $\mathfrak{R}/\mathfrak{i}$, corresponding primaries have the same index (see Th. VII).

Let $\mathfrak{i}$, $\mathfrak{j}$ be any two ideals in a ring $\mathfrak{R}$, where

$$\mathfrak{i} = 0 \ \ (\mathfrak{j}).$$

There is a homomorphism from $\mathfrak{R}$ to $\mathfrak{R}/\mathfrak{i}$, in which $\mathfrak{j}$ defines the ideal $\mathfrak{j}/\mathfrak{i}$ of $\mathfrak{R}/\mathfrak{i}$. In the same way we define a homomorphism from $\mathfrak{R}/\mathfrak{i}$ to $(\mathfrak{R}/\mathfrak{i})/(\mathfrak{j}/\mathfrak{i})$. Combining the two homomorphisms, we obtain a homomorphism of $\mathfrak{R}$ to $(\mathfrak{R}/\mathfrak{i})/(\mathfrak{j}/\mathfrak{i})$. An element α of $\mathfrak{R}$ maps on the zero of this ring if and only if it maps on an element of $\mathfrak{j}/\mathfrak{i}$ of $\mathfrak{R}/\mathfrak{i}$, that is, if and only if it is in $\mathfrak{j}$. Hence $(\mathfrak{R}/\mathfrak{i})/(\mathfrak{j}/\mathfrak{i})$ is isomorphic with the remainder-class ring of $\mathfrak{R}$ defined by $\mathfrak{j}$, that is, $\mathfrak{R}/\mathfrak{j}$. Hence we have

THEOREM VI. *If* $\mathfrak{i}$ *and* $\mathfrak{j}$ *are ideals of* $\mathfrak{R}$ *such that* $\mathfrak{i} \subseteq \mathfrak{j}$, *then* $\mathfrak{R}/\mathfrak{j}$ *is isomorphic with* $(\mathfrak{R}/\mathfrak{i})/(\mathfrak{j}/\mathfrak{i})$.

THEOREM VII. *If the Basis Theorem holds in* $\mathfrak{R}$, *it holds in* $\mathfrak{R}/\mathfrak{i}$.

Let $\bar{\mathfrak{j}}$ be any ideal of $\mathfrak{R}/\mathfrak{i}$, $\mathfrak{j}$ the largest ideal of $\mathfrak{R}$ which maps on $\bar{\mathfrak{j}}$, and let $\omega_1, \ldots, \omega_r$ be a basis for $\mathfrak{j}$. If $\bar{\alpha}$ is any element of $\bar{\mathfrak{j}}$, α an element of $\mathfrak{R}$ corresponding to $\bar{\alpha}$, then α is in $\mathfrak{j}$. Therefore

$$\alpha = \rho_1 \omega_1 + \ldots + \rho_r \omega_r + n_1 \omega_1 + \ldots + n_r \omega_r,$$

where $\rho_1, \ldots, \rho_r$ are in $\mathfrak{R}$, and $n_1, \ldots, n_r$ are integers. Therefore

$$\bar{\alpha} = \bar{\rho}_1 \bar{\omega}_1 + \ldots + \bar{\rho}_r \bar{\omega}_r + n_1 \bar{\omega}_1 + \ldots + n_r \bar{\omega}_r.$$

Hence $$\bar{\mathfrak{j}} \subseteq \mathfrak{R}/\mathfrak{i}\,.\,(\bar{\omega}_1, \ldots, \bar{\omega}_r).$$

But $\bar{\omega}_1, \ldots, \bar{\omega}_r$ belong to $\bar{\mathfrak{j}}$. Hence

$$\mathfrak{R}/\mathfrak{i}\,.\,(\bar{\omega}_1, \ldots, \bar{\omega}_r) \subseteq \bar{\mathfrak{j}}.$$

It follows that $\bar{\omega}_1, \ldots, \bar{\omega}_r$ is a basis for $\bar{\mathfrak{j}}$. Thus the theorem is proved.

The foregoing theorems enable us to examine the relations between properties of those ideals in $\mathfrak{R}$ which are factors of $\mathfrak{i}$, and properties of the corresponding ideals in $\mathfrak{R}/\mathfrak{i}$, and conversely. The problem is a very simple one, and we need not spend much time on it. The following are a few of the more useful results. The proofs may be left to the reader. Ideals in $\mathfrak{R}/\mathfrak{i}$ are represented by $\bar{\mathfrak{j}}, \bar{\mathfrak{k}}, \ldots$, and $\mathfrak{j}, \mathfrak{k}, \ldots$ are the largest ideals in $\mathfrak{R}$ which correspond to them.

(i) $\overline{(\mathfrak{j}, \mathfrak{k})} = (\bar{\mathfrak{j}}, \bar{\mathfrak{k}})$, and $(\mathfrak{j}, \mathfrak{k})$ is the largest ideal in $\mathfrak{R}$ corresponding to $(\bar{\mathfrak{j}}, \bar{\mathfrak{k}})$.

(ii) $\overline{[\mathfrak{j}, \mathfrak{k}]} = [\bar{\mathfrak{j}}, \bar{\mathfrak{k}}]$, and $[\mathfrak{j}, \mathfrak{k}]$ is the largest ideal in $\mathfrak{R}$ corresponding to $[\bar{\mathfrak{j}}, \bar{\mathfrak{k}}]$.

(iii) $\overline{\mathfrak{j}\mathfrak{k}} = \bar{\mathfrak{j}}\bar{\mathfrak{k}}$, $(\mathfrak{j}\mathfrak{k}, \mathfrak{i})$ is the largest ideal in $\mathfrak{R}$ corresponding to $\bar{\mathfrak{j}}\bar{\mathfrak{k}}$, and need not coincide with $\mathfrak{j}\mathfrak{k}$.

(iv) $\overline{\mathfrak{j} : \mathfrak{k}} = \bar{\mathfrak{j}} : \bar{\mathfrak{k}}$, and $\mathfrak{j} : \mathfrak{k}$ is the largest ideal in $\mathfrak{R}$ corresponding to $\bar{\mathfrak{j}} : \bar{\mathfrak{k}}$.

(v) If the Basis Theorem holds in $\mathfrak{R}$, and $[\mathfrak{q}_1, \ldots, \mathfrak{q}_k]$ is an uncontractible representation of an ideal $\mathfrak{j}$ which is a factor of $\mathfrak{i}$, then $[\bar{\mathfrak{q}}_1, \ldots, \bar{\mathfrak{q}}_k]$ is an uncontractible representation of $\bar{\mathfrak{j}}$.

4. Subrings and extension rings. In the following sections we shall often have to consider simultaneously two rings, one of which is a subring of the other, and it is convenient to collect together certain general results on ideals in rings which are so related. It will be sufficient for our purpose to consider the case in which the two rings have unity in common.

We shall denote the smaller of the rings by $\mathfrak{R}$, and ideals in it by German letters, and we shall denote the larger ring by $\mathfrak{R}^*$, and ideals in it by German letters with an asterisk attached. More generally, a symbol S will represent a set of elements in $\mathfrak{R}$, and a symbol S^* will denote a set of elements in $\mathfrak{R}^*$. Any set in $\mathfrak{R}$ is also a set in $\mathfrak{R}^*$, but an ideal in $\mathfrak{R}$ may not be an ideal in $\mathfrak{R}^*$. If $\mathfrak{i}$ and $\mathfrak{j}$ are ideals in $\mathfrak{R}$, the relation

$$\mathfrak{i} = 0 \quad (\mathfrak{j})$$

is a relation in $\mathfrak{R}$ between them, but

$$\mathfrak{i} \subseteq \mathfrak{j}$$

may be regarded either as a relation in $\mathfrak{R}$ or as a relation in $\mathfrak{R}^*$.

If S is any set of elements of $\mathfrak{R}$, we shall denote by $\mathfrak{R}^*.S$ the set of elements which can be written as *finite* sums $\sum_1^r \rho_i^* \alpha_i$, where ρ_i^* is any element of $\mathfrak{R}^*$, and $\alpha_1, \ldots, \alpha_r$ are elements of S. On the other hand, if S^* is any set in $\mathfrak{R}^*$, $\mathfrak{R}_\wedge S^*$ will denote the set of elements of $\mathfrak{R}$ which are in S^*.

THEOREM I. *If* $\mathfrak{i}$, $\mathfrak{j}$ *are ideals in* $\mathfrak{R}$, *then* $\mathfrak{R}^*.\mathfrak{i}$, $\mathfrak{R}^*.\mathfrak{j}$ *are ideals in* $\mathfrak{R}^*$, *and the following relations hold:*

(i) *if* $\quad \mathfrak{i} = 0 \ (\mathfrak{j}),$ *then* $\quad \mathfrak{R}^*.\mathfrak{i} = 0 \ (\mathfrak{R}^*.\mathfrak{j});$

(ii) $$\mathfrak{R}^*.(\mathfrak{i},\mathfrak{j}) = (\mathfrak{R}^*.\mathfrak{i}, \mathfrak{R}^*.\mathfrak{j});$$

(iii) $$\mathfrak{R}^*.[\mathfrak{i},\mathfrak{j}] \subseteq [\mathfrak{R}^*.\mathfrak{i}, \mathfrak{R}^*.\mathfrak{j}];$$

(iv) $$\mathfrak{R}^*.\mathfrak{i}\mathfrak{j} = (\mathfrak{R}^*.\mathfrak{i})(\mathfrak{R}^*.\mathfrak{j});$$

(v) $$\mathfrak{R}^*.(\mathfrak{i}:\mathfrak{j}) \subseteq (\mathfrak{R}^*.\mathfrak{i}):(\mathfrak{R}^*.\mathfrak{j}).$$

Let α^* and β^* be elements of $\mathfrak{R}^*.\mathfrak{i}$, and let ρ^* be any element of $\mathfrak{R}^*$. Then

$$\alpha^* = \sum_{i=1}^k \rho_i^* \alpha_i \quad \text{and} \quad \beta^* = \sum_{i=1}^s \sigma_i^* \beta_i,$$

where $\rho_1^*, \ldots, \rho_k^*, \sigma_1^*, \ldots, \sigma_s^*$ are in $\mathfrak{R}^*$, and $\alpha_1, \ldots, \alpha_k, \beta_1, \ldots, \beta_s$ are in $\mathfrak{i}$. Then it is clear that

$$\alpha^* - \beta^* = \sum_{i=1}^k \rho_i^* \alpha_i - \sum_{i=1}^s \sigma_i^* \beta_i,$$

and

$$\rho^* \alpha^* = \sum_{i=1}^k \rho^* \rho_i^* \alpha_i$$

are in $\mathfrak{R}^*.\mathfrak{i}$. Hence $\mathfrak{R}^*.\mathfrak{i}$ is an ideal in $\mathfrak{R}^*$.

(i) If α^* is any element of $\mathfrak{R}^*.\mathfrak{i}$,

$$\alpha^* = \sum_1^k \rho_i^* \alpha_i,$$

where $\alpha_i \in \mathfrak{i} \subseteq \mathfrak{j}$; hence $\alpha^* \in \mathfrak{R}^*.\mathfrak{j}$.

(ii) Any element of $(\mathfrak{i},\mathfrak{j})$ is of the form $\alpha+\beta$, where $\alpha \in \mathfrak{i}$ and $\beta \in \mathfrak{j}$. Hence, if $\gamma^* \in \mathfrak{R}^*.(\mathfrak{i},\mathfrak{j})$,

$$\gamma^* = \sum_1^k \rho_i^*(\alpha_i + \beta_i) = \sum_1^k \rho_i^* \alpha_i + \sum_1^k \rho_i^* \beta_i,$$

that is $$\gamma^* \in (\mathfrak{R}^*.\mathfrak{i}, \mathfrak{R}^*.\mathfrak{j}),$$

and therefore $$\mathfrak{R}^*.(\mathfrak{i}, \mathfrak{j}) \subseteq (\mathfrak{R}^*.\mathfrak{i}, \mathfrak{R}^*.\mathfrak{j}).$$

On the other hand, any element of $(\mathfrak{R}^*.\mathfrak{i}, \mathfrak{R}^*.\mathfrak{j})$ is of the form $\alpha^* + \beta^*$, where

$$\alpha^* = \sum_1^k \rho_i^* \alpha_i, \quad \beta^* = \sum_1^s \sigma_i^* \beta_i,$$

and $$\alpha_i = 0 \ (\mathfrak{i}), \quad \beta_i = 0 \ (\mathfrak{j}).$$

If $\gamma_i = \alpha_i \quad (i = 1, \ldots, k), \qquad \gamma_i = \beta_{i-k} \quad (i = k+1, \ldots, k+s),$

and $\tau_i^* = \rho_i^* \quad (i = 1, \ldots, k), \qquad \tau_i^* = \sigma_{i-k}^* \quad (i = k+1, \ldots, k+s),$

$$\alpha^* + \beta^* = \sum_1^{k+s} \tau_i^* \gamma_i,$$

where $$\gamma_i = 0 \ ((\mathfrak{i}, \mathfrak{j})).$$

Hence $\alpha^* + \beta^* \in \mathfrak{R}^*.(\mathfrak{i}, \mathfrak{j})$. Thus

$$(\mathfrak{R}^*.\mathfrak{i}, \mathfrak{R}^*.\mathfrak{j}) \subseteq \mathfrak{R}^*.(\mathfrak{i}, \mathfrak{j}).$$

From these two results we have

$$\mathfrak{R}^*.(\mathfrak{i}, \mathfrak{j}) = (\mathfrak{R}^*.\mathfrak{i}, \mathfrak{R}^*.\mathfrak{j}).$$

(iii) Any element γ^* of $\mathfrak{R}^*.[\mathfrak{i}, \mathfrak{j}]$ is of the form

$$\gamma^* = \sum_1^k \rho_i^* \gamma_i,$$

where $$\gamma_i = 0 \ (\mathfrak{i}) \quad \text{and} \quad \gamma_i = 0 \ (\mathfrak{j}).$$

Hence $$\gamma^* = 0 \ (\mathfrak{R}^*.\mathfrak{i}) \quad \text{and} \quad \gamma^* = 0 \ (\mathfrak{R}^*.\mathfrak{j}),$$

that is $$\mathfrak{R}^*.[\mathfrak{i}, \mathfrak{j}] \subseteq [\mathfrak{R}^*.\mathfrak{i}, \mathfrak{R}^*.\mathfrak{j}].$$

(iv) If $\gamma^* \in \mathfrak{R}^*.\mathfrak{i}\mathfrak{j}$,

$$\gamma^* = \sum_1^k \rho_i^* \alpha_i \beta_i = \sum_1^k (\rho_i^* \alpha_i) \beta_i,$$

where $$\alpha_i = 0 \ (\mathfrak{i}) \quad \text{and} \quad \beta_i = 0 \ (\mathfrak{j}).$$

Hence $$\rho_i^* \alpha_i \in \mathfrak{R}^*.\mathfrak{i} \quad \text{and} \quad \beta_i \in \mathfrak{R}^*.\mathfrak{j},$$

and therefore we have

$$\mathfrak{R}^*.\mathfrak{i}\mathfrak{j} \subseteq (\mathfrak{R}^*.\mathfrak{i})(\mathfrak{R}^*.\mathfrak{j}).$$

Conversely, if $$\gamma^* \in (\mathfrak{R}^*.\mathfrak{i})(\mathfrak{R}^*.\mathfrak{j}),$$

we have $$\gamma^* = \sum_{i=1}^{r} \alpha_i^* \beta_i^*,$$

where $$\alpha_i^* = \sum_{j=1}^{s} \rho_{ij}^* \alpha_j \quad \text{and} \quad \beta_i^* = \sum_{k=1}^{t} \sigma_{ik}^* \beta_k,$$

with $$\alpha_j = 0 \ (\mathfrak{i}) \quad \text{and} \quad \beta_j = 0 \ (\mathfrak{j}).$$

Hence $$\gamma^* = \sum_{i=1}^{r} \sum_{j=1}^{s} \sum_{k=1}^{t} \rho_{ij}^* \sigma_{ik}^* \alpha_j \beta_k \in \mathfrak{R}^*.\mathfrak{ij},$$

and therefore $$(\mathfrak{R}^*.\mathfrak{i})(\mathfrak{R}^*.\mathfrak{j}) \subseteq \mathfrak{R}^*.\mathfrak{ij}.$$

Combining the two results, we have

$$\mathfrak{R}^*.\mathfrak{ij} = (\mathfrak{R}^*.\mathfrak{i})(\mathfrak{R}^*.\mathfrak{j}).$$

(v) If $\gamma^* \in \mathfrak{R}^*.(\mathfrak{i}:\mathfrak{j})$, then

$$\gamma^* = \sum_{1}^{k} \rho_i^* \gamma_i,$$

where $$\gamma_i \mathfrak{j} = 0 \ (\mathfrak{i}).$$

Hence, by (i), $$\gamma_i \mathfrak{R}^*.\mathfrak{j} = 0 \ (\mathfrak{R}^*.\mathfrak{i}),$$

and therefore $$\gamma^* \mathfrak{R}^*.\mathfrak{j} = 0 \ (\mathfrak{R}^*.\mathfrak{i}).$$

Therefore $$\mathfrak{R}^*.(\mathfrak{i}:\mathfrak{j}) \subseteq (\mathfrak{R}^*.\mathfrak{i}):(\mathfrak{R}^*.\mathfrak{j}).$$

THEOREM II. *If* $\mathfrak{i}^*, \mathfrak{j}^*$ *are ideals in* $\mathfrak{R}^*$, *then* $\mathfrak{R}_\wedge \mathfrak{i}^*$, $\mathfrak{R}_\wedge \mathfrak{j}^*$ *are ideals in* $\mathfrak{R}$, *and we have*

(i) *if* $\mathfrak{i}^* = 0 \ (\mathfrak{j}^*)$, *then* $\mathfrak{R}_\wedge \mathfrak{i}^* = 0 \ (\mathfrak{R}_\wedge \mathfrak{j}^*)$;

(ii) $$\mathfrak{R}_\wedge (\mathfrak{i}^*, \mathfrak{j}^*) \supseteq (\mathfrak{R}_\wedge \mathfrak{i}^*, \mathfrak{R}_\wedge \mathfrak{j}^*);$$

(iii) $$\mathfrak{R}_\wedge [\mathfrak{i}^*, \mathfrak{j}^*] = [\mathfrak{R}_\wedge \mathfrak{i}^*, \mathfrak{R}_\wedge \mathfrak{j}^*];$$

(iv) $$\mathfrak{R}_\wedge (\mathfrak{i}^* \mathfrak{j}^*) \supseteq (\mathfrak{R}_\wedge \mathfrak{i}^*)(\mathfrak{R}_\wedge \mathfrak{j}^*);$$

(v) $$\mathfrak{R}_\wedge (\mathfrak{i}^* : \mathfrak{j}^*) \subseteq (\mathfrak{R}_\wedge \mathfrak{i}^*):(\mathfrak{R}_\wedge \mathfrak{j}^*).$$

Let α and β be elements of $\mathfrak{R}_\wedge \mathfrak{i}^*$, and let ρ be any element of $\mathfrak{R}$, and hence an element of $\mathfrak{R}^*$. Since $\mathfrak{i}^*$ is an ideal, $\alpha - \beta$ and $\rho\alpha$ are in $\mathfrak{i}^*$. But since they also lie in $\mathfrak{R}$, they belong to $\mathfrak{R}_\wedge \mathfrak{i}^*$, which is therefore an ideal.

(i) If $\alpha \in \mathfrak{R}_\wedge \mathfrak{i}^*$, $\alpha \in \mathfrak{i}^* \subseteq \mathfrak{j}^*$. Hence α is in $\mathfrak{j}^*$ and in $\mathfrak{R}$, and is therefore in $\mathfrak{R}_\wedge \mathfrak{j}^*$. Therefore

$$\mathfrak{R}_\wedge \mathfrak{i}^* = 0 \ (\mathfrak{R}_\wedge \mathfrak{j}^*).$$

(ii) If $\alpha \in (\mathfrak{R}_\wedge \mathfrak{i}^*, \mathfrak{R}_\wedge \mathfrak{j}^*)$, $\alpha \in (\mathfrak{i}^*, \mathfrak{j}^*)$. Since α is also in $\mathfrak{R}$, it is in $\mathfrak{R}_\wedge (\mathfrak{i}^*, \mathfrak{j}^*)$. Therefore

$$\mathfrak{R}_\wedge(\mathfrak{i}^*, \mathfrak{j}^*) \supseteq (\mathfrak{R}_\wedge \mathfrak{i}^*, \mathfrak{R}_\wedge \mathfrak{j}^*).$$

(iii) If $\alpha \in \mathfrak{R}_\wedge [\mathfrak{i}^*, \mathfrak{j}^*]$, we have

$$\alpha \in \mathfrak{i}^*, \quad \alpha \in \mathfrak{j}^*, \quad \alpha \in \mathfrak{R}.$$

Hence $$\alpha \in \mathfrak{R}_\wedge \mathfrak{i}^* \quad \text{and} \quad \alpha \in \mathfrak{R}_\wedge \mathfrak{j}^*.$$

Therefore $$\alpha \in [\mathfrak{R}_\wedge \mathfrak{i}^*, \mathfrak{R}_\wedge \mathfrak{j}^*],$$

and hence we have $$\mathfrak{R}_\wedge [\mathfrak{i}^*, \mathfrak{j}^*] \subseteq [\mathfrak{R}_\wedge \mathfrak{i}^*, \mathfrak{R}_\wedge \mathfrak{j}^*].$$

Conversely, we have $$\mathfrak{R}_\wedge \mathfrak{i}^* \subseteq \mathfrak{i}^*, \quad \mathfrak{R}_\wedge \mathfrak{j}^* \subseteq \mathfrak{j}^*.$$

Hence $$[\mathfrak{R}_\wedge \mathfrak{i}^*, \mathfrak{R}_\wedge \mathfrak{j}^*] \subseteq [\mathfrak{i}^*, \mathfrak{j}^*].$$

But $$[\mathfrak{R}_\wedge \mathfrak{i}^*, \mathfrak{R}_\wedge \mathfrak{j}^*] \subseteq \mathfrak{R};$$

hence $$[\mathfrak{R}_\wedge \mathfrak{i}^*, \mathfrak{R}_\wedge \mathfrak{j}^*] \subseteq \mathfrak{R}_\wedge [\mathfrak{i}^*, \mathfrak{j}^*].$$

Combining the two results, we have

$$\mathfrak{R}_\wedge [\mathfrak{i}^*, \mathfrak{j}^*] = [\mathfrak{R}_\wedge \mathfrak{i}^*, \mathfrak{R}_\wedge \mathfrak{j}^*].$$

(iv) Since $$\mathfrak{R}_\wedge \mathfrak{i}^* \subseteq \mathfrak{i}^* \quad \text{and} \quad \mathfrak{R}_\wedge \mathfrak{j}^* \subseteq \mathfrak{j}^*,$$

we have $$(\mathfrak{R}_\wedge \mathfrak{i}^*)(\mathfrak{R}_\wedge \mathfrak{j}^*) \subseteq \mathfrak{i}^*\mathfrak{j}^*.$$

Since $$(\mathfrak{R}_\wedge \mathfrak{i}^*)(\mathfrak{R}_\wedge \mathfrak{j}^*) \subseteq \mathfrak{R},$$

we obtain $$\mathfrak{R}_\wedge (\mathfrak{i}^*\mathfrak{j}^*) \supseteq (\mathfrak{R}_\wedge \mathfrak{i}^*)(\mathfrak{R}_\wedge \mathfrak{j}^*).$$

(v) If $$\gamma \in \mathfrak{R}_\wedge (\mathfrak{i}^* : \mathfrak{j}^*), \quad \text{then} \quad \gamma \in (\mathfrak{i}^* : \mathfrak{j}^*),$$

and hence $$\gamma \mathfrak{j}^* = 0 \quad (\mathfrak{i}^*).$$

By (i), we have $$\gamma \mathfrak{R}_\wedge \mathfrak{j}^* = 0 \quad (\mathfrak{R}_\wedge \mathfrak{i}^*),$$

hence $$\mathfrak{R}_\wedge (\mathfrak{i}^* : \mathfrak{j}^*) \subseteq (\mathfrak{R}_\wedge \mathfrak{i}^*) : (\mathfrak{R}_\wedge \mathfrak{j}^*).$$

THEOREM III. *If $\mathfrak{q}^*$ is a $\mathfrak{p}^*$-primary ideal in $\mathfrak{R}^*$, $\mathfrak{R}_\wedge \mathfrak{q}^*$ is $(\mathfrak{R}_\wedge \mathfrak{p}^*)$-primary in $\mathfrak{R}$. In particular, if $\mathfrak{p}^*$ is prime, $\mathfrak{R}_\wedge \mathfrak{p}^*$ is prime.*

(i) Let α, β be elements of $\mathfrak{R}$ such that

$$\alpha\beta = 0 \quad (\mathfrak{R}_\wedge \mathfrak{q}^*), \quad \text{while} \quad \alpha \neq 0 \quad (\mathfrak{R}_\wedge \mathfrak{q}^*).$$

Then $$\alpha\beta = 0 \quad (\mathfrak{q}^*),$$

while $$\alpha \neq 0 \quad (\mathfrak{q}^*).$$

Since $\mathfrak{q}^*$ is $\mathfrak{p}^*$-primary, it follows that $\beta \in \mathfrak{p}^*$, and since β is in $\mathfrak{R}$,

$$\beta = 0 \ (\mathfrak{R}_\wedge \mathfrak{p}^*).$$

(ii) Since
$$\mathfrak{q}^* = 0 \ (\mathfrak{p}^*),$$
it follows from Theorem II (i), that

$$\mathfrak{R}_\wedge \mathfrak{q}^* = 0 \ (\mathfrak{R}_\wedge \mathfrak{p}^*).$$

(iii) If
$$\beta = 0 \ (\mathfrak{R}_\wedge \mathfrak{p}^*),$$
we have
$$\beta = 0 \ (\mathfrak{p}^*).$$
Hence there exists an integer ρ such that

$$\beta^\rho = 0 \ (\mathfrak{q}^*).$$

But β^ρ is in $\mathfrak{R}$; hence
$$\beta^\rho = 0 \ (\mathfrak{R}_\wedge \mathfrak{q}^*).$$
From § 2, Th. I, it follows that $\mathfrak{R}_\wedge \mathfrak{q}^*$ is $(\mathfrak{R}_\wedge \mathfrak{p}^*)$-primary.

The second part of the theorem follows at once by taking $\mathfrak{q}^* = \mathfrak{p}^*$.

We now consider the effect of applying the operations $\mathfrak{R}^*.$ and $\mathfrak{R}_\wedge$ successively. Let $\mathfrak{i}$ be any ideal in $\mathfrak{R}$, and let $\mathfrak{i}^* = \mathfrak{R}^*.\mathfrak{i}$. We say that $\mathfrak{i}$ *extends* to $\mathfrak{i}^*$. If $\mathfrak{j} = \mathfrak{R}_\wedge \mathfrak{i}^*$, we have $\mathfrak{i} \subseteq \mathfrak{i}^*$, and hence

$$\mathfrak{i} = \mathfrak{R}_\wedge \mathfrak{i} \subseteq \mathfrak{R}_\wedge \mathfrak{i}^* = \mathfrak{j}.$$

Examples will be found in the next section which show that $\mathfrak{j}$ may be a proper factor of $\mathfrak{i}$. Since $\mathfrak{i} \subseteq \mathfrak{j}$, we have [Th. I (i)]

$$\mathfrak{i}^* = \mathfrak{R}^*.\mathfrak{i} \subseteq \mathfrak{R}^*.\mathfrak{j}.$$

But, on the other hand, we have $\mathfrak{j} \subseteq \mathfrak{i}^*$. Hence $\mathfrak{R}^*.\mathfrak{j} \subseteq \mathfrak{R}^*.\mathfrak{i}^* = \mathfrak{i}^*$, so that $\mathfrak{R}^*.\mathfrak{j} = \mathfrak{i}^*$. If $\mathfrak{i}'$ is any other ideal of $\mathfrak{R}$ such that

$$\mathfrak{R}^*.\mathfrak{i}' = \mathfrak{i}^*, \quad \text{then} \quad \mathfrak{i}' \subseteq \mathfrak{R}_\wedge \mathfrak{i}^* = \mathfrak{j}.$$

Hence $\mathfrak{j}$ is a factor of every ideal $\mathfrak{i}'$ which has the property $\mathfrak{R}^*.\mathfrak{i}' = \mathfrak{i}^*$. Thus we have

THEOREM IV. *If* $\mathfrak{i}$ *is any ideal of* $\mathfrak{R}$, *and* $\mathfrak{j} = \mathfrak{R}_\wedge(\mathfrak{R}^*.\mathfrak{i})$, *then* $\mathfrak{i} \subseteq \mathfrak{j}$ *and* $\mathfrak{R}^*.\mathfrak{i} = \mathfrak{R}^*.\mathfrak{j}$. *The ideal* $\mathfrak{j}$ *is the largest ideal in* $\mathfrak{R}$ *which extends to* $\mathfrak{R}^*.\mathfrak{i}$, *and it contains as a subideal every ideal of* $\mathfrak{R}$ *which extends to* $\mathfrak{R}^*.\mathfrak{i}$.

We may mention here a special case which will occur later in which we can prove that $\mathfrak{i} = \mathfrak{j}$ in Theorem IV.

THEOREM V. *If there exists a finite number of elements $\eta_1^*, \ldots, \eta_r^*$ of $\mathfrak{R}^*$ such that every element of $\mathfrak{R}^*$ can be written in the form $\sum_1^r \rho_i \eta_i^*$, where $\rho_1, \ldots, \rho_r$ are in $\mathfrak{R}$, and if $\mathfrak{p}$ is a prime ideal in $\mathfrak{R}$ with a finite basis, then $\mathfrak{p} = \mathfrak{R}_\wedge(\mathfrak{R}^*.\mathfrak{p})$.*

Let $\omega_1, \ldots, \omega_s$ be a basis for $\mathfrak{p}$. Then any element of $\mathfrak{R}^*.\mathfrak{p}$ is of the form $\sum_1^s \rho_i^* \omega_i$, where $\rho_1^*, \ldots, \rho_s^*$ are in $\mathfrak{R}^*$. Let ω be any element of $\mathfrak{R}_\wedge(\mathfrak{R}^*.\mathfrak{p})$. Then $\omega\eta_i^* \in \mathfrak{R}^*.\mathfrak{p}$, and hence

$$\omega\eta_i^* = \sum_{j=1}^s \rho_{ij}^* \omega_j \quad (i = 1, \ldots, r).$$

But
$$\rho_{ij}^* = \sum_{k=1}^r \rho_{ijk} \eta_k^*,$$

where ρ_{ijk} belongs to $\mathfrak{R}$. Hence

$$\omega\eta_i^* = \sum_{k=1}^r d_{ik} \eta_k^* \quad (i = 1, \ldots, r),$$

where
$$d_{ij} = 0 \ \ (\mathfrak{p}).$$

Eliminating $\eta_1^*, \ldots, \eta_r^*$, we have

$$| \delta_{ij}\omega - d_{ij} | = 0,$$

and hence
$$\omega^r = 0 \ \ (\mathfrak{p}).$$

Since $\mathfrak{p}$ is prime, we have $\omega = 0 \ \ (\mathfrak{p})$,

and hence
$$\mathfrak{R}_\wedge(\mathfrak{R}^*.\mathfrak{p}) \subseteq \mathfrak{p}.$$

Using Theorem IV, Theorem V follows at once.

The following corollary is also useful:

Corollary. If, in addition to the hypothesis of Theorem V, the Basis Theorem holds in $\mathfrak{R}^*$, and if $\mathfrak{R}^*.\mathfrak{p} = [\mathfrak{q}_1^*, \ldots, \mathfrak{q}_k^*]$, where $\mathfrak{q}_i^*$ is primary, then for at least one value of i we have $\mathfrak{R}_\wedge \mathfrak{q}_i^* = \mathfrak{p}$.

By Theorem II (iii) we have

$$\mathfrak{p} = \mathfrak{R}_\wedge(\mathfrak{R}^*.\mathfrak{p}) = [\mathfrak{R}_\wedge \mathfrak{q}_1^*, \ldots, \mathfrak{R}_\wedge \mathfrak{q}_k^*],$$

and by Theorem III, $\mathfrak{R}_\wedge \mathfrak{q}_i^*$ is primary. Let its radical be $\mathfrak{p}_i$. Then, since $\mathfrak{p} \subseteq \mathfrak{R}_\wedge \mathfrak{q}_i^* \subseteq \mathfrak{p}_i$,

$$\mathfrak{p} = 0 \ \ (\mathfrak{p}_i).$$

If each $\mathfrak{p}_i$ were a proper factor of $\mathfrak{p}$, we could find $\alpha_i \in \mathfrak{p}_i$, where α_i is not in $\mathfrak{p}$, for $i = 1, \ldots, k$. Hence, for a suitable integer ρ,

$$(\alpha_1 \ldots \alpha_k)^\rho = 0 \ \ ([\mathfrak{R}_\wedge \mathfrak{q}_1^*, \ldots, \mathfrak{R}_\wedge \mathfrak{q}_k^*]),$$

that is

$$(\alpha_1 \ldots \alpha_k)^\rho = 0 \ \ (\mathfrak{p}).$$

Since $\mathfrak{p}$ is prime, one at least of the elements α_i is in $\mathfrak{p}$, and we thus have a contradiction. Hence, for at least one i, $\mathfrak{p}_i = \mathfrak{p}$. But $\mathfrak{p} \subseteq \mathfrak{R}_\wedge \mathfrak{q}_i^* \subseteq \mathfrak{p}_i$. It follows that

$$\mathfrak{R}_\wedge \mathfrak{q}_i^* = \mathfrak{p}.$$

We now return to the general problem of the relation between ideals in $\mathfrak{R}$ and ideals in $\mathfrak{R}^*$. Let $\mathfrak{i}^*$ be an ideal in $\mathfrak{R}^*$, let $\mathfrak{i}$ be the ideal $\mathfrak{R}_\wedge \mathfrak{i}^*$, and let $\mathfrak{j}^* = \mathfrak{R}^*.\mathfrak{i}$. We say that $\mathfrak{i}^*$ *contracts* to $\mathfrak{R}_\wedge \mathfrak{i}^*$. Since $\mathfrak{i} \subseteq \mathfrak{i}^*$, $\mathfrak{j}^* = \mathfrak{R}^*.\mathfrak{i} \subseteq \mathfrak{R}^*.\mathfrak{i}^* = \mathfrak{i}^*$. The following example shows that $\mathfrak{j}^*$ may be a proper multiple of $\mathfrak{i}^*$. Let K be a commutative field, x an indeterminate over K, and let y be a root of the equation $y^2 = x^3$ in some extension of $K[x]$. We take $\mathfrak{R} = K[x, y]$, $\mathfrak{R}^* = K[x, y, yx^{-1}]$, and let

$$\mathfrak{i}^* = \mathfrak{R}^*.(x, y, yx^{-1}).$$

Then it can be verified that

$$\mathfrak{i} = \mathfrak{R}_\wedge \mathfrak{i}^* = \mathfrak{R}.(x, y) \quad \text{and} \quad \mathfrak{j}^* = \mathfrak{R}^*.\mathfrak{i} = \mathfrak{R}^*.(x, y) \neq \mathfrak{i}^*.$$

Since $\mathfrak{i} \subseteq \mathfrak{j}^*$, $\mathfrak{i} \subseteq \mathfrak{R}_\wedge \mathfrak{j}^*$, and since $\mathfrak{j}^* \subseteq \mathfrak{i}^*$, $\mathfrak{R}_\wedge \mathfrak{j}^* \subseteq \mathfrak{R}_\wedge \mathfrak{i}^* = \mathfrak{i}$. Hence $\mathfrak{R}_\wedge \mathfrak{j}^* = \mathfrak{R}_\wedge \mathfrak{i}^*$. We thus obtain

THEOREM VI. *If $\mathfrak{i}^*$ is any ideal of $\mathfrak{R}^*$, and if $\mathfrak{j}^* = \mathfrak{R}^*.(\mathfrak{R}_\wedge \mathfrak{i}^*)$, then $\mathfrak{j}^* \subseteq \mathfrak{i}^*$ and $\mathfrak{R}_\wedge \mathfrak{j}^* = \mathfrak{R}_\wedge \mathfrak{i}^*$; the ideal $\mathfrak{j}^*$ is the smallest ideal in $\mathfrak{R}^*$ which contracts to $\mathfrak{R}_\wedge \mathfrak{i}^*$, and is a multiple of every ideal of $\mathfrak{R}^*$ which contracts to $\mathfrak{R}_\wedge \mathfrak{i}^*$.*

Any ideal $\mathfrak{i}^*$ of $\mathfrak{R}^*$ which is equal to $\mathfrak{R}^*.\mathfrak{i}_1$, where $\mathfrak{i}_1$ is some ideal of $\mathfrak{R}$, is called an *extended ideal* in $\mathfrak{R}^*$. If $\mathfrak{i}_1^*$ is any ideal in $\mathfrak{R}^*$, $\mathfrak{i}^* = \mathfrak{R}^*.(\mathfrak{R}_\wedge \mathfrak{i}_1^*)$ is called the extended ideal *associated* with $\mathfrak{i}_1^*$. Similarly, an ideal $\mathfrak{i}$ of $\mathfrak{R}$ which is equal to $\mathfrak{R}_\wedge \mathfrak{i}_1^*$, where $\mathfrak{i}_1^*$ is some ideal of $\mathfrak{R}^*$, is called a *contracted ideal* of $\mathfrak{R}$, and if $\mathfrak{i}_1$ is any ideal of $\mathfrak{R}$, $\mathfrak{i} = \mathfrak{R}_\wedge (\mathfrak{R}^*.\mathfrak{i}_1)$ is called the contracted ideal of $\mathfrak{R}$ *associated* with $\mathfrak{i}_1$.

Every contracted ideal of $\mathfrak{R}$ is the contraction of exactly one extended ideal of $\mathfrak{R}^*$. Let $\mathfrak{i} = \mathfrak{R}_\wedge \mathfrak{i}_1^*$ be a contracted ideal of $\mathfrak{R}$.

If $\mathfrak{i}_2^* = \mathfrak{R}^*.\mathfrak{i}$, then [Th. VI] $\mathfrak{R}_\wedge \mathfrak{i}_2^* = \mathfrak{R}_\wedge \mathfrak{i}_1^* = \mathfrak{i}$, and $\mathfrak{i}$ is the contraction of the extended ideal $\mathfrak{i}_2^*$. If $\mathfrak{i}_3^* = \mathfrak{R}^*.\mathfrak{j}_1$ is any other extended ideal such that $\mathfrak{i} = \mathfrak{R}_\wedge \mathfrak{i}_3^*$, it follows from Theorem IV that $\mathfrak{R}^*.\mathfrak{j}_1 = \mathfrak{R}^*.\mathfrak{i}$, so that $\mathfrak{i}_3^* = \mathfrak{i}_2^*$. In exactly the same way we show that every extended ideal of $\mathfrak{R}^*$ is the extension of exactly one contracted ideal of $\mathfrak{R}$. Hence we have:

THEOREM VII. *There exists a one-to-one correspondence between the contracted ideals of* $\mathfrak{R}$ *and the extended ideals of* $\mathfrak{R}^*$. *If* $\mathfrak{i}$ *and* $\mathfrak{i}^*$ *correspond in this correspondence,* $\mathfrak{i}^* = \mathfrak{R}^*.\mathfrak{i}$ *and* $\mathfrak{i} = \mathfrak{R}_\wedge \mathfrak{i}^*$; $\mathfrak{i}$ *is the only contracted ideal which extends to* $\mathfrak{i}^*$, *and* $\mathfrak{i}^*$ *is the only extended ideal which contracts to* $\mathfrak{i}$.

THEOREM VIII.

(i) *If* $\mathfrak{i}, \mathfrak{j}$ *are contracted ideals of* $\mathfrak{R}$, *then* $[\mathfrak{i}, \mathfrak{j}]$ *and* $\mathfrak{i} : \mathfrak{j}$ *are contracted ideals;*

(ii) *If* $\mathfrak{i}^*, \mathfrak{j}^*$ *are extended ideals of* $\mathfrak{R}^*$, *then* $(\mathfrak{i}^*, \mathfrak{j}^*)$ *and* $\mathfrak{i}^*\mathfrak{j}^*$ *are extended ideals.*

(i) We may write $\mathfrak{i} = \mathfrak{R}_\wedge \mathfrak{i}^*$, $\mathfrak{j} = \mathfrak{R}_\wedge \mathfrak{j}^*$, where $\mathfrak{i}^* = \mathfrak{R}^*.\mathfrak{i}$, $\mathfrak{j}^* = \mathfrak{R}^*.\mathfrak{j}$. By Theorem II (iii),

$$[\mathfrak{i}, \mathfrak{j}] = \mathfrak{R}_\wedge [\mathfrak{i}^*, \mathfrak{j}^*],$$

and so $[\mathfrak{i}, \mathfrak{j}]$ is a contracted ideal.

By Theorem I (v), $\mathfrak{R}^*.(\mathfrak{i} : \mathfrak{j}) \subseteq \mathfrak{i}^* : \mathfrak{j}^*$, and, by Theorem II (v),

$$\mathfrak{R}_\wedge (\mathfrak{i}^* : \mathfrak{j}^*) \subseteq \mathfrak{i} : \mathfrak{j}.$$

Hence $$\mathfrak{R}_\wedge (\mathfrak{R}^*.(\mathfrak{i} : \mathfrak{j})) \subseteq \mathfrak{i} : \mathfrak{j}.$$

But the ideal on the left is the contracted ideal associated with $\mathfrak{i} : \mathfrak{j}$, and hence [Th. IV] it contains $\mathfrak{i} : \mathfrak{j}$. Therefore

$$\mathfrak{i} : \mathfrak{j} = \mathfrak{R}_\wedge (\mathfrak{R}^*.(\mathfrak{i} : \mathfrak{j})),$$

and it is therefore a contracted ideal.

(ii) If $\mathfrak{i}^* = \mathfrak{R}^*.\mathfrak{i}$ and $\mathfrak{j}^* = \mathfrak{R}^*.\mathfrak{j}$, it follows from Theorem I that

$$(\mathfrak{i}^*, \mathfrak{j}^*) = \mathfrak{R}^*.(\mathfrak{i}, \mathfrak{j}),$$

and $$\mathfrak{i}^*\mathfrak{j}^* = \mathfrak{R}^*.(\mathfrak{i}\mathfrak{j}).$$

Hence $(\mathfrak{i}^*, \mathfrak{j}^*)$ and $\mathfrak{i}^*\mathfrak{j}^*$ are extended ideals.

5. Quotient rings. In this section we consider a method, which has many applications to algebraic geometry, of passing from a ring $\mathfrak{R}$ to an extended ring $\mathfrak{R}^*$. Let $\mathfrak{R}$ be any commutative ring, and let S be a non-empty set of elements of $\mathfrak{R}$ with the following properties:

(i) S is multiplicatively closed; that is, if α and β are elements of $\mathfrak{R}$ in S, $\alpha\beta$ is in S;

(ii) S does not contain zero, or a divisor of zero, of $\mathfrak{R}$.

Thus if α and β are in S, $\alpha\beta$ is not zero, nor a divisor of zero.

The method by which $\mathfrak{R}^*$ is formed from S is reminiscent of the method of forming the quotient field of an integral domain [I, § 4]. We consider pairs of elements (α, β), in which α is any element of $\mathfrak{R}$ and β is any element of S. Two pairs (α, β) and (α', β') are said to be equivalent

$$(\alpha, \beta) \sim (\alpha', \beta') \quad \text{if} \quad \alpha\beta' = \alpha'\beta.$$

This relation is clearly reflexive and symmetric. To show that it is transitive, suppose that (α, β) is equivalent to (α', β'), and that (α', β') is equivalent to (α'', β''). Then

$$\alpha\beta' = \alpha'\beta \quad \text{and} \quad \alpha'\beta'' = \alpha''\beta'.$$

Therefore

$$\alpha\beta'\beta'' = \alpha'\beta\beta'' = \alpha'\beta''\beta = \alpha''\beta'\beta,$$

that is,

$$(\alpha\beta'' - \alpha''\beta)\beta' = 0,$$

and hence

$$\alpha\beta'' - \alpha''\beta = 0,$$

since β' is not a divisor of zero. Hence (α, β) is equivalent to (α'', β''). The equivalence is therefore a true equivalence relation [I, § 2], and it can be used to divide the pairs into mutually exclusive sets of equivalent pairs.

We define addition and multiplication of pairs just as in the definition of the quotient field of an integral domain:

$$(\alpha, \beta) + (\gamma, \delta) = (\alpha\delta + \beta\gamma, \beta\delta),$$

and

$$(\alpha, \beta)(\gamma, \delta) = (\alpha\gamma, \beta\delta).$$

Using the fact that S is multiplicatively closed and contains no divisors of zero, we prove, as in Chapter I, § 4, that if

$$(\alpha, \beta) \sim (\alpha', \beta') \quad \text{and} \quad (\gamma, \delta) \sim (\gamma', \delta'),$$

then

$$(\alpha, \beta) + (\gamma, \delta) \sim (\alpha', \beta') + (\gamma', \delta'),$$

and

$$(\alpha, \beta)(\gamma, \delta) \sim (\alpha', \beta')(\gamma', \delta').$$

Thus we can define addition and multiplication of sets of equivalent pairs, and as in the theory of quotient fields of integral domains, we prove that the sets, with these laws of addition and multiplication, form a ring $\mathfrak{R}^*$ whose zero is the set consisting of the equivalent pairs $(0, \beta)$, and which has as unity the set of equivalent pairs (α, α). To each element ξ of $\mathfrak{R}$ there corresponds an element ξ^* of $\mathfrak{R}^*$ consisting of the equivalent pairs $(\alpha\xi, \alpha)$. Now $(\alpha\xi, \alpha) \sim (\beta\eta, \beta)$ if and only if

$$\alpha\beta(\xi - \eta) = 0;$$

that is, if and only if $\xi = \eta$, since α and β belong to S and are therefore not divisors of zero. Since $(\xi+\eta)^* = \xi^*+\eta^*$, and $(\xi\eta)^* = \xi^*\eta^*$, the elements ξ^* therefore form a subring of $\mathfrak{R}^*$ which is isomorphic with $\mathfrak{R}$. We can then identify this subring with $\mathfrak{R}$, so that $\mathfrak{R}^*$ becomes an extension of $\mathfrak{R}$. When we do this, we usually represent any element of $\mathfrak{R}^*$ by α/β, where (α, β) is any pair of the corresponding set.

In the applications with which we are concerned the ring $\mathfrak{R}$ is always an integral domain, and *we shall henceforth assume that* $\mathfrak{R}$ *is such a domain.* Then $\mathfrak{R}^*$ is isomorphic with a ring containing $\mathfrak{R}$ and contained in the quotient field K of $\mathfrak{R}$, and will be identified with it. Since $\mathfrak{R}^*$ contains $\mathfrak{R}$, it has unity in common with $\mathfrak{R}$, and since it is contained in the field K, it has no divisors of zero. Hence it is an integral domain. Its quotient field clearly coincides with K. The sets S in $\mathfrak{R}$ with which we have to deal are obtained by selecting a prime ideal (not the unit ideal) $\mathfrak{p}$ in $\mathfrak{R}$, and taking S to be the set of elements of $\mathfrak{R}$ not in $\mathfrak{p}$. Since $\mathfrak{p}$ is prime, it is clear that S is multiplicatively closed and does not contain zero or any divisor of zero. The ring $\mathfrak{R}^*$ is then denoted by $\mathfrak{R}_\mathfrak{p}$, and is called the *quotient ring* of $\mathfrak{p}$. If $\mathfrak{p}$ is the zero ideal, $\mathfrak{R}_\mathfrak{p}$ is the quotient field of $\mathfrak{R}$, for the elements of $\mathfrak{R}$ not in $\mathfrak{p}$ are all the non-zero elements of $\mathfrak{R}$.

In considering the relations between ideals in $\mathfrak{R}$ and ideals in $\mathfrak{R}_\mathfrak{p}$, we can apply the results of §4. In the case of quotient rings $\mathfrak{R}_\mathfrak{p}$, however, it is possible to sharpen these results considerably, as the following theorems show.

THEOREM I. *Every ideal in* $\mathfrak{R}_\mathfrak{p}$ *is an extended ideal.*

Let $\mathfrak{i}^*$ be any ideal in $\mathfrak{R}_\mathfrak{p}$, and let $\mathfrak{i} = \mathfrak{R}_\wedge \mathfrak{i}^*$. If $\alpha/\beta \in \mathfrak{i}^*$, then $\alpha = \beta(\alpha/\beta) \in \mathfrak{i}^*$, and since α is in $\mathfrak{R}$, $\alpha \in \mathfrak{i}$. Conversely, if $\alpha \in \mathfrak{i}$ and β is not in $\mathfrak{p}$, we have $\alpha \in \mathfrak{i}^*$ and $\beta^{-1} \in \mathfrak{R}_\mathfrak{p}$. Hence $\alpha/\beta \in \mathfrak{i}^*$. Thus $\mathfrak{i}^*$

consists of the quotients α/β in which α belongs to $\mathfrak{i}$ and β is not in $\mathfrak{p}$. Any element of $\mathfrak{i}^*$ is therefore contained in $\mathfrak{R}_\mathfrak{p}.\mathfrak{i}$; that is, $\mathfrak{i}^* \subseteq \mathfrak{R}_\mathfrak{p}.\mathfrak{i}$. But [§ 4, Th. VI] $\mathfrak{R}_\mathfrak{p}.\mathfrak{i} \subseteq \mathfrak{i}^*$. Hence we have

$$\mathfrak{i}^* = \mathfrak{R}_\mathfrak{p}.\mathfrak{i}.$$

THEOREM II. *If* $\mathfrak{i}$ *is any ideal in* $\mathfrak{R}$, *and* $\mathfrak{i}^* = \mathfrak{R}_\mathfrak{p}.\mathfrak{i}$, *then* $\mathfrak{i}^*$ *is the unit ideal of* $\mathfrak{R}_\mathfrak{p}$ *if and only if* $\mathfrak{i} \neq 0 \ (\mathfrak{p})$.

Any element ξ of $\mathfrak{R}_\mathfrak{p}.\mathfrak{i}$ is of the form $\sum_1^r \frac{\alpha_i}{\beta_i}\gamma_i$, where $\alpha_i \in \mathfrak{R}$, $\gamma_i \in \mathfrak{i}$, $\beta_i \notin \mathfrak{p}$. Hence

$$\xi = \frac{\Sigma \alpha_i \gamma_i \beta_1 \ldots \beta_{i-1}\beta_{i+1} \ldots \beta_r}{\beta_1 \ldots \beta_r} = \frac{\gamma}{\beta},$$

where $\gamma \in \mathfrak{i}$ and $\beta \neq 0 \ (\mathfrak{p})$.

If $\mathfrak{i} \neq 0 \ (\mathfrak{p})$, there exists an element β of $\mathfrak{i}$ which is not in $\mathfrak{p}$. It follows that $1 = \beta/\beta$ is in $\mathfrak{R}_\mathfrak{p}.\mathfrak{i}$, and hence $\mathfrak{i}^*$ is the unit ideal. On the other hand, if $\mathfrak{i}^*$ is the unit ideal, it contains 1. Hence there exists an element β of $\mathfrak{R}$ not in $\mathfrak{p}$, and an element α in $\mathfrak{i}$ such that $\alpha/\beta = 1$. Hence $\alpha = \beta$ is in $\mathfrak{i}$ but not in $\mathfrak{p}$, that is, $\mathfrak{i} \neq 0 \ (\mathfrak{p})$.

THEOREM III. *If* $\mathfrak{P}$ *is a prime ideal of* $\mathfrak{R}$ *contained in* $\mathfrak{p}$, *and* $\mathfrak{Q}$ *is* $\mathfrak{P}$*-primary, then* $\mathfrak{q}^* = \mathfrak{R}_\mathfrak{p}.\mathfrak{Q}$ *is* $\mathfrak{p}^*$*-primary, where* $\mathfrak{p}^* = \mathfrak{R}_\mathfrak{p}.\mathfrak{P}$.

Any element of $\mathfrak{q}^* = \mathfrak{R}_\mathfrak{p}.\mathfrak{Q}$ can be written in the form α/β, where $\alpha \in \mathfrak{Q}$, and β is not in $\mathfrak{p}$. Similarly, any element of $\mathfrak{p}^*$ can be written in the form γ/δ, where $\gamma \in \mathfrak{P}$ and δ is not in $\mathfrak{p}$.

(i) Let α/β and γ/δ be two elements of $\mathfrak{R}_\mathfrak{p}$ such that

$$\frac{\alpha}{\beta}\cdot\frac{\gamma}{\delta} = 0 \ (\mathfrak{q}^*), \quad \text{while} \quad \frac{\alpha}{\beta} \neq 0 \ (\mathfrak{q}^*).$$

Then

$$\frac{\alpha\gamma}{\beta\delta} = \frac{\xi}{\eta},$$

where $\xi \in \mathfrak{Q}$ and η is not in $\mathfrak{p}$, and hence not in $\mathfrak{P}$, which is, by hypothesis, a multiple of $\mathfrak{p}$. Thus

$$\alpha\gamma\eta = \beta\delta\xi = 0 \ (\mathfrak{Q}),$$

and hence

$$\alpha\gamma = 0 \ (\mathfrak{Q}),$$

since η is not in $\mathfrak{P}$. Since α/β is not in $\mathfrak{q}^*$, α is not in $\mathfrak{Q}$, and, since $\mathfrak{Q}$ is $\mathfrak{P}$-primary, it follows that $\gamma \in \mathfrak{P}$, and hence that

$$\frac{\gamma}{\delta} = 0 \ (\mathfrak{p}^*).$$

(ii) Any element of $\mathfrak{q}^*$ is of the form α/β, where $\alpha \in \mathfrak{Q}$ and β is not in $\mathfrak{p}$. Hence $\alpha \in \mathfrak{P}$, and therefore $\alpha/\beta \in \mathfrak{p}^*$. Hence

$$\mathfrak{q}^* = 0 \quad (\mathfrak{p}^*).$$

(iii) Any element of $\mathfrak{p}^*$ is of the form α/β, where $\alpha \in \mathfrak{P}$. Then $\alpha^\rho \in \mathfrak{Q}$, for some integer ρ. Moreover, β is not in $\mathfrak{p}$, and, since $\mathfrak{p}$ is prime, β^ρ is not in $\mathfrak{p}$. Hence

$$\left(\frac{\alpha}{\beta}\right)^\rho = 0 \quad (\mathfrak{q}^*).$$

It follows from Theorem I of § 2 that $\mathfrak{q}^*$ is $\mathfrak{p}^*$-primary.

Using the notation of Theorem III, let α be an element of $\mathfrak{R}_\wedge \mathfrak{q}^*$. Then $\alpha \in \mathfrak{R}$, and also

$$\alpha = \frac{\beta}{\gamma},$$

where $\beta \in \mathfrak{Q}$ and γ is not in $\mathfrak{p}$. Hence $\gamma\alpha = \beta \in \mathfrak{Q}$, and since $\gamma \neq 0 \quad (\mathfrak{P})$, it follows that $\alpha \in \mathfrak{Q}$. Hence

$$\mathfrak{R}_\wedge \mathfrak{q}^* \subseteq \mathfrak{Q}.$$

But [§ 4, Th. IV] $$\mathfrak{Q} \subseteq \mathfrak{R}_\wedge \mathfrak{q}^*.$$

Hence $$\mathfrak{Q} = \mathfrak{R}_\wedge \mathfrak{q}^*.$$

A similar argument applies to $\mathfrak{R}_\wedge \mathfrak{p}^*$. From Theorems I and III we deduce

THEOREM IV. *There is a one-to-one correspondence between the primary ideals of* $\mathfrak{R}$ *whose radicals are multiples of* $\mathfrak{p}$ *and the primary ideals, other than the unit ideal, of* $\mathfrak{R}_\mathfrak{p}$, *the relation between corresponding ideals being that between corresponding contracted and extended ideals.*

Let us now consider the ideal $\mathfrak{R}_\mathfrak{p}.\mathfrak{p}$. Any element of $\mathfrak{R}_\mathfrak{p}$ not in this ideal can be written in the form α/β, where neither α nor β is in $\mathfrak{p}$. Hence β/α is in $\mathfrak{R}_\mathfrak{p}$, that is, the elements of $\mathfrak{R}_\mathfrak{p}$ not in $\mathfrak{R}_\mathfrak{p}.\mathfrak{p}$ are units of $\mathfrak{R}_\mathfrak{p}$. On the other hand, no element of $\mathfrak{R}_\mathfrak{p}.\mathfrak{p}$ can be a unit. For if $\mathfrak{R}_\mathfrak{p}.\mathfrak{p}$ contained a unit, it would be the unit ideal, and we should have a contradiction of Theorem II. Thus $\mathfrak{R}_\mathfrak{p}.\mathfrak{p}$ is an ideal whose elements are all the non-units of $\mathfrak{R}_\mathfrak{p}$. If $\mathfrak{i}^*$ is any ideal of $\mathfrak{R}_{\mathfrak{p}}$ different from the unit ideal, it cannot contain any unit of $\mathfrak{R}_{\mathfrak{p}}$. Hence it must be contained in $\mathfrak{R}_{\mathfrak{p}}.\mathfrak{p}$. Hence we have

THEOREM V. *The non-units of* $\mathfrak{R}_{\mathfrak{p}}$ *constitute the ideal* $\mathfrak{R}_{\mathfrak{p}}.\mathfrak{p}$. *Any ideal of* $\mathfrak{R}_{\mathfrak{p}}$ *other than the unit ideal is a multiple of* $\mathfrak{R}_{\mathfrak{p}}.\mathfrak{p}$.

Now let $\mathfrak{P}^*$ be any prime ideal of $\mathfrak{R}_{\mathfrak{p}}$, and let $\mathfrak{P} = \mathfrak{R}_\wedge \mathfrak{P}^*$. Then $\mathfrak{P}$ is prime, and, assuming that $\mathfrak{P}^*$ is not the unit ideal,

$$\mathfrak{P}^* = 0 \;\; (\mathfrak{R}_{\mathfrak{p}}.\mathfrak{p}), \quad \mathfrak{P} = 0 \;\; (\mathfrak{p}).$$

Let ζ be any element of $(\mathfrak{R}_{\mathfrak{p}})_{\mathfrak{P}^*}$. We can write it in the form ξ/η, where ξ, η are in $\mathfrak{R}_{\mathfrak{p}}$, and η is not in $\mathfrak{P}^*$. Then

$$\xi = \frac{\alpha}{\beta}, \quad \eta = \frac{\gamma}{\delta},$$

where neither β nor δ is in $\mathfrak{p}$, and γ is not in $\mathfrak{P}$. Then

$$\zeta = \frac{\alpha\delta}{\beta\gamma},$$

and since $\mathfrak{P} \subseteq \mathfrak{p}$, $\beta\gamma$ is not in $\mathfrak{P}$. Hence ζ is in $\mathfrak{R}_{\mathfrak{P}}$, that is,

$$(\mathfrak{R}_{\mathfrak{p}})_{\mathfrak{P}^*} \subseteq \mathfrak{R}_{\mathfrak{P}}.$$

Conversely, if ξ is any element of $\mathfrak{R}_{\mathfrak{P}}$, write $\xi = \alpha/\beta$, where α and β are in $\mathfrak{R}$ and β is not in $\mathfrak{P}$. Then α and β are in $\mathfrak{R}_{\mathfrak{p}}$, and β is not in $\mathfrak{P}^*$. Therefore ξ is in $(\mathfrak{R}_{\mathfrak{p}})_{\mathfrak{P}^*}$, that is,

$$\mathfrak{R}_{\mathfrak{P}} \subseteq (\mathfrak{R}_{\mathfrak{p}})_{\mathfrak{P}^*}.$$

Taking the two results together, we have

$$(\mathfrak{R}_{\mathfrak{p}})_{\mathfrak{P}^*} = \mathfrak{R}_{\mathfrak{P}}.$$

Moreover, if $\mathfrak{P}$ is a proper multiple of $\mathfrak{p}$, $\mathfrak{P}^*$ is a proper multiple of $\mathfrak{R}_{\mathfrak{p}}.\mathfrak{P}$, and is therefore a proper ideal of $\mathfrak{R}_{\mathfrak{p}}$, and there are non-units of $\mathfrak{R}_{\mathfrak{p}}$ which are not in $\mathfrak{P}^*$. Then

$$\mathfrak{R}_{\mathfrak{p}} \subset (\mathfrak{R}_{\mathfrak{p}})_{\mathfrak{P}^*} = \mathfrak{R}_{\mathfrak{P}}.$$

Next suppose that $\mathfrak{P}$ is any ideal of $\mathfrak{R}$ such that

$$\mathfrak{R}_{\mathfrak{p}} \subseteq \mathfrak{R}_{\mathfrak{P}}.$$

Then every unit of $\mathfrak{R}_{\mathfrak{p}}$ is a unit of $\mathfrak{R}_{\mathfrak{P}}$. In particular, any element of $\mathfrak{R}$ which is not in $\mathfrak{p}$ is not in $\mathfrak{P}$, and hence $\mathfrak{P} \subseteq \mathfrak{p}$. Thus, summing up, we have

THEOREM VI. *If* $\mathfrak{p}$ *and* $\mathfrak{P}$ *are two prime ideals of* $\mathfrak{R}$, *then* $\mathfrak{R}_{\mathfrak{p}} \subseteq \mathfrak{R}_{\mathfrak{P}}$ *if and only if* $\mathfrak{P} \subseteq \mathfrak{p}$. *In particular,* $\mathfrak{R}_{\mathfrak{p}} = \mathfrak{R}_{\mathfrak{P}}$ *if and only if* $\mathfrak{p} = \mathfrak{P}$. *Moreover, if* $\mathfrak{P} \subseteq \mathfrak{p}$, $\mathfrak{R}_{\mathfrak{P}} = (\mathfrak{R}_{\mathfrak{p}})_{\mathfrak{R}_{\mathfrak{p}}.\mathfrak{P}}$.

THEOREM VII. *If the Basis Theorem holds in* $\mathfrak{R}$, *it holds in* $\mathfrak{R}_{\mathfrak{p}}$.

Let $\mathfrak{i}^*$ be any ideal in $\mathfrak{R}_{\mathfrak{p}}$, and let $\mathfrak{i} = \mathfrak{R}_\wedge \mathfrak{i}^*$. The ideal $\mathfrak{i}$ is in $\mathfrak{R}$, and has a finite basis, say, $\omega_1, \ldots, \omega_r$. Since $\omega_j \in \mathfrak{i}$, we have $\omega_j \in \mathfrak{i}^*$. Now, any element α of $\mathfrak{i}$ is of the form $\sum_1^r \rho_i \omega_i$, where ρ_i is in $\mathfrak{R}$. But [Th. I] any element α^* of $\mathfrak{i}^*$ is of the form α/β, where $\alpha \in \mathfrak{i}$ and $\beta \neq 0$ $(\mathfrak{p})$. Hence

$$\alpha^* = \sum_1^r \frac{\rho_i}{\beta} \omega_i,$$

and therefore $\omega_1, \ldots, \omega_r$ is a basis for $\mathfrak{i}^*$. This proves the theorem.

For the remainder of this section, *we assume that the Basis Theorem holds in* $\mathfrak{R}$.

In view of Theorem VII, the Basis Theorem also holds in $\mathfrak{R}_{\mathfrak{p}}$, and we can then assign indices to primary ideals in $\mathfrak{R}$ and in $\mathfrak{R}_{\mathfrak{p}}$. We now prove

THEOREM VIII. *Primary ideals in* $\mathfrak{R}$ *and* $\mathfrak{R}_{\mathfrak{p}}$ *which correspond as in Theorem IV have the same indices.*

Let $\mathfrak{q}^*$ be a $\mathfrak{p}^*$-primary ideal in $\mathfrak{R}_{\mathfrak{p}}$, and let $\mathfrak{Q} = \mathfrak{R}_\wedge \mathfrak{q}^*$ be the corresponding primary of $\mathfrak{R}$. Its radical is $\mathfrak{P} = \mathfrak{R}_\wedge \mathfrak{p}^*$.

(i) Let ρ be the index of $\mathfrak{Q}$. Then

$$\mathfrak{P}^\rho = 0 \ \ (\mathfrak{Q}).$$

By § 4, Theorem I (iv),

$$\mathfrak{R}_{\mathfrak{p}} . \mathfrak{P}^\rho = (\mathfrak{R}_{\mathfrak{p}} . \mathfrak{P})^\rho = (\mathfrak{p}^*)^\rho.$$

Since $\mathfrak{q}^* = \mathfrak{R}_{\mathfrak{p}} . \mathfrak{Q}$, we have from § 4, Theorem I (i),

$$(\mathfrak{p}^*)^\rho = 0 \ \ (\mathfrak{q}^*).$$

Hence the index of $\mathfrak{q}^*$ cannot exceed ρ.

(ii) Let σ be the index of $\mathfrak{q}^*$. Then, using § 4, Theorem II (i) and (iv),

$$\mathfrak{P}^\sigma = (\mathfrak{R}_\wedge \mathfrak{p}^*)^\sigma \subseteq \mathfrak{R}_\wedge \mathfrak{p}^{*\sigma} \subseteq \mathfrak{R}_\wedge \mathfrak{q}^* = \mathfrak{Q}.$$

Hence the index of $\mathfrak{Q}$ cannot exceed σ.

Taking these two results together, we obtain $\rho = \sigma$, the required result.

Now consider any ideal $\mathfrak{i}$ in $\mathfrak{R}$, and let

$$\mathfrak{i} = [\mathfrak{q}_1, \ldots, \mathfrak{q}_s, \mathfrak{q}'_{s+1}, \ldots, \mathfrak{q}'_t]$$

be a representation of it as an uncontractible intersection of primaries, where the radical of $\mathfrak{q}_i$, and therefore $\mathfrak{q}_i$ itself, is contained in $\mathfrak{p}$ $(i = 1, \ldots, s)$, while $\mathfrak{q}'_i \neq 0 \ (\mathfrak{p})$ $(i = s+1, \ldots, t)$. By Theorem I (iii) of § 4

$$\begin{aligned}\mathfrak{R}_\mathfrak{p} . \mathfrak{i} &\subseteq [\mathfrak{R}_\mathfrak{p} . \mathfrak{q}_1, \ldots, \mathfrak{R}_\mathfrak{p} . \mathfrak{q}_s, \mathfrak{R}_\mathfrak{p} . \mathfrak{q}'_{s+1}, \ldots, \mathfrak{R}_\mathfrak{p} . \mathfrak{q}'_t] \\ &= [\mathfrak{R}_\mathfrak{p} . \mathfrak{q}_1, \ldots, \mathfrak{R}_\mathfrak{p} . \mathfrak{q}_s],\end{aligned}$$

since, by Theorem II,

$$\mathfrak{R}_\mathfrak{p} . \mathfrak{q}'_i = \mathfrak{R}_\mathfrak{p} \quad (i = s+1, \ldots, t).$$

Let ξ be any element of $[\mathfrak{R}_\mathfrak{p} . \mathfrak{q}_1, \ldots, \mathfrak{R}_\mathfrak{p} . \mathfrak{q}_s]$. Then, for $i = 1, \ldots, s$, we may write

$$\xi = \frac{\alpha_i}{\beta_i},$$

where $\alpha_i \in \mathfrak{q}_i$, and β_i is not in $\mathfrak{p}$. Then

$$\alpha_i \beta_j = \alpha_j \beta_i = 0 \ (\mathfrak{q}_j).$$

Since the radical of $\mathfrak{q}_j$ is a multiple of $\mathfrak{p}$, and β_j is not in $\mathfrak{p}$, β_j is not in the radical of $\mathfrak{q}_j$, hence $\alpha_i \in \mathfrak{q}_j$. Since $\mathfrak{q}'_j$ is not contained in $\mathfrak{p}$, we can find a β'_j in $\mathfrak{q}'_j$ which is not in $\mathfrak{p}$, for $j = s+1, \ldots, t$. Hence

$$\xi = \frac{\alpha_1 \beta'_{s+1} \cdots \beta'_t}{\beta_1 \beta'_{s+1} \cdots \beta'_t} = \frac{\alpha}{\beta},$$

where α is in $[\mathfrak{q}_1, \ldots, \mathfrak{q}_s, \mathfrak{q}'_{s+1}, \ldots, \mathfrak{q}'_t] = \mathfrak{i}$, and β is not in $\mathfrak{p}$. Hence [Th. I]

$$[\mathfrak{R}_\mathfrak{p} . \mathfrak{q}_1, \ldots, \mathfrak{R}_\mathfrak{p} . \mathfrak{q}_s] \subseteq \mathfrak{R}_\mathfrak{p} . \mathfrak{i}.$$

It follows that $\mathfrak{R}_\mathfrak{p} . \mathfrak{i} = [\mathfrak{R}_\mathfrak{p} . \mathfrak{q}_1, \ldots, \mathfrak{R}_\mathfrak{p} . \mathfrak{q}_s]$.

This representation of $\mathfrak{R}_\mathfrak{p} . \mathfrak{i}$ is uncontractible. For if, say,

$$\mathfrak{R}_\mathfrak{p} . \mathfrak{q}_i \supseteq [\mathfrak{R}_\mathfrak{p} . \mathfrak{q}_1, \ldots, \mathfrak{R}_\mathfrak{p} . \mathfrak{q}_{i-1}, \mathfrak{R}_\mathfrak{p} . \mathfrak{q}_{i+1}, \ldots, \mathfrak{R}_\mathfrak{p} . \mathfrak{q}_s],$$

we have [§ 4, Th. II (iii) and Th. IV]

$$\begin{aligned}\mathfrak{q}_i = \mathfrak{R}_\wedge(\mathfrak{R}_\mathfrak{p} . \mathfrak{q}_i) \supseteq [&\mathfrak{R}_\wedge(\mathfrak{R}_\mathfrak{p} . \mathfrak{q}_1), \ldots, \mathfrak{R}_\wedge(\mathfrak{R}_\mathfrak{p} . \mathfrak{q}_{i-1}), \\ &\mathfrak{R}_\wedge(\mathfrak{R}_\mathfrak{p} . \mathfrak{q}_{i+1}), \ldots, \mathfrak{R}_\wedge(\mathfrak{R}_\mathfrak{p} . \mathfrak{q}_s)] \\ = [\mathfrak{q}_1, \ldots, \mathfrak{q}_{i-1}, \mathfrak{q}_{i+1}, \ldots, \mathfrak{q}_s] &\supseteq [\mathfrak{q}_1, \ldots, \mathfrak{q}_{i-1}, \mathfrak{q}_{i+1}, \ldots, \mathfrak{q}_s, \mathfrak{q}'_{s+1}, \ldots, \mathfrak{q}'_t],\end{aligned}$$

and it would follow that $\mathfrak{q}_i$ is irrelevant in the representation of $\mathfrak{i}$, contrary to hypothesis.

If, on the other hand, we start from an ideal $\mathfrak{i}^*$ of $\mathfrak{R}_\mathfrak{p}$ and an uncontractible representation $\mathfrak{i}^* = [\mathfrak{q}_1^*, \ldots, \mathfrak{q}_r^*]$ of $\mathfrak{i}^*$, Theorem II (iii) of § 4 tells us that

$$\mathfrak{R}_\wedge \mathfrak{i}^* = [\mathfrak{R}_\wedge \mathfrak{q}_1^*, \ldots, \mathfrak{R}_\wedge \mathfrak{q}_r^*],$$

and an argument similar to the above tells us that this is an uncontractible representation of $\mathfrak{R}_\wedge \mathfrak{i}^*$.

We thus have:

THEOREM IX. *The (proper) contracted ideals of $\mathfrak{R}$ are those which can be represented as the intersections of primaries whose radicals are multiples of $\mathfrak{p}$. The contracted ideal associated with $\mathfrak{i} = [\mathfrak{q}_1, \ldots, \mathfrak{q}_t]$ is obtained by omitting those primaries whose radicals are not multiples of $\mathfrak{p}$. If $\mathfrak{i} = [\mathfrak{q}_1, \ldots, \mathfrak{q}_r]$ is an uncontractible representation of a contracted ideal of $\mathfrak{R}$, $[\mathfrak{R}_\mathfrak{p}.\mathfrak{q}_1, \ldots, \mathfrak{R}_\mathfrak{p}.\mathfrak{q}_r]$ is an uncontractible representation of $\mathfrak{R}_\mathfrak{p}.\mathfrak{i}$; and if $[\mathfrak{q}_1^*, \ldots, \mathfrak{q}_k^*]$ is an uncontractible representation of an ideal $\mathfrak{i}^*$ of $\mathfrak{R}_\mathfrak{p}$, $[\mathfrak{R}_\wedge \mathfrak{q}_1^*, \ldots, \mathfrak{R}_\wedge \mathfrak{q}_k^*]$ is an uncontractible representation of $\mathfrak{R}_\wedge \mathfrak{i}^*$.*

This result illustrates one of the advantages of using quotient rings. In many problems, particularly when we apply ideal theory to algebraic geometry, we are only concerned with those components of ideals which are multiples of a given prime ideal $\mathfrak{p}$, and the use of quotient rings enables us to eliminate all the components with which we are not concerned. In later chapters many examples of the use of this device will appear.

The remaining two theorems of this section are of frequent use in the applications of ideal theory to algebraic geometry. The first assumes that the Basis Theorem holds in $\mathfrak{R}$, the second does not.

THEOREM X. *The integral domain $\mathfrak{R}$ is the intersection of quotient rings of its prime ideals, and, indeed, is the intersection of the quotient rings of its maximal ideals.*

By the intersection of the quotient rings, we mean the aggregate of those elements of the quotient field of $\mathfrak{R}$ which are in all the quotient rings.

We first show that it is only necessary to consider the maximal ideals in $\mathfrak{R}$. These are prime, since $\mathfrak{R}$ has unity. Let $\mathfrak{p}$ be any prime ideal of $\mathfrak{R}$. If $\mathfrak{p}$ is not maximal, it is a proper multiple of some proper ideal $\mathfrak{i}$ of $\mathfrak{R}$, and hence of any primary ideal $\mathfrak{q}'$ which is a component of $\mathfrak{i}$. The radical $\mathfrak{p}'$ of $\mathfrak{q}'$ is not the unit ideal. For if $\mathfrak{p}' = \mathfrak{R}$, $\mathfrak{p}'^\rho$, for any integer ρ, is also equal to $\mathfrak{R}$, since $\mathfrak{R}$ has unity. But since $\mathfrak{q}'$ has a finite index, there exists a value of ρ such that $\mathfrak{p}'^\rho \subseteq \mathfrak{q}'$, so that $\mathfrak{q}'$ is the unit ideal, contrary to the hypothesis that $\mathfrak{q}'$ is a component of a proper ideal of $\mathfrak{R}$. We now have $\mathfrak{p} \subset \mathfrak{q}' \subseteq \mathfrak{p}'$, so that there is a proper factor of $\mathfrak{p}$ which is a proper prime ideal of $\mathfrak{R}$. If $\mathfrak{p}'$ is not maximal, it has, in the same way, a proper factor $\mathfrak{p}''$ which is a proper prime ideal in $\mathfrak{R}$. If $\mathfrak{p}''$ is not maximal, we proceed as before. We thus obtain a sequence of ideals

$$\mathfrak{p} \subset \mathfrak{p}' \subset \mathfrak{p}'' \subset \dots.$$

By Theorem VI of § 1, this sequence is finite, and hence after a finite number of steps we reach an ideal $\mathfrak{p}_1$ which is maximal and is a factor of $\mathfrak{p}$. By Theorem VI, $\mathfrak{R}_{\mathfrak{p}_1} \subseteq \mathfrak{R}_{\mathfrak{p}}$, and hence any element of the quotient field of $\mathfrak{R}$ which is in $\mathfrak{R}_{\mathfrak{p}_1}$ is in $\mathfrak{R}_{\mathfrak{p}}$. Hence, in considering the elements of the quotient field of $\mathfrak{R}$ which are in the quotient rings of prime ideals of $\mathfrak{R}$, we need not consider $\mathfrak{R}_{\mathfrak{p}}$, since if the elements are restricted to lie in $\mathfrak{R}_{\mathfrak{p}_1}$, they necessarily lie in $\mathfrak{R}_{\mathfrak{p}}$. Hence we need only consider the quotient rings of the maximal ideals in $\mathfrak{R}$.

Clearly, every element of $\mathfrak{R}$ is in every quotient ring. Now let ξ be any element of the quotient field of $\mathfrak{R}$ which is in the quotient ring of every prime ideal of $\mathfrak{R}$. Let $\xi = \alpha_0/\beta_0$, where α_0 and β_0 are in $\mathfrak{R}$. If β_0 is a unit of $\mathfrak{R}$, ξ is in $\mathfrak{R}$. If β_0 is not a unit of $\mathfrak{R}$, $\mathfrak{R}.(\beta_0)$ is a proper ideal of $\mathfrak{R}$. Write

$$\mathfrak{R}.(\beta_0) = [\mathfrak{q}_1, \dots, \mathfrak{q}_r],$$

where $\mathfrak{q}_i$ is $\mathfrak{p}_i$-primary. ξ is in $\mathfrak{R}_{\mathfrak{p}_1}$, by hypothesis; hence $\xi = \alpha_1/\beta_1$, where α_1 and β_1 are in $\mathfrak{R}$ and β_1 is not in $\mathfrak{p}_1$. Hence β_1 is not in $\mathfrak{q}_1$, and therefore not in $\mathfrak{R}.(\beta_0)$. Therefore

$$\mathfrak{R}.(\beta_0) \subset \mathfrak{R}.(\beta_0, \beta_1).$$

If $\mathfrak{R}.(\beta_0, \beta_1)$ is not the unit ideal, we write

$$\mathfrak{R}.(\beta_0, \beta_1) = [\mathfrak{q}'_1, \dots, \mathfrak{q}'_s],$$

where $\mathfrak{q}'_i$ is $\mathfrak{p}'_i$-primary, ξ is in the quotient ring of $\mathfrak{p}'_1$, and we may write $\xi = \alpha_2/\beta_2$, where α_2 and β_2 are in $\mathfrak{R}$ and β_2 is not in $\mathfrak{p}'_1$. Hence, as before, we obtain

$$\mathfrak{R}.(\beta_0, \beta_1) \subset \mathfrak{R}.(\beta_0, \beta_1, \beta_2).$$

We thus construct a sequence of ideals

$$\mathfrak{R}.(\beta_0) \subset \mathfrak{R}.(\beta_0, \beta_1) \subset \mathfrak{R}.(\beta_0, \beta_1, \beta_2) \subset \dots,$$

which only terminates when we reach the unit ideal. But, by Theorem VI of § 1, the sequence must terminate; hence there exists an integer r such that

$$\mathfrak{R}.(\beta_0, \ldots, \beta_r) = \mathfrak{R}.$$

There must therefore exist elements $\rho_0, \ldots, \rho_r$ of $\mathfrak{R}$ such that

$$\rho_0\beta_0 + \ldots + \rho_r\beta_r = 1.$$

But $\xi\beta_i = \alpha_i$ $(i = 0, \ldots, r)$. Hence

$$\xi = \rho_0\xi\beta_0 + \ldots + \rho_r\xi\beta_r$$

$$= \rho_0\alpha_0 + \ldots + \rho_r\alpha_r;$$

that is, ξ is contained in $\mathfrak{R}$. This proves the theorem.

Our final theorem concerns the permutability of the operations of forming remainder-class rings and quotient rings. Let $\mathfrak{R}$ be an integral domain, $\mathfrak{p}$ a prime ideal in it, and let $\mathfrak{i}$ be any ideal in $\mathfrak{R}$ which is contained in $\mathfrak{p}$. Then [§ 3, Th. V], when we pass to the remainder-class ring $\mathfrak{R}/\mathfrak{i}$, $\mathfrak{p}/\mathfrak{i}$ is prime, and we can form the quotient ring $(\mathfrak{R}/\mathfrak{i})_{\mathfrak{p}/\mathfrak{i}}$. Again, since $\mathfrak{i} \subseteq \mathfrak{p}$, $\mathfrak{R}_\mathfrak{p}.\mathfrak{i}$ is an ideal of $\mathfrak{R}_\mathfrak{p}$ different from the unit ideal [Th. II]. Hence we can form the remainder-class ring $\mathfrak{R}_\mathfrak{p}/\mathfrak{R}_\mathfrak{p}.\mathfrak{i}$. We now prove

THEOREM XI. *The rings* $(\mathfrak{R}/\mathfrak{i})_{\mathfrak{p}/\mathfrak{i}}$ *and* $\mathfrak{R}_\mathfrak{p}/\mathfrak{R}_\mathfrak{p}.\mathfrak{i}$ *are isomorphic.*

Let $\bar{\xi}/\bar{\eta}$ be any element of $(\mathfrak{R}/\mathfrak{i})_{\mathfrak{p}/\mathfrak{i}}$, where $\bar{\xi}$ and $\bar{\eta}$ are in $\mathfrak{R}/\mathfrak{i}$, and $\bar{\eta}$ is not in $\mathfrak{p}/\mathfrak{i}$. If $\xi, \xi', \ldots$ are elements of $\mathfrak{R}$ in the remainder-class $\bar{\xi}$, and $\eta, \eta', \ldots$ are elements of $\mathfrak{R}$ in the remainder-class $\bar{\eta}$, the elements $\eta, \eta', \ldots$ are not in $\mathfrak{p}$, while

$$\xi - \xi' = 0 \ (\mathfrak{i}), \quad \eta - \eta' = 0 \ (\mathfrak{i}).$$

Since $\eta, \eta', \ldots$ are not in $\mathfrak{p}$, ξ/η and ξ'/η' are in $\mathfrak{R}_\mathfrak{p}$. Again,

$$\frac{\xi}{\eta} - \frac{\xi'}{\eta'} = \frac{(\xi - \xi')\eta' - \xi'(\eta - \eta')}{\eta\eta'} = 0 \quad (\mathfrak{R}_\mathfrak{p}.\mathfrak{i});$$

hence $\xi/\eta, \xi'/\eta', \ldots$ belong to the same remainder-class of $\mathfrak{R}_\mathfrak{p}$, modulo $\mathfrak{R}_\mathfrak{p}.\mathfrak{i}$. Therefore $\bar{\xi}/\bar{\eta}$ determines a unique element of $\mathfrak{R}_\mathfrak{p}/\mathfrak{R}_\mathfrak{p}.\mathfrak{i}$. We thus have a mapping of the elements of $(\mathfrak{R}/\mathfrak{i})_{\mathfrak{p}/\mathfrak{i}}$ on $\mathfrak{R}_\mathfrak{p}/\mathfrak{R}_\mathfrak{p}.\mathfrak{i}$, which we denote by H.

Conversely, let us consider an element ζ^* of $\mathfrak{R}_\mathfrak{p}/\mathfrak{R}_\mathfrak{p}.\mathfrak{i}$. Let $\xi/\eta, \xi'/\eta', \ldots$ be elements of $\mathfrak{R}_\mathfrak{p}$ in the remainder-class ζ^*, where $\xi, \xi', \ldots$ are in $\mathfrak{R}$, and $\eta, \eta', \ldots$ are in $\mathfrak{R}$, but not in $\mathfrak{p}$. Since

$$\frac{\xi}{\eta} - \frac{\xi'}{\eta'} = 0 \ \ (\mathfrak{R}_\mathfrak{p}.\mathfrak{i}),$$

we have
$$\xi\eta' - \xi'\eta = 0 \ \ (\mathfrak{i}).$$

Pass to the remainder-class ring $\mathfrak{R}/\mathfrak{i}$. If $\bar{\xi}, \bar{\xi}', \ldots, \bar{\eta}, \bar{\eta}', \ldots$ denote the elements of $\mathfrak{R}/\mathfrak{i}$ corresponding to $\xi, \xi', \ldots, \eta, \eta', \ldots,$

$$\bar{\xi}\bar{\eta}' - \bar{\xi}'\bar{\eta} = 0,$$

and, since η, η' are not in $\mathfrak{p}$, $\bar{\eta}, \bar{\eta}'$ are not in $\mathfrak{p}/\mathfrak{i}$. Hence $\bar{\xi}/\bar{\eta} = \bar{\xi}'/\bar{\eta}'$ is in $(\mathfrak{R}/\mathfrak{i})_{\mathfrak{p}/\mathfrak{i}}$, so that ζ^* determines a unique element $\bar{\xi}/\bar{\eta}$ of $(\mathfrak{R}/\mathfrak{i})_{\mathfrak{p}/\mathfrak{i}}$. It is clear that this element $\bar{\xi}/\bar{\eta}$ determines ζ^* in the mapping H of $(\mathfrak{R}/\mathfrak{i})_{\mathfrak{p}/\mathfrak{i}}$ on $\mathfrak{R}_\mathfrak{p}/\mathfrak{R}_\mathfrak{p}.\mathfrak{i}$. Thus we have a one-to-one correspondence between the elements of the two rings, and it is seen at once that this correspondence preserves sums and products. It is therefore an isomorphism, and Theorem XI is proved.

Corollary. If $\mathfrak{i} = \mathfrak{p}$, $(\mathfrak{R}/\mathfrak{p})_{\mathfrak{p}/\mathfrak{p}}$ *is the quotient field of* $\mathfrak{R}/\mathfrak{p}$. *Hence* $\mathfrak{R}_\mathfrak{p}/\mathfrak{R}_\mathfrak{p}.\mathfrak{p}$ *is isomorphic with the quotient field of* $\mathfrak{R}/\mathfrak{p}$.

6. Modules. As a preliminary to the development of a multiplicative theory of ideals, we give a brief discussion of some properties of modules over a commutative ring.

Let $\mathfrak{R}$ be any commutative ring, and let $\mathfrak{M}$ be any commutative group, written as an additive group. Suppose, now, that there exists a multiplication between the elements of $\mathfrak{R}$ and the elements of $\mathfrak{M}$ such that, if α is in $\mathfrak{R}$ and ξ is in $\mathfrak{M}$, then $\alpha\xi = \xi\alpha$ is in $\mathfrak{M}$. $\mathfrak{M}$ is called an $\mathfrak{R}$-*module*, sometimes merely a *module*, if the following properties hold:

(i) $$(\alpha+\beta)\,\xi = \alpha\xi+\beta\xi,$$

(ii) $$\alpha(\xi+\eta) = \alpha\xi+\alpha\eta,$$

(iii) $$\alpha(\beta\xi) = (\alpha\beta)\,\xi.$$

In the case in which $\mathfrak{R}$ is a field, a linear set [II, § 1] is an $\mathfrak{R}$-module. Again, any ideal in $\mathfrak{R}$ is an $\mathfrak{R}$-module.

If η is the zero of $\mathfrak{M}$, and ξ is any element of $\mathfrak{M}$, we have, for any element α of $\mathfrak{R}$,

$$\alpha\xi = \alpha(\xi+\eta) = \alpha\xi+\alpha\eta;$$

hence, since $\mathfrak{M}$ is an additive group,

$$\alpha\eta = 0.$$

Similarly, if α is the zero of $\mathfrak{R}$, and ξ is any element of $\mathfrak{M}$, we prove that

$$\alpha\xi = 0.$$

Consider the elements α of $\mathfrak{R}$ such that $\alpha\xi = 0$ for all ξ in $\mathfrak{M}$. If α and β are two such elements of $\mathfrak{R}$, and ρ is any element of $\mathfrak{R}$, we have

$$(\alpha-\beta)\xi = \alpha\xi-\beta\xi = 0 \quad \text{and} \quad (\rho\alpha)\xi = \rho(\alpha\xi) = 0.$$

Hence the elements of $\mathfrak{R}$ in question form an ideal $\mathfrak{i}$. Again, if α and β are any two elements of $\mathfrak{R}$ such that $\alpha\xi = \beta\xi$ for all elements ξ of $\mathfrak{M}$, we have $(\alpha-\beta)\xi = 0$, and therefore $\alpha-\beta \in \mathfrak{i}$. Hence α and β belong to the same remainder-class of $\mathfrak{R}/\mathfrak{i}$. Conversely, we see at once that if α and β belong to the same remainder-class, $\alpha\xi = \beta\xi$ for all ξ in $\mathfrak{M}$. Hence, in considering $\mathfrak{M}$ as an $\mathfrak{R}$-module, we may replace $\mathfrak{R}$ by $\mathfrak{R}/\mathfrak{i}$, and the properties of $\mathfrak{M}$ can be obtained by considering $\mathfrak{M}$ as an $\mathfrak{R}/\mathfrak{i}$-module, and then using the results of § 3 for the relations between $\mathfrak{R}$ and $\mathfrak{R}/\mathfrak{i}$. It will therefore be sufficient to consider $\mathfrak{M}$ as an $\mathfrak{R}/\mathfrak{i}$-module, or, in other words, to assume that the equation $\alpha\xi = 0$ for all elements ξ of $\mathfrak{M}$ implies $\alpha = 0$. *We make this assumption in what follows*, and we shall also assume *that* $\mathfrak{R}$ *has unity*.

In an exactly similar manner, we show that the elements ξ of $\mathfrak{M}$ which have the property

$$\alpha\xi = 0$$

for all α in $\mathfrak{R}$ form a subgroup $\mathfrak{M}_1$ of $\mathfrak{M}$. Now suppose that e is the unity of $\mathfrak{R}$. Let ξ be any element of $\mathfrak{M}$, and let $\bar{\xi} = e\xi$. Then

$$e\bar{\xi} = e^2\xi = e\xi = \bar{\xi},$$

and hence

$$e(\xi-\bar{\xi}) = e\xi-e\bar{\xi} = 0.$$

If we write $\bar{\eta} = \xi-\bar{\xi}$, and take any element α of $\mathfrak{R}$, we have

$$\alpha\bar{\eta} = (\alpha e)\bar{\eta} = \alpha(e\bar{\eta}) = 0.$$

Hence $\bar{\eta} \in \mathfrak{M}_1$. Therefore any element ξ of $\mathfrak{M}$ can be written in the form

$$\xi = \bar{\xi}+\bar{\eta},$$

where

$$e\bar{\xi} = \bar{\xi} \quad \text{and} \quad e\bar{\eta} = 0.$$

Moreover, this representation is unique. For if

$$\bar{\xi}+\bar{\eta} = \bar{\xi}'+\bar{\eta}',$$

where $$e\bar{\xi}' = \bar{\xi}' \quad \text{and} \quad e\bar{\eta}' = 0,$$

we have $$\bar{\xi}-\bar{\xi}' = \bar{\eta}'-\bar{\eta}.$$

Hence $$\bar{\xi}-\bar{\xi}' = e(\bar{\xi}-\bar{\xi}') = e\bar{\eta}'-e\bar{\eta} = 0;$$

that is, $\bar{\xi} = \bar{\xi}'$, and hence $\bar{\eta} = \bar{\eta}'$.

We can prove at once that the elements $\bar{\xi}$ of $\mathfrak{M}$ which have the property $e\bar{\xi} = \bar{\xi}$ form a subgroup $\mathfrak{M}_2$. For

$$e(\bar{\xi}+\bar{\eta}) = e\bar{\xi}+e\bar{\eta} = \bar{\xi}+\bar{\eta}.$$

Thus $\mathfrak{M}$ is the 'direct sum' of the subgroups $\mathfrak{M}_1$ and $\mathfrak{M}_2$, and it is obvious that we can obtain all the properties of $\mathfrak{M}$ as an $\mathfrak{R}$-module from the properties of $\mathfrak{M}_2$ as an $\mathfrak{R}$-module. Since we shall only be concerned with rings $\mathfrak{R}$ with unity, we may therefore confine our attention to modules $\mathfrak{M}$ with the property that if the equation

$$\alpha\xi = 0$$

holds for all α in $\mathfrak{R}$ then ξ is zero. *We shall assume this property in what follows.* This implies that $e\xi = \xi$ for all ξ in $\mathfrak{M}$.

The theory of $\mathfrak{R}$-modules, even under the restrictions which we have imposed, is a large one, and a full development of the theory would take us far beyond our objective. We therefore confine ourselves to a few results which will be required later.

The $\mathfrak{R}$-module $\mathfrak{M}$ is said to have a finite $\mathfrak{R}$-basis if there exists in $\mathfrak{M}$ a finite set of elements $\omega_1, \ldots, \omega_r$ such that $\mathfrak{M}$ consists of the elements which can be written in the form

$$\alpha_1\omega_1 + \ldots + \alpha_r\omega_r,$$

where $\alpha_1, \ldots, \alpha_r$ are elements of $\mathfrak{R}$. An ideal in $\mathfrak{R}$ with a finite basis is a module with a finite $\mathfrak{R}$-basis, and, following our custom for ideals, we denote the module with the finite $\mathfrak{R}$-basis $\omega_1, \ldots, \omega_r$ by $\mathfrak{R}.(\omega_1, \ldots, \omega_r)$.

Any subgroup of $\mathfrak{M}$ which is closed under the operation of multiplication by elements of $\mathfrak{R}$ is called a submodule. We now prove

THEOREM I. *If the Basis Theorem holds in $\mathfrak{R}$, and if $\mathfrak{M}$ has a finite $\mathfrak{R}$-basis, then any submodule $\mathfrak{M}'$ of $\mathfrak{M}$ has a finite $\mathfrak{R}$-basis.*

Let $\mathfrak{M} = \mathfrak{R}.(\omega_1, \ldots, \omega_r)$. We prove our theorem by induction on the number r of elements in the basis. Consider first the case $r = 1$. Any element of $\mathfrak{M}'$ is of the form $\alpha\omega_1$, where α is in $\mathfrak{R}$. If $\alpha\omega_1$ and $\beta\omega_1$ are elements of $\mathfrak{M}'$, and ρ is any element of $\mathfrak{R}$,

$$(\alpha - \beta)\,\omega_1 = \alpha\omega_1 - \beta\omega_1$$

is in $\mathfrak{M}'$, since $\mathfrak{M}'$ is a group, and

$$(\rho\alpha)\,\omega_1 = \rho(\alpha\omega_1)$$

is in $\mathfrak{M}'$, since $\mathfrak{M}'$ is closed under multiplication by elements of $\mathfrak{R}$. It follows that the elements α of $\mathfrak{R}$ which appear as coefficients of ω_1 in elements of $\mathfrak{M}'$ form an ideal $\mathfrak{i}$ in $\mathfrak{R}$. This ideal has a finite basis, say $\alpha_1, \ldots, \alpha_k$. Since $\alpha_i \in \mathfrak{i}$, $\alpha_i\omega_1$ is in $\mathfrak{M}'$. Any element $\alpha\omega_1$ of $\mathfrak{M}'$ is of the form

$$\left(\sum_1^k \rho_i \alpha_i\right)\omega_1 = \sum_1^k \rho_i(\alpha_i\omega_1),$$

where $\rho_1, \ldots, \rho_k$ are in $\mathfrak{R}$, and conversely, any element of this form is in $\mathfrak{M}'$, since $\sum_1^k \rho_i\alpha_i \in \mathfrak{i}$. Hence $\mathfrak{M}' = \mathfrak{R}.(\alpha_1\omega_1, \ldots, \alpha_k\omega_1)$.

Now suppose that the theorem is true for all $\mathfrak{R}$-modules which have an $\mathfrak{R}$-basis consisting of less than r elements, and let

$$\mathfrak{M}_1 = \mathfrak{R}.(\omega_1, \ldots, \omega_{r-1}).$$

If ξ and η are any two elements of $\mathfrak{M} = \mathfrak{R}.(\omega_1, \ldots, \omega_r)$ which are in $\mathfrak{M}'$ and in $\mathfrak{M}_1$, and if α is any element of $\mathfrak{R}$, it is clear that $\xi - \eta$ and $\alpha\xi$ lie in $\mathfrak{M}'$ and in $\mathfrak{M}_1$. Hence the elements of $\mathfrak{M}'$ which lie in $\mathfrak{M}_1$ form a submodule $\mathfrak{M}_1'$ of $\mathfrak{M}_1$. By the hypothesis of induction, $\mathfrak{M}_1'$ has a finite $\mathfrak{R}$-basis, say, $\mathfrak{M}_1' = \mathfrak{R}.(\nu_1, \ldots, \nu_s)$.

Consider the elements α of $\mathfrak{R}$ such that there exist elements $\alpha_1, \ldots, \alpha_{r-1}$ in $\mathfrak{R}$ so that $\alpha_1\omega_1 + \ldots + \alpha_{r-1}\omega_{r-1} + \alpha\omega_r$ is in $\mathfrak{M}'$. As in the case $r = 1$, we show that these elements α form an ideal in $\mathfrak{R}$, say, $\mathfrak{R}.(\beta_1, \ldots, \beta_t)$. Then there exist elements $\nu_{s+1}, \ldots, \nu_{s+t}$ of $\mathfrak{M}'$ which can be written as

$$\nu_i = \sum_{j=1}^{r-1} \alpha_{ij}\omega_j + \beta_{i-s}\omega_r \quad (i = s+1, \ldots, s+t).$$

Let
$$\omega = \sum_{i=1}^{r} \alpha_i\omega_i$$

be any element of $\mathfrak{M}'$. We know that $\alpha_r \in \mathfrak{R}.(\beta_1, \ldots, \beta_t)$, say,

$$\alpha_r = \sum_{i=1}^{t} \rho_{s+i}\beta_i.$$

Then
$$\omega - \sum_{i=s+1}^{s+t} \rho_i \nu_i$$
is in $\mathfrak{M}'_1$. Hence it is equal to $\sum_{i=1}^{s} \rho_i \nu_i$, for some choice of $\rho_1, \ldots, \rho_s$ in $\mathfrak{R}$. Therefore
$$\omega = \sum_{i=1}^{s+t} \rho_i \nu_i,$$
that is,
$$\omega \in \mathfrak{R}.(\nu_1, \ldots, \nu_{s+t}).$$

On the other hand, it follows from the construction that $\nu_1, \ldots, \nu_{s+t}$ are in $\mathfrak{M}'$, so that
$$\mathfrak{R}.(\nu_1, \ldots, \nu_{s+t}) \subseteq \mathfrak{M}'.$$
We therefore conclude that
$$\mathfrak{M}' = \mathfrak{R}.(\nu_1, \ldots, \nu_{s+t}),$$
and the theorem is proved.

Corollary. It will be observed that if $\mathfrak{R}$ is a principal ideal ring, $t = 0$ or 1, and hence it is easily shown that if $\mathfrak{M}$ has a basis of r elements, $\mathfrak{M}'$ has a basis consisting of r elements at most.

7. Multiplicative theory of ideals. We are now in a position to consider some multiplicative properties of ideals in an integral domain. Let $\mathfrak{R}$ be an integral domain, K its quotient field. We shall assume that the Basis Theorem holds in $\mathfrak{R}$. An $\mathfrak{R}$-module $\mathfrak{M}$ is called a *fractional ideal* of $\mathfrak{R}$ when (i) every element of $\mathfrak{M}$ is in K; (ii) the product of an element α of $\mathfrak{R}$ and an element ξ of $\mathfrak{M}$ is equal to the product $\alpha\xi$ in K; (iii) $\mathfrak{M}$ has a finite $\mathfrak{R}$-basis. Any ideal in $\mathfrak{R}$ is thus a special case of a fractional ideal. An ideal in $\mathfrak{R}$ will be referred to as an *integral* ideal when we wish to distinguish it from a more general fractional ideal, and the term ideal will, in this section, include fractional ideals. As a preliminary to considering a multiplicative theory of integral ideals we develop some properties of the more general fractional ideals.

If $\mathfrak{M}_1$ and $\mathfrak{M}_2$ are two fractional ideals of $\mathfrak{R}$, we define their *join* $(\mathfrak{M}_1, \mathfrak{M}_2)$ as the aggregate of elements of K which can be written in the form $\xi_1 + \xi_2$, where $\xi_1 \in \mathfrak{M}_1$ and $\xi_2 \in \mathfrak{M}_2$. It can be verified at once that this is a fractional ideal. Again, the *intersection* of $\mathfrak{M}_1$ and $\mathfrak{M}_2$, written $[\mathfrak{M}_1, \mathfrak{M}_2]$, is the set of elements of K which lie in $\mathfrak{M}_1$ and in $\mathfrak{M}_2$. It is clearly a module, and it is a submodule of $\mathfrak{M}_1$ and of $\mathfrak{M}_2$. Hence [§ 6, Th. I] it has a finite $\mathfrak{R}$-basis, and it is therefore a fractional ideal. Finally, the *product* $\mathfrak{M}_1\mathfrak{M}_2$ of $\mathfrak{M}_1$ and $\mathfrak{M}_2$ is the set of elements of K which can be written as $\Sigma\xi_i\eta_i$, where ξ_i is in $\mathfrak{M}_1$ and η_i is in $\mathfrak{M}_2$. Again, it can be shown that this is an $\mathfrak{R}$-module, and finally that it is a fractional ideal.

Let $\mathfrak{M}$ be any non-zero fractional ideal of $\mathfrak{R}$, and consider the set S of elements ξ in K such that $\xi\eta$ is in $\mathfrak{R}$ for every element η of $\mathfrak{M}$.

If ξ and ξ' are any two elements of S,

$$\xi\eta\in\mathfrak{R} \quad \text{and} \quad \xi'\eta\in\mathfrak{R}.$$

Hence $$(\xi+\xi')\,\eta\in\mathfrak{R} \quad \text{and} \quad \alpha\xi\eta\in\mathfrak{R}$$

for every element η of $\mathfrak{M}$, where α is any element of $\mathfrak{R}$. Hence S is an $\mathfrak{R}$-module. Moreover, if η is any element of $\mathfrak{M}$ different from zero, and ξ is any element of S, $\xi\eta$ is in $\mathfrak{R}$, and therefore ξ is in $\mathfrak{R}.(\eta^{-1})$. Thus S is a submodule of the module $\mathfrak{R}.(\eta^{-1})$, which has a finite $\mathfrak{R}$-basis. Hence [§ 6, Th. I] S has a finite $\mathfrak{R}$-basis. S is therefore a fractional ideal, which we denote by $\mathfrak{M}^{-1}$. (We may note that if $\mathfrak{M}$ is the zero module, that is, if it consists of zero only, S is then equal to K, but in general this has not a finite $\mathfrak{R}$-basis.)

$\mathfrak{M}^{-1}$ is called the *inverse* of the fractional ideal $\mathfrak{M}$. If $\mathfrak{M} = \mathfrak{R}.(\zeta)$, where ζ is a non-zero element of K, and ξ is any element of $\mathfrak{M}^{-1}$, then $\xi\zeta\in\mathfrak{R}$, so that $\xi\in\mathfrak{R}.(\zeta^{-1})$; and if $\xi\in\mathfrak{R}.(\zeta^{-1})$, $\xi = \alpha/\zeta$, where α is in $\mathfrak{R}$. Hence $\xi\,(\beta\zeta) = \alpha\beta\in\mathfrak{R}$, where $\beta\zeta$ is any element of $\mathfrak{M}$. Therefore $\mathfrak{M}^{-1} = \mathfrak{R}.(\zeta^{-1})$. In particular, if $\mathfrak{M} = \mathfrak{R}$, $\mathfrak{M}^{-1} = \mathfrak{R}$. $\mathfrak{M} = \mathfrak{R}$, regarded as a fractional ideal, plays an important part in what follows, and it is convenient to denote it by **1**, since it plays the part of unity in our factorization theory.

I. If $\mathfrak{M}_1 \subseteq \mathfrak{M}_2$, any element ζ of $\mathfrak{M}_2^{-1}$ multiplied by any element of $\mathfrak{M}_1$ (which is also an element of $\mathfrak{M}_2$) must be in $\mathfrak{R}$. Hence $\mathfrak{M}_2^{-1} \subseteq \mathfrak{M}_1^{-1}$. The converse of this is not, however, always true. For example, let $\mathfrak{R}$ be the ring $K_1[x, y]$ of polynomials in two indeterminates over the ground field K_1, and let $\mathfrak{M}_1 = \mathbf{1}$, $\mathfrak{M}_2 = \mathfrak{R}.(x, y)$. Let $\alpha(x, y)/\beta(x, y)$ be any element of $\mathfrak{M}_2^{-1}$, where $\alpha(x, y)$ and $\beta(x, y)$ are polynomials in $\mathfrak{R}$ without common factor. Then

$$\frac{x\alpha(x,y)}{\beta(x,y)} \quad \text{and} \quad \frac{y\alpha(x,y)}{\beta(x,y)}$$

are in $\mathfrak{R}$. It follows from the unique factorization property of $\mathfrak{R}$ that $\beta(x, y)$ must be in K_1; that is, $\alpha(x, y)/\beta(x, y)$ is in $\mathfrak{R}$.

Conversely, it is clear that if $\alpha(x,y) \in \mathfrak{R}$, $\alpha(x,y) \in \mathfrak{M}_2^{-1}$. Hence $\mathfrak{M}_2^{-1} = \mathbf{1} = \mathfrak{M}_1^{-1}$. But $\mathfrak{M}_1 \not\subseteq \mathfrak{M}_2$.

II. Two fractional ideals $\mathfrak{M}_1$ and $\mathfrak{M}_2$ are said to be *quasi-equal;* $\mathfrak{M}_1 \sim \mathfrak{M}_2$, if $\mathfrak{M}_1^{-1} = \mathfrak{M}_2^{-1}$. The example given above shows that quasi-equality does not imply equality. It follows from the definition that quasi-equality is a true equivalence relation.

From the definition of $\mathfrak{M}_1^{-1}$, $\mathfrak{M}_1\mathfrak{M}_1^{-1} \subseteq \mathfrak{R}$; hence

$$\mathfrak{M}_1 \subseteq (\mathfrak{M}_1^{-1})^{-1} = \mathfrak{M}_1^*, \text{ say.}$$

Therefore, by I above, $\quad (\mathfrak{M}_1^*)^{-1} \subseteq \mathfrak{M}_1^{-1}$.

On the other hand, we have

$$\mathfrak{M}_1^{-1}\mathfrak{M}_1^* = \mathfrak{M}_1^{-1}(\mathfrak{M}_1^{-1})^{-1} \subseteq \mathfrak{R},$$

and so $$\mathfrak{M}_1^{-1} \subseteq (\mathfrak{M}_1^*)^{-1}.$$

Therefore $$(\mathfrak{M}_1^*)^{-1} = \mathfrak{M}_1^{-1},$$

that is, $$\mathfrak{M}_1^* \sim \mathfrak{M}_1.$$

Similarly, if $\mathfrak{M}_2 \sim \mathfrak{M}_1$,

$$\mathfrak{M}_2 \subseteq (\mathfrak{M}_2^{-1})^{-1} = (\mathfrak{M}_1^{-1})^{-1} = \mathfrak{M}_1^*,$$

so that $$\mathfrak{M}_2 \subseteq \mathfrak{M}_1^*.$$

Hence $\mathfrak{M}_1^*$ is the largest fractional ideal quasi-equal to $\mathfrak{M}_1$, and contains every fractional ideal quasi-equal to $\mathfrak{M}_1$. We shall call it the *reduced* fractional ideal obtained from $\mathfrak{M}_1$. Any fractional ideal written with an asterisk will denote a reduced fractional ideal (but note that $\mathfrak{M}^*\mathfrak{N}^*$ may not be reduced).

We note the following immediate consequences of the definition of reduced ideals.

(i) If $\mathfrak{M}_1 \sim \mathfrak{M}_2$, $$\mathfrak{M}_1 \sim \mathfrak{M}_2 \sim \mathfrak{M}_2^*,$$

and hence $$\mathfrak{M}_2^* \subseteq \mathfrak{M}_1^*.$$

Similarly, $$\mathfrak{M}_1^* \subseteq \mathfrak{M}_2^*,$$

so that $$\mathfrak{M}_1^* = \mathfrak{M}_2^*.$$

Conversely, if $$\mathfrak{M}_1^* = \mathfrak{M}_2^*,$$

we have $$\mathfrak{M}_1 \sim \mathfrak{M}_1^* = \mathfrak{M}_2^* \sim \mathfrak{M}_2,$$

and so $$\mathfrak{M}_1 \sim \mathfrak{M}_2.$$

(ii) In (i) take $\mathfrak{M}_2 = \mathfrak{M}_1^*$. We then have

$$(\mathfrak{M}_1^*)^* = \mathfrak{M}_1^*.$$

We also note that if $\mathfrak{M} = \mathfrak{R}.(\zeta)$, where ζ is any non-zero element of K, $\mathfrak{M}^{-1} = \mathfrak{R}.(\zeta^{-1})$, and hence $\mathfrak{M}^* = (\mathfrak{M}^{-1})^{-1} = \mathfrak{R}.(\zeta) = \mathfrak{M}$. Also, if $\mathfrak{M}_2^{-1} \subseteq \mathfrak{M}_1^{-1}$, then $(\mathfrak{M}_1^{-1})^{-1} \subseteq (\mathfrak{M}_2^{-1})^{-1}$ [see I above]. Hence, although $\mathfrak{M}_2^{-1} \subseteq \mathfrak{M}_1^{-1}$ does not imply $\mathfrak{M}_1 \subseteq \mathfrak{M}_2$, it does imply $\mathfrak{M}_1^* \subseteq \mathfrak{M}_2^*$. But, by I, $\mathfrak{M}_1 \subseteq \mathfrak{M}_2$ implies $\mathfrak{M}_2^{-1} \subseteq \mathfrak{M}_1^{-1}$. Hence we see that if

$$\mathfrak{M}_1 \subseteq \mathfrak{M}_2,$$

then

$$\mathfrak{M}_1^* \subseteq \mathfrak{M}_2^*.$$

A particular application of this result concerns the case in which $\mathfrak{M} = \mathfrak{i}$ is an integral ideal. Then

$$\mathfrak{i} \subseteq \mathbf{1},$$

so

$$\mathfrak{i}^* \subseteq \mathbf{1}^* = \mathbf{1};$$

that is, if $\mathfrak{i}$ is integral, so is $\mathfrak{i}^*$.

Also, we note that $(\mathfrak{M}^*)^{-1} = ((\mathfrak{M}^{-1})^{-1})^{-1} = (\mathfrak{M}^{-1})^*$.

III. We shall say that $\mathfrak{M}$ is a *quasi-multiple* of $\mathfrak{N}$ (or that $\mathfrak{N}$ is a *quasi-factor* of $\mathfrak{M}$), and write $\mathfrak{M} \geqslant \mathfrak{N}$ (or $\mathfrak{N} \leqslant \mathfrak{M}$) if $\mathfrak{M}\mathfrak{N}^{-1}$ is contained in $\mathfrak{R}$. Since $\mathfrak{M}^{-1}$ contains all elements of K whose product with every element of $\mathfrak{M}$ is in $\mathfrak{R}$, this condition is equivalent to $\mathfrak{N}^{-1} \subseteq \mathfrak{M}^{-1}$. We note that the simultaneous relations

$$\mathfrak{M} \geqslant \mathfrak{N} \quad \text{and} \quad \mathfrak{N} \geqslant \mathfrak{M}$$

imply

$$\mathfrak{M} \sim \mathfrak{N}.$$

For the two relations imply, respectively,

$$\mathfrak{N}^{-1} \subseteq \mathfrak{M}^{-1} \quad \text{and} \quad \mathfrak{M}^{-1} \subseteq \mathfrak{N}^{-1},$$

so that

$$\mathfrak{N}^{-1} = \mathfrak{M}^{-1}.$$

From I we see that if $\mathfrak{M} \subseteq \mathfrak{N}$, $\mathfrak{M}^{-1} \supseteq \mathfrak{N}^{-1}$, and therefore $\mathfrak{M} \geqslant \mathfrak{N}$; and from II we see that if $\mathfrak{M} \geqslant \mathfrak{N}$, $\mathfrak{M}^* \subseteq \mathfrak{N}^*$. Therefore $\mathfrak{M} \geqslant \mathbf{1}$ implies $\mathfrak{M} \subseteq \mathfrak{M}^* \subseteq \mathbf{1}^* = \mathfrak{R}$; hence $\mathfrak{M} \geqslant \mathbf{1}$ implies that $\mathfrak{M}$ is an integral ideal. Conversely, if $\mathfrak{M}$ is integral, $\mathfrak{M} \subseteq \mathbf{1}$, and therefore $\mathfrak{M} \geqslant \mathbf{1}$.

Now let $\mathfrak{M}$ and $\mathfrak{N}$ be two non-zero fractional ideals such that $\mathfrak{M} \leqslant \mathfrak{N}$, and let $\mathfrak{L}$ be any other non-zero fractional ideal. Then $\mathfrak{L}\mathfrak{M}$ is not the zero ideal, since K has no divisors of zero, and hence we can define $(\mathfrak{L}\mathfrak{M})^{-1}$. $(\mathfrak{L}\mathfrak{M})^{-1}(\mathfrak{L}\mathfrak{M})$ is contained in $\mathfrak{R}$, by the definition of $(\mathfrak{L}\mathfrak{M})^{-1}$. Hence

$$(\mathfrak{L}\mathfrak{M})^{-1}\mathfrak{L} \subseteq \mathfrak{M}^{-1} \subseteq \mathfrak{N}^{-1},$$

and therefore $$(\mathfrak{L}\mathfrak{M})^{-1}\mathfrak{L}\mathfrak{N} \subseteq \mathfrak{R}.$$

Hence we have $$\mathfrak{L}\mathfrak{M} \leqslant \mathfrak{L}\mathfrak{N}.$$

If $\mathfrak{M} \sim \mathfrak{N}$, we have

$$\mathfrak{M} \geqslant \mathfrak{N} \quad \text{and} \quad \mathfrak{N} \geqslant \mathfrak{M};$$

hence, by the result just proved,

$$\mathfrak{L}\mathfrak{M} \geqslant \mathfrak{L}\mathfrak{N} \quad \text{and} \quad \mathfrak{L}\mathfrak{N} \geqslant \mathfrak{L}\mathfrak{M}.$$

Therefore $$\mathfrak{L}\mathfrak{M} \sim \mathfrak{L}\mathfrak{N}.$$

From this we can deduce the more general result that if

$$\mathfrak{M} \sim \mathfrak{M}' \quad \text{and} \quad \mathfrak{N} \sim \mathfrak{N}',$$

then $$\mathfrak{M}\mathfrak{N} \sim \mathfrak{M}\mathfrak{N}' \sim \mathfrak{M}'\mathfrak{N}'.$$

Let us consider the particular case of this in which $\mathfrak{N} = \mathfrak{M}^{-1}$, $\mathfrak{M}' = \mathfrak{M}^*$, and $\mathfrak{N}' = \mathfrak{N}^* = (\mathfrak{M}^*)^{-1}$. Then we see that

$$\mathfrak{M}\mathfrak{M}^{-1} \sim \mathfrak{M}^*(\mathfrak{M}^*)^{-1}.$$

To proceed further with the properties of ideals under the relation of quasi-equality we require to know whether

$$\mathfrak{M}\mathfrak{M}^{-1} \sim \mathbf{1}.$$

This, however, is not true in general, as the following example will show. Let K_1 be a field, x an indeterminate over K_1, and y a solution of the equation

$$y^2 = x^3$$

in some extension of $K_1[x]$. Take $\mathfrak{R} = K_1[x, y]$ and $\mathfrak{M} = \mathfrak{R}.(x, y)$. Any element of the quotient field K of $\mathfrak{R}$ is of the form $\alpha + \beta y$, where α and β are rational functions of x over K_1. If $\xi = \alpha + \beta y$ is in $\mathfrak{M}^{-1}$,

$$\alpha x + \beta x y \quad \text{and} \quad \beta x^3 + \alpha y$$

are in $\mathfrak{R}$. We conclude that α and βx must lie in $\mathfrak{R}$. Conversely, if α and βx lie in $\mathfrak{R}$, $\xi = \alpha + \beta y$ is in $\mathfrak{M}^{-1}$. Hence

$$\mathfrak{M}^{-1} = \mathfrak{R}.(1, y/x).$$

Then

$$\mathfrak{M}\mathfrak{M}^{-1} = \mathfrak{R}.(x, y, y, y^2/x) = \mathfrak{R}.(x, y, x^2) = \mathfrak{R}.(x, y) = \mathfrak{M}.$$

Now $\mathfrak{M}^{-1} \nsubseteq \mathfrak{R}$, hence

$$\mathfrak{M}^{-1} \neq \mathbf{1} = \mathbf{1}^{-1},$$

that is, $\mathfrak{M}$ is not quasi-equal to $\mathbf{1}$. Therefore $\mathfrak{M}\mathfrak{M}^{-1}$ is not quasi-equal to $\mathbf{1}$.

However, for some important classes of integral domains it can be shown that $\mathfrak{M}\mathfrak{M}^{-1} \sim \mathbf{1}$ for all non-zero fractional ideals. An example will be given in § 8. It is therefore worth while considering further the properties of fractional ideals in rings $\mathfrak{R}$ for which $\mathfrak{M}\mathfrak{M}^{-1}$ is quasi-equal to the unit ideal for all non-zero fractional ideals $\mathfrak{M}$ of $\mathfrak{R}$. *For the remainder of this section we assume that*

$$\mathfrak{M}\mathfrak{M}^{-1} \sim \mathbf{1}$$

for all non-zero fractional ideals of $\mathfrak{R}$.

This assumption, together with the properties of fractional ideals already proved, enables us to develop a complete factorisation theory of *integral* ideals of $\mathfrak{R}$, and indeed it is for this reason that we introduce the idea of a fractional ideal.

An immediate consequence of our new assumption is that if $\mathfrak{M}$ is any fractional ideal of $\mathfrak{R}$, $\mathfrak{M} \sim \mathfrak{i}\mathfrak{j}^{-1}$, where $\mathfrak{i}$ and $\mathfrak{j}$ are integral ideals. Let

$$\mathfrak{M} = \mathfrak{R}.(\zeta_1, \ldots, \zeta_r),$$

where

$$\zeta_i = \alpha_i/\beta_i,$$

and α_i and β_i are in $\mathfrak{R}$. Let $\beta = \beta_1 \ldots \beta_r$, and let

$$\mathfrak{j} = \mathfrak{R}.(\beta).$$

Then

$$\mathfrak{j}\mathfrak{M} = \mathfrak{R}.(\beta\zeta_1, \ldots, \beta\zeta_r) \subseteq \mathfrak{R}.$$

Hence $\mathfrak{j}\mathfrak{M}$ is an integral ideal, $\mathfrak{i}$, say. Then

$$\mathfrak{i}\mathfrak{j}^{-1} = \mathfrak{j}^{-1}\mathfrak{j}\mathfrak{M} \sim \mathbf{1}\mathfrak{M} \sim \mathfrak{M}.$$

THEOREM I. *The classes of quasi-equal non-zero fractional ideals of* $\mathfrak{R}$ *form a commutative group.*

If

$$\mathfrak{M} \sim \mathfrak{M}' \quad \text{and} \quad \mathfrak{N} \sim \mathfrak{N}',$$

then, by III above,

$$\mathfrak{M}\mathfrak{N} \sim \mathfrak{M}'\mathfrak{N}'.$$

Hence the multiplication of classes can be defined uniquely. Since the associative and commutative laws of multiplication hold for fractional ideals, they must hold for the multiplication of classes. Also, if $\mathfrak{M}$ and $\mathfrak{N}$ are non-zero fractional ideals, the equation

$$\mathfrak{X}\mathfrak{M} \sim \mathfrak{N}$$

has the solution

$$\mathfrak{X} = \mathfrak{M}^{-1}\mathfrak{N},$$

since

$$\mathfrak{M}^{-1}\mathfrak{N}\mathfrak{M} = \mathfrak{M}^{-1}\mathfrak{M}\mathfrak{N} \sim \mathbf{1}\mathfrak{N} = \mathfrak{N}.$$

The theorem then follows. The class of quasi-equal ideals containing $\mathbf{1}$ is the unity of the group.

Since we shall be mainly concerned with our earlier concept of an (integral) ideal in $\mathfrak{R}$, we revert to our practice of denoting integral ideals by small letters in German type, such as $\mathfrak{i}, \mathfrak{j}, \ldots$.

Now let $\mathfrak{i}$ and $\mathfrak{j}$ be non-zero *integral* ideals such that

$$\mathfrak{i}\mathfrak{j} \sim \mathbf{1}.$$

Then
$$\mathfrak{i}^{-1} = \mathfrak{i}^{-1}\mathbf{1} \sim \mathfrak{i}^{-1}\mathfrak{i}\mathfrak{j} \sim \mathfrak{j} \geqslant \mathbf{1}.$$

Hence $\mathfrak{i}^{-1}$ is an integral ideal. It follows that $\mathfrak{i}^{-1} \subseteq \mathbf{1} = \mathbf{1}^{-1}$, and therefore $\mathfrak{i} \leqslant \mathbf{1}$. On the other hand, $\mathfrak{i}$ is integral, and therefore $\mathfrak{i} \geqslant \mathbf{1}$. It follows that $\mathfrak{i} \sim \mathbf{1}$. Similarly $\mathfrak{j} \sim \mathbf{1}$. Hence we have

THEOREM II. *If* $\mathfrak{i}$ *and* $\mathfrak{j}$ *are (integral) ideals in* $\mathfrak{R}$ *with* $\mathfrak{i}\mathfrak{j} \sim \mathbf{1}$, *then* $\mathfrak{i} \sim \mathbf{1}, \mathfrak{j} \sim \mathbf{1}$.

Next, suppose that $\mathfrak{i} \geqslant \mathfrak{j}$. Then $\mathfrak{i}\mathfrak{j}^{-1} = \mathfrak{k}$ is integral, and

$$\mathfrak{j}\mathfrak{k} = \mathfrak{i}(\mathfrak{j}^{-1}\mathfrak{j}) \sim \mathfrak{i}\mathbf{1} \sim \mathfrak{i},$$

and since $\mathfrak{j}^{-1}\mathfrak{j}$ is contained in $\mathfrak{R}$, $\mathfrak{j}\mathfrak{k} = \mathfrak{i}(\mathfrak{j}^{-1}\mathfrak{j}) \subseteq \mathfrak{i}$. Hence we have

THEOREM III. *If* $\mathfrak{i}$ *and* $\mathfrak{j}$ *are ideals in* $\mathfrak{R}$ *and* $\mathfrak{i} \geqslant \mathfrak{j}$, *there exists an ideal* $\mathfrak{k}$ *in* $\mathfrak{R}$ *such that*

$$\mathfrak{i} \sim \mathfrak{j}\mathfrak{k}, \quad \mathfrak{j}\mathfrak{k} = 0 \ (\mathfrak{i}).$$

An ideal $\mathfrak{i}$ in $\mathfrak{R}$ is said to be *quasi-irreducible* if the quasi-equality $\mathfrak{i} \sim \mathfrak{j}\mathfrak{k}$ implies either $\mathfrak{j} \sim \mathbf{1}$ or $\mathfrak{k} \sim \mathbf{1}$; otherwise $\mathfrak{i}$ is *quasi-reducible*. Since $\mathfrak{i} \sim \mathfrak{i}^*$, $\mathfrak{i}$ is quasi-irreducible if and only if $\mathfrak{i}^*$ is quasi-irreducible.

THEOREM IV. *If* $\mathfrak{i}$ *is quasi-irreducible, then* $\mathfrak{i}^*$ *is prime, and, conversely, if* $\mathfrak{i}^*$ *is prime then* $\mathfrak{i}$ *is quasi-irreducible.*

(i) Suppose that $\mathfrak{i}$ is quasi-irreducible. If $\mathfrak{i}^* = \mathbf{1}$, it is prime. Suppose that $\mathfrak{i}^* \neq \mathbf{1}$, and that $\mathfrak{a}$ and $\mathfrak{b}$ are two ideals in $\mathfrak{R}$ such that

$$\mathfrak{a}\mathfrak{b} = 0 \ (\mathfrak{i}^*), \quad \mathfrak{b} \neq 0 \ (\mathfrak{i}^*).$$

Let
$$\mathfrak{c} = (\mathfrak{b}, \mathfrak{i}^*).$$

Then $\mathfrak{c}$ is a proper factor of $\mathfrak{i}^*$. Hence, $\mathfrak{i}^* \geqslant \mathfrak{c}$, and therefore [Th. III] there exists an ideal $\mathfrak{d}$ in $\mathfrak{R}$ such that

$$\mathfrak{i}^* \sim \mathfrak{c}\mathfrak{d}.$$

Since $\mathfrak{i}^*$ is quasi-irreducible, either $\mathfrak{c} \sim \mathbf{1}$ or $\mathfrak{d} \sim \mathbf{1}$. If $\mathfrak{d} \sim \mathbf{1}$, $\mathfrak{i}^* \sim \mathfrak{c}$, and therefore $\mathfrak{c}^* = \mathfrak{i}^*$. But $\mathfrak{c} \subseteq \mathfrak{c}^* = \mathfrak{i}^*$, and this contradicts the fact that $\mathfrak{c}$ is a proper factor of $\mathfrak{i}^*$. Hence we must have $\mathfrak{c} \sim \mathbf{1}$. But we also have

$$\mathfrak{a} = \mathfrak{R}.\mathfrak{a} \sim \mathfrak{c}\mathfrak{a} = (\mathfrak{b}\mathfrak{a}, \mathfrak{i}^*\mathfrak{a}) \subseteq \mathfrak{i}^*.$$

Hence $\mathfrak{a} = 0 \ (\mathfrak{i}^*)$, since $(\mathfrak{c}\mathfrak{a})^* \subseteq (\mathfrak{i}^*)^* = \mathfrak{i}^*$ and $\mathfrak{a} \subseteq (\mathfrak{c}\mathfrak{a})^*$. It follows that $\mathfrak{i}^*$ is prime.

(ii) Suppose that $\mathfrak{i}^*$ is prime. If $\mathfrak{i}^* = \mathbf{1}$, it is quasi-irreducible, by Theorem II.

Suppose that $\mathfrak{i}^* \neq \mathbf{1}$, and that it is quasi-reducible, that is, there exist reduced ideals $\mathfrak{j}^*$, $\mathfrak{k}^*$ in $\mathfrak{R}$ not quasi-equal to $\mathbf{1}$ such that

$$\mathfrak{i}^* \sim \mathfrak{j}^*\mathfrak{k}^*.$$

Since $\mathfrak{i}^*$ is the reduced ideal obtained from $\mathfrak{j}^*\mathfrak{k}^*$,

$$\mathfrak{j}^*\mathfrak{k}^* = 0 \ (\mathfrak{i}^*).$$

Also $\mathfrak{i}^*(\mathfrak{j}^*)^{-1} \sim \mathfrak{k}^* \subseteq \mathfrak{R}$; that is, $\mathfrak{i}^* \geqslant \mathfrak{j}^*$, and hence $\mathfrak{i}^* \subseteq \mathfrak{j}^*$; similarly $\mathfrak{i}^* \subseteq \mathfrak{k}^*$. If $\mathfrak{i}^* = \mathfrak{j}^*$, $\quad \mathfrak{k}^* \sim \mathfrak{i}^*(\mathfrak{j}^*)^{-1} = \mathfrak{i}^*(\mathfrak{i}^*)^{-1} \sim \mathbf{1}$,

contrary to hypothesis. Hence

$$\mathfrak{i}^* \subset \mathfrak{j}^*, \quad \text{and similarly} \quad \mathfrak{i}^* \subset \mathfrak{k}^*.$$

Therefore there exists an element α of $\mathfrak{j}^*$ which is not in $\mathfrak{i}^*$, and an element β in $\mathfrak{k}^*$ which is not in $\mathfrak{i}^*$. Then $\alpha\beta \in \mathfrak{j}^*\mathfrak{k}^* \subseteq \mathfrak{i}^*$. Hence there exist two elements α and β of $\mathfrak{R}$ such that

$$\alpha\beta = 0 \ (\mathfrak{i}^*), \quad \alpha \neq 0 \ (\mathfrak{i}^*), \quad \beta \neq 0 \ (\mathfrak{i}^*).$$

Therefore $\mathfrak{i}^*$ is not prime, and we have a contradiction. Hence it follows that if $\mathfrak{i}^*$ is prime, $\mathfrak{i}$ is quasi-irreducible.

Corollary. If $\mathfrak{i}$ *is quasi-irreducible, and* $\mathfrak{i}\mathfrak{j} \sim \mathfrak{a}\mathfrak{b}$, *then* $\mathfrak{i}$ *is a quasi-factor of* $\mathfrak{a}$ *or of* $\mathfrak{b}$.

For $\qquad \mathfrak{a}\mathfrak{b}\mathfrak{i}^{-1} \sim \mathfrak{j} \subseteq \mathbf{1}$,

so that $\qquad \mathfrak{a}\mathfrak{b} \geqslant \mathfrak{i}$.

Therefore $\qquad \mathfrak{a}\mathfrak{b} \subseteq (\mathfrak{a}\mathfrak{b})^* \subseteq \mathfrak{i}^*$;

that is, $\qquad \mathfrak{a}\mathfrak{b} = 0 \ (\mathfrak{i}^*)$.

Since $\mathfrak{i}^*$ is quasi-irreducible and therefore prime, we then have

$$\mathfrak{a} = 0 \ (\mathfrak{i}^*) \quad \text{or} \quad \mathfrak{b} = 0 \ (\mathfrak{i}^*).$$

Hence $\qquad \mathfrak{i}^{-1} = (\mathfrak{i}^*)^{-1} \subseteq \mathfrak{a}^{-1} \quad \text{or} \quad \mathfrak{i}^{-1} = (\mathfrak{i}^*)^{-1} \subseteq \mathfrak{b}^{-1}$,

and therefore $\qquad \mathfrak{a} \geqslant \mathfrak{i} \quad \text{or} \quad \mathfrak{b} \geqslant \mathfrak{i}$.

Now let $\mathfrak{i}$ be any ideal in $\mathfrak{R}$. We wish to show that $\mathfrak{i}$ is quasi-equal to the product of a finite number of quasi-irreducible ideals in $\mathfrak{R}$. Suppose that this is not the case. Then $\mathfrak{i}$ cannot itself be quasi-irreducible, and hence

$$\mathfrak{i} \sim \mathfrak{i}_1 \mathfrak{j}_1,$$

where $\mathfrak{i}_1$ and $\mathfrak{j}_1$ are not quasi-equal to $\mathbf{1}$. Then

$$\mathfrak{i}\mathfrak{i}_1^{-1} \sim \mathfrak{j}_1 \subseteq \mathfrak{R}.$$

Hence
$$\mathfrak{i}^* \geqslant \mathfrak{i}_1^*.$$

If $\mathfrak{i}^* = \mathfrak{i}_1^*$,
$$\mathfrak{j}_1 \sim \mathfrak{i}\mathfrak{i}_1^{-1} \sim \mathfrak{i}^*(\mathfrak{i}_1^*)^{-1} \sim \mathbf{1},$$

contrary to hypothesis. Hence

$$\mathfrak{i}^* \subset \mathfrak{i}_1^*, \quad \text{and similarly,} \quad \mathfrak{i}^* \subset \mathfrak{j}_1^*.$$

If $\mathfrak{i}_1$ and $\mathfrak{j}_1$ were each quasi-equal to the product of a finite number of quasi-irreducible ideals, the same would be true for $\mathfrak{i} \sim \mathfrak{i}_1\mathfrak{j}_1$. Hence, since we are assuming that such a representation is not possible for $\mathfrak{i}$, it is not possible for one of $\mathfrak{i}_1$ and $\mathfrak{j}_1$, say $\mathfrak{i}_1$. Then, as above,

$$\mathfrak{i}_1 \sim \mathfrak{i}_2 \mathfrak{j}_2,$$

where
$$\mathfrak{i}_1^* \subset \mathfrak{i}_2^*, \quad \mathfrak{i}_1^* \subset \mathfrak{j}_2^*,$$

and $\mathfrak{i}_2$ and $\mathfrak{j}_2$ are not both quasi-equal to finite products of quasi-irreducible ideals. We suppose that $\mathfrak{i}_2$ is not so expressible, and proceed as before. It is then clear that unless $\mathfrak{i}$ is quasi-equal to the product of a finite number of quasi-irreducible ideals, we can obtain an infinite sequence of ideals $\mathfrak{i}^*, \mathfrak{i}_1^*, \mathfrak{i}_2^*, \ldots$, such that

$$\mathfrak{i}^* \subset \mathfrak{i}_1^* \subset \mathfrak{i}_2^* \subset \ldots.$$

But by § 1, Theorem VI, this is impossible, since the Basis Theorem holds in $\mathfrak{R}$. Hence, using Theorem IV, we can write

$$\mathfrak{i} \sim \mathfrak{p}_1^* \mathfrak{p}_2^* \ldots \mathfrak{p}_r^*,$$

where $\mathfrak{p}_1^*, \ldots, \mathfrak{p}_r^*$ are prime, and hence quasi-irreducible, and not quasi-equal to $\mathbf{1}$.

Suppose that we have another representation of $\mathfrak{i}$ in which $\mathfrak{i}$ is quasi-equal to a product of quasi-irreducible ideals. Then we can write

$$\mathfrak{i} \sim \mathfrak{p}_1^* \mathfrak{p}_2^* \ldots \mathfrak{p}_r^* \sim \mathfrak{p}_1'^* \mathfrak{p}_2'^* \ldots \mathfrak{p}_s'^*.$$

By the corollary to Theorem IV, $\mathfrak{p}_1^*$ is a factor of $\mathfrak{p}_1'^*$ or of $\mathfrak{p}_2'^* \ldots \mathfrak{p}_s'^*$. If it is a factor of the latter, it is a factor of $\mathfrak{p}_2'^*$, or of $\mathfrak{p}_3'^* \ldots \mathfrak{p}_s'^*$.

Proceeding in this way, we see that $\mathfrak{p}_1^*$ is a factor of some $\mathfrak{p}_i'^*$, and by arranging the factors suitably, we may suppose that it is a factor of $\mathfrak{p}_1'^*$. Then, by Theorem III, we have

$$\mathfrak{p}_1'^* \sim \mathfrak{p}_1^* \mathfrak{j}.$$

But, by hypothesis, $\mathfrak{p}_1'^*$ is quasi-irreducible, but $\mathfrak{p}_1^*$ is not quasi-equal to $\mathbf{1}$. Hence $\mathfrak{j} \sim \mathbf{1}$, that is,

$$\mathfrak{p}_1'^* \sim \mathfrak{p}_1^*.$$

Since they are both reduced ideals, we therefore have

$$\mathfrak{p}_1^* = \mathfrak{p}_1'^*.$$

We then have

$$\mathfrak{p}_2^* \dots \mathfrak{p}_r^* \sim \mathfrak{p}_1^{*-1} \mathfrak{p}_1'^* \mathfrak{p}_2'^* \dots \mathfrak{p}_s'^* \sim \mathfrak{p}_2'^* \dots \mathfrak{p}_s'^*.$$

Hence, following the usual induction argument, as used for instance, in proving Theorem VIII of § 2, we deduce that $r = s$, and that if the factors are suitably arranged,

$$\mathfrak{p}_i^* = \mathfrak{p}_i'^* \quad (i = 1, \dots, r).$$

We therefore have *the Unique Factorisation Theorem:*

THEOREM V. *Any ideal* $\mathfrak{i}$ *in* $\mathfrak{R}$ *not quasi-equal to* $\mathbf{1}$ *satisfies a quasi-equality*

$$\mathfrak{i} \sim \mathfrak{p}_1^* \dots \mathfrak{p}_r^*,$$

where $\mathfrak{p}_1^*, \dots, \mathfrak{p}_r^*$ *are quasi-irreducible and not quasi-equal to* $\mathbf{1}$, *and the quasi-factors* $\mathfrak{p}_i^*$ *are uniquely determined, save for order.*

Our next problem is to determine the prime ideals of $\mathfrak{R}$ which are quasi-equal to $\mathbf{1}$. Let $\mathfrak{p}$ be any prime ideal which is quasi-equal to $\mathbf{1}$, and let $\alpha \in \mathfrak{p}$. If $\mathfrak{i} = \mathfrak{R}.(\alpha)$, $\mathfrak{i}^{-1} = \mathfrak{R}.(\alpha^{-1})$, and hence $\mathfrak{i}$ is not quasi-equal to $\mathbf{1}$. Also

$$\mathfrak{i} \sim \mathfrak{p}_1^* \dots \mathfrak{p}_r^*,$$

where $\mathfrak{p}_1^*, \dots, \mathfrak{p}_r^*$ are quasi-irreducible and not quasi-equal to $\mathbf{1}$ [Th. V]. Since $\mathfrak{i}^* = \mathfrak{i}$ ($\mathfrak{i}$ being a principal ideal),

$$\mathfrak{p}_1^* \dots \mathfrak{p}_r^* \subseteq \mathfrak{i} \subseteq \mathfrak{p},$$

and hence, for some value of $\mathfrak{j}$,

$$\mathfrak{p}_j^* = 0 \ (\mathfrak{p}).$$

Since $$\mathfrak{p}_j^* \neq \mathbf{1} \quad \text{and} \quad \mathfrak{p}^* = \mathbf{1},$$

$\mathfrak{p}_j^*$ is a prime ideal which is a proper multiple of $\mathfrak{p}$. Thus any prime ideal of $\mathfrak{R}$ which is quasi-equal to **1** has a proper multiple which is prime.

We next show that if $\mathfrak{p}^*$ is a quasi-irreducible (and therefore a prime) ideal not quasi-equal to **1**, it cannot have a proper multiple which is prime. Let $\mathfrak{i}$ be any prime ideal which is a multiple of $\mathfrak{p}^*$:

$$\mathfrak{i} = 0 \ \ (\mathfrak{p}^*).$$

Then $\mathfrak{i} \geqslant \mathfrak{p}^*$, and hence [Th. III] there exists an ideal $\mathfrak{q}$ such that

$$\mathfrak{i} \sim \mathfrak{p}^*\mathfrak{q}, \quad \mathfrak{p}^*\mathfrak{q} = 0 \ \ (\mathfrak{i}).$$

If $\mathfrak{q} \subseteq \mathfrak{i}$, we can, similarly, find an ideal $\mathfrak{r}$ such that

$$\mathfrak{q} \sim \mathfrak{i}\mathfrak{r}.$$

Hence $$\mathfrak{i} \sim \mathfrak{i}\mathfrak{p}^*\mathfrak{r},$$

and therefore $$\mathfrak{p}^*\mathfrak{r} \sim \mathbf{1}.$$

By Theorem II this implies $\mathfrak{p}^* \sim \mathbf{1}$, which is contrary to hypothesis.

We therefore have

$$\mathfrak{p}^*\mathfrak{q} = 0 \ \ (\mathfrak{i}), \quad \mathfrak{q} \neq 0 \ \ (\mathfrak{i});$$

hence, since $\mathfrak{i}$ is prime, $$\mathfrak{p}^* = 0 \ \ (\mathfrak{i}).$$

Therefore $$\mathfrak{i} = \mathfrak{p}^*.$$

Finally, we consider any prime ideal $\mathfrak{p}$ in $\mathfrak{R}$ which is not quasi-equal to **1**. Then

$$\mathfrak{p} \sim \mathfrak{p}_1^* \ldots \mathfrak{p}_r^*,$$

where the $\mathfrak{p}_i^*$ are quasi-irreducible, and not quasi-equal to **1**. Then

$$\mathfrak{p}(\mathfrak{p}_i^*)^{-1} \sim \mathfrak{p}_1^* \ldots \mathfrak{p}_{i-1}^*\mathfrak{p}_{i+1}^* \ldots \mathfrak{p}_r^* \subseteq \mathbf{1},$$

so that $\mathfrak{p} \geqslant \mathfrak{p}_i^*$, and therefore $\mathfrak{p} \subseteq \mathfrak{p}^* \subseteq \mathfrak{p}_i^*$. $\mathfrak{p}$ is thus a prime ideal which is a multiple of $\mathfrak{p}_i^*$. Hence, by the result proved above, $\mathfrak{p} = \mathfrak{p}_i^*$, and it further follows that $r = 1$.

Summing up these results, we have

THEOREM VI. *A prime ideal $\mathfrak{p}$ in $\mathfrak{R}$ is either quasi-equal to* **1**, *or is quasi-irreducible and equal to the associated reduced ideal. $\mathfrak{p}$ is quasi-equal to* **1** *if and only if there exists a proper multiple of $\mathfrak{p}$ which is prime.*

We next consider the primary ideals in $\mathfrak{R}$. Let $\mathfrak{q}$ be a $\mathfrak{p}$-primary ideal in $\mathfrak{R}$, ρ its index. Suppose, first, that $\mathfrak{p} \sim \mathbf{1}$. Then we have

$$\mathfrak{p}^\rho = 0 \ \ (\mathfrak{q}),$$

and hence [Th. III] there exists an ideal $\mathfrak{i}$ in $\mathfrak{R}$ such that

$$\mathfrak{p}^\rho \sim \mathfrak{q}\mathfrak{i}.$$

Since $\mathfrak{p} \sim \mathbf{1}$, we have $\qquad \mathfrak{q}\mathfrak{i} \sim \mathbf{1},$

and hence [Th. II] $\qquad \mathfrak{q} \sim \mathbf{1}.$

Again, since $\mathfrak{q} = 0 \ (\mathfrak{p})$, there exists an ideal $\mathfrak{j}$ in $\mathfrak{R}$ such that

$$\mathfrak{q} \sim \mathfrak{p}\mathfrak{j}.$$

Hence, if $\mathfrak{q} \sim \mathbf{1}$, we must have $\mathfrak{p} \sim \mathbf{1}$.

We next consider the case in which $\mathfrak{p}$ is not quasi-equal to $\mathbf{1}$. Then $\mathfrak{p} = \mathfrak{p}^*$. We first prove the

Lemma. If $\mathfrak{a}$ *and* $\mathfrak{b}$ *are quasi-equal ideals in* $\mathfrak{R}$, *there exist ideals* $\mathfrak{c}$ *and* $\mathfrak{d}$ *in* $\mathfrak{R}$, *each quasi-equal to* $\mathbf{1}$, *such that* $\mathfrak{a}\mathfrak{c} = \mathfrak{b}\mathfrak{d}$.

Since $\mathfrak{a} \sim \mathfrak{b}$, we have $\mathfrak{a}^{-1} = \mathfrak{b}^{-1}$. Hence

$$\mathfrak{a}(\mathfrak{b}^{-1}\mathfrak{b}) = (\mathfrak{a}\mathfrak{a}^{-1})\,\mathfrak{b}.$$

But $\mathfrak{c} = \mathfrak{b}^{-1}\mathfrak{b}$ and $\mathfrak{d} = \mathfrak{a}\mathfrak{a}^{-1}$ are ideals in $\mathfrak{R}$ each quasi-equal to $\mathbf{1}$. They therefore satisfy the requirements of the lemma.

We apply the lemma to the quasi-equal ideals $\mathfrak{q}$ and $\mathfrak{q}^*$. Then there exist ideals $\mathfrak{c}$ and $\mathfrak{d}$ such that

$$\mathfrak{q}\mathfrak{c} = \mathfrak{q}^*\mathfrak{d}, \quad \mathfrak{c} \sim \mathbf{1}, \quad \mathfrak{d} \sim \mathbf{1}.$$

If $\mathfrak{d} \subseteq \mathfrak{p}$ there exists an ideal $\mathfrak{i}$ in $\mathfrak{R}$ such that

$$\mathfrak{d} \sim \mathfrak{p}\mathfrak{i}.$$

If this is so $\qquad \mathfrak{p}\mathfrak{i} \sim \mathbf{1},$

and hence [Th. II] $\qquad \mathfrak{p} \sim \mathbf{1},$

contrary to hypothesis. It follows that $\mathfrak{d}$ is not contained in $\mathfrak{p}$. But

$$\mathfrak{q}^*\mathfrak{d} = \mathfrak{q}\mathfrak{c} = 0 \ (\mathfrak{q}),$$

and $\mathfrak{q}$ is $\mathfrak{p}$-primary. Since $\mathfrak{d}$ is not contained in $\mathfrak{p}$, we have

$$\mathfrak{q}^* = 0 \ (\mathfrak{q}).$$

Since $\mathfrak{q}^*$ is the reduced ideal obtained from $\mathfrak{q}$, we have

$$\mathfrak{q} = 0 \ (\mathfrak{q}^*).$$

Therefore $\qquad \mathfrak{q} = \mathfrak{q}^*.$

Thus we have

THEOREM VII. *If* $\mathfrak{q}$ *is a* $\mathfrak{p}$*-primary ideal in* $\mathfrak{R}$, $\mathfrak{q} \sim \mathbf{1}$ *if and only if* $\mathfrak{p} \sim \mathbf{1}$, *and if* $\mathfrak{p}$ *is not quasi-equal to* $\mathbf{1}$, $\mathfrak{q} = \mathfrak{q}^*$.

Now consider any ideal $\mathfrak{i}$ in $\mathfrak{R}$ which is not quasi-equal to $\mathbf{1}$. Let $[\mathfrak{q}_1, \dots \mathfrak{q}_r]$ be an uncontractible representation of $\mathfrak{i}$ as an intersection of primaries. We may suppose that $\mathfrak{q}_1$ is an isolated component of $\mathfrak{i}$ and for convenience we write $\mathfrak{q}_1 = \mathfrak{q}$, $[\mathfrak{q}_2, \ldots, \mathfrak{q}_r] = \mathfrak{k}$, and let $\mathfrak{p}$ be the radical of $\mathfrak{q}$. Let $\mathfrak{j} = (\mathfrak{q}, \mathfrak{k})$. We prove that $\mathfrak{j} \sim \mathbf{1}$.

Suppose that $\mathfrak{j}$ is not quasi-equal to $\mathbf{1}$, and let

$$\mathfrak{j} \sim \mathfrak{p}_1 \ldots \mathfrak{p}_s \quad (s \geqslant 1),$$

where $\mathfrak{p}_1, \ldots, \mathfrak{p}_s$ are prime ideals not quasi-equal to $\mathbf{1}$. If ρ is the index of $\mathfrak{q}$,

$$\mathfrak{p}^\rho \subseteq \mathfrak{q} \subseteq \mathfrak{j};$$

hence

$$\mathfrak{p}^\rho \sim \mathfrak{p}_1 \ldots \mathfrak{p}_s \mathfrak{r},$$

where $\mathfrak{r}$ is some ideal in $\mathfrak{R}$ [Th. III]. By the unique factorisation theorem we must have $\mathfrak{p}_i \sim \mathfrak{p}$, and hence

$$\mathfrak{j} \sim \mathfrak{p}^s.$$

It follows that (i) $\mathfrak{p}$ is not quasi-equal to $\mathbf{1}$, and (ii) $\mathfrak{j} \subseteq \mathfrak{p}$. Since $\mathfrak{q} = \mathfrak{q}_1$ is an isolated component of $\mathfrak{i}$, we can find an element α_i in the radical of $\mathfrak{q}_i$ $(i > 1)$ which is not in $\mathfrak{p}$. There then exists an integer σ such that

$$(\alpha_2 \ldots \alpha_r)^\sigma \in \mathfrak{k}$$

but is not in $\mathfrak{p}$. Hence $(\alpha_2 \ldots \alpha_r)^\sigma$ is contained in $\mathfrak{j}$ but not in $\mathfrak{p}$, which contradicts the result (ii) proved above. It follows that our assumption that $\mathfrak{j}$ is not quasi-equal to $\mathbf{1}$ is incorrect. Therefore we have

$$\mathfrak{j} \sim \mathbf{1}.$$

Since

$$\mathfrak{i} = [\mathfrak{q}, \mathfrak{k}] \quad \text{and} \quad \mathfrak{j} = (\mathfrak{q}, \mathfrak{k}),$$

we have [§ 1, Th. IV]

$$\mathfrak{i}\mathfrak{j} = 0 \ \ (\mathfrak{q}\mathfrak{k}).$$

Therefore

$$\mathfrak{i}\mathfrak{j} \geqslant \mathfrak{q}\mathfrak{k},$$

and hence

$$\mathfrak{i}^*\mathfrak{j}^* = \mathfrak{i}^* \geqslant \mathfrak{q}^*\mathfrak{k}^*.$$

On the other hand, we have

$$\mathfrak{q}\mathfrak{k} = 0 \ \ (\mathfrak{i}),$$

and therefore

$$\mathfrak{q}^*\mathfrak{k}^* = 0 \ \ (\mathfrak{i}^*),$$

that is

$$\mathfrak{q}^*\mathfrak{k}^* \geqslant \mathfrak{i}^*.$$

It follows that

$$\mathfrak{i}^* \sim \mathfrak{q}^*\mathfrak{k}^*.$$

Since an ideal is quasi-equal to its reduced ideal, and from the above, we have

$$[\mathfrak{q}^*, \mathfrak{k}^*] \sim [\mathfrak{q}^*, \mathfrak{k}^*]^* \sim \mathfrak{q}^{**}\mathfrak{k}^{**} = \mathfrak{q}^*\mathfrak{k}^* \sim \mathfrak{i}^*.$$

Hence $$\mathfrak{i}^* \sim [\mathfrak{q}^*, \mathfrak{k}^*].$$

Since $\mathfrak{i}^*$ contains every ideal quasi-equal to it,

$$[\mathfrak{q}^*, \mathfrak{k}^*] = 0 \ \ (\mathfrak{i}^*);$$

on the other hand, we have

$$\mathfrak{i} \subseteq \mathfrak{q} \quad \text{and} \quad \mathfrak{i} \subseteq \mathfrak{k},$$

so that $$\mathfrak{i}^* \subseteq \mathfrak{q}^* \quad \text{and} \quad \mathfrak{i}^* \subseteq \mathfrak{k}^*.$$

Therefore $$\mathfrak{i}^* \subseteq [\mathfrak{q}^*, \mathfrak{k}^*].$$

Taking the two results together, we obtain

$$\mathfrak{i}^* = [\mathfrak{q}^*, \mathfrak{k}^*].$$

Treating $\mathfrak{k} = [\mathfrak{q}_2, \ldots, \mathfrak{q}_r]$ as we treated $[\mathfrak{q}_1, \ldots, \mathfrak{q}_r]$, and proceeding in this way, we obtain, finally,

$$\mathfrak{i}^* = [\mathfrak{q}_1^*, \ldots, \mathfrak{q}_r^*].$$

Returning now to Theorem VII, $\mathfrak{q}_i^* \sim \mathbf{1}$ if and only if the radical $\mathfrak{p}_i$ of $\mathfrak{q}_i$ is quasi-equal to $\mathbf{1}$, and if the radical is not quasi-equal to $\mathbf{1}$, $\mathfrak{q}_i^* = \mathfrak{q}_i$. Thus if $\mathfrak{i} = [\mathfrak{q}_1, \ldots, \mathfrak{q}_r]$, we obtain $\mathfrak{i}^*$ by omitting all the primaries whose radicals are quasi-equal to $\mathbf{1}$. We notice, further, that if the radical $\mathfrak{p}_i$ of $\mathfrak{q}_i$ is not quasi-equal to $\mathbf{1}$, then $\mathfrak{p}_i$ has no proper multiple which is prime [Th. VI]. This means that $\mathfrak{q}_i$ is isolated. Hence we obtain

THEOREM VIII. *If* $[\mathfrak{q}_1, \ldots, \mathfrak{q}_r]$ *is an uncontractible representation of an ideal* $\mathfrak{i}$ *of* $\mathfrak{R}$, *and* $\mathfrak{q}_i = \mathfrak{q}_i^*$ $(i = 1, \ldots, s)$, $\mathfrak{q}_i \sim \mathbf{1}$ $(i = s+1, \ldots, r)$, *then* $[\mathfrak{q}_1, \ldots, \mathfrak{q}_s]$ *is an uncontractible representation of* $\mathfrak{i}^*$ *as an intersection of primaries, and* $\mathfrak{q}_1, \ldots, \mathfrak{q}_s$ *are isolated components of* $\mathfrak{i}$ *and of* $\mathfrak{i}^*$.

Let us apply this theorem to the powers of a prime ideal $\mathfrak{p}$ of $\mathfrak{R}$ [p. 23]. If $\mathfrak{p} \sim \mathbf{1}$, $\mathfrak{p}^\rho \sim \mathbf{1}$ for all values of ρ. If $\mathfrak{p}$ is not quasi-equal to $\mathbf{1}$, neither is $\mathfrak{p}^\rho$, for any value of ρ greater than zero. Let

$$\mathfrak{p}^\rho = [\mathfrak{p}^{(\rho)}, \mathfrak{q}_1, \ldots, \mathfrak{q}_k],$$

where $\mathfrak{p}^{(\rho)}$ is the symbolic ρth power of $\mathfrak{p}$, and $\mathfrak{q}_1, \ldots, \mathfrak{q}_k$ are embedded components of $\mathfrak{p}^\rho$. If $\mathfrak{q}_i$ is $\mathfrak{p}_i$-primary, $\mathfrak{p}$ is a prime ideal which is a proper multiple of $\mathfrak{p}_i$, hence $\mathfrak{p}_i \sim \mathbf{1}$ [Th. VI], and therefore [Th. VII] $\mathfrak{q}_i \sim \mathbf{1}$. Therefore [Th. VIII]

$$(\mathfrak{p}^\rho)^* = \mathfrak{p}^{(\rho)}.$$

Now consider any primary ideal $\mathfrak{q}$ of $\mathfrak{R}$, which is not quasi-equal to **1**. Let $\mathfrak{p}$ be the radical of $\mathfrak{q}$, and σ its index. We have

$$\mathfrak{q} = 0 \ (\mathfrak{p}),$$

and therefore there exists an ideal $\mathfrak{i}$ such that

$$\mathfrak{q} \sim \mathfrak{p}\mathfrak{i}.$$

We also have
$$\mathfrak{p}^\sigma = 0 \ (\mathfrak{q}),$$

and hence, similarly, there exists an ideal $\mathfrak{j}$ such that

$$\mathfrak{p}^\sigma \sim \mathfrak{q}\mathfrak{j}.$$

Therefore
$$\mathfrak{p}^\sigma \sim \mathfrak{p}\mathfrak{i}\mathfrak{j},$$

or, equivalently,
$$\mathfrak{p}^{\sigma-1} \sim \mathfrak{i}\mathfrak{j}.$$

By the unique factorisation theorem, it follows that

$$\mathfrak{i} \sim \mathfrak{p}^{\tau-1}$$

for some integer τ ($\tau \leqslant \sigma$), and therefore

$$\mathfrak{q} \sim \mathfrak{p}^\tau.$$

Since $\mathfrak{q} = \mathfrak{q}^*$,
$$\mathfrak{p}^\tau \subseteq (\mathfrak{p}^\tau)^* = \mathfrak{q}^* = \mathfrak{q}.$$

Hence $\tau \geqslant \sigma$. Therefore we conclude that $\tau = \sigma$, and

$$\mathfrak{q} = \mathfrak{q}^* = (\mathfrak{p}^\sigma)^* = \mathfrak{p}^{(\sigma)}.$$

Hence we have

THEOREM IX. *The only primary ideals of $\mathfrak{R}$ not quasi-equal to* **1** *are symbolic powers of primes.*

We conclude with a useful result on principal ideals in $\mathfrak{R}$. Let $\mathfrak{i} = \mathfrak{R}.(\alpha)$ be a principal ideal in $\mathfrak{R}$. Then, as we have seen, $\mathfrak{i}^* = \mathfrak{i}$. Now if

$$\mathfrak{i} = [\mathfrak{q}_1, \ldots, \mathfrak{q}_r],$$

$$\mathfrak{i} = \mathfrak{i}^* = [\mathfrak{p}_1^{(\rho_1)}, \ldots, \mathfrak{p}_s^{(\rho_s)}],$$

where $\mathfrak{p}_i^{(\rho_i)}$ is a symbolic power of a prime $\mathfrak{p}_i$ which has no proper multiple which is prime. No $\mathfrak{p}_i^{(\rho_i)}$, therefore, is embedded, and we have

THEOREM X. *All the components of a principal ideal in $\mathfrak{R}$ are isolated.*

8. Integral dependence. Let $\mathfrak{R}$ be any ring, and let $\mathfrak{S}$ be any ring which contains $\mathfrak{R}$ as a subring. An element ξ of $\mathfrak{S}$ will be said to be *integrally dependent on* $\mathfrak{R}$ if it satisfies an equation of the form

$$\xi^n + a_1\xi^{n-1} + \ldots + a_n + m_1\xi^{n-1} + \ldots + m_{n-1}\xi = 0 \tag{1}$$

for some integer n, where $a_1, \ldots, a_n$ are elements of $\mathfrak{R}$ and $m_1, \ldots, m_{n-1}$ are integers. If ξ is an element a of $\mathfrak{R}$,

$$\xi - a = 0;$$

hence any element of $\mathfrak{R}$ is integrally dependent on $\mathfrak{R}$. It is clear from (1) that if ξ is integrally dependent on $\mathfrak{R}$, satisfying (1), the powers $\xi, \xi^2, \xi^3, \ldots$ of ξ are contained in the finite $\mathfrak{R}$-module $\mathfrak{R}.(\xi, \ldots, \xi^{n-1})$. The reader should be warned that in many books on Modern Algebra this fact is used as a basis for the definition of integral dependence; an element ξ of $\mathfrak{S}$ is said to be integrally dependent on $\mathfrak{R}$ if the powers $\xi, \xi^2, \xi^3, \ldots$ of ξ belong to a finite $\mathfrak{R}$-module. These definitions are not the same; as we have seen, integral dependence in our sense implies integral dependence as defined in terms of $\mathfrak{R}$-modules, but the converse is not in general true. We shall, in fact, prove that the two definitions are equivalent in the case in which the Basis Theorem holds in $\mathfrak{R}$, but it is not possible to limit ourselves to rings $\mathfrak{R}$ in which the Basis Theorem holds; for later we shall have to consider properties of integral dependence in connection with valuation rings, in which the Basis Theorem may not hold. Our definition is therefore a narrower one, but it is more convenient for the geometrical applications which we have in view. In order to develop the theory of integral dependence on our narrower definition, we shall shortly have to introduce restrictions on $\mathfrak{R}$ and $\mathfrak{S}$ which would not be necessary with the wider definition, but these restrictions are, in fact, always satisfied by the rings which appear in the geometrical problems to which we apply our theory.

We first prove the theorem mentioned above. In all that follows, integral dependence is to be understood in our narrower sense.

THEOREM I. *If the Basis Theorem holds in* $\mathfrak{R}$, *and* ξ *is any element of* $\mathfrak{S}$ *such that* $\xi, \xi^2, \xi^3, \ldots$ *are contained in a finite* $\mathfrak{R}$*-module* $\mathfrak{M}$, *then* ξ *is integrally dependent on* $\mathfrak{R}$.

Since $\xi, \xi^2, \xi^3, \ldots$ are contained in $\mathfrak{M}$, any polynomial

$$a_1\xi + a_2\xi^2 + \ldots + a_n\xi^n + m_1\xi + \ldots + m_n\xi^n,$$

where $a_i \in \mathfrak{R}$ and m_i is an integer, is contained in $\mathfrak{M}$. These polynomials clearly form an $\mathfrak{R}$-module $\mathfrak{M}_1$, and $\mathfrak{M}_1 \subseteq \mathfrak{M}$. Hence, since $\mathfrak{M}$ is a finite $\mathfrak{R}$-module and the Basis Theorem holds in $\mathfrak{R}$, it follows from Theorem I of § 6 that $\mathfrak{M}_1$ is a finite $\mathfrak{R}$-module:

$$\mathfrak{M}_1 = \mathfrak{R}.(\eta_1, \ldots, \eta_s).$$

Since η_i is in $\mathfrak{M}_1$, $$\eta_i = \sum_{j=1}^{n_i} a_{ij}\xi^j + \sum_{j=1}^{n_i} m_{ij}\xi^j.$$

Let $n = \max[n_1, \ldots, n_s]$. Then

$$\mathfrak{M}_1 \subseteq \mathfrak{R}.(\xi, \ldots, \xi^n);$$

and since $\xi^i \subseteq \mathfrak{M}_1$, $$\mathfrak{R}.(\xi, \ldots, \xi^n) \subseteq \mathfrak{M}_1.$$

Hence $$\mathfrak{M}_1 = \mathfrak{R}.(\xi, \ldots, \xi^n).$$

But ξ^{n+1} is in $\mathfrak{M}_1$. Hence

$$\xi^{n+1} = a_n\xi^n + \ldots + a_1\xi + m_n\xi^n + \ldots + m_1\xi,$$

where $a_i \in \mathfrak{R}$, and m_i is an integer. Hence ξ is integrally dependent on $\mathfrak{R}$.

The only cases with which we shall be concerned are those in which $\mathfrak{R}$ and $\mathfrak{S}$ are both integral domains. Since $\mathfrak{S}$ is an integral domain, it has unity, and since $\mathfrak{R} \subseteq \mathfrak{S}$, the unity of $\mathfrak{S}$ is the unity of $\mathfrak{R}$. For the rest of this section we therefore assume that $\mathfrak{R}$ and $\mathfrak{S}$ are integral domains. In (1) we can write b_i for $a_i + m_i$. Then if ξ is integrally dependent on $\mathfrak{R}$

$$\xi^n + b_1\xi^{n-1} + \ldots + b_n = 0 \quad (b_i \in \mathfrak{R}).$$

The ring $\mathfrak{R}$ is said to be *integrally closed in* $\mathfrak{S}$ if every element of $\mathfrak{S}$ which is integrally dependent on $\mathfrak{R}$ is in $\mathfrak{R}$. We often have to consider the case in which $\mathfrak{S}$ is the quotient field of $\mathfrak{R}$; and if we say that an integral domain is integrally closed, without specifying in which ring $\mathfrak{S}$, we shall understand that $\mathfrak{R}$ is integrally closed in its quotient field. More generally, the set $\mathfrak{R}^*$ of elements in $\mathfrak{S}$ which are integrally dependent on $\mathfrak{R}$ is called the *integral closure of* $\mathfrak{R}$ *in* $\mathfrak{S}$.

THEOREM II. *The integral closure* $\mathfrak{R}^*$ *of* $\mathfrak{R}$ *in* $\mathfrak{S}$ *is an integral domain, and* $\mathfrak{R}^*$ *is integrally closed in* $\mathfrak{S}$.

To prove that $\mathfrak{R}^*$ is a ring, we have to show that if ξ and η are any elements of $\mathfrak{R}^*$, then $\xi \pm \eta$ and $\xi\eta$ are in $\mathfrak{R}^*$. We therefore prove that if $f(x, y)$ is any polynomial in x, y, with coefficients in $\mathfrak{R}$, then $\zeta = f(\xi, \eta)$ is integrally dependent on $\mathfrak{R}$. Let

$$\xi^n + a_1 \xi^{n-1} + \ldots + a_n = 0 \quad (a_i \in \mathfrak{R})$$

be the equation which expresses the integral dependence of ξ on $\mathfrak{R}$, and, similarly, let

$$\eta^m + b_1 \eta^{m-1} + \ldots + b_m = 0 \quad (b_i \in \mathfrak{R})$$

be the equation expressing the integral dependence of η on $\mathfrak{R}$. Let $\omega_1, \ldots, \omega_r$ be the mn products $\xi^i \eta^j$ $(0 \leqslant i \leqslant n-1;\ 0 \leqslant j \leqslant m-1)$, arranged in some order. Then $\zeta\omega_i$ is a polynomial in ξ, η, and, using the equations for ξ and η written above, we see that

$$\zeta\omega_i = \sum_{j=1}^{r} c_{ij}\omega_j \quad (i = 1, \ldots, r),$$

where $c_{ij} \in \mathfrak{R}$. Hence, by the usual elimination process,

$$| \zeta\delta_{ij} - c_{ij} | \, \omega_k = 0,$$

where δ_{ij} is the Kronecker delta, and since at least one ω_k is different from zero, and is not a divisor of zero ($\mathfrak{S}$ being an integral domain), we have

$$| \zeta\delta_{ij} - c_{ij} | = 0,$$

which expresses the fact that ζ is integrally dependent on $\mathfrak{R}$. Hence ζ is in $\mathfrak{R}^*$, and $\mathfrak{R}^*$ is therefore a ring. Since $\mathfrak{R}^* \supseteq \mathfrak{R}$ and $\mathfrak{R}$ contains the unity of $\mathfrak{S}$, $\mathfrak{R}^*$ has unity, and since $\mathfrak{R}^* \subseteq \mathfrak{S}$, $\mathfrak{R}^*$ has no divisors of zero. Hence $\mathfrak{R}^*$ is an integral domain.

Next, let ξ be any element of $\mathfrak{S}$ which is integrally dependent on $\mathfrak{R}^*$. Then

$$\xi^n + \alpha_1 \xi^{n-1} + \ldots + \alpha_n = 0,$$

where $\alpha_1, \ldots, \alpha_n$ are in $\mathfrak{R}^*$. Since α_i is integrally dependent on $\mathfrak{R}$, we have equations

$$\alpha_i^{m_i} + a_{i1}\alpha_i^{m_i-1} + \ldots + a_{im_i} = 0,$$

where $a_{ij} \in \mathfrak{R}$ $(i = 1, \ldots, n)$. Let $\omega_1, \ldots, \omega_s$ be the products $\alpha_1^{k_1} \ldots \alpha_n^{k_n} \xi^k$ $(0 \leqslant k_i \leqslant m_i - 1;\ 0 \leqslant k \leqslant n-1)$, in some order. Just as above, we see that

$$\xi\omega_i = \sum_{j=1}^{s} b_{ij}\omega_j \quad (i = 1, \ldots, s),$$

and hence

$$| \xi\delta_{ij} - b_{ij} | \, \omega_k = 0,$$

and, since ω_k is not a divisor of zero of $\mathfrak{S}$,

$$|\xi\delta_{ij} - b_{ij}| = 0,$$

and hence ξ is integrally dependent on $\mathfrak{R}$. Hence ξ is in $\mathfrak{R}^*$, and therefore $\mathfrak{R}^*$ is integrally closed in $\mathfrak{S}$. This completes the proof of Theorem II.

It should be noted that the second part of Theorem II is equivalent to saying that *the integral dependence of an element of $\mathfrak{S}$ on a subring of $\mathfrak{S}$ is a transitive property.*

THEOREM III. *If $\mathfrak{R}$ is an integral domain which is integrally closed in its quotient field, and if the Basis Theorem holds in $\mathfrak{R}$, then $\mathfrak{M}\mathfrak{M}^{-1} \sim \mathbf{1}$, for any non-zero fractional ideal $\mathfrak{M}$ of $\mathfrak{R}$.*

If $\mathfrak{M}$ is any non-zero fractional ideal in $\mathfrak{R}$, we know, from the definition of $\mathfrak{M}^{-1}$, that

$$\mathfrak{M}\mathfrak{M}^{-1} \subseteq \mathfrak{R},$$

that is, that

$$\mathfrak{M}\mathfrak{M}^{-1} \geqslant \mathbf{1}.$$

We have to prove that

$$\mathfrak{M}\mathfrak{M}^{-1} \leqslant \mathbf{1}.$$

It is sufficient to show that $(\mathfrak{M}\mathfrak{M}^{-1})^{-1} \subseteq \mathfrak{R}$. Let λ be any element of $(\mathfrak{M}\mathfrak{M}^{-1})^{-1}$; then $\lambda\mathfrak{M}\mathfrak{M}^{-1} \subseteq \mathfrak{R}$. Therefore $\lambda\mathfrak{M}^{-1} \subseteq \mathfrak{M}^{-1}$. Thus we have

$$\lambda\mathfrak{M}^{-1} \subseteq \mathfrak{M}^{-1},$$

$$\lambda^2\mathfrak{M}^{-1} \subseteq \lambda\mathfrak{M}^{-1} \subseteq \mathfrak{M}^{-1},$$

$$\cdots\cdots\cdots\cdots\cdots\cdots$$

$$\lambda^r\mathfrak{M}^{-1} \subseteq \mathfrak{M}^{-1},$$

for any positive integer r. Hence

$$\lambda^r\mathfrak{M}\mathfrak{M}^{-1} \subseteq \mathfrak{M}\mathfrak{M}^{-1} \subseteq \mathfrak{R}.$$

Let μ be any non-zero element of $\mathfrak{M}\mathfrak{M}^{-1}$. Then $\lambda^r\mu \in \mathfrak{R}$, for all values of r. Hence all powers of λ belong to the finite $\mathfrak{R}$-module $\mathfrak{R}.(\mu^{-1})$. By Theorem I it follows that λ is integrally dependent on $\mathfrak{R}$, and hence, since $\mathfrak{R}$ is, by hypothesis, integrally closed, λ must be in $\mathfrak{R}$. Therefore

$$(\mathfrak{M}\mathfrak{M}^{-1})^{-1} \subseteq \mathfrak{R}.$$

Hence we have

$$\mathfrak{M}\mathfrak{M}^{-1} \leqslant \mathbf{1},$$

and since also

$$\mathfrak{M}\mathfrak{M}^{-1} \geqslant \mathbf{1},$$

we have

$$\mathfrak{M}\mathfrak{M}^{-1} \sim \mathbf{1}.$$

It follows from this result that the multiplicative theory of ideals developed in § 7 is applicable to any integral domain for which the Basis Theorem holds if the integral domain is integrally closed in its quotient field. This is one of the reasons why, in the applications of ideal theory to algebraic geometry which we make later in this volume, we pay so much attention to integrally closed integral domains.

THEOREM IV. *Let $\mathfrak{R}$ be an integral domain, and $\mathfrak{R}^*$ its integral closure. If $\mathfrak{p}$ is any prime ideal in $\mathfrak{R}$, a necessary and sufficient condition that the quotient ring $\mathfrak{R}_\mathfrak{p}$ be integrally closed is that $\mathfrak{R}^* \subseteq \mathfrak{R}_\mathfrak{p}$.*

Suppose first that $\mathfrak{R}_\mathfrak{p}$ is integrally closed. If ξ is any element of the quotient field of $\mathfrak{R}$ (which is also the quotient field of $\mathfrak{R}_\mathfrak{p}$) which is integrally dependent on $\mathfrak{R}$, then ξ is integrally dependent on $\mathfrak{R}_\mathfrak{p}$, since $\mathfrak{R} \subseteq \mathfrak{R}_\mathfrak{p}$. Since $\mathfrak{R}_\mathfrak{p}$ is integrally closed, ξ is therefore contained in $\mathfrak{R}_\mathfrak{p}$. But ξ is any element of $\mathfrak{R}^*$. Hence

$$\mathfrak{R}^* \subseteq \mathfrak{R}_\mathfrak{p}.$$

Conversely, suppose that $\mathfrak{R}^* \subseteq \mathfrak{R}_\mathfrak{p}$. Let ξ be any element of the quotient field of $\mathfrak{R}$ which is integrally dependent on $\mathfrak{R}_\mathfrak{p}$. Then

$$\xi^n + \alpha_1 \xi^{n-1} + \dots + \alpha_n = 0,$$

where $\alpha_i \in \mathfrak{R}_\mathfrak{p}$; that is, $\alpha_i = a_i/b_i$, where a_i and b_i are elements of $\mathfrak{R}$, and b_i is not contained in $\mathfrak{p}$. Multiplying by $(b_1 \dots b_n)^n$, we obtain an equation

$$(b_1 \dots b_n \xi)^n + a_1 b_2 \dots b_n (b_1 \dots b_n \xi)^{n-1} + \dots + a_n b_1^n \dots b_{n-1}^n b_n^{n-1} = 0,$$

which expresses the fact that $b_1 \dots b_n \xi$ is integrally dependent on $\mathfrak{R}$. Hence

$$b_1 \dots b_n \xi \in \mathfrak{R}^* \subseteq \mathfrak{R}_\mathfrak{p}.$$

Since $\mathfrak{p}$ is prime, and $b_i \neq 0$ $(\mathfrak{p})$ $(i = 1, \dots, n)$, $b_1 \dots b_n \neq 0$ $(\mathfrak{p})$; that is, $b_1 \dots b_n$ is a unit of $\mathfrak{R}_\mathfrak{p}$. Hence ξ is in $\mathfrak{R}_\mathfrak{p}$, and therefore $\mathfrak{R}_\mathfrak{p}$ is integrally closed.

An immediate consequence of this result is

THEOREM V. *If $\mathfrak{R}$ is an integral domain for which the Basis Theorem holds, a necessary and sufficient condition that $\mathfrak{R}$ be integrally closed is that the quotient rings of all its prime ideals be integrally closed. In applying the sufficiency criterion, we need only consider the maximal ideals of $\mathfrak{R}$.*

Suppose, first, that $\mathfrak{R}$ is integrally closed. If $\mathfrak{p}$ is any prime ideal of $\mathfrak{R}$, we have $\mathfrak{R}^* = \mathfrak{R} \subseteq \mathfrak{R}_\mathfrak{p}$, and $\mathfrak{R}_\mathfrak{p}$ is integrally closed, by Theorem IV. On the other hand, suppose that the quotient rings of all maximal ideals of $\mathfrak{R}$ are integrally closed. We know [§ 5, Th. X] that $\mathfrak{R}$ is the intersection of these quotient rings. Any element of $\mathfrak{R}^*$, the integral closure of $\mathfrak{R}$, is contained in the quotient ring of each maximal ideal of $\mathfrak{R}$, by Theorem IV, and is therefore contained in the intersection of these rings, that is, in $\mathfrak{R}$. Hence $\mathfrak{R}^* = \mathfrak{R}$.

The next two theorems deal with integral closure properties of the kinds of ring with which we are mainly concerned in geometrical applications.

THEOREM VI. *If K is any field, and $x_1, \ldots, x_r$ are independent indeterminates over K, then $\mathfrak{R} = K[x_1, \ldots, x_r]$ is integrally closed (in its quotient field).*

$\mathfrak{R}$ is a unique factorisation domain. Hence any element ξ of $K(x_1, \ldots, x_r)$ can be written in the form

$$\xi = \frac{a(x_1, \ldots, x_r)}{b(x_1, \ldots, x_r)} = \frac{a}{b},$$

where a and b are polynomials in $x_1, \ldots, x_r$, without common factor. If ξ is integrally dependent on $\mathfrak{R}$, we have an equation

$$\xi^n + c_1\xi^{n-1} + \ldots + c_n = 0,$$

where $c_i = c_i\,(x_1, \ldots, x_r)$ is a polynomial in $x_1, \ldots, x_r$ with coefficients in K, and so

$$a^n + c_1ba^{n-1} + \ldots + c_nb^n = 0,$$

that is

$$a^n = b(-c_1a^{n-1} - \ldots - c_nb^{n-1}).$$

b is therefore a factor of a^n. Since $\mathfrak{R}$ is a unique factorisation domain, and a and b are without common factors, b must be in K, and hence ξ is in $\mathfrak{R}$. This proves the theorem.

Now let $\mathfrak{R}$ be an integral domain (in which the Basis Theorem holds), and let Σ be its quotient field. Assume that Σ is of characteristic zero, and that $\mathfrak{R}$ is integrally closed in Σ. Let Σ^* be a finite (and hence a simple) algebraic extension of Σ. Let $\mathfrak{R}^*$ be the integral closure of $\mathfrak{R}$ in Σ^*. We prove

THEOREM VII. *Σ^* is the quotient field of $\mathfrak{R}^*$, and $\mathfrak{R}^*$ is a finite $\mathfrak{R}$-module. The Basis Theorem holds in $\mathfrak{R}^*$.*

Σ^* is a simple extension of Σ. Hence there exists in Σ^* an element ξ which (i) satisfies an equation

$$\xi^n + \alpha_1\xi^{n-1} + \ldots + \alpha_n = 0, \qquad (2)$$

where $\alpha_1, \ldots, \alpha_n$ are in Σ and $x^n+\alpha_1 x^{n-1}+\ldots+\alpha_n$ is an irreducible polynomial over Σ, and (ii) has the further property that any element of Σ^* can be written as

$$\beta_0+\beta_1\xi+\ldots+\beta_{n-1}\xi^{n-1},$$

where $\beta_0, \ldots, \beta_{n-1}$ are in Σ. The element ξ is what we have called a *primitive element* of Σ^* over Σ. Since α_i is in Σ, $\alpha_i = a_i/b_i$, where a_i and b_i are in $\mathfrak{R}$. From (2) we have

$$(b_1 \ldots b_n \xi)^n + a_1 b_2 \ldots b_n (b_1 \ldots b_n \xi)^{n-1} + \ldots + a_n b_1^n \ldots b_{n-1}^n b_n^{n-1} = 0.$$

Hence $\eta = b_1 \ldots b_n \xi$ is an element of Σ^* which is integrally dependent on $\mathfrak{R}$; that is, $\eta \in \mathfrak{R}^*$. Clearly, η is a primitive element of Σ^* over Σ; thus we can choose the primitive element of Σ^* in $\mathfrak{R}^*$.

Any element ζ of Σ^* is of the form

$$\zeta = \gamma_0+\gamma_1\eta+\ldots+\gamma_{n-1}\eta^{n-1},$$

where $\gamma_i = c_i/d_i$ is in Σ. Thus we can write

$$\zeta = \frac{\sigma}{d},$$

where $\sigma \in \mathfrak{R}[\eta]$; that is, σ is in the ring of polynomials in η with coefficients in $\mathfrak{R}$. Hence Σ^* is the quotient field of $\mathfrak{R}[\eta]$. But, since η is integrally dependent on $\mathfrak{R}$, $\mathfrak{R}[\eta] \subseteq \mathfrak{R}^*$. Hence Σ^* is the quotient field of $\mathfrak{R}^*$. This proves the first part of the theorem.

Let $\phi(x)$ be the characteristic polynomial of η, with respect to Σ. If Σ_1 is the root field of $\phi(x)$, and $\eta_1, \ldots, \eta_n$ the zeros of $\phi(x)$ in Σ_1, we may put $\eta = \eta_1$. By our choice of η, $\phi(x)$ has coefficients in $\mathfrak{R}$ and its leading coefficient is 1. Then, since $\phi(\eta_i) = 0$, $\eta_1, \ldots, \eta_n$ are all integrally dependent on $\mathfrak{R}$ (which is a subring of $\Sigma^* = \Sigma(\eta_1) \subseteq \Sigma_1$). Let ζ be any element of Σ^* which is integrally dependent on $\mathfrak{R}$. Then

$$\zeta = \rho_0+\rho_1\eta+\ldots+\rho_{n-1}\eta^{n-1},$$

where $\rho_i \in \Sigma$, and

$$\zeta_i = \rho_0+\rho_1\eta_i+\ldots+\rho_{n-1}\eta_i^{n-1} \tag{3}$$

is integrally dependent on $\mathfrak{R}$ $(i = 1, \ldots, n)$. Let

$$D = \det(\eta_j^i) = \prod_{i<j} (\eta_i - \eta_j) \neq 0.$$

Since each η_i is integrally dependent on $\mathfrak{R}$, D is integrally dependent on $\mathfrak{R}$. Moreover, D^2 is invariant under the automorphisms of Σ_1 over Σ. Hence, by X, § 10, Lemma V, D^2 is in Σ. But $\mathfrak{R}$ is integrally closed in Σ; therefore D^2 belongs to $\mathfrak{R}$.

If we solve the equations (3), we obtain

$$D\rho_i = \sum_{j=1}^{n} S_{ij}\zeta_j,$$

where S_{ij} is in Σ_1 and is integrally dependent on $\mathfrak{R}$. Hence $s_i = D^2\rho_i$ is integrally dependent on $\mathfrak{R}$. But it is in Σ; therefore it is in $\mathfrak{R}$.

Thus $$\zeta = \frac{s_0 + s_1\eta + \ldots + s_{n-1}\eta^{n-1}}{D^2}.$$

Now ζ is any element of Σ^* which is integrally dependent on $\mathfrak{R}$, and D^2 does not depend on our choice of ζ. Hence

$$\mathfrak{R}^* \subseteq \mathfrak{R}.(D^{-2}, D^{-2}\eta, \ldots, D^{-2}\eta^{n-1}).$$

$\mathfrak{R}^*$ is therefore a submodule of a finite $\mathfrak{R}$-module. Since the Basis Theorem holds in $\mathfrak{R}$ it follows [§ 6, Th. I] that $\mathfrak{R}^*$ is a finite $\mathfrak{R}$-module.

Finally, any ideal $\mathfrak{i}^*$ in $\mathfrak{R}^*$ is a submodule of $\mathfrak{R}^*$ over $\mathfrak{R}$. Hence it has a finite $\mathfrak{R}$-basis. This basis is also a basis of $\mathfrak{i}^*$ as an ideal in $\mathfrak{R}^*$. Hence the Basis Theorem holds in $\mathfrak{R}^*$. This completes the proof of Theorem VII.

The theory of § 4 can be applied to obtain relations between ideals in an integral domain $\mathfrak{R}$ and in its integral closure $\mathfrak{R}^*$. We now assume that $\mathfrak{R}^*$ is the integral closure of $\mathfrak{R}$ in its quotient field, that $\mathfrak{R}^*$ is a finite $\mathfrak{R}$-module, and that the Basis Theorem holds in $\mathfrak{R}$. By the proof just given, it then also holds in $\mathfrak{R}^*$. We prove that if $\mathfrak{i}$ is a proper ideal of $\mathfrak{R}$, $\mathfrak{R}^*.\mathfrak{i}$ is a proper ideal of $\mathfrak{R}^*$, and that if $\mathfrak{i}^*$ is a proper ideal of $\mathfrak{R}^*$, $\mathfrak{R}_\wedge \mathfrak{i}^*$ is a proper ideal of $\mathfrak{R}$.

Let $\mathfrak{i}$ be a proper ideal of $\mathfrak{R}$. Since $\mathfrak{i} \subseteq \mathfrak{R}^*.\mathfrak{i}$, $\mathfrak{R}^*.\mathfrak{i}$ is not the zero ideal of $\mathfrak{R}^*$. We now show that it is not the unit ideal of $\mathfrak{R}^*$. By hypothesis, $\mathfrak{R}^*$ is a finite $\mathfrak{R}$-module; let $\zeta_1, \ldots, \zeta_t$ be an $\mathfrak{R}$-basis for it. Then the elements of $\mathfrak{R}^*.\mathfrak{i}$ are of the form $\Sigma c_j\zeta_j$, where $c_j \in \mathfrak{i}$. If $\mathfrak{R}^*.\mathfrak{i} = \mathfrak{R}^*$, we can find elements c_{ij} in $\mathfrak{i}$ such that

$$\sum_{j=1}^{t} c_{ij}\zeta_j = \zeta_i \quad (i = 1, \ldots, t).$$

Eliminating $\zeta_1, \ldots, \zeta_t$ we obtain the equation

$$|c_{ij} - \delta_{ij}| = 0,$$

and since $c_{ij} \in \mathfrak{i}$, we have $$1 = 0 \ (\mathfrak{i}),$$

so that $\mathfrak{i} = \mathfrak{R}$. Since we assumed that $\mathfrak{i}$ is a proper ideal of $\mathfrak{R}$, we have a contradiction. It follows that $\mathfrak{R}^*.\mathfrak{i}$ is not the unit ideal. It is therefore a proper ideal.

Now let $\mathfrak{i}^*$ be a proper ideal of $\mathfrak{R}^*$. Let ζ be a non-zero element of it. Then ζ is integrally dependent on $\mathfrak{R}$, so we have an equation

$$\zeta^n + a_1 \zeta^{n-1} + \ldots + a_n = 0,$$

where $a_i \in \mathfrak{R}$. If $a_{m+1} = a_{m+2} = \ldots = a_n = 0$, but $a_m \neq 0$, we have

$$\zeta^{n-m}(\zeta^m + a_1 \zeta^{m-1} + \ldots + a_m) = 0,$$

and hence, as ζ is not zero and R^* is an integral domain, we have

$$\zeta^m + a_1 \zeta^{m-1} + \ldots + a_m = 0.$$

Hence we may assume that $a_n \neq 0$. Since

$$\zeta^n + a_1 \zeta^{n-1} + \ldots + a_{n-1}\zeta \in \mathfrak{i}^*,$$

a_n belongs to $\mathfrak{i}^*$. Therefore $\mathfrak{R}_\wedge \mathfrak{i}^*$ is not the zero ideal of $\mathfrak{R}$. Suppose that $\mathfrak{R}_\wedge \mathfrak{i}^* = \mathfrak{R}$. Then

$$\mathfrak{R}^*.(\mathfrak{R}_\wedge \mathfrak{i}^*) = \mathfrak{R}^*.$$

But [§ 4, Th. VI]

$$\mathfrak{R}^*.(\mathfrak{R}_\wedge \mathfrak{i}^*) \subseteq \mathfrak{i}^*.$$

Hence $\mathfrak{i}^* = \mathfrak{R}^*$, contrary to hypothesis. We thus conclude that $\mathfrak{R}_\wedge \mathfrak{i}^*$ is a proper ideal of $\mathfrak{R}$.

Since $\mathfrak{R}^*$ is assumed to be a finite $\mathfrak{R}$-module, we can apply Theorem V of § 4 and its corollary. By these results, we know that if $\mathfrak{p}$ is any prime ideal of $\mathfrak{R}$,

$$\mathfrak{p} = \mathfrak{R}_\wedge (\mathfrak{R}^*.\mathfrak{p});$$

and that if

$$\mathfrak{R}^*.\mathfrak{p} = [\mathfrak{q}_1^*, \ldots, \mathfrak{q}_r^*],$$

where $\mathfrak{q}_i^*$ is $\mathfrak{p}_i^*$-primary, then $\mathfrak{R}_\wedge \mathfrak{p}_i^*$ is a factor of $\mathfrak{p}$, and for at least one value of i, $\mathfrak{R}_\wedge \mathfrak{q}_i^* = \mathfrak{R}_\wedge \mathfrak{p}_i^* = \mathfrak{p}$. We can make this last result more precise in the case in which $\mathfrak{R}_\mathfrak{p}$ is integrally closed in its quotient field.

Suppose that $\mathfrak{R}_\mathfrak{p}$ is integrally closed. Then $\mathfrak{R}^* \subseteq \mathfrak{R}_\mathfrak{p}$ [Th. IV]. Let $\mathfrak{q}_i^*$ be a component of $\mathfrak{R}^*.\mathfrak{p}$ whose radical $\mathfrak{p}_i^*$ contracts to $\mathfrak{p}$, that is,

$$\mathfrak{R}_\wedge \mathfrak{p}_i^* = \mathfrak{p}.$$

We show that

$$\mathfrak{R}^*_{\mathfrak{p}_i^*} = \mathfrak{R}_\mathfrak{p}.$$

Any element of $\mathfrak{R}_\mathfrak{p}$ is of the form α/β, where α and β are in $\mathfrak{R}$, and hence in $\mathfrak{R}^*$, and β is not in $\mathfrak{p}$. Hence β cannot be in $\mathfrak{p}_i^*$, for this would imply that β is in $\mathfrak{R}_\wedge \mathfrak{p}_i^* = \mathfrak{p}$. Hence α/β is in $\mathfrak{R}^*_{\mathfrak{p}_i^*}$, and therefore

$$\mathfrak{R}_\mathfrak{p} \subseteq \mathfrak{R}^*_{\mathfrak{p}_i^*}. \tag{4}$$

Now consider an element of $\mathfrak{R}^*_{\mathfrak{p}_i^*}$. This is equal to α^*/β^*, where α^* and β^* lie in $\mathfrak{R}^*$, and hence in $\mathfrak{R}_\mathfrak{p}$, and β^* is not in $\mathfrak{p}_i^*$. Now β^* is integrally dependent on $\mathfrak{R}$ and therefore satisfies an equation

$$\beta^{*n}+a_1\beta^{*n-1}+\ldots+a_n=0,$$

where $a_1, \ldots, a_n$ are in $\mathfrak{R}$. These coefficients cannot, however, all lie in $\mathfrak{p}$, for if they did we should have

$$\beta^{*n}=-a_1\beta^{*n-1}-\ldots-a_n\in\mathfrak{R}^*.\mathfrak{p}\subseteq\mathfrak{p}_i^*,$$

and hence, since $\mathfrak{p}_i^*$ is prime, $\beta^*=0\ (\mathfrak{p}_i^*)$, contrary to what was said above.

Suppose that m $(m\geqslant 1)$ is such that

$$a_m\neq 0\ (\mathfrak{p}),\quad a_{m+i}=0\ (\mathfrak{p})\quad (i=1,\ldots,n-m).$$

Then
$$\beta^{*n-m}(\beta^{*m}+a_1\beta^{*m-1}+\ldots+a_m)=0\ (\mathfrak{p}_i^*),$$

and, since $\mathfrak{p}_i^*$ is prime and does not contain β^*, we have

$$\zeta^*\equiv\beta^{*m}+a_1\beta^{*m-1}+\ldots+a_m=0\ (\mathfrak{p}_i^*).$$

Then
$$\frac{\alpha^*}{\beta^*}=\frac{\alpha^*(\beta^{*m-1}+a_1\beta^{*m-2}+\ldots+a_{m-1})}{\zeta^*-a_m}.$$

Now the non-units of $\mathfrak{R}_\mathfrak{p}$ form the ideal $\mathfrak{R}_\mathfrak{p}.\mathfrak{p}$ [§ 5, Th. V]. Since a_m is not contained in $\mathfrak{p}$, it is a unit of $\mathfrak{R}_\mathfrak{p}$. If ζ^* were a unit of $\mathfrak{R}_\mathfrak{p}$, we could find γ and δ in $\mathfrak{R}$, with δ not in $\mathfrak{p}$, such that

$$\gamma\zeta^*=\delta.$$

Then $\gamma\zeta^*\in\mathfrak{R}$, and since $\zeta^*\in\mathfrak{p}_i^*$, $\delta=\gamma\zeta^*\in\mathfrak{R}\wedge\mathfrak{p}_i^*=\mathfrak{p}$, which is a contradiction. Thus ζ^* is a non-unit of $\mathfrak{R}_\mathfrak{p}$. Therefore ζ^*-a_m is a unit, for if it lay in $\mathfrak{R}_\mathfrak{p}.\mathfrak{p}$, then $a_m=\zeta^*-(\zeta^*-a_m)$ would be in this ideal, whereas we know that a_m is a unit of $\mathfrak{R}_\mathfrak{p}$. Thus

$$\alpha^*(\beta^{*m-1}+a_1\beta^{*m-2}+\ldots+a_{m-1})\quad\text{and}\quad\zeta^*-a_m$$

are in $\mathfrak{R}^*$, which is contained in $\mathfrak{R}_\mathfrak{p}$, and ζ^*-a_m is a unit. Therefore $\alpha^*/\beta^*\in\mathfrak{R}_\mathfrak{p}$, and hence

$$\mathfrak{R}^*_{\mathfrak{p}_i^*}\subseteq\mathfrak{R}_\mathfrak{p}.$$

Taking this together with equation (4), we have

$$\mathfrak{R}_\mathfrak{p}=\mathfrak{R}^*_{\mathfrak{p}_i^*}.$$

Since
$$\mathfrak{R}^*.\mathfrak{p}=[\mathfrak{q}_1^*,\ldots,\mathfrak{q}_r^*]\quad\text{and}\quad\mathfrak{R}^*\subseteq\mathfrak{R}_\mathfrak{p},$$

we have, further,

$$\mathfrak{R}_{\mathfrak{p}}.(\mathfrak{R}^*.\mathfrak{p}) = \mathfrak{R}_{\mathfrak{p}}.\mathfrak{p} = \mathfrak{R}^*_{\mathfrak{p}_i^*}.[\mathfrak{q}_1^*, \ldots, \mathfrak{q}_r^*]$$
$$= [\mathfrak{R}^*_{\mathfrak{p}_i^*}.\mathfrak{q}_1^*, \ldots, \mathfrak{R}^*_{\mathfrak{p}_i^*}.\mathfrak{q}_r^*]$$

[§ 5, Th. IX]. Now $\mathfrak{R}_{\mathfrak{p}}.\mathfrak{p}$ is a maximal ideal in $\mathfrak{R}_{\mathfrak{p}} = \mathfrak{R}^*_{\mathfrak{p}_i^*}$ [§ 5, Th. V]. But, clearly, $\bar{\mathfrak{j}} = [\mathfrak{R}^*_{\mathfrak{p}_i^*}.\mathfrak{q}_1^*, \ldots, \mathfrak{R}^*_{\mathfrak{p}_i^*}.\mathfrak{q}_r^*]$ can only be maximal if, for each value of k, either

$$\mathfrak{R}_{\mathfrak{p}}.\mathfrak{q}_k^* = \mathfrak{R}^*_{\mathfrak{p}_i^*}.\mathfrak{q}_k^* = \mathfrak{R}^*_{\mathfrak{p}_i^*},$$

or

$$\mathfrak{R}^*_{\mathfrak{p}_i^*}.\mathfrak{q}_k^* = \bar{\mathfrak{j}}.$$

In the particular case $k = i$, $\mathfrak{R}^*_{\mathfrak{p}_i^*}.\mathfrak{q}_i^*$ is not the unit ideal [§ 5, Th. II]; hence

$$\mathfrak{R}_{\mathfrak{p}}.\mathfrak{q}_i^* = \mathfrak{R}_{\mathfrak{p}}.\mathfrak{p}.$$

This ideal is prime; hence [§ 5, Th. VIII] $\mathfrak{q}_i^*$ must be prime; that is, $\mathfrak{q}_i^* = \mathfrak{p}_i^*$.

Again, [§ 4, Th. V],

$$\mathfrak{p}_i^* = \mathfrak{R}^*{}_\wedge(\mathfrak{R}^*_{\mathfrak{p}_i^*}.\mathfrak{p}_i^*) = \mathfrak{R}^*{}_\wedge(\mathfrak{R}_{\mathfrak{p}}.\mathfrak{p}).$$

It follows that the radical of the primary component of $\mathfrak{R}^*.\mathfrak{p}$ which contracts to $\mathfrak{p}$ is uniquely determined, and hence that if $[\mathfrak{q}_1^*, \ldots, \mathfrak{q}_r^*]$ is an uncontractible representation of $\mathfrak{R}^*.\mathfrak{p}$, only one of the components $\mathfrak{q}_i^*$ has a radical contracting to $\mathfrak{p}$. Summing up we have:

THEOREM VIII. *If $\mathfrak{p}$ is a prime ideal of $\mathfrak{R}$ such that $\mathfrak{R}_{\mathfrak{p}}$ is integrally closed, then $\mathfrak{R}^*.\mathfrak{p}$ can be represented as an uncontractible intersection of primary ideals in the form*

$$\mathfrak{R}^*.\mathfrak{p} = [\mathfrak{p}^*, \mathfrak{q}_1^*, \ldots, \mathfrak{q}_s^*],$$

where

$$\mathfrak{R}_\wedge\mathfrak{p}^* = \mathfrak{p},$$

and $\mathfrak{R}_\wedge\mathfrak{p}_i^*$ is a proper factor of $\mathfrak{p}$, $\mathfrak{p}_i^*$ being the radical of $\mathfrak{q}_i^*$. Moreover,

$$\mathfrak{R}_{\mathfrak{p}} = \mathfrak{R}^*_{\mathfrak{p}^*}.$$

The following corollary is of particular importance in geometrical applications.

Corollary. If $\mathfrak{p}$ is maximal in $\mathfrak{R}$ and $\mathfrak{R}_{\mathfrak{p}}$ is integrally closed, then

$$\mathfrak{R}^*.\mathfrak{p} = \mathfrak{p}^*$$

is prime, and $\mathfrak{p}$ and $\mathfrak{p}^*$ are corresponding contracted and extended ideals.

By Theorem VIII we have

$$\mathfrak{R}^*.\mathfrak{p} = [\mathfrak{p}^*, \mathfrak{q}_1^*, \ldots, \mathfrak{q}_s^*].$$

If $s > 0$, $\mathfrak{R}_\wedge \mathfrak{p}_i^*$ is a proper factor of $\mathfrak{p}$. But $\mathfrak{p}$ is maximal. Hence

$$\mathfrak{R}_\wedge \mathfrak{p}_i^* = \mathfrak{R}.$$

But, as we have seen, this implies that $\mathfrak{p}_i^*$ is the unit ideal. Since $\mathfrak{R}$ is an integral domain in which the Basis Theorem holds, it follows that $\mathfrak{q}_i^*$ is the unit ideal. Thus in the uncontractible representation $s = 0$.

CHAPTER XVI

THE ARITHMETIC THEORY OF VARIETIES

1. Algebraic varieties in affine space. We are now in a position to apply the methods of ideal theory to the geometry of algebraic varieties. In making these applications, it is usually convenient to use non-homogeneous coordinates, and some preliminary remarks on these are necessary.

The ground field K may be any field of characteristic zero, and we consider a space S_n over K. To introduce non-homogeneous coordinates [X, § 3] we select a prime Π which is to be the 'prime at infinity'. Our choice of Π may be completely arbitrary, or may be limited by the problem under consideration; for instance, if we wish to study a point P on an algebraic variety V in S_n, we must select Π so that it does not contain P, and hence does not contain V entirely. We then choose our homogeneous coordinates in S_n so that Π has the equation

$$x_0 = 0,$$

and take the non-homogeneous coordinates $(x'_1, \ldots, x'_n)$ which are given by the equations

$$x'_i = x_i/x_0 \quad (i = 1, \ldots, n).$$

To each point P of S_n which is not in Π there corresponds a unique set of coordinates $(x'_1, \ldots, x'_n)$, and conversely. If $(x'_1, \ldots, x'_n)$ are the non-homogeneous coordinates of P, the homogeneous coordinates of P are $(x_0, x_0x'_1, \ldots, x_0x'_n)$, where x_0 is any non-zero element of a field which is an extension of K. The points of S_n which do not lie in Π constitute what is called an *affine space* of n dimensions over K, which we denote by A_n.

Consider an allowable transformation of coordinates in S_n:

$$y_i = \sum_{j=0}^{n} a_{ij}x_j \quad (i = 0, \ldots, n).$$

If, in the new coordinate system, Π has the equation $y_0 = 0$, we must have

$$a_{01} = a_{02} = \ldots = a_{0n} = 0,$$

and since the matrix of the transformation is non-singular, we must therefore have a_{00} different from zero. Then

$$y'_i = \frac{y_i}{y_0} = \frac{a_{i0}}{a_{00}} + \sum_{j=1}^{n} \frac{a_{ij}x'_j}{a_{00}} \quad (i = 1, \ldots, n),$$

which we write in the form

$$y'_i = b_i + \sum_{j=1}^{n} b_{ij}x'_j \quad (i = 1, \ldots, n).$$

We note further that since

$$\begin{vmatrix} a_{00} & a_{01} & . & a_{0n} \\ a_{10} & a_{11} & . & a_{1n} \\ . & . & . & . \\ a_{n0} & a_{n1} & . & a_{nn} \end{vmatrix} = a_{00}^{n+1} \begin{vmatrix} a_{11}/a_{00} & . & . & a_{1n}/a_{00} \\ . & . & . & . \\ . & . & . & . \\ a_{n1}/a_{00} & . & . & a_{nn}/a_{00} \end{vmatrix},$$

we have

$$\begin{vmatrix} b_{11} & . & b_{1n} \\ . & . & . \\ b_{n1} & . & b_{nn} \end{vmatrix} \neq 0.$$

Conversely, we see at once that if

$$y_i = \sum_{j=0}^{n} a_{ij}x_j \quad (i = 0, \ldots, n)$$

is any allowable transformation in S_n for which $a_{01} = \ldots = a_{0n} = 0$, Π has, in the new coordinate system, the equation $y_0 = 0$; and that if

$$y'_i = b_i + \sum_{j=1}^{n} b_{ij}x'_j \quad (i = 1, \ldots, n), \tag{1}$$

where $|b_{ij}| \neq 0$, is a transformation of the non-homogeneous coordinates in A_n, this can be obtained as above from the allowable transformation $y = Ax$ in S_n, where

$$A = \begin{Vmatrix} 1 & 0 & . & . & 0 \\ b_1 & b_{11} & . & . & b_{1n} \\ b_2 & b_{21} & . & . & b_{2n} \\ . & . & . & . & . \\ b_n & b_{n1} & . & . & b_{nn} \end{Vmatrix}.$$

Thus we may define the *allowable transformations* of coordinates in A_n to be the transformations of the type (1).

The reader will find it an interesting exercise to define an affine space of n dimensions by the methods which were used in Chapter V to define a projective space. Since, however, we shall only meet in this book those affine spaces which are derived from projective spaces by choosing a prime at infinity, we shall find it convenient to think of them in this way, instead of defining them abstractly. For our purpose, an affine space is just a projective space with a prime removed, and the allowable transformations of coordinates are just those of type (1).

Logically, we ought to define varieties in affine space, and develop their properties, but as the theory is very similar to that of varieties in projective space, it will only be necessary at this stage to indicate a few of the salient properties. A variety V in A_n is the set of points which (in a given allowable coordinate system) satisfies a set of (non-homogeneous) equations

$$f_i(x_1, \ldots, x_n) = 0 \quad (i = 1, 2, \ldots).$$

Using Hilbert's Basis Theorem [IV, § 2, Th. I] we can show, as in the case of varieties in projective space, that any variety in A_n is given by a finite set of equations

$$f_i(x_1, \ldots, x_n) = 0 \quad (i = 1, \ldots, r).$$

If V_1 and V_2 are two varieties in A_n, their join $V_1 + V_2$ is defined as the set of points lying on V_1 or V_2, and the intersection $V_1 \wedge V_2$ is defined as the set of points lying simultaneously on V_1 and V_2. The associative, commutative and distributive properties of the operations '$+$' and '$\wedge$', proved in projective space, hold also in affine space.

A variety V in A_n is said to be *irreducible* if an equation

$$V = V_1 + V_2$$

necessarily implies either

$$V = V_1 \quad \text{or} \quad V = V_2,$$

and as in Chapter X, § 2, we can show that a necessary and sufficient condition that a variety be reducible is that there exist two polynomials $f(x_1, \ldots, x_n)$ and $g(x_1, \ldots, x_n)$, neither of which vanishes on V (that is, neither vanishes at all points of V), such that

$$f(x_1, \ldots, x_n)\, g(x_1, \ldots, x_n) = 0$$

on V.

A point $\xi = (\xi_1, \ldots, \xi_n)$ in A_n is called a generic point of a variety V (over the ground field K) if (1) ξ lies on V, and (2) if $f(x_1, \ldots, x_n)$ is any polynomial in $x_1, \ldots, x_n$ with coefficients in K such that

$$f(\xi_1, \ldots, \xi_n) = 0,$$

then

$$f(x_1, \ldots, x_n) = 0$$

on V. As in § 3 of Chapter X, we show that if a variety V has a generic point, then it is irreducible. To show that if V is irreducible it has a generic point, we consider the rational functions

$$f(x_1, \ldots, x_n)/g(x_1, \ldots, x_n)$$

of $x_1, \ldots, x_n$ over K for which $g(x_1, \ldots, x_n)$ is not zero on V and define an equivalence relation

$$\frac{f(x_1, \ldots, x_n)}{g(x_1, \ldots, x_n)} \sim \frac{f'(x_1, \ldots, x_n)}{g'(x_1, \ldots, x_n)}$$

if and only if

$$f(x_1, \ldots, x_n)\, g'(x_1, \ldots, x_n) - f'(x_1, \ldots, x_n)\, g(x_1, \ldots, x_n) = 0$$

on V. As in the case of varieties in projective space, we show that this is a true equivalence relation when V is irreducible, and go on to define addition and multiplication of equivalent classes. We thus arrive at a field Σ, the *function field* of V, which contains a subfield isomorphic with K. Identifying this with K, we get an extension of K. If ξ_i is the element of this extension corresponding to the class which contains $x_i/1$, we then show that $(\xi_1, \ldots, \xi_n)$ is a generic point of V. In the next section we shall introduce a simpler method of obtaining the generic point of an irreducible variety in A_n, using the methods of ideal theory.

The use of non-homogeneous coordinates allows us to introduce the notion of a proper specialisation of an element η of the function field $K(\xi_1, \ldots, \xi_n)$ of an algebraic variety V. If $\eta \in K(\xi_1, \ldots, \xi_n)$, the pair (ξ, η) is the generic point of a correspondence in which the object-variety is V, and the image-variety is a variety V' which is a line if η is transcendental over K, and a zero-dimensional variety if η is algebraic over K. If x' is any point of V, there corresponds to it at least one point y' of V'. The non-homogeneous coordinate y' of any such point (which may be at infinity) is called *a proper specialisation of* η *corresponding to the specialisation* $\xi \to x'$.

Using these results as a starting point, many of the other results established in Chapters X and XI for projective varieties can be established for affine varieties. It is not, however, necessary to do so, because we are now in a position to examine the relation between a variety in S_n and a variety derived from it in an affine space obtained by choosing a prime at infinity in S_n. This will enable us to translate any theorem on a variety in projective space into a theorem on a variety in affine space, and conversely.

Let S_n be an n-dimensional projective space over the ground field K, and let $(x_0, \ldots, x_n)$ be an allowable coordinate system in it. Let A_n be the affine space obtained from it by taking $x_0 = 0$ as the prime at infinity. Since points of S_n not in this prime have $x_0 \neq 0$, we may suppose the coordinates of all such points normalised so that $x_0 = 1$. Then $(x_1, \ldots, x_n)$ are the non-homogeneous coordinates of the point $(1, x_1, \ldots, x_n)$, regarded as a point of S_n.

Let
$$V = V_1 + V_2 + \ldots + V_k + V_{k+1} + \ldots + V_s$$
be any variety in S_n, where $V_1, \ldots, V_s$ are irreducible components of V, and suppose that $V_1, \ldots, V_k$ do not lie in $x_0 = 0$, while $V_{k+1}, \ldots, V_s$ lie in this prime. We shall denote by $(1, \xi_1^{(i)}, \ldots, \xi_n^{(i)})$ the generic point of V_i $(i = 1, \ldots, k)$, and by $(0, \xi_1^{(i)}, \ldots, \xi_n^{(i)})$ the generic point of V_i $(i > k)$. Further, let
$$f_i(x_0, \ldots, x_n) = 0 \quad (i = 1, \ldots, r) \tag{2}$$
be the equations of V.

Let V'_i be the irreducible variety in A_n whose generic point is $(\xi_1^{(i)}, \ldots, \xi_n^{(i)})$ $(i = 1, \ldots, k)$, and let
$$V' = V'_1 + \ldots + V'_k.$$
Since
$$f_i(x_0, \ldots, x_n) = 0$$
on V, for $i = 1, \ldots, r$, we have
$$f_i(1, \xi_1^{(j)}, \ldots, \xi_n^{(j)}) = 0 \quad (i = 1, \ldots, r; j = 1, \ldots, k).$$
Hence
$$f_i(1, x_1, \ldots, x_n) = 0$$
on V'_j, for $i = 1, \ldots, r$, for all values of j $(1 \leqslant j \leqslant k)$. Hence V' satisfies the equations
$$f_i(1, x_1, \ldots, x_n) = 0 \quad (i = 1, \ldots, r). \tag{3}$$

Conversely, let $f(x_1, \ldots, x_n)$ be any polynomial over K which vanishes on V', and let $F(x_0, \ldots, x_n)$ be a homogeneous polynomial such that

$$F(1, x_1, \ldots, x_n) = f(x_1, \ldots, x_n).$$

Then $\quad F(1, \xi_1^{(j)}, \ldots, \xi_n^{(j)}) = f(\xi_1^{(j)}, \ldots, \xi_n^{(j)}) = 0 \quad (j = 1, \ldots, k).$

Hence

$$F(x_0, \ldots, x_n) = 0$$

on $V_1 + \ldots + V_k$. It follows that $(x_1', \ldots, x_n')$ lies on V' if and only if $(1, x_1', \ldots, x_n')$ lies on V, and therefore equations (3) are the equations of V'. Thus the points of V not in $x_0 = 0$ form an algebraic variety in affine space, and this is irreducible if and only if V has a single component not lying in $x_0 = 0$.

Conversely, a variety

$$V' = V_1' + \ldots + V_k'$$

in A_n, where V_i' is irreducible and has generic point $(\xi_1^{(i)}, \ldots, \xi_n^{(i)})$, determines a variety

$$\bar{V} = V_1 + \ldots + V_k$$

in S_n, where V_i has the generic point $(1, \xi_1^{(i)}, \ldots, \xi_n^{(i)})$. It follows, as above, that $(x_1', \ldots, x_n')$ lies on V' if and only if $(1, x_1', \ldots, x_n')$ lies on $\bar{V}$. We note that if V' is the variety originally obtained from the variety V of S_n, V and $\bar{V}$ differ only by components lying in $x_0 = 0$, and also that if we begin with a variety V' in A_n, and pass to the variety $\bar{V}$ of S_n and then back to a variety of A_n, we return to V'. Further, as we saw above, if $f(x_1, \ldots, x_n)$ is any polynomial over K which vanishes on V', and $F(x_0, \ldots, x_n)$ is a homogeneous polynomial which reduces to $f(x_1, \ldots, x_n)$ when x_0 is replaced by 1, $F(x_0, \ldots, x_n)$ vanishes on $\bar{V}$. But it is to be observed that if

$$f_i(x_1, \ldots, x_n) = 0 \quad (i = 1, \ldots, r)$$

are the equations of V', and $F_i(x_0, \ldots, x_n)$ is a homogeneous polynomial reducing to $f_i(x_1, \ldots, x_n)$ when x_0 is replaced by 1, the equations

$$F_i(x_0, \ldots, x_n) = 0 \quad (i = 1, \ldots, r)$$

are satisfied by $\bar{V}$, but may define a variety differing from $\bar{V}$ by components lying in $x_0 = 0$. For instance, if V' is given in A_2 by the equations

$$x_2 = 0, \quad x_1 x_2 = 0,$$

these equations, now regarded as homogeneous equations, determine the variety in S_2 with a single component

$$x_2 = 0.$$

V' can also be given by

$$(x_1+1)x_2 = 0, \quad x_1x_2 = 0,$$

and the variety in S_2 obtained by making these equations homogeneous necessarily has the additional component consisting of the point $(0, 0, 1)$.

This ambiguity in the determination of the variety in S_n corresponding to a variety in A_n will not, however, arise in practice, since we shall always be concerned, in the first instance, with a variety in S_n, and shall choose our prime at infinity to contain no component of this. If we agree to exclude all varieties having components in S_n which lie in the prime at infinity, the correspondence between this restricted class of varieties in S_n and varieties in A_n is one-to-one without exception.

Before proceeding to develop an arithmetic theory of varieties in affine space by means of ideal theory, we make a few preliminary remarks, which may prove helpful, on the choice of coordinate system in the affine space. The results which we seek must not depend on the coordinate system chosen, but proofs may often be simplified by a judicious choice of coordinate system. The reader is aware of many examples of this in elementary geometry. In proving a theorem it will often be convenient to assume that our coordinate system is 'sufficiently general'. This description relates to the problem in question, and implies that we choose our coordinate system so as to avoid some awkward configurations, or to simplify certain calculations. For instance, in dealing with the generic point of a variety of dimension d, we have often found it convenient to suppose that the coordinates can be chosen so that the coordinates $(\xi_0, \ldots, \xi_n)$ of the generic point are normalised with $\xi_0 = 1$, and $\xi_1, \ldots, \xi_d$ are algebraically independent over the ground field. More precisely, if we make a transformation

$$y_i = b_i + \sum_{j=1}^{n} b_{ij}x_j \quad (i = 1, \ldots, n) \tag{4}$$

of the affine coordinates we may, in a given problem, wish to exclude certain choices of coordinates which will arise when the b_i and b_{ij} satisfy any one of a finite set of algebraic relations. A transformation of coordinates in which none of these relations is satisfied is said to be a *sufficiently general* transformation (for the problem in question). If the identity transformation is sufficiently general, then we say that the coordinate system $(x_1, \ldots, x_n)$ is itself a *sufficiently general* coordinate system. A few examples of sufficiently general coordinate systems will make clearer what we mean.

(1) There exists a tangent space at a simple point of a variety of dimension d, given by a set of equations

$$a_i + \sum_{j=1}^{n} a_{ij} x_j = 0 \quad (i = 1, 2, \ldots, n-d),$$

where the matrix $(a_i\ a_{ij})$ is of rank $n-d$. Not all of the $(n-d)$-rowed square submatrices of (a_{ij}) need be of rank $n-d$. If we make a transformation (4) with generic coefficients, all the $(n-d)$-rowed square submatrices of the derived matrix of coefficients of the equation of the tangent space are of rank $n-d$, and when we specialise the transformation, these matrices will have rank less than $n-d$ only if the coefficients in the transformation (4) satisfy certain algebraic relations. In many problems the transformation is sufficiently general if none of these relations is satisfied; or, in other words, the coordinates $(x_1, \ldots, x_n)$ are *sufficiently general for the problem* if every matrix of $n-d$ rows and columns found from (a_{ij}) is of rank $n-d$. Then the tangent space can be written in the form

$$x_{d+i} = c_{d+i} + \sum_{j=1}^{d} c_{ij} x_j \quad (i = 1, \ldots, n-d).$$

The reader may satisfy himself that, in this particular example, the b_i are not restricted in any way, so that if the point in question is rational, the coordinate system may be chosen so that this point has coordinates $(0, \ldots, 0)$, and still be sufficiently general.

(2) Suppose that $(\xi_1, \ldots, \xi_n)$ are the coordinates of a generic point of a variety V' in A_n of dimension d. Let u_{ij} $(i, j = 1, \ldots, n)$ be n^2 indeterminates. If

$$\zeta_i = \sum_{j=1}^{n} u_{ij} \xi_j \quad (i = 1, \ldots, n),$$

we know [X, § 6] that $\zeta_1, \ldots, \zeta_d$ are algebraically independent over $K(u_{ij})$, and that there exists an irreducible algebraic relation

$$f_i(u_{ij}, \zeta_1, \ldots, \zeta_d, \zeta_{d+i}) = 0 \quad (i = 1, \ldots, n-d).$$

If g is the order of V', this relation is of degree g in $\zeta_1, \ldots, \zeta_{d+i}$, and there is present a term ζ_{d+i}^g which has a coefficient which is a non-zero polynomial $a_i(u)$ in the u_{ij}. Now consider the transformation

$$y_i = \sum_{j=1}^{n} b_{ij} x_j \quad (i = 1, \ldots, n).$$

This will be sufficiently general for many problems if $a_i(b) \neq 0$ $(i = 1, \ldots, n-d)$, and $\sum_j b_{ij} \xi_j$ $(i = 1, \ldots, d)$ are algebraically independent over K. Then the coordinate system $(x_1, \ldots, x_n)$ is sufficiently general if $\xi_1, \ldots, \xi_d$ are algebraically independent over K, and $\xi_{d+1}, \ldots, \xi_n$ are integrally dependent on $K[\xi_1, \ldots, \xi_d]$.

(3) If $\xi_1, \ldots, \xi_d$ are algebraically independent over K, we can find elements $a_1, \ldots, a_n$ of K such that $\sum_1^n a_i \xi_i$ is a primitive element of $K(\xi_1, \ldots, \xi_n)$ over $K(\xi_1, \ldots, \xi_d)$. Then there exists a sufficiently general coordinate system in which $\xi_1, \ldots, \xi_d$ are algebraically independent over K and ξ_{d+1} is a primitive element of $K(\xi_1, \ldots, \xi_n)$. The proof may be left to the reader.

(4) In any given problem there is only a finite number of algebraic relations connecting the coefficients of an allowable transformation which must be avoided if the new coordinate system is to be sufficiently general. It is therefore possible simultaneously to combine a finite number of sets of restrictions in order to obtain coordinate systems which are sufficiently general for a number of purposes. A situation which will often arise is that in which we are considering the geometry of an irreducible variety V of dimension d in the neighbourhood of a point. We shall then require

(i) that the tangent space at the point can be written in the form

$$x_{d+i} = c_{d+i} + \sum_{j=1}^{d} c_{ij} x_j \quad (i = 1, \ldots, n-d);$$

(ii) that, if $(\xi_1, \ldots, \xi_n)$ is a generic point, $\xi_1, \ldots, \xi_d$ be algebraically independent over K;

(iii) that ξ_{d+1} be a primitive element of the function field of V over $K(\xi_1, \ldots, \xi_d)$;

(iv) that $\xi_{d+1}, \ldots, \xi_n$ be integrally dependent on $K[\xi_1, \ldots, \xi_d]$.

It is clear from what has been said above that the coordinate system can be chosen sufficiently generally to satisfy all these conditions.

After this explanation of the term 'sufficiently general' it will be possible to leave to the reader the proof that the coordinate system may be chosen sufficiently generally for the purposes of any proof in the sequel.

2. Ideals and varieties in affine space. Let A_n be an affine space of n dimensions defined over the ground field K, and let $(x_1, \ldots, x_n)$ be an allowable coordinate system in A_n. We consider the integral domain $\mathfrak{R} = K[x_1, \ldots, x_n]$. We know [IV, § 2, Th. I] that the Basis Theorem holds in $\mathfrak{R}$. Let $\mathfrak{i}$ be any ideal in $\mathfrak{R}$ and suppose that $f_i(x_1, \ldots, x_n)$ $(i = 1, \ldots, r)$ is a basis for $\mathfrak{i}$. Then the equations

$$f_i(x_1, \ldots, x_n) = 0 \quad (i = 1, \ldots, r) \tag{1}$$

determine an algebraic variety U in A_n, and since any polynomial in $\mathfrak{i}$ is of the form $\Sigma a_i(x_1, \ldots, x_n) f_i(x_1, \ldots, x_n)$, any polynomial in $\mathfrak{i}$ vanishes on U. If $\mathfrak{i}$ is the zero ideal, U is the whole space A_n, and conversely if U is the whole space, the equations (1) must be satisfied by all points of the space, and hence

$$f_i(x_1, \ldots, x_n) \equiv 0 \quad (i = 1, \ldots, r);$$

that is, $\mathfrak{i}$ is the zero ideal. On the other hand, if $\mathfrak{i}$ is the unit ideal, and x' is any point of A_n, there exists a polynomial in $\mathfrak{i}$ which does not vanish at x'. Therefore x' is not on U, and we conclude that U is the empty set of points. Conversely, if U is empty, it follows that equations (1) have no solutions, and hence [IV, § 7, Th. I] there exist polynomials $a_i(x_1, \ldots, x_n)$ such that

$$\Sigma a_i(x_1, \ldots, x_n) f_i(x_1, \ldots, x_n) = 1.$$

Hence $\mathfrak{i}$ contains the unity of $\mathfrak{R}$, and is therefore the unit ideal.

Two different ideals in $\mathfrak{R}$ may determine the same variety U. Let $\mathfrak{j}$ be any other ideal of $\mathfrak{R}$ determining the same variety U as $\mathfrak{i}$, and let $g_i(x_1, \ldots, x_n)$ $(i = 1, \ldots, s)$ be a basis for $\mathfrak{j}$. Since $g_i(x_1, \ldots, x_n)$ vanishes on U, it follows from Hilbert's Zero Theorem that there exists an integer σ_i such that $g_i^{\sigma_i}$ belongs to $\mathfrak{i}$. If $\sigma = \Sigma(\sigma_i - 1) + 1$, it then follows that

$$\mathfrak{j}^{\sigma} = 0 \ (\mathfrak{i}).$$

Similarly, there exists an integer ρ such that

$$\mathfrak{i}^{\rho} = 0 \ (\mathfrak{j}).$$

Conversely, if the two ideals $\mathfrak{i}$ and $\mathfrak{j}$ in $\mathfrak{R}$ are related by the congruences

$$\mathfrak{j}^\sigma = 0 \ (\mathfrak{i}) \quad \text{and} \quad \mathfrak{i}^\rho = 0 \ (\mathfrak{j}),$$

they determine the same variety. For the first congruence tells us that

$$g_i^\sigma(x_1, \ldots, x_n) = \Sigma a_{ij}(x_1, \ldots, x_n) f_j(x_1, \ldots, x_n) \quad (i = 1, \ldots, s),$$

and hence $g_i(x_1, \ldots, x_n)$ vanishes at every point of the variety determined by $\mathfrak{i}$. Therefore the variety determined by $\mathfrak{j}$ contains the variety determined by $\mathfrak{i}$. The second congruence likewise tells us that the variety determined by $\mathfrak{i}$ contains that determined by $\mathfrak{j}$. Hence the two varieties coincide.

The same proof tells us that, if $\mathfrak{i}$ and $\mathfrak{j}$ are two ideals of $\mathfrak{R}$ determining, respectively, varieties U and V, then a necessary and sufficient condition that $U \subseteq V$ is that there exists an integer σ such that $\mathfrak{j}^\sigma \subseteq \mathfrak{i}$.

If, on the other hand, we are given a variety U in A_n, it is obvious that the polynomials in $\mathfrak{R}$ which vanish on U form an ideal. If $\mathfrak{i}^*$ is the ideal determined by U in this way, and $\mathfrak{i}$ is any ideal of $\mathfrak{R}$ which determines U, then $\mathfrak{i} \subseteq \mathfrak{i}^*$, since all polynomials of $\mathfrak{i}$ vanish on U, and hence belong to $\mathfrak{i}^*$. Thus $\mathfrak{i}^*$ is the largest ideal in $\mathfrak{R}$ which determines U, and it contains every other ideal which determines U. Let us now see how we may pass from $\mathfrak{i}$ to $\mathfrak{i}^*$.

Let

$$\mathfrak{i} = [\mathfrak{q}_1, \ldots, \mathfrak{q}_k, \mathfrak{q}_{k+1}, \ldots, \mathfrak{q}_t]$$

be an uncontractible representation of $\mathfrak{i}$ as the intersection of primary ideals, and suppose that $\mathfrak{q}_1, \ldots, \mathfrak{q}_k$ are isolated components, while $\mathfrak{q}_{k+1}, \ldots, \mathfrak{q}_t$ are embedded. Denote the radical of $\mathfrak{q}_i$ by $\mathfrak{p}_i$. If $f \equiv f(x_1, \ldots, x_n)$ is any polynomial which vanishes on U, the variety determined by the principal ideal $\mathfrak{j} = \mathfrak{R}.f$ contains U, and hence there exists an integer ρ such that $\mathfrak{j}^\rho \subseteq \mathfrak{i}$. Therefore

$$f^\rho = 0 \ (\mathfrak{i}),$$

and hence

$$f^\rho \in \mathfrak{q}_i \subseteq \mathfrak{p}_i \quad (i = 1, \ldots, t).$$

Hence

$$f = 0 \ (\mathfrak{p}_i) \quad (i = 1, \ldots, t).$$

Therefore we have

$$f \in [\mathfrak{p}_1, \ldots, \mathfrak{p}_t] = [\mathfrak{p}_1, \ldots, \mathfrak{p}_k],$$

since for each value of a greater than k there exists an integer b not greater than k such that $\mathfrak{p}_b \subseteq \mathfrak{p}_a$.

Conversely, if $f \in [\mathfrak{p}_1, \ldots, \mathfrak{p}_k]$, we have

$$f = 0 \ (\mathfrak{p}_i) \quad (i = 1, \ldots, t),$$

and there exists an integer ρ such that

$$f^\rho = 0 \ (\mathfrak{q}_i) \quad (i = 1, \ldots, t),$$

and therefore

$$f^\rho = 0 \ (\mathfrak{i}).$$

Hence $f(x_1, \ldots, x_n)$ vanishes on U. It follows that

$$\mathfrak{i}^* = [\mathfrak{p}_1, \ldots, \mathfrak{p}_k].$$

It is clear that if $\mathfrak{i}^*$ and $\mathfrak{j}^*$ are the ideals formed by the polynomials which vanish on the varieties U and V, respectively, $U \subseteq V$ if and only if $\mathfrak{j}^* \subseteq \mathfrak{i}^*$.

Let U be a variety in A_n, $\mathfrak{i}^*$ the ideal formed by the polynomials which vanish on U. A necessary and sufficient condition that U be irreducible is that if $f(x_1, \ldots, x_n)$ and $g(x_1, \ldots, x_n)$ are two polynomials whose product vanishes on U, then either $f(x_1, \ldots, x_n)$ or $g(x_1, \ldots, x_n)$ vanishes on U. This is the same as saying that if

$$f(x_1, \ldots, x_n)\, g(x_1, \ldots, x_n) = 0 \ (\mathfrak{i}^*),$$

then either

$$f(x_1, \ldots, x_n) = 0 \ (\mathfrak{i}^*)$$

or

$$g(x_1, \ldots, x_n) = 0 \ (\mathfrak{i}^*);$$

that is, that $\mathfrak{i}^*$ is prime. Hence

THEOREM I. *A necessary and sufficient condition that a variety U in A_n be irreducible is that the ideal of polynomials which vanish on U be prime.*

Next, let $\mathfrak{i}$ be any ideal defining U, and let

$$\mathfrak{i} = [\mathfrak{q}_1, \ldots, \mathfrak{q}_t]$$

be an uncontractible representation of $\mathfrak{i}$ as the intersection of primaries. Then, as we have seen,

$$\mathfrak{i}^* = [\mathfrak{p}_1, \ldots, \mathfrak{p}_k],$$

where $\mathfrak{q}_1, \ldots, \mathfrak{q}_k$ are the isolated components of $\mathfrak{i}$, and $\mathfrak{p}_i$ is the radical of $\mathfrak{q}_i$. Since U is irreducible, $\mathfrak{i}^*$ is prime. Hence $k = 1$. Thus we have

THEOREM II. *A necessary and sufficient condition that the variety defined by an ideal $\mathfrak{i}$ in $\mathfrak{R}$ be irreducible is that all the components of $\mathfrak{i}$ except one be embedded.*

Let us now consider the relationship between joins and intersections of ideals and the joins and intersections of the varieties which they define. Let $\mathfrak{i}$ be an ideal in $\mathfrak{R}$, $[\mathfrak{q}_1, \ldots, \mathfrak{q}_t]$ an uncontractible representation of $\mathfrak{i}$, where $\mathfrak{q}_1, \ldots, \mathfrak{q}_k$ are the isolated components of $\mathfrak{i}$ and $\mathfrak{p}_i$ is the radical of $\mathfrak{q}_i$. If $\mathfrak{i}$ defines the variety U, U is also defined by the ideal $\mathfrak{i}^* = [\mathfrak{p}_1, \ldots, \mathfrak{p}_k]$. Let U_i be the irreducible variety defined by $\mathfrak{p}_i$. A necessary and sufficient condition that $f(x_1, \ldots, x_n)$ should vanish on U is that it belong to $\mathfrak{p}_1, \mathfrak{p}_2, \ldots, \mathfrak{p}_k$ and hence that it vanish on $U_1 \dotplus \ldots \dotplus U_k$. Therefore

$$U = U_1 \dotplus U_2 \dotplus \ldots \dotplus U_k.$$

If $\mathfrak{i} = [\mathfrak{q}_1, \ldots, \mathfrak{q}_t]$ is an ideal defining the variety U, and

$$\mathfrak{j} = [\mathfrak{q}'_1, \ldots, \mathfrak{q}'_s]$$

is an ideal defining the variety V, then, since

$$[\mathfrak{i}, \mathfrak{j}] = [\mathfrak{q}_1, \ldots, \mathfrak{q}_t, \mathfrak{q}'_1, \ldots, \mathfrak{q}'_s],$$

it follows that $[\mathfrak{i}, \mathfrak{j}]$ defines the variety $U \dotplus V$.

Let $f_i(x_1, \ldots, x_n)$ $(i = 1, \ldots, r)$ be a basis for $\mathfrak{i}$, and let $g_i(x_1, \ldots, x_n)$ $(i = 1, \ldots, s)$ be a basis for $\mathfrak{j}$. Then

$$f_i(x_1, \ldots, x_n)\, g_j(x_1, \ldots, x_n) \quad (i = 1, \ldots, r;\ j = 1, \ldots, s)$$

is a basis for $\mathfrak{i}\mathfrak{j}$. Hence it follows that $\mathfrak{i}\mathfrak{j}$ also defines $U \dotplus V$.

Using the same notation, we see that the set of $r+s$ polynomials $f_i(x_1, \ldots, x_n)$ $(i = 1, \ldots, r)$ and $g_i(x_1, \ldots, x_n)$ $(i = 1, \ldots, s)$ is a basis for $(\mathfrak{i}, \mathfrak{j})$. Hence $(\mathfrak{i}, \mathfrak{j})$ determines the intersection $U_\wedge V$.

If $\mathfrak{R}$ is any ring which contains a field K, and $\mathfrak{i}$ is any ideal in $\mathfrak{R}$, $\mathfrak{i}$ is said to be *of dimension* r over K if (i) there exist r elements $\xi_1, \ldots, \xi_r$ of $\mathfrak{R}$ such that if $f(x_1, \ldots, x_r)$ is a polynomial in $K[x_1, \ldots, x_r]$ with the property that

$$f(\xi_1, \ldots, \xi_r) = 0 \ (\mathfrak{i}),$$

then

$$f(x_1, \ldots, x_r) \equiv 0;$$

and (ii) if $\xi_1, \ldots, \xi_{r+1}$ are *any* $r+1$ elements of $\mathfrak{R}$, then there exists a non-zero polynomial $f(x_1, \ldots, x_{r+1})$ with coefficients in K such that

$$f(\xi_1, \ldots, \xi_{r+1}) = 0 \ (\mathfrak{i}).$$

We apply this idea to the integral domain $\mathfrak{R} = K[x_1, \ldots, x_n]$ which we have been considering in connection with the affine space A_n. Let $\mathfrak{i}$ be any ideal in $\mathfrak{R}$, defining a variety U, and let $[\mathfrak{q}_1, \ldots, \mathfrak{q}_t]$ be an uncontractible representation of $\mathfrak{i}$ as the intersection of primary ideals, $\mathfrak{q}_1, \ldots, \mathfrak{q}_k$ being the isolated components, and $\mathfrak{p}_i$ denoting the radical of $\mathfrak{q}_i$. If $\xi_1, \ldots, \xi_r$ denote elements of $\mathfrak{R}$, and $f(x_1, \ldots, x_r)$ is any polynomial over K, the relation

$$f(\xi_1, \ldots, \xi_r) = 0 \ (\mathfrak{i})$$

implies $$f(\xi_1, \ldots, \xi_r) \in \mathfrak{q}_i \subseteq \mathfrak{p}_i \quad (i = 1, \ldots, k).$$

Conversely, if there exist polynomials $f_i(x_1, \ldots, x_r)$ $(i = 1, \ldots, k)$ such that

$$f_i(\xi_1, \ldots, \xi_r) \subseteq \mathfrak{p}_i \quad (i = 1, \ldots, k),$$

then there exist polynomials $f_i(x_1, \ldots, x_r)$ $(i = 1, \ldots, t)$ such that

$$f_i(\xi_1, \ldots, \xi_r) \subseteq \mathfrak{p}_i \quad (i = 1, \ldots, t),$$

since for any j greater than k, there exists a value of i not exceeding k such that $\mathfrak{p}_i \subseteq \mathfrak{p}_j$, and for $f_j(x_1, \ldots, x_r)$ we need only take the corresponding $f_i(x_1, \ldots, x_r)$. If ρ_i is the index of $\mathfrak{q}_i$, and

$$f(x_1, \ldots, x_r) = \prod_{i=1}^{t} f_i^{\rho_i}(x_1, \ldots, x_r),$$

then $$f(\xi_1, \ldots, \xi_r) = 0 \ (\mathfrak{i}).$$

From this we prove that the dimension of $\mathfrak{i}$ is equal to the greatest of the dimensions of $\mathfrak{p}_1, \ldots, \mathfrak{p}_k$. Let r be the dimension of $\mathfrak{i}$, r_i that of $\mathfrak{p}_i$. Then any $r+1$ elements of $\mathfrak{R}$ are algebraically dependent modulo $\mathfrak{i}$, and hence are algebraically dependent modulo $\mathfrak{p}_i$. Hence

$$r \geqslant \max [r_1, \ldots, r_k].$$

If, on the other hand, $s > r_i$ $(i = 1, \ldots, k)$, any s elements of $\mathfrak{R}$ are algebraically dependent modulo $\mathfrak{p}_i$ $(i = 1, \ldots, k)$, and are therefore algebraically dependent modulo $\mathfrak{i}$. Hence $r \leqslant \max [r_1, \ldots, r_k]$, and therefore

$$r = \max [r_1, \ldots, r_k].$$

Now let $\mathfrak{p}$ be a prime ideal, formed by the polynomials which vanish on an irreducible variety U. Let $(\eta_1, \ldots, \eta_n)$ be a generic point of U. If U is of dimension d, as defined in Chapter X, there exist d of the η_i, say $\eta_{i_1}, \ldots, \eta_{i_d}$, which are algebraically independent over K. Then if $f(x_{i_1}, \ldots, x_{i_d}) \in \mathfrak{p}$,

$$f(\eta_{i_1}, \ldots, \eta_{i_d}) = 0,$$

and hence $$f(x_{i_1}, \ldots, x_{i_d}) \equiv 0.$$

Therefore the dimension of $\mathfrak{p}$ is not less than d. On the other hand, if $a_i(x)$ $(i = 1, \ldots, d+1)$ are any $d+1$ elements of $\mathfrak{R}$, there exists a non-zero polynomial $F(y_1, \ldots, y_{d+1})$ such that

$$F(a_1(\eta), \ldots, a_{d+1}(\eta)) = 0.$$

Hence $$F(a_1(x), \ldots, a_{d+1}(x)) = 0 \ (\mathfrak{p}).$$

It follows that the dimension of $\mathfrak{p}$ is equal to the dimension of U.

We conclude that the dimension of any ideal $\mathfrak{i}$ in $\mathfrak{R}$ is equal to the dimension of the components of greatest dimension of the variety which it defines, that is, it is equal to the dimension of the variety it defines. Thus for our purposes the dimension of an ideal is not a new concept, but we shall sometimes find it useful to use the ideal-theoretic method of defining the dimension of a variety.

The foregoing remarks are preliminary to the study of the geometry of an irreducible variety. Let V be an irreducible variety in A_n. This may be the whole space A_n, but we shall assume that it is not vacuous.

Let $\mathfrak{P}$ be the prime ideal formed by the polynomials in

$$\mathfrak{R} = K[x_1, \ldots, x_n]$$

which vanish on V. We consider the remainder-class ring $\mathfrak{R}/\mathfrak{P}$, which, since $\mathfrak{P}$ is prime, is an integral domain [XV, § 3, p. 30], and in which the Basis Theorem holds, since it holds in $\mathfrak{R}$ [XV, § 3, Th. VII]. Let a and b be any two elements of K, and let $\bar{a}$ and $\bar{b}$ be the elements of $\mathfrak{R}/\mathfrak{P}$ on which they map. Then $\bar{a} = \bar{b}$ if and only if $a-b$ is in $\mathfrak{P}$, and since $\mathfrak{P}$ is not the unit ideal, this can only happen when $a = b$. Hence $\mathfrak{R}/\mathfrak{P}$ contains a subring isomorphic with K. We may identify this with K, and regard $\mathfrak{R}/\mathfrak{P}$ as an extension of K. We then denote it by $\mathfrak{J}$. Let ξ_i be the element of $\mathfrak{J}$ which corresponds to the element x_i of $\mathfrak{R}$. If $f(x_1, \ldots, x_n)$ is any polynomial in $\mathfrak{R}$, it maps on $f(\xi_1, \ldots, \xi_n)$ of $\mathfrak{J}$, and $f(x_1, \ldots, x_n)$ belongs to $\mathfrak{P}$ if and only if $f(\xi_1, \ldots, \xi_n) = 0$. Hence $(\xi_1, \ldots, \xi_n)$ is a generic point of V. Moreover, $\mathfrak{J} = K[\xi_1, \ldots, \xi_n]$, the ring of polynomials in $\xi_1, \ldots, \xi_n$ with coefficients in K. If $(\eta_1, \ldots, \eta_n)$ is any generic point of V, $K[\xi_1, \ldots, \xi_n]$ and $K[\eta_1, \ldots, \eta_n]$ are equivalent extensions of K in which ξ_i and η_i correspond. The ring $\mathfrak{J}$ is thus uniquely determined to within isomorphism by a generic point of V. $\mathfrak{J}$ (or any isomorph of it) is called the *integral domain of V in A_n*.

The integral domain $\mathfrak{I}$ does not depend on the allowable coordinate system chosen in A_n. Let $(x'_1, \ldots, x'_n)$ be any other allowable coordinate system. Then

$$x'_i = b_i + \Sigma b_{ij} x_j \quad (i = 1, \ldots, n),$$

and

$$x_i = c_i + \Sigma c_{ij} x'_j \quad (i = 1, \ldots, n),$$

where b_i, b_{ij}, c_i, c_{ij} are in K. If $(\xi_1, \ldots, \xi_n)$ are the coordinates of a generic point of V in the original system, the coordinates $(\xi'_1, \ldots, \xi'_n)$ of this point in the new coordinate system satisfy the equations

$$\xi'_i = b_i + \Sigma b_{ij} \xi_j \quad (i = 1, \ldots, n),$$

and

$$\xi_i = c_i + \Sigma c_{ij} \xi'_j \quad (i = 1, \ldots, n).$$

From these relations it follows that

$$K[\xi_1, \ldots, \xi_n] = K[\xi'_1, \ldots, \xi'_n].$$

In Chapter XV, § 3, we considered the relations between ideals in a ring and ideals in a remainder-class ring. We apply this to the ideals in $\mathfrak{R}$ and $\mathfrak{I}$. To each ideal $\mathfrak{K}$ in $\mathfrak{R}$ there corresponds a unique ideal $\mathfrak{k}$ in $\mathfrak{I}$, and the largest ideal in $\mathfrak{R}$ which corresponds to $\mathfrak{k}$ is $(\mathfrak{K}, \mathfrak{P})$. To each ideal $\mathfrak{k}$ of $\mathfrak{I}$ there corresponds one ideal $\mathfrak{K}_1$ of $\mathfrak{R}$ which is a factor of $\mathfrak{P}$, and any ideal of $\mathfrak{R}$ which corresponds to $\mathfrak{k}$ is a multiple of this. If $\mathfrak{K}$ is any ideal of $\mathfrak{R}$ defining a variety U of A_n, $(\mathfrak{K}, \mathfrak{P})$ defines $U_\wedge V$. The variety $U_\wedge V$ is vacuous if and only if $(\mathfrak{K}, \mathfrak{P})$ is the unit ideal, that is, if and only if the corresponding ideal $\mathfrak{k}$ of $\mathfrak{I}$ is the unit ideal. Thus any ideal $\mathfrak{i}$ in $\mathfrak{I}$ determines a variety contained in V, as follows. If $f_i(x_1, \ldots, x_n)$ $(i = 1, \ldots, r)$ is a basis for $\mathfrak{P}$, and $g_i(\xi_1, \ldots, \xi_n)$ $(i = 1, \ldots, s)$ is a basis for $\mathfrak{i}$, then $f_i(x_1, \ldots, x_n)$ $(i = 1, \ldots, r)$ and $g_i(x_1, \ldots, x_n)$ $(i = 1, \ldots, s)$ is a basis for the corresponding ideal of $\mathfrak{R}$ which is a factor of $\mathfrak{P}$. The variety determined by $\mathfrak{i}$ has the equations

$$\left.\begin{aligned} f_i(x_1, \ldots, x_n) &= 0 \quad (i = 1, \ldots, r), \\ g_j(x_1, \ldots, x_n) &= 0 \quad (j = 1, \ldots, s). \end{aligned}\right\}$$

If, conversely, U is any variety on V, and $g_i(x_1, \ldots, x_n)$ $(i = 1, \ldots, s)$ is a basis for the ideal of polynomials in $\mathfrak{R}$ which vanish on U, then $g_i(\xi_1, \ldots, \xi_n)$ $(i = 1, \ldots, s)$ is a basis for the largest ideal in $\mathfrak{I}$ which determines U.

The results proved at the beginning of this section concerning the relation between ideals in $\mathfrak{R}$ and varieties in A_n can now be translated at once into results relating ideals in $\mathfrak{J}$ with varieties on V. Indeed, the earlier results correspond to the special case in which $\mathfrak{P}$ is the zero ideal. It is not necessary to give the details of this translation, which follow at once from the results of Chapter XV, § 3. The most important result is that if U is any variety on V, U is irreducible if and only if the largest ideal of $\mathfrak{J}$ which defines U is prime. We note that if $\mathfrak{k}$ is any ideal of $\mathfrak{J}$, and if $\mathfrak{K}$ is the ideal of $\mathfrak{R}$ corresponding to it which is a factor of $\mathfrak{P}$, the dimension of $\mathfrak{k}$ over K, as an ideal in $\mathfrak{J}$, is equal to the maximum number of elements of $\mathfrak{J}$ which are algebraically independent modulo $\mathfrak{k}$, and hence to the maximum number of elements of $\mathfrak{R}$ which are algebraically independent modulo $\mathfrak{K}$; that is, the dimension of $\mathfrak{k}$ is equal to the dimension of $\mathfrak{K}$. Hence the dimension of $\mathfrak{k}$ over K is equal to the dimension of the variety defined by $\mathfrak{k}$.

Since $\mathfrak{J}$ is an integral domain, its maximal ideals are all prime [XV, § 3, Th. III]. Let $\mathfrak{p}$ be a maximal ideal of $\mathfrak{J}$, U the irreducible variety defined by it. If U is of dimension greater than zero, there exists an irreducible variety U' properly contained in it. If $\mathfrak{p}'$ is the largest ideal defining U', $\mathfrak{p}'$ is not the unit ideal, since there are points on U', and from the relation $U' \subset U$, we have $\mathfrak{p} \subset \mathfrak{p}'$, which contradicts the fact that $\mathfrak{p}$ is maximal in $\mathfrak{J}$. Hence U is of dimension zero. Conversely, if $\mathfrak{p}$ is the prime ideal defining a variety U of dimension zero, $\mathfrak{p}$ must be maximal. For if it were not, there would exist a prime ideal $\mathfrak{p}'$ different from the unit ideal such that $\mathfrak{p} \subset \mathfrak{p}'$. Then the variety defined by $\mathfrak{p}'$ is properly contained in U, which is impossible since U is of dimension zero. Hence we have

THEOREM III. *The maximal ideals of $\mathfrak{J}$ correspond to the irreducible varieties of dimension zero on V.*

Let $\mathfrak{p}$ be any prime ideal in $\mathfrak{J}$, and let U be the irreducible variety on V defined by it. The quotient ring $\mathfrak{J}_{\mathfrak{p}}$ of $\mathfrak{p}$ is called the quotient ring of U, and is often denoted by $Q(U)$. By XV, § 5, Th. X, $\mathfrak{J}$ is the intersection of the quotient rings of the irreducible varieties U contained in V, and, indeed, it is the intersection of the quotient rings of the irreducible varieties of dimension zero contained in V, since the irreducible varieties of dimension zero are defined by the maximal ideals of $\mathfrak{J}$.

Let U_1 and U_2 be two irreducible varieties contained in V, and let $\mathfrak{p}_1$ and $\mathfrak{p}_2$ be the corresponding prime ideals of $\mathfrak{J}$. Then, as we have seen, $U_1 \subseteq U_2$ if and only if $\mathfrak{p}_2 \subseteq \mathfrak{p}_1$. But [XV, § 5, Th. VI]

$\mathfrak{p}_2 \subseteq \mathfrak{p}_1$ if and only if $\mathfrak{J}_{\mathfrak{p}_1} \subseteq \mathfrak{J}_{\mathfrak{p}_2}$. Hence we conclude that U_1 is contained in U_2 if and only if $Q(U_1)$ is contained in $Q(U_2)$. In particular, U_1 is a proper subvariety of U_2 ($U_1 \subset U_2$) if and only if $Q(U_1)$ is a proper subring of $Q(U_2)$, for in this case U_2 is not contained in U_1, and hence $Q(U_2)$ is not contained in $Q(U_1)$. Thus we have

THEOREM IV. *If U_1 and U_2 are irreducible varieties contained in V,*
(i) *$U_1 \subseteq U_2$ if and only if $Q(U_1) \subseteq Q(U_2)$;*
(ii) *$U_1 \subset U_2$ if and only if $Q(U_1) \subset Q(U_2)$.*

It is to be recalled that the results of this section refer to varieties in affine space. If V is derived from a variety $\bar{V}$ in projective space, the integral domain $\mathfrak{J}$ depends on the choice of the prime at infinity. For example, if $\bar{V}$ is the projective space $(x_0, \ldots, x_n)$, and we choose $x_0 = 0$ to be the prime at infinity, we take the generic point of $\bar{V}$ to be $(1, \xi_1, \ldots, \xi_n)$, where $\xi_1, \ldots, \xi_n$ are independent indeterminates over K, and $\mathfrak{J} = K[\xi_1, \ldots, \xi_n]$. Using the same function field for $\bar{V}$, we must take $(\xi_n^{-1}, \xi_n^{-1}\xi_1, \ldots, \xi_n^{-1}\xi_{n-1}, 1)$ as the generic point when $x_n = 0$ is taken as the prime at infinity, and the integral domain for the corresponding affine variety is $K[\xi_n^{-1}, \xi_n^{-1}\xi_1, \ldots, \xi_n^{-1}\xi_{n-1}]$, which is different from $\mathfrak{J}$. It should also be observed that in the theorem which says that $\mathfrak{J}$ is the intersection of the quotient rings of the irreducible subvarieties of V of dimension zero, we only consider subvarieties in the affine space.

However, the quotient ring of an irreducible subvariety of V is a projective concept. Let $\bar{V}$ be an irreducible variety, not lying in $x_0 = 0$, in a projective space S_n with the coordinate system $(x_0, \ldots, x_n)$ and let V be the corresponding variety in the affine space A_n obtained from S_n by taking $x_0 = 0$ to be the prime at infinity. We recall how we obtained the function field of $\bar{V}$ in Chapter X, § 3. We considered the ratios

$$\frac{f(x)}{g(x)} = \frac{f(x_0, \ldots, x_n)}{g(x_0, \ldots, x_n)}$$

of homogeneous polynomials of like degree in $K[x_0, \ldots, x_n]$, where $g(x_0, \ldots, x_n)$ does not vanish on $\bar{V}$, and we defined the ratios $f(x)/g(x)$ and $f'(x)/g'(x)$ to be equivalent if

$$f(x)\,g'(x) - f'(x)\,g(x) = 0$$

on $\bar{V}$. By suitably defining addition and multiplication for classes of equivalent ratios, we obtained a field isomorphic with the function field of $\bar{V}$. Let Σ denote an extension of K isomorphic with the field of equivalent classes. Now let $\bar{U}$ be an irreducible variety on $\bar{V}$ not lying in $x_0 = 0$, and let U be the corresponding variety in A_n. We consider the elements of Σ which correspond to ratios $f(x)/g(x)$ in which $g(x)$ does not vanish on $\bar{U}$. If $f(x)/g(x)$ and $f'(x)/g'(x)$ are two ratios whose denominators do not vanish on $\bar{U}$, then

$$\frac{f(x)}{g(x)}+\frac{f'(x)}{g'(x)}=\frac{f(x)\,g'(x)+f'(x)\,g(x)}{g(x)\,g'(x)}$$

and

$$\frac{f(x)}{g(x)}\cdot\frac{f'(x)}{g'(x)}=\frac{f(x)\,f'(x)}{g(x)\,g'(x)}$$

are ratios whose denominators do not vanish on $\bar{U}$. Hence the corresponding elements of Σ form a ring $\mathfrak{R}$. It is clear that $\mathfrak{R}$ contains K, and hence has unity, and since Σ has no divisors of zero, $\mathfrak{R}$ has none. Therefore $\mathfrak{R}$ is an integral domain. Moreover, Σ is the quotient field of $\mathfrak{R}$.

Let ξ_i be the element of $\mathfrak{R}$ (and hence of Σ) which corresponds to the class x_i/x_0 $(i = 1, \ldots, n)$. Then $(1, \xi_1, \ldots, \xi_n)$ is a generic point of $\bar{V}$, and $(\xi_1, \ldots, \xi_n)$ is a generic point of V. The ring $\mathfrak{J} = K[\xi_1, \ldots, \xi_n]$ is the integral domain of V. Let $\mathfrak{p}$ be the prime ideal of $\mathfrak{J}$ corresponding to U. Any element ζ of Σ is of the form

$$\zeta = \frac{f(1, \xi_1, \ldots, \xi_n)}{g(1, \xi_1, \ldots, \xi_n)},$$

where $f(x_0, \ldots, x_n)$ and $g(x_0, \ldots, x_n)$ are homogeneous polynomials over K of the same degree, the second not vanishing on $\bar{V}$. ζ belongs to $\mathfrak{J}_\mathfrak{p}$ if and only if we can find $f(x)$ and $g(x)$ such that $g(1, \xi_1, \ldots, \xi_n)$ does not belong to $\mathfrak{p}$. But, by the definition of $\mathfrak{p}$, this is simply the condition that $g(1, x_1, \ldots, x_n)$ does not vanish on U, that is, that $g(x_0, \ldots, x_n)$ does not vanish on $\bar{U}$. Hence ζ belongs to $\mathfrak{J}_\mathfrak{p}$ if and only if it is contained in $\mathfrak{R}$. Thus $\mathfrak{J}_\mathfrak{p} = \mathfrak{R}$.

The quotient ring of U can be defined without reference to the function field of V, simply by considering only the ratios $f(x)/g(x)$ in which $g(x)$ does not vanish on U, and defining equivalent ratios as before. Introducing addition and multiplication of classes of equivalent ratios in the usual manner, we then show that these classes form an integral domain, which is, by what we have seen, isomorphic with the quotient ring of U. But as we usually have to consider the quotient rings of several subvarieties of V simultaneously, we must consider them as subrings of the same representation of the function field of V.

These considerations show that the quotient ring of an irreducible subvariety of $\bar{V}$, a variety in projective space, is determined by the function field of $\bar{V}$ and the subvariety and does not depend on the selection of a prime at infinity. For this reason we shall try, in the following sections, to express the properties of varieties in affine space in terms of quotient rings, so that they may be interpreted as properties of varieties in projective space.

We conclude this section with some remarks on the problem of finding a basis for the ideal determined by an irreducible variety on V. This is not a problem which arises much in practice, since in general it is sufficient to know that a basis exists, and this we know from the Basis Theorem, but we give here one or two results which will be useful later.

Let U be an irreducible subvariety of V, of e dimensions, and let $F(u_0, \ldots, u_e)$ be its Cayley form (that is, the Cayley form of U when A_n is 'completed' to form a projective space S_n). Then if $S^0, \ldots, S^e$ are skew-symmetric matrices of $n+1$ rows and columns whose elements are independent indeterminates over K, the equations of U are obtained by equating to zero the coefficients of the independent power-products of the s^i_{jk} in

$$F(S^0x, \ldots, S^ex),$$

where x stands for $(1, x_1, \ldots, x_n)$, written as a column vector. It follows that if $\xi = (1, \xi_1, \ldots, \xi_n)$, the coefficients of the power-products of s^i_{jk} in

$$F(S^0\xi, \ldots, S^e\xi)$$

generate an ideal $\mathfrak{i}$ in $\mathfrak{J}$, which determines the variety U. Let $\mathfrak{i} = [\mathfrak{q}_1, \ldots, \mathfrak{q}_r]$ be a representation of $\mathfrak{i}$ as an uncontractible intersection of primary ideals; and let $\mathfrak{p}_i$ be the radical of $\mathfrak{q}_i$, U_i the irreducible variety on V defined by $\mathfrak{p}_i$. Then

$$U = U_1 \dotplus \ldots \dotplus U_r.$$

Since U is irreducible, one of the components U_i, say U_1, must coincide with U, and the others must be embedded. Hence $\mathfrak{p}_1$ is the prime ideal of U, and the components $\mathfrak{q}_2, \ldots, \mathfrak{q}_r$ of $\mathfrak{i}$ are embedded.

There is one case, which will be of importance later, in which we can go further. This is the case in which U is an irreducible subvariety of V, not consisting entirely of multiple points of V, of dimension $d-1$, d being the dimension of V. In this case, we can show that $\mathfrak{q}_1$ is a prime ideal, and hence the prime ideal of U. We prove

THEOREM V. *Let V be an irreducible variety of dimension d in affine space, and let $\mathfrak{J}$ be its integral domain. Let U be an irreducible subvariety of V of dimension $d-1$ which is simple on V, and let $F(u_0, \ldots, u_{d-1})$ be its Cayley form. Then if $(\xi_1, \ldots, \xi_n)$ $(\xi_i \in \mathfrak{J})$ is a generic point of V, and S^i $(i = 0, \ldots, d-1)$ are d independent indeterminate skew-symmetric matrices, the prime ideal of U in $\mathfrak{J}$ is an isolated component of the ideal generated by the coefficients of the power-products of s^i_{jk} in $F(S^0\xi, \ldots, S^{d-1}\xi)$.*

It will be convenient to revert to the projective space S_n derived from A_n; V is then the variety whose generic point is $(1, \xi_1, \ldots, \xi_n)$ in the coordinate system $(x_0, \ldots, x_n)$.

Let $W(x) = W(x_0, \ldots, x_n)$ be any homogeneous form with indeterminate coefficients $w_1, \ldots, w_t$. We first show how we may obtain the Cayley form of the intersection of V with the primal $W(x) = 0$.

Let
$$f_i(x) = 0 \quad (i = 1, \ldots, s) \tag{2}$$
be the homogeneous equations of V, and let
$$\sum_{j=0}^{n} u_{ij} x_j = 0 \quad (i = 0, \ldots, d-1) \tag{3}$$
be d independent generic primes in S_n. The equations (2) and (3) have a finite number of solutions $\xi^{(\nu)} = (1, \xi_1^{(\nu)}, \ldots, \xi_n^{(\nu)})$ $(\nu = 1, \ldots, g)$. These g points are conjugate over $K(u_0, \ldots, u_{d-1})$, and each of them is a generic point of V over K [X, §7]. From this we draw the following conclusions:

(i) $\prod_{\nu=1}^{g} W(\xi^{(\nu)})$ is irreducible as a form in $w_1, \ldots, w_t$, with coefficients in $K(u_0, \ldots, u_{d-1})$;

(ii) $\prod_{\nu=2}^{g} W(\xi^{(\nu)})$ is a form in $w_1, \ldots, w_t$, with coefficients in
$$K(u_0, \ldots, u_{d-1}; \xi_1^{(1)}, \ldots, \xi_n^{(1)}).$$

On account of (i), we can write
$$\prod_{\nu=1}^{g} W(\xi^{(\nu)}) = \frac{A(u_0, \ldots, u_{d-1})}{H(u_0, \ldots, u_{d-1})} B(w, u_0, \ldots, u_{d-1}),$$
where $A(u_0, \ldots, u_{d-1})$ and $H(u_0, \ldots, u_{d-1})$ are in $K[u_0, \ldots, u_{d-1}]$, and may be assumed to be without common factor, and $B(w, u_0, \ldots, u_{d-1})$ is an irreducible polynomial in $K[w, u_0, \ldots, u_{d-1}]$.

For any specialisation $W'(x)$ of $W(x)$, the equations (2) and (3) and

$$W'(x) = 0$$

have a solution if and only if $B(w', u_0, \ldots, u_{d-1})$ is zero. Hence $B(w, u_0, \ldots, u_{d-1})$ is the u_0-resultant of equations (2) and the equations

$$\sum_{j=0}^{n} u_{ij} x_j = 0 \quad (i = 1, \ldots, d-1),$$

and

$$W(x) = 0.$$

Hence [vol. II, p. 41] it is the Cayley form of the intersection of V with $W(x) = 0$.

Let $\mathfrak{p}$ be the prime ideal of U in $\mathfrak{J}$, and let $W'(1, \xi_1, \ldots, \xi_n) = W'(\xi)$ be any non-zero element of $\mathfrak{p}$, where $W'(x_0, \ldots, x_n)$ is homogeneous. Then

$$B(w', u_0, \ldots, u_{d-1}) = F(u_0, \ldots, u_{d-1}) F'(u_0, \ldots, u_{d-1}),$$

where $F(u_0, \ldots, u_{d-1})$ is the Cayley form of U, and $F'(u_0, \ldots, u_{d-1})$ is also a Cayley form, namely, the Cayley form of the variety which is the residual intersection of V and $W'(x) = 0$, possibly together with U itself. Hence we have

$$H(u_0, \ldots, u_{d-1}) \prod_{\nu=1}^{g} W'(\xi^{(\nu)})$$
$$= A(u_0, \ldots, u_{d-1}) F'(u_0, \ldots, u_{d-1}) F(u_0, \ldots, u_{d-1}).$$

Using (ii) above, we can write

$$\prod_{\nu=2}^{g} W'(\xi^{(\nu)}) = \frac{C(u_0, \ldots, u_{d-1}, \xi^{(1)})}{C'(u_0, \ldots, u_{d-1}, \xi^{(1)})},$$

where $C(u_0, \ldots, u_{d-1}, x)$ and $C'(u_0, \ldots, u_{d-1}, x)$ are homogeneous in $x_0, \ldots, x_n$, with coefficients in $K[u_0, \ldots, u_{d-1}]$. Then we have

$$H(u_0, \ldots, u_{d-1}) C(u_0, \ldots, u_{d-1}, \xi^{(1)}) W'(\xi^{(1)})$$
$$= A(u_0, \ldots, u_{d-1}) C'(u_0, \ldots, u_{d-1}, \xi^{(1)}) F'(u_0, \ldots, u_{d-1}) F(u_0, \ldots, u_{d-1}).$$

Now $\xi^{(1)}$ is a generic point of V, and

$$\sum_{j=0}^{n} u_{ij} x_j = 0$$

is a generic prime through it. But ξ is also a generic point of V, and d independent generic primes through it can be written in the form

$$\sum_{j,k} s^i_{jk} \xi_j x_k = 0 \quad (i = 0, \ldots, d-1),$$

where $S^0, \ldots, S^{d-1}$ are d independent indeterminate skew-symmetric matrices. Hence

$$H(S^0\xi, \ldots, S^{d-1}\xi)\, C(S^0\xi, \ldots, S^{d-1}\xi, \xi)\, W'(\xi)$$
$$= A(S^0\xi, \ldots, S^{d-1}\xi)\, C'(S^0\xi, \ldots, S^{d-1}\xi, \xi)\, F'(S^0\xi, \ldots, S^{d-1}\xi)$$
$$\times F(S^0\xi, \ldots, S^{d-1}\xi). \quad (4)$$

Over the field $K(S^0, \ldots, S^{d-1})$, V is irreducible [X, § 5, Th. VI], and a point x' of U which is simple for V over K is also simple for V over $K(S^0, \ldots, S^{d-1})$. Now

$$\zeta = \frac{C(S^0\xi, \ldots, S^{d-1}\xi, \xi)}{C'(S^0\xi, \ldots, S^{d-1}\xi, \xi)} = \prod_{\nu=2}^{g} W'(\eta^{(\nu)}),$$

where $\xi, \eta^{(2)}, \ldots, \eta^{(g)}$ are the g distinct points of V which lie in the $(n-d)$-space

$$\sum_{j,k} s^i_{jk}\xi_j x_k = 0 \quad (i = 0, \ldots, d-1).$$

If we specialise ξ to x', a point of U simple for V,

$$(\xi, \eta^{(2)}, \ldots, \eta^{(g)}) \rightarrow (x', y^{(2)}, \ldots, y^{(g)}),$$

where $x', y^{(2)}, \ldots, y^{(g)}$ are all distinct and at a finite distance. It follows that as $\xi \rightarrow x'$, ζ has a unique specialisation, different from zero, since a generic $(n-d)$-space over K through x' does not meet $W'(x) = 0$ on V, except at x'. By a result to be proved later [XVIII, § 1, Th. VII, Cor.], this implies that ζ is a unit of the quotient ring of the variety of which x' is a generic point. Hence, without loss of generality we may assume that

$$C(S^0\xi, \ldots, S^{d-1}\xi, \xi) \quad \text{and} \quad C'(S^0\xi, \ldots, S^{d-1}\xi, \xi)$$

do not vanish for the specialisation $\xi \rightarrow x'$, and therefore are not in $K(S^0, \ldots, S^{d-1}).\mathfrak{p}$, where $\mathfrak{p}$ is the prime ideal of U.

Again, $H(u_0, \ldots, u_{d-1})$ vanishes only when the $(n-d)$-space

$$\sum_j u_{ij} x_j = 0 \quad (i = 0, \ldots, d-1)$$

has an intersection with V at infinity. Hence $H(S^0x', \ldots, S^{d-1}x') \neq 0$, and therefore

$$H(S^0\xi, \ldots, S^{d-1}\xi) \neq 0 \quad (K(S^0, \ldots, S^{d-1}).\mathfrak{p}).$$

We can then determine specialisations $T^0, \ldots, T^{d-1}$ of $S^0, \ldots, S^{d-1}$ in K so that

$$[C(S^0\xi, \ldots, S^{d-1}\xi, \xi),\, C'(S^0\xi, \ldots, S^{d-1}\xi, \xi),\, H(S^0\xi, \ldots, S^{d-1}\xi)]$$
$$\rightarrow [C(\xi), C'(\xi), H(\xi)],$$

where $C(\xi)$, $C'(\xi)$, $H(\xi)$ are not contained in $\mathfrak{p}$.

Let $\mathfrak{i}$ be the ideal generated by the coefficients of $F(S^0\xi, \ldots, S^{d-1}\xi)$. Then, as we saw above,

$$\mathfrak{i} = [\mathfrak{q}_1, \mathfrak{q}_2, \ldots, \mathfrak{q}_r],$$

where $\mathfrak{q}_1$ has $\mathfrak{p}$ as radical, and $\mathfrak{q}_2, \ldots, \mathfrak{q}_r$ are embedded components. When we specialise $S^0, \ldots, S^{d-1}$ to $T^0, \ldots, T^{d-1}$, we conclude from (4) that

$$H(\xi)\, C(\xi)\, W'(\xi) \subseteq \mathfrak{i} \subseteq \mathfrak{q}_1,$$

and since $H(\xi)$ and $C(\xi)$ are not in the radical of $\mathfrak{q}_1$ we have

$$W'(\xi) = 0 \ \ (\mathfrak{q}_1).$$

But $W'(\xi)$ is any element of $\mathfrak{p}$; hence we have

$$\mathfrak{p} = 0 \ \ (\mathfrak{q}_1).$$

But since $\mathfrak{p}$ is the radical of $\mathfrak{q}_1$, this implies that $\mathfrak{p} = \mathfrak{q}_1$, which is the required result.

3. **Simple points.** In this section we investigate in some detail the properties of a point P on an irreducible variety V of dimension d in S_n, defined over the ground field K, which is (i) a rational point, (ii) a simple point of V. In the next section we shall extend our results to algebraic points. We take the prime at infinity $x_0 = 0$ to be any prime not passing through P, and consider P to be on a variety in affine space A_n. Since the point is rational it has, in the coordinate system $(x_1, \ldots, x_n)$, coordinates $(\alpha_1, \ldots, \alpha_n)$, where α_i $(i = 1, \ldots, n)$ is in K. If $(\xi_1, \ldots, \xi_n)$ is a generic point of V, and $\mathfrak{J} = K[\xi_1, \ldots, \xi_n]$ is the integral domain of the variety, the prime ideal defined by P is $\mathfrak{p} = \mathfrak{J}\cdot(\xi_1 - \alpha_1, \ldots, \xi_n - \alpha_n)$, whether P is simple or not. Our object is to express the fact that P is simple in terms of the ideal $\mathfrak{p}$.

Our investigations will be simplified by the choice of an allowable coordinate system in A_n. If we make the transformation of coordinates

$$x'_i = x_i - \alpha_i \quad (i = 1, \ldots, n),$$

P is at the 'origin' of the new coordinate system. We now suppose the coordinate system chosen so that P is the origin, and is otherwise sufficiently general to ensure that if $(\xi_1, \ldots, \xi_n)$ is a generic point of V, then

(i) $\xi_1, \ldots, \xi_d$ are algebraically independent over K;

(ii) ξ_{d+1} is a primitive element of the function field

$$\Sigma = K(\xi_1, \ldots, \xi_n)$$

of V over $K(\xi_1, \ldots, \xi_d)$;

(iii) $\xi_{d+1}, \ldots, \xi_n$ are integrally dependent on $K[\xi_1, \ldots, \xi_d]$;

(iv) the tangent space to V at P is given by equations of the form

$$x_{d+i} = \sum_{j=1}^{d} a_{ij} x_j \quad (i = 1, \ldots, n-d);$$

(v) if
$$f(x_1, \ldots, x_{d+1}) = 0$$

is the irreducible equation satisfied by $\xi_1, \ldots, \xi_{d+1}$, the equation

$$f(0, \ldots, 0, x) = 0$$

has simple roots.

While it is not always necessary to assume that all these conditions are satisfied, all have their uses, and in dealing with simple points it is usually best to prepare the ground in this way.

We saw in § 1 that it is possible to choose the coordinate system sufficiently generally to ensure that conditions (i) ... (iv) are satisfied. If we can show that we can choose the coordinate system sufficiently generally to satisfy (v), it will follow that the coordinate system can be chosen to satisfy (i), ..., (v) simultaneously.

To show that the coordinate system can be chosen to satisfy (v), let
$$F(u_0, \ldots, u_d) = f(u_{ij};\, u_{00}, \ldots, u_{d0})$$

be the Cayley form of V. It can be factorised in the form

$$F(u_0, \ldots, u_d) = A(u_1, \ldots, u_d) \prod_{\nu=1}^{g} \left(u_{00} + \sum_{i=1}^{n} u_{0i} \xi_i^{(\nu)}\right),$$

where g is the order of V, and $(\xi_1^{(\nu)}, \ldots, \xi_n^{(\nu)})$ $(\nu = 1, \ldots, g)$ are the points in which V_d is met by the $(n-d)$-space

$$u_{i0} + \sum_{j=1}^{n} u_{ij} x_j = 0 \quad (i = 1, \ldots, d).$$

If
$$u'_{i0} + \sum_{j=1}^{n} u'_{ij} x_j = 0 \quad (i = 1, \ldots, d)$$

is any $(n-d)$-space whose points of intersection with V are finite in number and at a finite distance, $F(u_0, u'_1, \ldots, u'_d)$, as a form in $u_{00}, \ldots, u_{0n}$, is the Cayley form of the intersection.

Consider, in particular, the case in which $u'_{i0} = 0$ $(i = 1, \ldots, d)$ and u'_{ij} $(i = 1, \ldots, d;\, j = 1, \ldots, n)$ are independent indeterminates. Then $F(u_0, u'_1, \ldots, u'_d)$ is the Cayley form of the intersection of V and a generic $(n-d)$-space through P. Since P is simple, we know

[XI, § 10, Ths. I and III] that this intersection consists of P, counted once, and $g-1$ generic points $\eta^{(2)}, \ldots, \eta^{(g)}$ of V which are conjugate over $K(u'_1, \ldots, u'_d)$, none of them being at infinity. Then we have

$$f(u'_{ij}; u_{00}, 0, \ldots, 0) = F(u_0, u'_1, \ldots, u'_d) = Au_{00} \prod_{\nu=2}^{g} \left(u_{00} + \sum_1^n u_{0i} \eta_i^{(\nu)}\right).$$

Hence
$$f(u'_{ij}; z, 0, \ldots, 0) = 0$$

has simple roots. Also, if

$$-\zeta_0 = \sum_1^n u_{0j} \xi_j,$$

and
$$-\zeta_i = \sum_1^n u_{ij} \xi_j \quad (i = 1, \ldots, d),$$

then
$$f(u'_{ij}; \zeta_0, \ldots, \zeta_d) = 0$$

is the irreducible equation connecting $\zeta_0, \ldots, \zeta_d$ [X, § 6].

For a specialisation $u'_{ij} \to a_{ij}$, the equation has less than g distinct finite roots if and only if the a_{ij} satisfy a certain finite number of algebraic relations. Changing our notation by writing a_{d+1j} for a_{0j} $(j = 1, \ldots, n)$, we deduce that for sufficiently general values of a_{ij} $(i = 1, \ldots, d+1; j = 1, \ldots, n)$, the irreducible equation

$$f(\eta_1, \ldots, \eta_{d+1}) = 0$$

connecting
$$-\eta_i = \sum_{j=1}^n a_{ij} \xi_j \quad (i = 1, \ldots, d+1)$$

is such that the equation

$$f(0, 0, \ldots, 0, x) = 0$$

has g simple roots. Hence if

$$x'_i = \sum_{j=1}^n a_{ij} x_j \quad (i = 1, \ldots, n)$$

is a sufficiently general transformation of coordinates preserving the origin at P, the new coordinate system satisfies the requirement (v) above.

It will be observed that the condition (v) is equivalent to saying that if g is the order of V (in projective space), the $(n-d)$-space S'_{n-d} given by

$$x_i = 0 \quad (i = 1, \ldots, d)$$

meets V in g points (one of which is the origin), none of them lying in $x_0 = 0$. If any one of these were a multiple point of V, an

$(n-d)$-space through it would meet V either in an infinite number of points or in less than g points. Hence the g points in which S'_{n-d} meets V are all simple. Again, if S'_{n-d} were tangent to V at any of the points, it would meet V in less than g points. Hence the effect of (v) is to ensure that the $(n-d)$-space

$$x_i = 0 \quad (i = 1, \ldots, d)$$

meets V in g (finite) simple points, and does not touch V at any of them.

THEOREM I. *The quotient ring of a simple point is integrally closed.*

We assume that the coordinate system in A_n is sufficiently general in the sense explained above. In order to prove our result it is sufficient to show that the integral closure of $\mathfrak{J} = K[\xi_1, \ldots, \xi_n]$ is contained in $\mathfrak{J}_\mathfrak{p}$, where $\mathfrak{p} = \mathfrak{J}.(\xi_1, \ldots, \xi_n)$ is the prime ideal of the point [XV, § 8, Th. IV]. Let ζ be any element of the quotient field of $\mathfrak{J}$ which is integrally dependent on $\mathfrak{J}$. Since $\xi_{d+1}, \ldots, \xi_n$ are integrally dependent on $K[\xi_1, \ldots, \xi_d]$, it follows from XV, § 8, Th. II that ζ is integrally dependent on $K[\xi_1, \ldots, \xi_d]$. Since ξ_{d+1} is a primitive element over $K(\xi_1, \ldots, \xi_d)$ of the function field of V, it follows from XV, § 8, Th. VII that

$$\zeta = \frac{r_0 + r_1\xi_{d+1} + \ldots + r_{g-1}\xi_{d+1}^{g-1}}{D^2},$$

where $r_0, \ldots, r_{g-1}$ are in $K[\xi_1, \ldots, \xi_d]$, and D^2 is the resultant of

$$f(\xi_1, \ldots, \xi_d, x) \quad \text{and} \quad \frac{\partial f}{\partial x}(\xi_1, \ldots, \xi_d, x),$$

$$f(x_1, \ldots, x_{d+1}) = 0$$

being the irreducible equation satisfied by $\xi_1, \ldots, \xi_{d+1}$. By condition (v) on our coordinate system, it follows that $D^2 \neq 0 \ (\mathfrak{p})$, and hence $\zeta \in \mathfrak{J}_\mathfrak{p}$. Hence the integral closure of $\mathfrak{J}$ is contained in $\mathfrak{J}_\mathfrak{p}$, that is, $\mathfrak{J}_\mathfrak{p}$ is integrally closed.

It is as well to point out here that the quotient ring of a point may be integrally closed even when the point is not simple. Con sider the surface in A_3 whose equation is

$$x_2x_3 = x_1^2.$$

It can be verified at once that the origin is not a simple point. We shall show that its quotient ring is integrally closed by showing that the ring $\mathfrak{I}$ for the surface is integrally closed, and applying XV, § 8, Th. V.

A generic point of the surface is $(u, v, u^2/v)$, where u and v are independent indeterminates over the ground field K. Then

$$\mathfrak{I} = K[u, v, u^2/v],$$

and any element of $\mathfrak{I}$ is of the form $\Sigma a_{ij} u^i v^j$, where a_{ij} is in K and $i \geqslant 0$, $i+2j \geqslant 0$. Conversely, we show that, if $i \geqslant 0$, and $i+2j \geqslant 0$, then $u^i v^j \in \mathfrak{I}$. If $j \geqslant 0$, this is obvious; if $j = -k < 0$, we have

$$u^i v^j = u^i/v^k = u^{2k} u^{i-2k}/v^k = u^{i+2j}(u^2/v)^k \in \mathfrak{I}.$$

It is evident that the quotient field of $\mathfrak{I}$ is $K(u, v)$. Let

$$\zeta = \frac{p(u,v)}{q(u,v)} = \frac{p}{q}$$

be any element of this which is integrally dependent on $\mathfrak{I}$, where $p = p(u, v)$ and $q = q(u, v)$ are polynomials in u and v without common factor. Then we have an equation

$$\zeta^n + a_1 \zeta^{n-1} + \ldots + a_n = 0,$$

where $a_i \in \mathfrak{I}$, and hence $a_i = b_i/v^{k_i}$, where $b_i \in K[u, v]$. Hence there exists an integer k such that $v^k \zeta$ is integrally dependent on $K[u, v]$, and hence [XV, § 8, Th. VI] $v^k \zeta$ is contained in $K[u, v]$. Hence

$$\zeta = \Sigma a_{ij} u^i v^j \quad (i \geqslant 0),$$

where $a_{ij} \in K$. Since ζ is integrally dependent on $\mathfrak{I}$, $\Sigma a_{ij} \alpha^i \beta^j t^{i+2j}$ is integrally dependent on $K[\alpha t, \beta t^2, (\alpha^2/\beta)] = K[t]$, where α and β are non-zero elements of K. But [XV, § 8, Th. VI] $K[t]$ is integrally closed; hence, for each term in $\Sigma a_{ij} \alpha^i \beta^j t^{i+2j}$, $i + 2j \geqslant 0$, and it follows from what was said above on the form of elements in $\mathfrak{I}$, that ζ is in $\mathfrak{I}$. Therefore $\mathfrak{I}$ is integrally closed.

We now consider certain properties of a simple point of a variety which lead to a characterisation of a simple point in terms of its quotient ring. (Since the quotient ring is a projective concept, this characterisation is a projective one.) We suppose that P is defined by the prime ideal $\mathfrak{p} = \mathfrak{I}.(\xi_1, \ldots, \xi_n)$, and that the coordinate system is sufficiently general, as described above. The tangent space at P can therefore be written in the form

$$x_{d+i} = \sum_{j=1}^{d} a_{ij} x_j \quad (i = 1, \ldots, n-d).$$

Hence, among the equations of V we can find $n-d$ equations

$$f_i(x_1, \ldots, x_n) = 0 \quad (i = 1, \ldots, n-d)$$

such that

$$f_i(x_1, \ldots, x_n) \equiv x_{d+i} - \sum_{j=1}^{d} a_{ij} x_j - f_i^*(x_1, \ldots, x_n) \quad (i = 1, \ldots, n-d),$$

where $f_i^*(x_1, \ldots, x_n)$ has no terms of degree less than two. Since

$$f_i(\xi_1, \ldots, \xi_n) = 0 \quad (i = 1, \ldots, n-d),$$

we have
$$\xi_{d+i} = \sum_{j=1}^{d} a_{ij} \xi_j + f_i^*(\xi_1, \ldots, \xi_n),$$

that is
$$\xi_{d+i} = \sum_{j=1}^{d} a_{ij} \xi_j \ (\mathfrak{p}^2) \quad (i = 1, \ldots, n-d). \tag{1}$$

If ω is any element of $\mathfrak{J}$,

$$\omega = a + \sum_{1}^{n} a_i \xi_i + f^*(\xi_1, \ldots, \xi_n),$$

where no term of $f^*(x_1, \ldots, x_n)$ is of degree less than two. Hence $f^*(\xi_1, \ldots, \xi_n) \in \mathfrak{p}^2$, and therefore, using (1),

$$\omega = a + \sum_{1}^{d} a_i' \xi_i \ (\mathfrak{p}^2),$$

where $a, a_1', \ldots, a_d'$ are in K, and $a = 0$ if and only if ω belongs to $\mathfrak{p}$.

Conversely, suppose that there exist d' elements, $\zeta_1, \ldots, \zeta_{d'}$ of the ideal $\mathfrak{p} = \mathfrak{J}.(\xi_1, \ldots, \xi_n)$ of a point on V such that any element ω of $\mathfrak{J}$ satisfies a relation

$$\omega = a + \sum_{1}^{d'} a_i \zeta_i \ (\mathfrak{p}^2),$$

where $a, a_1, \ldots, a_{d'}$ belong to K. We show that $d' \geqslant d$, and that if $d' = d$, then P is simple. Since $\xi_1, \ldots, \xi_n$ are in $\mathfrak{p}$,

$$\xi_i = \sum_{j=1}^{d'} b_{ij} \zeta_j \ (\mathfrak{p}^2) \quad (i = 1, \ldots, n),$$

where $b_{ij} \in K$. The rank d'' of the matrix (b_{ij}) is at most equal to d'. We therefore obtain $n-d'' \geqslant n-d'$ linearly independent equations

$$\sum_{j=1}^{n} c_{ij} \xi_j = 0 \ (\mathfrak{p}^2).$$

We can now construct polynomials

$$\phi_i(x_1, \ldots, x_n) \equiv \sum_{j=1}^{n} c_{ij} x_j + \phi_i^*(x_1, \ldots, x_n),$$

where $\phi_i^*(x_1, \ldots, x_n)$ has no terms of degree less than two, which vanish at ξ and therefore on V. Hence if

$$f_i(x_0, \ldots, x_n) = 0 \quad (i = 1, \ldots, r)$$

is a basis for the homogeneous equations of V, the matrix

$$\left(\frac{\partial f_i}{\partial x_j}\right)$$

has rank $n - d''$ at least at P. Since V is of dimension d, and P is on V, the rank of the matrix is at most $n - d$ [X, § 14, Th. I]. Therefore,

$$d' \geqslant d'' \geqslant d.$$

If $d' = d, d'' = d$, and V has a tangent space at P. Then P is a simple point of V.

If $\mathfrak{p}$ is the prime ideal of a point of V, and there exist d elements $\zeta_1, \ldots, \zeta_d$ of $\mathfrak{p}$ such that any element ω of $\mathfrak{J}$ satisfies a relation

$$\omega = a + \sum_1^d a_i \zeta_i \ (\mathfrak{p}^2),$$

the variety is said to possess *uniformising parameters* at P, and $\zeta_1, \ldots, \zeta_d$ are called uniformising parameters at P. Hence we have

THEOREM II. *A necessary and sufficient condition that P be a simple point of V is that there exist a set of uniformising parameters at P.*

We note that if the coordinate system is sufficiently general and have P as origin, $\xi_1, \ldots, \xi_d$ are uniformising parameters at P. Let $\zeta_1, \ldots, \zeta_d$ be any other set of uniformising parameters at P. Then, since $\xi_1, \ldots, \xi_d$ and $\zeta_1, \ldots, \zeta_d$ are both sets of parameters, and therefore lie in $\mathfrak{p}$, we have relations

$$\zeta_i = \sum_1^d a_{ij} \xi_j \ (\mathfrak{p}^2) \quad (i = 1, \ldots, d),$$

and

$$\xi_i = \sum_1^d b_{ij} \zeta_j \ (\mathfrak{p}^2) \quad (i = 1, \ldots, d),$$

where a_{ij} and b_{ij} belong to K. Hence

$$\xi_i = \sum_j \sum_k b_{ij} a_{jk} \xi_k \quad (\mathfrak{p}^2).$$

If the matrix product $(b_{ij})(a_{jk})$ is not the unit matrix, it follows that there exists a relation

$$\sum_1^d a_i \xi_i = 0 \quad (\mathfrak{p}^2),$$

in which $a_1, \ldots, a_d$ are elements of K not all zero. If $a_k \neq 0$, we may, without loss of generality, take it to be unity. If ω is any element of $\mathfrak{J}$, we have a relation

$$\omega = b + \sum_1^d b_i \xi_i = b + \sum_1^d (b_i - b_k a_i) \xi_i \quad (\mathfrak{p}^2),$$

and hence any element of $\mathfrak{J}$ is congruent to a linear expression in $\xi_1, \ldots, \xi_{k-1}, \xi_{k+1}, \ldots, \xi_d$ modulo $\mathfrak{p}^2$, which we know to be impossible. Hence $(b_{ij})(a_{jk})$ is the unit matrix, and therefore (a_{ij}) is non-singular.

Conversely, if (c_{ij}) is a non-singular $d \times d$ matrix, and $\tau_1, \ldots, \tau_d$ are elements of $\mathfrak{p}$ such that

$$\tau_i = \sum_{j=1}^d c_{ij} \xi_j \quad (\mathfrak{p}^2),$$

then we can solve these equations and write

$$\xi_i = \sum_1^d d_{ij} \tau_j \quad (\mathfrak{p}^2),$$

and it follows at once that $\tau_1, \ldots, \tau_d$ are uniformising parameters. Thus we see how to get any set of uniformising parameters at P from any given set.

THEOREM III. *If P is a simple point of an irreducible variety V of d dimensions, given by the prime ideal $\mathfrak{p}$ in the integral domain $\mathfrak{J}$ of V, and $\zeta_1, \ldots, \zeta_d$ are uniformising parameters at P, then $\mathfrak{p}$ is an isolated component of the ideal $\mathfrak{i} = \mathfrak{J}.(\zeta_1, \ldots, \zeta_d)$. Conversely, if P is a point of V, given by the prime ideal $\mathfrak{p}$ of $\mathfrak{J}$, and there exist d elements $\zeta_1, \ldots, \zeta_d$ of $\mathfrak{p}$ such that $\mathfrak{p}$ is an isolated component of*

$$\mathfrak{i} = \mathfrak{J}.(\zeta_1, \ldots, \zeta_d),$$

then P is a simple point of V, and $\zeta_1, \ldots, \zeta_d$ are uniformising parameters at P.

We consider first the case in which we are given that P is a simple point, and that $\zeta_1, \ldots, \zeta_d$ are uniformising parameters at P. We shall make use of the lemma proved at the end of XV, § 2. In the first place, if ω is any element of $\mathfrak{p}$, we have a relation

$$\omega = \sum_1^d a_i \zeta_i \quad (\mathfrak{p}^2),$$

where $a_1, \ldots, a_d$ are in K. Hence

$$\mathfrak{p} \subseteq (\mathfrak{i}, \mathfrak{p}^2).$$

Let
$$\mathfrak{i} = [\mathfrak{q}_1, \ldots, \mathfrak{q}_r]$$

be a representation of $\mathfrak{i}$ as an uncontractible intersection of primary ideals, and let $\mathfrak{p}_i$ be the radical of $\mathfrak{q}_i$. Since $\mathfrak{i} \subseteq \mathfrak{q}_i$, we see that $(\mathfrak{i}, \mathfrak{p}^2) \subseteq (\mathfrak{q}_i, \mathfrak{p}^2)$, and hence

$$\mathfrak{p} \subseteq (\mathfrak{i}, \mathfrak{p}^2) \subseteq [(\mathfrak{q}_1, \mathfrak{p}^2), \ldots, (\mathfrak{q}_r, \mathfrak{p}^2)].$$

Since $\mathfrak{p}$ is a maximal ideal of $\mathfrak{J}$, the ideal on the right is either equal to $\mathfrak{p}$ or to the unit ideal. It can only be the unit ideal if $(\mathfrak{q}_i, \mathfrak{p}^2)$ is the unit ideal for $i = 1, \ldots, r$, and hence, by the lemma quoted, it can only be the unit ideal if $\mathfrak{p}_i \nsubseteq \mathfrak{p}$ for each value of i. Suppose, first, that this is the case. If η_i is an element of $\mathfrak{p}_i$ not in $\mathfrak{p}$, and σ_i is the index of $\mathfrak{q}_i$, then $\eta_1^{\sigma_1} \ldots \eta_r^{\sigma_r}$ belongs to $\mathfrak{i}$ but not to $\mathfrak{p}$. But since $\zeta_i \in \mathfrak{p}$, $\mathfrak{i} \subseteq \mathfrak{p}$, and we have a contradiction. Hence

$$\mathfrak{p} = [(\mathfrak{q}_1, \mathfrak{p}^2), \ldots, (\mathfrak{q}_r, \mathfrak{p}^2)].$$

Using the lemma once more, $(\mathfrak{q}_i, \mathfrak{p}^2)$ is $\mathfrak{p}$-primary if and only if $\mathfrak{p}_i \subseteq \mathfrak{p}$, and is equal to $\mathfrak{p}$ if and only if $\mathfrak{q}_i = \mathfrak{p}$. Since $\mathfrak{p}_1, \ldots, \mathfrak{p}_r$ are all different, one $\mathfrak{q}_i$, say $\mathfrak{q}_1$, is equal to $\mathfrak{p}$, and for the other components of $\mathfrak{i}$, $(\mathfrak{q}_j, \mathfrak{p}^2)$ is the unit ideal, that is, $\mathfrak{q}_j$ is not contained in $\mathfrak{p}$. Hence

$$\mathfrak{i} = [\mathfrak{p}, \mathfrak{q}_2, \ldots, \mathfrak{q}_r],$$

and $\mathfrak{p}$ is an isolated component of $\mathfrak{i}$.

It may be noted in passing that in the case in which the ground field K is algebraically closed, the various components of

$$\mathfrak{J} \cdot (\xi_1, \ldots, \xi_d)$$

define the g points in which the S_{n-d} given by $x_1 = \ldots = x_d = 0$ meets V, and, in view of the restriction (v) on our coordinate system, all these are simple points of V, and $\xi_1, \ldots, \xi_d$ are uniformising parameters at each of them. We conclude that in this case $r = g$, and $\mathfrak{q}_2, \ldots, \mathfrak{q}_g$ are prime, defining the remaining $g-1$ points in which S_{n-d} meets V.

Returning to the theorem, suppose, conversely, that $\mathfrak{p}$ is an isolated component of $\mathfrak{i} = \mathfrak{J}.(\zeta_1, \ldots, \zeta_d)$, say

$$\mathfrak{i} = [\mathfrak{p}, \mathfrak{q}_2, \ldots, \mathfrak{q}_r].$$

If ω is any element of $\mathfrak{J}$, we have

$$\omega = a + \sum_1^n a_i \xi_i + \text{terms of higher degree},$$

where $a, a_1, \ldots, a_n$ are in K. Since $\xi_i \in \mathfrak{p}$ $(i = 1, \ldots, n)$, $\omega - a \in \mathfrak{p}$. Hence for every element ω of $\mathfrak{J}$ there exists an element a of K such that

$$\omega = a \quad (\mathfrak{p}),$$

where a is zero if and only if $\omega \in \mathfrak{p}$.

Since $\mathfrak{p}$ is an isolated component of $\mathfrak{i}$, there exists an element α of $\mathfrak{J}$ which is in $[\mathfrak{q}_2, \ldots, \mathfrak{q}_r]$ but not in $\mathfrak{p}$. Let ω be any element of $\mathfrak{J}$, and let

$$\omega = a \quad (\mathfrak{p}).$$

Then $\alpha(\omega - a)$ is in $\mathfrak{p}$ and in $[\mathfrak{q}_2, \ldots, \mathfrak{q}_r]$; hence it is in $\mathfrak{i}$. We can therefore find elements $\alpha_1, \ldots, \alpha_d$ in $\mathfrak{J}$ such that

$$\alpha(\omega - a) = \sum_1^d \alpha_i \zeta_i.$$

Let
$$\alpha = c \quad (\mathfrak{p}),$$
and
$$\alpha_i = c_i \quad (\mathfrak{p}) \quad (i = 1, \ldots, d),$$

where $c, c_1, \ldots, c_d$ are in K. Since α is not in $\mathfrak{p}$, c is not zero. If $a_i = c_i/c$, we deduce that

$$\omega = a + \sum_1^d a_i \zeta_i \quad (\mathfrak{p}^2),$$

and hence $\zeta_1, \ldots, \zeta_d$ are uniformising parameters at P, which is therefore a simple point.

Now, in the case in which $\zeta_1, \ldots, \zeta_d$ are uniformising parameters at the simple point P, let us pass to the quotient ring $\mathfrak{J}_\mathfrak{p}$ of P. If $\mathfrak{i} = \mathfrak{J}.(\zeta_1, \ldots, \zeta_d)$,

$$\mathfrak{i} = [\mathfrak{p}, \mathfrak{q}_2, \ldots, \mathfrak{q}_r],$$

where $\mathfrak{q}_i$ is not contained in $\mathfrak{p}$. Hence [XV, § 5, Th. IX],

$$\mathfrak{J}_\mathfrak{p}.\mathfrak{i} = \mathfrak{J}_\mathfrak{p}.\mathfrak{p},$$

that is, the ideal of non-units of $\mathfrak{J}_\mathfrak{p}$ is $\mathfrak{J}_\mathfrak{p}.(\zeta_1, \ldots, \zeta_d)$.

Conversely, let the quotient ring $\mathfrak{J}_\mathfrak{p}$ of a point of V have the property that its ideal of non-units has a basis consisting of d elements $\omega_1, \ldots, \omega_d$. We can write

$$\omega_i = \zeta_i/\zeta_0,$$

where $\zeta_0, \ldots, \zeta_d$ are in $\mathfrak{J}$ and ζ_0 is not in $\mathfrak{p}$, and is therefore a unit of $\mathfrak{J}_\mathfrak{p}$. Hence $\zeta_1, \ldots, \zeta_d$ is a basis for the ideal of non-units of $\mathfrak{J}_\mathfrak{p}$. This ideal is

$$\mathfrak{J}_\mathfrak{p} \cdot \mathfrak{p} = \mathfrak{J}_\mathfrak{p} \cdot (\zeta_1, \ldots, \zeta_d).$$

Let
$$\mathfrak{i} = \mathfrak{J} \cdot (\zeta_1, \ldots, \zeta_d) = [\mathfrak{q}_1, \ldots, \mathfrak{q}_r].$$

Since
$$\mathfrak{J}_\mathfrak{p} \cdot \mathfrak{p} = \mathfrak{J}_\mathfrak{p} \cdot \mathfrak{i},$$

it follows from XV, § 5, Th. IX that one $\mathfrak{q}_i$, say $\mathfrak{q}_1$, is equal to $\mathfrak{p}$, and for $i > 1$

$$\mathfrak{q}_i \neq 0 \quad (\mathfrak{p}).$$

Hence $\mathfrak{p}$ is an isolated component of $\mathfrak{i}$. Therefore, by Theorem III, P is a simple point of V, and $\zeta_1, \ldots, \zeta_d$ are uniformising parameters of V at P. Thus we have

THEOREM IV. *A necessary and sufficient condition that a point P of V be simple is that the ideal of non-units of its quotient ring has a basis consisting of d members. If $\zeta_1, \ldots, \zeta_d$ form a basis for the ideal and lie in $\mathfrak{J}$, $\zeta_1, \ldots, \zeta_d$ are uniformising parameters at P.*

We may also note that if P is *any* rational point of V, and $\zeta_1, \ldots, \zeta_{d'}$ is a basis for the ideal of non-units of its quotient ring, then $d' \geqslant d$. For if

$$\mathfrak{J}_\mathfrak{p} \cdot \mathfrak{p} = \mathfrak{J}_\mathfrak{p} \cdot (\zeta_1, \ldots, \zeta_{d'}),$$

we can, as above, suppose that $\zeta_1, \ldots, \zeta_{d'}$ lie in $\mathfrak{J}$. Then $\mathfrak{p}$ is an isolated component of $\mathfrak{J} \cdot (\zeta_1, \ldots, \zeta_{d'})$. The second stage of the proof of Theorem III then shows us that if ω is any element of $\mathfrak{J}$,

$$\omega = a + \sum_1^{d'} a_i \zeta_i \quad (\mathfrak{p}^2),$$

and hence, as we have seen, we must have $d' \geqslant d$.

The next result which we prove helps to explain the use of the term 'uniformising parameters'. We consider a simple point P of an irreducible variety V of d dimensions, and, as usual, use sufficiently general coordinates which have origin at P. $(\xi_1, \ldots, \xi_n)$ are the non-homogeneous coordinates of a generic point of V.

THEOREM V. *Let P be a simple point of V, given by the prime ideal $\mathfrak{p}$ of the integral domain $\mathfrak{J} = K[\xi_1, \ldots, \xi_n]$, and let $\zeta_1, \ldots, \zeta_d$ be a basis* *for the ideal of non-units of the quotient ring* $\mathfrak{I}_\mathfrak{p}$. *Given any element* ω *of* $\mathfrak{I}_\mathfrak{p}$, *there exists a unique sequence* $\psi_0, \psi_1, \psi_2, \ldots$, *where* ψ_m *is a homogeneous polynomial of degree* m *in* $\zeta_1, \ldots, \zeta_d$, *with coefficients in* K, *such that* $\omega - \sum_0^{m-1} \psi_i \in (\mathfrak{I}_\mathfrak{p} . \mathfrak{p})^m$, *for* $m = 1, 2, \ldots$. *Moreover, the mapping*

$$\omega \to \psi_0 + \psi_1 + \ldots$$

of the elements of $\mathfrak{I}_\mathfrak{p}$ *on the ring* $K\{\zeta_1, \ldots, \zeta_d\}$ *of formal power series in* $\zeta_1, \ldots, \zeta_d$ *is an isomorphism of* $\mathfrak{I}_\mathfrak{p}$ *on a subring of* $K\{\zeta_1, \ldots, \zeta_d\}$.

We shall begin by proving the first part of the theorem in the case in which

(i) $\zeta_i = \xi_i$ $(i = 1, \ldots, d)$;

(ii) ω belongs to $\mathfrak{I}$.

In this case we replace $\mathfrak{I}_\mathfrak{p} . \mathfrak{p}$ by $\mathfrak{p}$ in the enunciation of the theorem.

The existence of the constant ψ_0 and its uniqueness have already been proved. The existence of ψ_1 has also been proved, and it is a trivial exercise to prove its uniqueness.

We assume the truth of the theorem when $m = 0$, and prove the theorem generally by induction on m. Suppose that we know that there exist uniquely determined polynomials $\psi_0, \ldots, \psi_{m-1}$ such that

$$\omega = \psi_0 + \psi_1 + \ldots + \psi_{i-1} \quad (\mathfrak{p}^i) \quad (i = 1, \ldots, m).$$

We proceed to construct ψ_m and prove its uniqueness. Since

$$\omega - \sum_0^{m-1} \psi_i = 0 \quad (\mathfrak{p}^m),$$

there exist elements $\alpha_{i_1 \ldots i_n}$ in $\mathfrak{I}$, where $i_1, \ldots, i_n$ take all integral non-negative values subject to the condition

$$i_1 + \ldots + i_n = m,$$

such that

$$\omega - \sum_0^{m-1} \psi_i = \sum \alpha_{i_1 \ldots i_n} \xi_1^{i_1} \ldots \xi_n^{i_n}.$$

From the case $m = 0$, already proved, we know that there exist elements $a_{i_1 \ldots i_n}$ in K such that

$$\alpha_{i_1 \ldots i_n} = a_{i_1 \ldots i_n} \quad (\mathfrak{p}).$$

Hence

$$\omega - \sum_0^{m-1} \psi_i = \sum a_{i_1 \ldots i_n} \xi_1^{i_1} \ldots \xi_n^{i_n} \quad (\mathfrak{p}^{m+1}), \tag{2}$$

since $\xi_i \in \mathfrak{p}$. Moreover, there exist constants a_{ij} such that

$$\xi_{d+i} = \sum_{j=1}^{d} a_{ij}\xi_j \quad (\mathfrak{p}^2) \quad (i = 1, \ldots, n-d).$$

Substituting in (2) we obtain the relation

$$\omega - \sum_{0}^{m-1} \psi_i = \sum b_{i_1\ldots i_d} \xi_1^{i_1} \ldots \xi_d^{i_d} \quad (\mathfrak{p}^{m+1}),$$

where $b_{i_1\ldots i_d}$ is in K, and the summation is over non-negative values of $i_1, \ldots, i_d$ such that

$$i_1 + \ldots + i_d = m.$$

If

$$\psi_m = \Sigma b_{i_1\ldots i_d} \xi_1^{i_1} \ldots \xi_d^{i_d},$$

then, clearly,

$$\omega = \psi_0 + \psi_1 + \ldots + \psi_m \quad (\mathfrak{p}^{m+1}).$$

We have now to prove the uniqueness of ψ_m. Suppose that there exists another homogeneous polynomial ψ'_m of degree m such that

$$\omega = \psi_0 + \psi_1 + \ldots + \psi_{m-1} + \psi'_m \quad (\mathfrak{p}^{m+1}).$$

Then $\chi(\xi_1, \ldots, \xi_d) = \psi_m - \psi'_m$ is a homogeneous polynomial of degree m in $\xi_1, \ldots, \xi_d$ belonging to $\mathfrak{p}^{m+1}$.

Let $\mathfrak{J}.(\xi_1, \ldots, \xi_d) = \mathfrak{i} = [\mathfrak{p}, \mathfrak{q}_2, \ldots, \mathfrak{q}_r]$. We have seen that $\mathfrak{q}_2, \ldots, \mathfrak{q}_r$ are, in fact, the prime ideals of the zero-dimensional varieties other than P in which the $(n-d)$-space

$$x_1 = \ldots = x_d = 0$$

meets V. Hence they are all maximal ideals of $\mathfrak{J}$. Since $(\mathfrak{q}_i, \mathfrak{q}_j)$ and $(\mathfrak{p}, \mathfrak{q}_i)$ are therefore equal to the unit ideal, $\mathfrak{i} = \mathfrak{p}\mathfrak{q}_2 \ldots \mathfrak{q}_r$ [XV, § 1, Th. IV, Cor. I]. Then

$$\chi(\xi_1, \ldots, \xi_d) \in \mathfrak{p}^m \mathfrak{q}_2^m \ldots \mathfrak{q}_r^m.$$

But, by hypothesis, $\chi \in \mathfrak{p}^{m+1}$. Hence $\chi \in \mathfrak{p}^{m+1}\mathfrak{q}_2^m \ldots \mathfrak{q}_r^m = \mathfrak{p}\mathfrak{i}^m$. Therefore

$$\chi = \sum_i \gamma_{i_1\ldots i_d} \xi_1^{i_1} \ldots \xi_d^{i_d},$$

where the summation is over values of $i_1, \ldots, i_d$ whose sum is m, and $\gamma_{i_1\ldots i_d} \in \mathfrak{p}$. Since $\xi_{d+1}, \ldots, \xi_n$ are integrally dependent on $K[\xi_1, \ldots, \xi_d]$, $\mathfrak{J}$ is integrally dependent on this ring, and hence the $\gamma_{i_1\ldots i_d}$ are integrally dependent on $K[\xi_1, \ldots, \xi_d]$. They belong to an algebraic extension of $K(\xi_1, \ldots, \xi_d)$, and their conjugates

$$\gamma^{(1)}_{i_1\ldots i_d} = \gamma_{i_1\ldots i_d}, \quad \gamma^{(2)}_{i_1\ldots i_d}, \quad \ldots, \quad \gamma^{(g)}_{i_1\ldots i_d}$$

over this field are also integrally dependent on $K[\xi_1, \ldots, \xi_d]$. Therefore $\Phi = \prod_{\nu=1}^{g} \Sigma \gamma^{(\nu)}_{i_1 \ldots i_d} \xi_1^{i_1} \ldots \xi_d^{i_d}$ is a polynomial of degree mg in $\xi_1, \ldots, \xi_d$ whose coefficients are integrally dependent on $K[\xi_1, \ldots, \xi_d]$. But these coefficients are symmetric functions of the roots of an algebraic equation over $K(\xi_1, \ldots, \xi_d)$, which is of characteristic zero; hence they lie in this field. Also $K[\xi_1, \ldots, \xi_d]$ is integrally closed in $K(\xi_1, \ldots, \xi_d)$, since $\xi_1, \ldots, \xi_d$ are algebraically independent over K [XV, § 8, Th. VI]. Hence Φ is a polynomial of degree mg in $\xi_1, \ldots, \xi_d$, with coefficients in $K[\xi_1, \ldots, \xi_d]$. Moreover, since $\gamma^{(1)}_{i_1 \ldots i_d} = 0$ $(\mathfrak{p})$, these coefficients are in $\mathfrak{i}$. Therefore Φ is a polynomial in $\xi_1, \ldots, \xi_d$ having no term of degree less than $mg+1$. But since

$$\Sigma \gamma_{i_1 \ldots i_d} \xi_1^{i_1} \ldots \xi_d^{i_d} = \chi(\xi_1, \ldots, \xi_d),$$

we have, taking conjugates over $K(\xi_1, \ldots, \xi_d)$,

$$\Sigma \gamma^{(\nu)}_{i_1 \ldots i_d} \xi_1^{i_1} \ldots \xi_d^{i_d} = \chi(\xi_1, \ldots, \xi_d) \quad (i = 1, \ldots, g),$$

and hence

$$\Phi = [\chi(\xi_1, \ldots, \xi_d)]^g.$$

Since $\chi(\xi_1, \ldots, \xi_d)$ is of degree m exactly, and Φ is a polynomial in $\xi_1, \ldots, \xi_d$ having no term of degree less than $mg+1$, this gives a proper relation over K connecting $\xi_1, \ldots, \xi_d$, unless Φ and χ are both zero. But $\xi_1, \ldots, \xi_d$ are algebraically independent over K. Hence $\chi(\xi_1, \ldots, \xi_d) = 0$. This proves the uniqueness of ψ_m.

Now let ω and ν be two elements of $\mathfrak{J}$, and let $\psi_0, \psi_1, \ldots$ be the sequence obtained, as above, for ω, and let $\phi_0, \phi_1, \ldots$ be the sequence similarly obtained for ν. Then, since

$$\omega = \psi_0 + \ldots + \psi_{m-1} \ (\mathfrak{p}^m),$$

and

$$\nu = \phi_0 + \ldots + \phi_{m-1} \ (\mathfrak{p}^m),$$

$$\omega + \nu = (\psi_0 + \phi_0) + \ldots + (\psi_{m-1} + \phi_{m-1}) \ (\mathfrak{p}^m),$$

and since $\psi_i + \phi_i$ is a homogeneous polynomial of degree i in $\xi_1, \ldots, \xi_d$, the sequence $\psi_0 + \phi_0, \psi_1 + \phi_1, \ldots$ corresponds to $\omega + \nu$.

Also
$$\omega\nu = \psi_0\phi_0 + (\psi_0\phi_1 + \psi_1\phi_0) + \ldots + \psi_{m-1}\phi_{m-1} \ (\mathfrak{p}^m).$$

But $\psi_i \phi_j$ is a polynomial in $\xi_1, \ldots, \xi_d$ of degree $i+j$, and hence lies in $\mathfrak{p}^m$ if $i+j \geqslant m$. Therefore, if

$$\theta_i = \psi_0\phi_i + \psi_1\phi_{i-1} + \ldots + \psi_i\phi_0,$$

then

$$\omega\nu = \theta_0 + \theta_1 + \ldots + \theta_{m-1} \ (\mathfrak{p}^m).$$

Since θ_i is a homogeneous polynomial of degree i in $\xi_1, \ldots, \xi_d$, $\theta_0, \theta_1, \ldots$ is the sequence determined by $\omega\nu$.

Let $K\{\xi_1, \ldots, \xi_d\}$ denote the ring of all formal power series in $\xi_1, \ldots, \xi_d$. We then map the element ω of $\mathfrak{I}$ on the element

$$\overline{\omega} = \psi_0 + \psi_1 + \ldots$$

of $K\{\xi_1, \ldots, \xi_d\}$, where $\psi_0, \psi_1, \ldots$ is the sequence obtained from ω, as above. The result just proved shows that if ω, ν map on the elements $\overline{\omega}$, $\overline{\nu}$ of $K\{\xi_1, \ldots, \xi_d\}$, $\omega + \nu$ and $\omega\nu$ map on $\overline{\omega} + \overline{\nu}$ and $\overline{\omega}\overline{\nu}$, respectively. Hence the mapping of $\mathfrak{I}$ in $K\{\xi_1, \ldots, \xi_d\}$ is a homomorphism. It is an isomorphism if the only element of $\mathfrak{I}$ which maps on zero is $\omega = 0$. If ω maps on zero, the sequence $\psi_0, \psi_1, \ldots$ must consist of zero only. Hence

$$\omega = 0 \ (\mathfrak{p}^m) \quad (m = 1, 2, \ldots).$$

But [XV, § 2, Th. X] the only element common to $\mathfrak{p}, \mathfrak{p}^2, \mathfrak{p}^3, \ldots$ is zero. Hence if ω maps on the zero of $K\{\xi_1, \ldots, \xi_d\}$, $\omega = 0$. Hence the mapping of $\mathfrak{I}$ in $K\{\xi_1, \ldots, \xi_d\}$ is an isomorphism. The images of the elements of $\mathfrak{I}$ form a subring $\boldsymbol{R}$ of $K\{\xi_1, \ldots, \xi_d\}$ isomorphic with $\mathfrak{I}$.

The elements of $K\{\xi_1, \ldots, \xi_d\}$ which have no constant term form a prime ideal $\boldsymbol{\Pi}$ of this ring, as can be verified at once. Hence $\boldsymbol{\pi} = \boldsymbol{R}_\wedge \boldsymbol{\Pi}$ is a prime ideal of $\boldsymbol{R}$ [XV, § 4, Th. III]. An element ω of $\mathfrak{I}$ maps on an element of $\boldsymbol{\pi}$ if and only if

$$\omega = 0 \ (\mathfrak{p});$$

hence $\boldsymbol{\pi}$ is the image of $\mathfrak{p}$ in the isomorphism between $\mathfrak{I}$ and $\boldsymbol{R}$. Therefore $\mathfrak{I}_\mathfrak{p}$ is isomorphic with $\boldsymbol{R}_{\boldsymbol{\pi}}$. But, by an elementary property of formal power series, an element $\overline{\omega}$ of $K\{\xi_1, \ldots, \xi_d\}$ not in $\boldsymbol{\Pi}$ has an inverse in $K\{\xi_1, \ldots, \xi_d\}$; therefore $\boldsymbol{R}_{\boldsymbol{\pi}}$ is a subring $\boldsymbol{R}^*$ of $K\{\xi_1, \ldots, \xi_d\}$ containing $\boldsymbol{R}$. Hence there is an isomorphic mapping of $\mathfrak{I}_\mathfrak{p}$ on a subring $\boldsymbol{R}^*$ of $K\{\xi_1, \ldots, \xi_d\}$. If ω is any element of $\mathfrak{I}_\mathfrak{p}$ and

$$\overline{\omega} = \omega_0 + \omega_1 + \ldots,$$

its image in $\boldsymbol{R}^*$, $\omega - \sum_0^{m-1} \omega_i$ maps on $\omega_m + \omega_{m+1} + \ldots$, and this is in $\boldsymbol{\Pi}^m$. Since it is also in $\boldsymbol{R}^*$, it is contained in

$$\boldsymbol{R}^*{}_\wedge \boldsymbol{\Pi}^m = (\boldsymbol{R}^*{}_\wedge \boldsymbol{\Pi})^m = (\boldsymbol{R}_{\boldsymbol{\pi}} . \boldsymbol{\pi})^m.$$

Hence
$$\omega - \sum_0^{m-1} \omega_i \in (\mathfrak{I}_\mathfrak{p} . \mathfrak{p})^m.$$

Thus $\omega_0, \omega_1, \ldots$ is a sequence corresponding to ω, as required by our theorem. The uniqueness of this sequence in $\mathfrak{J}$ is an immediate consequence of the uniqueness of the sequences corresponding to elements in $\mathfrak{J}$.

Finally, let $\zeta_1, \ldots, \zeta_d$ be any basis for the ideal of non-units of $\mathfrak{J}_\mathfrak{p}$. Then, as we have seen, we can write

$$\zeta_i = \eta_i/\eta_0,$$

where $\eta_0, \ldots, \eta_d$ are elements of $\mathfrak{J}$, and $\eta_0 \neq 0 \ (\mathfrak{p})$, while $\eta_1, \ldots, \eta_d$ are uniformising parameters at P. Then, by a property already proved for uniformising parameters,

$$\eta_i = \sum_{j=1}^{d} c_{ij}\xi_j \ (\mathfrak{p}^2) \quad (i = 1, \ldots, d),$$

where $|c_{ij}| \neq 0$. If $c \neq 0$ is the element of K such that $\eta_0 = c \ (\mathfrak{p})$ and $a_{ij} = c_{ij}/c$, we have

$$\zeta_i = \sum_{j=1}^{d} a_{ij}\xi_j \ ((\mathfrak{J}_\mathfrak{p} . \mathfrak{p})^2).$$

Hence if ζ_i maps on the series

$$\psi_{i0} + \psi_{i1} + \cdots \tag{3}$$

of $\boldsymbol{R}^*$, we have $\psi_{i0} = 0 \ (i = 1, \ldots, d)$ and

$$\psi_{i1} = \sum_{1}^{d} a_{ij}\xi_j \ (i = 1, \ldots, d),$$

where $|a_{ij}| \neq 0$. The series (3) may therefore be inverted to express ξ_i as a power series in $\zeta_1, \ldots, \zeta_d$. Hence we obtain an isomorphic mapping of $\boldsymbol{R}^*$ on a subring $\boldsymbol{S}^*$ of $K\{\zeta_1, \ldots, \zeta_d\}$. It follows easily from this that we obtain an isomorphic mapping of $\mathfrak{J}_\mathfrak{p}$ on $\boldsymbol{S}^*$ which has all the required properties. This completes the proof of Theorem V.

Another useful characterisation of a simple point in terms of the prime ideal $\mathfrak{p}$ in $\mathfrak{J}$ which defines it is given by

THEOREM VI. *If P is a rational point on V given by the prime ideal $\mathfrak{p}$, P is simple if and only if $\mathfrak{p}/\mathfrak{p}^2$ is a K-module with a basis of d elements.*

Suppose, first, that P is simple, and take $\xi_1, \ldots, \xi_d$ as the uniformising parameters. If ω is any element of $\mathfrak{p}$, we have

$$\omega = \sum_{1}^{d} a_i\xi_i \ (\mathfrak{p}^2),$$

where $a_1, \ldots, a_d$ are in K. Now consider the mapping of $\mathfrak{J}$ on $\mathfrak{J}/\mathfrak{p}^2$. If $\omega, \xi_1, \ldots, \xi_d$ map on $\bar{\omega}, \bar{\xi}_1, \ldots, \bar{\xi}_d$ we have

$$\bar{\omega} = \sum_1^d a_i \bar{\xi}_i,$$

that is, $\bar{\xi}_1, \ldots, \bar{\xi}_d$ is a K-basis for $\mathfrak{p}/\mathfrak{p}^2$.

Suppose, conversely, that $\bar{\zeta}_1, \ldots, \bar{\zeta}_d$ is a K-basis for $\mathfrak{p}/\mathfrak{p}^2$, and let $\zeta_1, \ldots, \zeta_d$ be elements of $\mathfrak{J}$ which map on $\bar{\zeta}_1, \ldots, \bar{\zeta}_d$ respectively. If ω is any element of $\mathfrak{J}$, there exists an element a of K such that $\omega - a \in \mathfrak{p}$. The corresponding element of $\mathfrak{p}/\mathfrak{p}^2$ is of the form $\sum_1^d a_i \bar{\zeta}_i$. Hence

$$\omega = a + a_1 \zeta_1 + \ldots + a_d \zeta_d \quad (\mathfrak{p}^2).$$

Therefore $\zeta_1, \ldots, \zeta_d$ are uniformising parameters at P, and P must be a simple point.

Since we have seen that for any point P, simple or not, we cannot find d' elements $\eta_1, \ldots, \eta_{d'}$ (with $d' < d$) such that any element ω of $\mathfrak{J}$ satisfies a relation

$$\omega = a + \sum_1^{d'} a_i \eta_i \quad (\mathfrak{p}^2),$$

it follows that the module $\mathfrak{p}/\mathfrak{p}^2$ can never have a K-basis of less than d elements.

4. Irreducible subvarieties of V_d. The investigations of the preceding section deal only with *rational* simple points on a variety V, that is, with points whose non-homogeneous coordinates lie in the ground field K. In the case in which K is algebraically closed, this includes all algebraic points, and is really sufficient for most purposes. But when we pass to consider varieties of dimension greater than zero on V, we find that the natural procedure is to deduce the properties of an irreducible simple variety of dimension s on V from a study of an irreducible variety of dimension zero on a new variety V_1, defined over a ground field K_1, which, however, may not be algebraically closed, even though the original ground field K is closed. Thus it becomes necessary to consider irreducible varieties of dimension zero on a variety V defined over a ground field which is not algebraically closed. We therefore begin by extending the results of § 3 to varieties of dimension zero which are not simply points, but sets of conjugate points over the ground field.

We are again concerned with a variety V, of d dimensions in affine space A_n, which is irreducible over the ground field K. As in § 3, we shall find it convenient in our investigations to assume that the coordinate system in A_n is 'sufficiently general'; this concept will by this time be sufficiently familiar to the reader, and we shall usually leave it to him to verify that the coordinate system can be chosen so as to justify the assertion that stated properties hold when the coordinate system is sufficiently general.

If V is not absolutely irreducible over K, there exists [X, § 11, Th. II] an algebraic extension K^* of K over which V splits up into a finite number of absolutely irreducible varieties, each of d dimensions:

$$V = V^{(1)} + \ldots + V^{(k)}.$$

An algebraic extension K^* of K with this property will be called a *splitting extension* for V. It is easy to see that a point P of V is a simple point of V if and only if it lies on just one component $V^{(i)}$ of V and is a simple point of this component. We prove this in the projective case, from which the affine case follows at once. We recall [XI, § 10, Th. II, IV] that P is a simple point of V if and only if a generic $(n-d)$-space through P meets V simply at P. If $F(u_0, \ldots, u_d)$ is the Cayley form of V_d, irreducible over K, and if

$$\sum_{j=0}^{n} v_{ij} x_j = 0 \quad (i = 1, \ldots, d)$$

is a generic $(n-d)$-space through $P\,(x'_0, \ldots, x'_n)$, it follows that P is simple if and only if $\Sigma u_{0i} x'_i$ is a simple factor of $F(u_0, v_1, \ldots, v_d)$, regarded as a form in $u_{00}, \ldots, u_{0n}$. If K^* is a splitting extension for V, we have, over K^*,

$$F(u_0, \ldots, u_d) = \prod_{i=1}^{k} F_i^*(u_0, \ldots, u_d),$$

where $F_i^*(u_0, \ldots, u_d)$ is the (irreducible) Cayley form of $V^{(i)}$. It follows that $\Sigma u_{0i} x'_i$ is a simple factor of $F(u_0, v_1, \ldots, v_d)$ if and only if it is a factor of just one $F_i^*(u_0, v_1, \ldots, v_d)$, and if it is a simple factor of that. The result follows immediately.

In this section it is important to recognise when a variety U on V consists entirely of points lying on more than one absolutely irreducible component of V. The following Theorem I provides the most convenient form of answer to this question. In a convenient coordinate system in A_n we consider a generic point $(\xi_1, \ldots, \xi_n)$ of

V, and, as usual, let $\mathfrak{J} = K[\xi_1, \ldots, \xi_n]$ be the integral domain of V. Then V is absolutely irreducible if and only if every element of the function field $K(\xi_1, \ldots, \xi_n)$ of V which is algebraic over K lies in K [X, § 11, Th. IV]. If V is not absolutely irreducible, the elements of $K(\xi_1, \ldots, \xi_n)$ which are algebraic over K form a finite algebraic extension of K, which is called the *algebraic closure* of K in $K(\xi_1, \ldots, \xi_n)$.

THEOREM I. *Let U be an irreducible subvariety of V.* (i) *If the algebraic closure K' of K in $K(\xi_1, \ldots, \xi_n)$ is contained in the quotient ring $Q(U)$ of U, then U contains points which lie on only one absolutely irreducible component of V.* (ii) *If U is simple on V, then $K' \subseteq Q(U)$.*

(i) Let α be a primitive element of K' over K. Since $K' \subseteq Q(U)$, we can write $\alpha = a(\xi)/b(\xi)$, where $a(x)$ and $b(x)$ are polynomials over K, and $b(x)$ is not zero on U. Since

$$a(\xi) - \alpha b(\xi) = 0,$$

it follows that $a(x) - \alpha b(x) = 0$ on a component V^* over K^* (any normal splitting extension of K which contains K'), but not on V. Let $\alpha_1, \ldots, \alpha_k$ be the conjugates of α in K^*. Then [X, § 11] V has k absolutely irreducible components $V_1, \ldots, V_k$, and on any one of them at least one of the forms $g_i(x) = a(x) - \alpha_i b(x)$ vanishes; also $g_i(x)$ vanishes on some V_j, for each value of i. Suppose that it vanishes on more than one component. Then on some component V_j two of the forms $g_p(x)$ and $g_q(x)$ vanish. Since $\alpha_p \neq \alpha_q$, this implies that $b(x)$ vanishes on V_j. Since $b(x) \in K[x_1, \ldots, x_n]$, and $\dim V_j = \dim V$, $b(x)$ vanishes on V and hence on U, which is contrary to hypothesis. Hence each $g_i(x)$ vanishes on just one of the components V_i of V.

Since $b(x)$ does not vanish on U we can find a point x' of U such that $b(x') \neq 0$. Suppose that x' lay on two components V_i and V_j. Then

$$g_i(x') = 0, \quad g_j(x') = 0,$$

and hence, since $\alpha_i \neq \alpha_j$, $b(x') = 0$. We thus have a contradiction. Hence x' can only lie on one absolutely irreducible component of V, and the first part of the theorem is proved.

(ii) Let U be of dimension s. Since U is simple on V, the points of U which are multiple for V form a subvariety of U of dimension s' $(s' < s)$. This contains the points of U which lie on more than one absolutely irreducible component of V. A sufficiently general $(n-s)$-

space in A_n therefore meets U in a finite set of algebraic points which are simple for V. Hence U must contain a set of conjugate points $x^{(1)}, \ldots, x^{(r)}$ which are simple for V (and hence each lies on exactly one irreducible component of V). Let U_0 be the variety over K formed by $x^{(1)}, \ldots, x^{(r)}$. Let $\alpha \in K(\xi_1, \ldots, \xi_n)$ be a primitive element of K' over K, and let $\alpha_1, \ldots, \alpha_k$ be its conjugates over K. We can then find polynomials $f_0(x), \ldots, f_{k-1}(x)$ such that

$$f_0(x) + f_1(x)\,\alpha_i + \ldots + f_{k-1}(x)\,\alpha_i^{k-1}$$

vanishes on the absolutely irreducible component V_i, but does not vanish at any point $x^{(j)}$ which is not on V_i. Suppose that $x^{(1)}$ lies on V_1. Then the equations

$$f_0(x^{(1)}) + f_1(x^{(1)})\,z + \ldots + f_{k-1}(x^{(1)})\,z^{k-1} = 0,$$

$$\prod_{i=1}^{k} (z - \alpha_i) = 0,$$

have a simple root α_1 in common, which can be found by the division algorithm. It follows that when $\xi \to x^{(1)}$, $\alpha \in K(\xi_1, \ldots, \xi_n)$ has a unique specialisation. Since $x^{(1)}$ is simple, its quotient ring is integrally closed. Hence, by a result to be proved later [XVIII, § 1, Th. VII, Cor.], $\alpha \in Q(x^{(1)})$. Since $x^{(1)}$ is a generic point of U_0 over K, it follows that $\alpha \subseteq Q(U_0) \subseteq Q(U)$. Therefore $K' \subseteq Q(U)$.

We note in passing that in (ii) it is only necessary to assume that U contains a subvariety, defined over some extension of K, which lies only on one component of V and whose quotient ring is integrally closed, in order to apply the foregoing method to establish the relation $K' \subseteq Q(U)$. Even with this generalisation, the result given is an incomplete one, but Theorem I as stated contains all that we shall require.

We next consider some relations concerning maximal ideals in the integral domain $\mathfrak{J} = K[\xi_1, \ldots, \xi_n]$ of the variety V. Let $\mathfrak{p}$ be a maximal ideal in $\mathfrak{J}$, necessarily prime since $\mathfrak{J}$ is an integral domain. $\mathfrak{p}$ is the ideal of an irreducible variety U of dimension zero on V. U therefore consists of a set of points $x^{(1)}, \ldots, x^{(r)}$, conjugate over K. We consider a normal extension K^* of K, which we assume is a splitting extension of K and contains the non-homogeneous coordinates of the points $x^{(i)}$. The ring $\mathfrak{J}^* = K^*[\xi_1, \ldots, \xi_n]$ is the integral domain (over K^*) of one of the absolutely irreducible components V^* of V [X, § 12]. There are points of U on each component of V, and we may suppose the points $x^{(i)}$ so arranged that $x^{(1)}, \ldots, x^{(k)}$ lie on V^*, while $x^{(k+1)}, \ldots, x^{(r)}$ do not lie on V^*. Let $\mathfrak{p}_i^*$ be the prime ideal of $x^{(i)}$ in $\mathfrak{J}^*$ $(i = 1, \ldots, k)$. $\mathfrak{J}^*$ is an extension of $\mathfrak{J}$; hence any ideal in $\mathfrak{J}$ defines an extended ideal in $\mathfrak{J}^*$. We prove

THEOREM II. $$\mathfrak{J}^*.\mathfrak{p} = \mathfrak{p}_1^* \ldots \mathfrak{p}_k^*.$$

Let $$f_i(x) = 0 \quad (i = 1, \ldots, s)$$

be a basis for the equations of U in A_n. Then $f_i(\xi)$ $(i = 1, \ldots, s)$ is a basis for $\mathfrak{p}$ in $\mathfrak{J}$. These elements of $\mathfrak{J}$ form a basis for the ideal $\mathfrak{J}^*.\mathfrak{p}$ in $\mathfrak{J}^*$. Hence the variety on V^* defined by $\mathfrak{J}^*.\mathfrak{p}$ is the intersection of V^* and the variety defined by the equations

$$f_i(x) = 0 \quad (i = 1, \ldots, s).$$

Hence it must be the variety consisting of $x^{(1)}, \ldots, x^{(k)}$. Hence

$$\mathfrak{J}^*.\mathfrak{p} = [\mathfrak{q}_1^*, \ldots, \mathfrak{q}_k^*],$$

where $\mathfrak{q}_i^*$ is $\mathfrak{p}_i^*$-primary. Since $\mathfrak{p}_i^*$ is a maximal ideal in $\mathfrak{J}^*$, and $\mathfrak{p}_1^*, \ldots, \mathfrak{p}_k^*$ are all different, $(\mathfrak{q}_i^*, \mathfrak{q}_j^*) = \mathfrak{J}^*$ whenever $i \neq j$, and hence

$$[\mathfrak{q}_1^*, \ldots, \mathfrak{q}_k^*] = \mathfrak{q}_1^* \mathfrak{q}_2^* \ldots \mathfrak{q}_k^*.$$

Now let $f^*(x_1, \ldots, x_n)$ be any polynomial over K^* such that $f^*(\xi_1, \ldots, \xi_n) \in \mathfrak{p}_i^*$ $(i = 1, \ldots, k)$. Then $f^*(x_1, \ldots, x_n)$ vanishes at $x^{(1)}, \ldots, x^{(k)}$. We can find a polynomial $g^*(x_1, \ldots, x_n)$ which vanishes at $x^{(k+1)}, \ldots, x^{(r)}$ but does not vanish at $x^{(1)}, \ldots, x^{(k)}$. Then $f^*(x)\, g^*(x)$ vanishes at $x^{(1)}, \ldots, x^{(r)}$. Let θ be a primitive element of K^* over K, and let $\theta_1 = \theta, \theta_2, \ldots, \theta_t$ be its conjugates. We can write

$$f^*(x)\, g^*(x) = h_0(x) + h_1(x)\,\theta + \ldots + h_{t-1}(x)\,\theta^{t-1},$$

where $h_i(x)$ is in $K[x_1, \ldots, x_n]$. Consider the automorphism of K^* over K which takes θ into θ_i. It takes $x^{(1)}, \ldots, x^{(r)}$ into the same set of points rearranged. Hence

$$h_0(x) + h_1(x)\,\theta_i + \ldots + h_{t-1}(x)\,\theta_i^{t-1}$$

vanishes at $x^{(1)}, \ldots, x^{(r)}$. This holds for $i = 1, \ldots, t$. Since the determinant $|\theta_i^{j-1}|$ is not zero, we deduce that $h_i(x) = 0$ at $x^{(1)}, \ldots, x^{(r)}$ for $i = 1, \ldots, t$. Hence $h_i(\xi) \in \mathfrak{p}$, and therefore

$$f^*(\xi)\, g^*(\xi) = 0 \quad ([\mathfrak{q}_1^*, \mathfrak{q}_2^*, \ldots, \mathfrak{q}_k^*]).$$

But we have arranged that $g^*(\xi) \neq 0 \ (\mathfrak{p}_i^*) \ (i = 1, \ldots, k)$; hence [XV, § 2, Th. I, Cor. I]

$$f^*(\xi) = 0 \quad ([\mathfrak{q}_1^*, \mathfrak{q}_2^*, \ldots, \mathfrak{q}_k^*]).$$

$f^*(\xi)$ was, however, any element of $[\mathfrak{p}_1^*, \ldots, \mathfrak{p}_k^*]$. Therefore

$$[\mathfrak{p}_1^*, \ldots, \mathfrak{p}_k^*] \subseteq [\mathfrak{q}_1^*, \ldots, \mathfrak{q}_k^*].$$

This implies that $\mathfrak{q}_i^* = \mathfrak{p}_i^* \ (i = 1, \ldots, k)$. Hence

$$\mathfrak{J}^*.\mathfrak{p} = \mathfrak{p}_1^* \ldots \mathfrak{p}_k^*.$$

Now let $\mathfrak{Q}^*$ be the integral domain formed by elements of $K^*(\xi_1, \ldots, \xi_n)$ which can be written in the form α^*/β, where α^* is in $\mathfrak{J}^*$, and β is in $\mathfrak{J}$ but not in $\mathfrak{p}$. $\mathfrak{Q}^*$ is an extension of $\mathfrak{J}$; hence any ideal $\mathfrak{i}$ in $\mathfrak{J}$ defines an extended ideal $\mathfrak{Q}^*.\mathfrak{i}$ in $\mathfrak{Q}^*$. We now prove

THEOREM III. $\mathfrak{Q}^*.\mathfrak{i} = (\mathfrak{J}_{\mathfrak{p}_1^*}^*.\mathfrak{i}) \wedge (\mathfrak{J}_{\mathfrak{p}_2^*}^*.\mathfrak{i}) \wedge \cdots \wedge (\mathfrak{J}_{\mathfrak{p}_k^*}^*.\mathfrak{i})$,

where $\mathfrak{J}_{\mathfrak{p}_h^*}^*$ *is the quotient ring of* $\mathfrak{p}_h^*$ in $\mathfrak{J}^*$.

Any element of $\mathfrak{Q}^*.\mathfrak{i}$ can be written in the form $\alpha^*(\xi)/\beta(\xi)$, where $\alpha^*(x)$ is in $K^*[x_1, \ldots, x_n]$ and $\beta(x)$ is in $K[x_1, \ldots, x_n]$, and $\alpha^*(\xi)$ is in $\mathfrak{J}^*.\mathfrak{i}$, while $\beta(\xi)$ is not in $\mathfrak{p}$. Hence $\beta(x)$ does not vanish on U, and therefore does not contain any of the points $x^{(1)}, \ldots, x^{(k)}$. Hence $\beta(\xi) \neq 0 \ (\mathfrak{p}_h^*) \ (h = 1, \ldots, k)$, so that $\alpha^*(\xi)/\beta(\xi)$ is in $\mathfrak{J}_{\mathfrak{p}_h^*}^*.\mathfrak{i}$ $(h = 1, \ldots, k)$. Therefore

$$\mathfrak{Q}^*.\mathfrak{i} \subseteq (\mathfrak{J}_{\mathfrak{p}_1^*}^*.\mathfrak{i}) \wedge \cdots \wedge (\mathfrak{J}_{\mathfrak{p}_k^*}^*.\mathfrak{i}). \tag{1}$$

Next, we can find an element ω_h^* in $\mathfrak{p}_h^*$ which is not in $\mathfrak{p}_j^*$ if $j \neq h$.

Let
$$\nu_h^* = \prod_{i \neq h} \omega_i^*.$$

Then
$$\nu_h^* = 0 \ (\mathfrak{p}_i^*) \quad (i \neq h),$$

and
$$\nu_h^* = k_h^* \neq 0 \ (\mathfrak{p}_h^*),$$

where k_h^* is some element of K^*. Let $\lambda_h^* = (k_h^*)^{-1} \nu_h^*$. Then, if δ_{hi} is the Kronecker delta,

$$\lambda_h^* = \delta_{hi} \ (\mathfrak{p}_i^*).$$

Let ξ^* be any element common to the k ideals $\mathfrak{J}_{\mathfrak{p}_h^*}^*.\mathfrak{i}$ $(h = 1, \ldots, k)$. Then we can write

$$\xi^* = \alpha_h^*/\beta_h^*, \quad \text{where} \quad \alpha_h^* \in \mathfrak{J}^*.\mathfrak{i} \quad \text{and} \quad \beta_h^* \neq 0 \ (\mathfrak{p}_h^*).$$

Therefore $$\xi^* = (\Sigma\lambda_h^*\alpha_h^*)/(\Sigma\lambda_h^*\beta_h^*) = \alpha^*/\beta^*,$$

where $$\alpha^* \in \mathfrak{J}^*.\mathfrak{i} \quad \text{and} \quad \beta^* \neq 0 \ (\mathfrak{p}_h^*) \quad (h = 1, \ldots, k).$$

The ideal $\mathfrak{J}^*.(\mathfrak{p}, \beta^*)$ in $\mathfrak{J}^*$ defines a vacuous variety on V^*, since the variety defined by it is the intersection of the variety defined by $\mathfrak{J}^*.\mathfrak{p}$, that is, the set of points $x^{(1)}, \ldots, x^{(k)}$, and the variety defined by β^*, where $\beta^* \neq 0$ $(\mathfrak{p}_h^*)$ for any value of h. Hence $\mathfrak{J}^*.(\mathfrak{p}, \beta^*)$ must be the unit ideal of $\mathfrak{J}^*$. It follows that there must exist an element η^* in $\mathfrak{J}^*.\mathfrak{p}$, and an element ϵ^* of $\mathfrak{J}^*$, such that

$$\eta^* + \epsilon^*\beta^* = 1.$$

Since η^* is in $\mathfrak{J}^*.\mathfrak{p}$, it can be written in the form

$$\eta^* = \sum_{i=0}^{t-1} \eta_i \theta^i,$$

where θ is a primitive element of K^* over K, and $\eta_0, \ldots, \eta_{t-1}$ are in $\mathfrak{p}$. If $\theta_1 = \theta$, $\theta_2, \ldots, \theta_t$ are the conjugates of θ over K, they all lie in K^*, since K^* is a normal extension. Hence if

$$\eta_j^* = \sum_{i=0}^{t-1} \eta_i \theta_j^i,$$

$\eta_1^*, \ldots, \eta_t^*$ are in $\mathfrak{J}^*.\mathfrak{p}$, and

$$\prod_{j=1}^{t} (1 - \eta_j^*) = 1 + \zeta,$$

where ζ belongs to $\mathfrak{p}$. Now we have

$$\xi^* = \frac{\epsilon^*\alpha^* \prod_2^t (1 - \eta_j^*)}{(1 - \eta^*) \prod_2^t (1 - \eta_j^*)} = \frac{\gamma^*}{1 + \zeta},$$

where $\gamma^* \in \mathfrak{J}^*.\mathfrak{i}$ and $1 + \zeta$ is in $\mathfrak{J}$ but not in $\mathfrak{p}$. Hence ξ^* belongs to $\mathfrak{O}^*.\mathfrak{i}$. Therefore $$(\mathfrak{J}_{\mathfrak{p}_1^*}^*.\mathfrak{i}) \wedge \cdots \wedge (\mathfrak{J}_{\mathfrak{p}_k^*}^*.\mathfrak{i}) \subseteq \mathfrak{O}^*.\mathfrak{i}.$$

Taking this result with equation (1), Theorem III follows immediately.

Corollary I. If $\mathfrak{i}$ is the unit ideal it contains the unity of $\mathfrak{J}^*$. Hence $$\mathfrak{O}^*.\mathfrak{i} = \mathfrak{O}^*, \quad \mathfrak{J}_{\mathfrak{p}_h^*}^*.\mathfrak{i} = \mathfrak{J}_{\mathfrak{p}_h^*}^*.$$

Therefore $$\mathfrak{Q}^* = \mathfrak{J}^*_{\mathfrak{p}_1^*} \wedge \cdots \wedge \mathfrak{J}^*_{\mathfrak{p}_k^*}.$$

Corollary II. If $\mathfrak{i} = \mathfrak{p}$,

$$\begin{aligned} \mathfrak{J}^*_{\mathfrak{p}_h^*}.\mathfrak{p} &= \mathfrak{J}^*_{\mathfrak{p}_h^*}.[\mathfrak{p}_1^*, \ldots, \mathfrak{p}_k^*] \quad \text{[Th. II]} \\ &= \mathfrak{J}^*_{\mathfrak{p}_h^*}.\mathfrak{p}_h^* \quad \text{[XV, § 5, Th. II].} \end{aligned}$$

Hence $\mathfrak{Q}^*.\mathfrak{p}$ is the intersection of the ideals of non-units of the quotient rings $\mathfrak{J}^*_{\mathfrak{p}_h^*}$.

Before passing to the next theorem we require the following lemma:

Lemma. If $\mathfrak{i}^*$ *is an ideal in* $\mathfrak{Q}^*$ *which contains a unit of* $\mathfrak{J}^*_{\mathfrak{p}_h^*}$ *for each value of h, then* $\mathfrak{i}^*$ *is the unit ideal.*

Let α_h^*/β_h^* be a unit of $\mathfrak{J}^*_{\mathfrak{p}_h^*}$ which is in $\mathfrak{i}^*$, where α_h^* and β_h^* are in $\mathfrak{J}^*$ and not in $\mathfrak{p}_h^*$. Since α_h^*/β_h^* is in $\mathfrak{Q}^*$, we can write

$$\frac{\alpha_h^*}{\beta_h^*} = \frac{\gamma_h^*}{\delta_h},$$

where γ_h^* is in $\mathfrak{J}^*$ and δ_h is in $\mathfrak{J}$ but not in $\mathfrak{p}$. Then δ_h is not in $\mathfrak{p}_j^*$ for any value of j. We show that γ_h^* is not in $\mathfrak{p}_h^*$. We have

$$\delta_h \alpha_h^* = \beta_h^* \gamma_h^*;$$

hence $\delta_h \alpha_h^*$ is in $\mathfrak{p}_h^*$ if γ_h^* belongs to this ideal. Since $\mathfrak{p}_h^*$ is prime, this implies that δ_h or α_h^* is in $\mathfrak{p}_h^*$, which is not the case. Thus γ_h^* is not in $\mathfrak{p}_h^*$. We then consider an element $\xi_h^* = \gamma_h^*/\delta_h$ which is in $\mathfrak{i}^*$ and is a unit of $\mathfrak{J}^*_{\mathfrak{p}_h^*}$, for each value of h. We can bring these elements to the same common denominator $\delta = \delta_1 \ldots \delta_k$, and write

$$\xi_h^* = \epsilon_h^*/\delta,$$

where ϵ_h^* is not in $\mathfrak{p}_h^*$. If $\lambda_1^*, \ldots, \lambda_k^*$ are the elements defined in the proof of Theorem III, $\xi^* = \Sigma \lambda_h^* \xi_h^* = (\Sigma \lambda_h^* \epsilon_h^*)/\delta = \epsilon^*/\delta$ and is in $\mathfrak{i}^*$, and

$$\epsilon^* = \Sigma \lambda_j^* \epsilon_j^* \neq 0 \ \ (\mathfrak{p}_h^*) \quad (h = 1, \ldots, k).$$

Hence $(\xi^*)^{-1}$ is in $\mathfrak{J}^*_{\mathfrak{p}_h^*}$ for each value of h, and is therefore in $\mathfrak{Q}^*$. Hence $\mathfrak{i}^*$ contains a unit of $\mathfrak{Q}^*$, and $\mathfrak{i}^*$ is therefore the unit ideal.

THEOREM IV. $\mathfrak{Q}^*$ *contains k maximal ideals, namely,*

$$\pi_h^* = \mathfrak{J}^*_{\mathfrak{p}_1^*} \wedge \cdots \wedge \mathfrak{J}^*_{\mathfrak{p}_{h-1}^*} \wedge (\mathfrak{J}^*_{\mathfrak{p}_h^*}.\mathfrak{p}_h^*) \wedge \mathfrak{J}^*_{\mathfrak{p}_{h+1}^*} \wedge \cdots \wedge \mathfrak{J}^*_{\mathfrak{p}_k^*} \quad (h = 1, \ldots, k).$$

Clearly, π_h^*, defined in the enunciation, is a proper ideal of $\mathfrak{O}^*$. It is a maximal ideal, since any larger ideal must contain a unit of each quotient ring $\mathfrak{J}_{\mathfrak{p}_h^*}^*$, and hence, by the foregoing lemma, it must be the unit ideal of $\mathfrak{O}^*$. On the other hand, any proper ideal π^* of $\mathfrak{O}^*$ must be such that its elements are all non-units of some $\mathfrak{J}_{\mathfrak{p}_h^*}^*$, and therefore $\pi^* \subseteq \pi_h^*$ for some value of h. If π^* is maximal, it follows that $\pi^* = \pi_h^*$. Hence $\pi_1^*, \ldots, \pi_k^*$ are the only maximal ideals of $\mathfrak{O}^*$. The fact that these k ideals are all distinct is an immediate consequence of the next theorem.

THEOREM V. *The quotient ring $\mathfrak{O}_{\pi_h^*}^*$ is equal to the quotient ring $\mathfrak{J}_{\mathfrak{p}_h^*}^*$.*

We first note that

$$\mathfrak{J}^* \wedge \pi_h^* = (\mathfrak{J}^* \wedge \mathfrak{J}_{\mathfrak{p}_1^*}^*) \wedge \cdots \wedge (\mathfrak{J}^* \wedge \mathfrak{J}_{\mathfrak{p}_h^*}^* . \mathfrak{p}_h^*) \wedge \cdots \wedge (\mathfrak{J}^* \wedge \mathfrak{J}_{\mathfrak{p}_k^*}^*) = \mathfrak{p}_h^*.$$

We also have $$\mathfrak{O}^* \subseteq \mathfrak{J}_{\mathfrak{p}_h^*}^*, \quad \mathfrak{p}_h^* \subseteq \pi_h^*,$$

so that $$\mathfrak{O}_{\pi_h^*}^* \subseteq \mathfrak{J}_{\mathfrak{p}_h^*}^*.$$

On the other hand, if α^*/β^* is an element of $\mathfrak{J}_{\mathfrak{p}_h^*}^*$, where α^* and β^* are in $\mathfrak{J}^*$ and β^* is not in $\mathfrak{p}_h^*$, β^* is not in π_h^*, otherwise it would be in $\mathfrak{J}^* \wedge \pi_h^* = \mathfrak{p}_h^*$. Hence $\alpha^*/\beta^* \in \mathfrak{O}_{\pi_h^*}^*$, that is,

$$\mathfrak{J}_{\mathfrak{p}_h^*}^* \subseteq \mathfrak{O}_{\pi_h^*}^*.$$

The result follows.

THEOREM VI. *If the algebraic closure K' of K in $K(\xi, \ldots, \xi_n)$ is in $\mathfrak{J}_\mathfrak{p}$, and $\mathfrak{i}$ is any ideal in $\mathfrak{J}$, then*

$$\mathfrak{J}_\mathfrak{p} \wedge (\mathfrak{O}^* . \mathfrak{i}) = \mathfrak{J}_\mathfrak{p} . \mathfrak{i}.$$

Let α be a primitive element of K' over K, and let θ be a primitive element of K^* over K'. Then any element η of $\mathfrak{O}^* . \mathfrak{i}$ is of the form $\sum_{i,j} \eta_{ij} \alpha^i \theta^j$, where $\eta_{ij} \in \mathfrak{J}_\mathfrak{p} . \mathfrak{i}$.

Suppose that η is also in $\mathfrak{J}_\mathfrak{p}$, and choose β ($\beta \neq 0$) in $\mathfrak{J}$ so that

$$\beta\eta = \zeta(\xi) \quad \text{and} \quad \beta\eta_{ij} = \zeta_{ij}(\xi)$$

are in $\mathfrak{J}$. Then $$\zeta(x) - \sum_{i,j} \zeta_{ij}(x)\, \alpha^i \theta^j$$

vanishes on V^*. V^* is absolutely irreducible, and can be defined over $K' = K(\alpha)$. Hence if $\theta_1, \ldots, \theta_s$ are the conjugates of θ over K',

$$\zeta(x) - \sum_{i,j} \zeta_{ij}(x)\, \alpha^i \theta_t^j = 0$$

on V^*. Since the determinant $|\theta^i_k|$ is not zero, it follows that

$$\zeta(x) - \Sigma\zeta_{i0}(x)\,\alpha^i = 0$$

on V^*. Hence
$$\eta = \Sigma\eta_{i0}\alpha^i.$$

But since $K' \subseteq \mathfrak{J}_\mathfrak{p}$, $\alpha \in \mathfrak{J}_\mathfrak{p}$. Hence $\Sigma\eta_{i0}\alpha^i \in \mathfrak{J}_\mathfrak{p}.\mathfrak{i}$. Therefore

$$\mathfrak{J}_\mathfrak{p} \wedge (\mathfrak{Q}^*.\mathfrak{i}) \subseteq \mathfrak{J}_\mathfrak{p}.\mathfrak{i}.$$

On the other hand, since $\mathfrak{J}_\mathfrak{p} \subseteq \mathfrak{Q}^*$,

$$\mathfrak{J}_\mathfrak{p}.\mathfrak{i} \subseteq \mathfrak{Q}^*.\mathfrak{i},$$

and the result follows.

Theorems II–V hold for any maximal ideal $\mathfrak{p}$ in $\mathfrak{J}$, that is, for the prime ideal of any variety U of dimension zero on V which is irreducible over K. Theorem VI requires that $K' \subseteq \mathfrak{J}_\mathfrak{p}$, but otherwise holds for any U. In what follows we consider the case in which U is simple on V, that is, when U contains at least one point which is simple for V. If it contains such a point, U is not contained in the subvariety of multiple points of V. Since U has no proper subvarieties, being of dimension zero, it cannot intersect this variety of multiple points. Hence all its points are simple on V.

If U is simple on V, then $x^{(h)} = (x^{(h)}_1, \ldots, x^{(h)}_n)$ is simple on V^*, for $h = 1, \ldots, k$. We may suppose that the coordinate system in A_n is sufficiently generally chosen so that $\xi_1 - x^{(h)}_1, \ldots, \xi_d - x^{(h)}_d$ are uniformising parameters of V^* at $x^{(h)}$ $(h = 1, \ldots, k)$, and that the equations of U over K can be written, in non-homogeneous coordinates, in the form

$$\phi_1(x) = 0,$$

$$\phi_i(x) \equiv x_i - \psi_i(x_1) = 0 \quad (i = 2, \ldots, n),$$

where $\phi_1(x)$ is an irreducible polynomial in x_1, over K, of degree r, and $\psi_i(x_1)$ is a polynomial in x_1. Then

$$\phi_1(\xi) = \phantom{-\psi'_i(x^{(h)}_1)} \phi'_1(x^{(h)}_1)(\xi_1 - x^{(h)}_1) \quad (\mathfrak{p}^{*2}_h),$$
$$\phi_i(\xi) = -\psi'_i(x^{(h)}_1)(\xi_1 - x^{(h)}_1) + (\xi_i - x^{(h)}_i) \quad (\mathfrak{p}^{*2}_h) \quad (i = 2, \ldots, n),$$

where $\phi'_1(x^{(h)}_1)$ is not zero, since $\phi_1(x)$ is irreducible and $\phi_1(x^{(h)}_1) = 0$. The $d \times d$ matrix

$$\begin{pmatrix} \phi'_1(x^{(h)}_1) & 0 & . & 0 \\ -\psi'_2(x^{(h)}_1) & 1 & . & 0 \\ . & . & . & . \\ -\psi'_d(x^{(h)}_1) & . & . & 1 \end{pmatrix}$$

is non-singular; hence [cf. § 3, Th. II] $\eta_i = \phi_i(\xi)$ $(i = 1, ..., d)$ is a set of uniformising parameters on V^* at P_h^* $(h = 1, ..., k)$. It follows that

$$\mathfrak{J}^*_{\mathfrak{p}_h^*} . \mathfrak{p}_h^* = \mathfrak{J}^*_{\mathfrak{p}_h^*} . (\eta_1, ..., \eta_d).$$

Now
$$\mathfrak{Q}^* . \mathfrak{p} = (\mathfrak{J}^*_{\mathfrak{p}_1^*} . \mathfrak{p}_1^*) \wedge \dots \wedge (\mathfrak{J}^*_{\mathfrak{p}_k^*} . \mathfrak{p}_k^*)$$

$$= (\mathfrak{J}^*_{\mathfrak{p}_1^*} \wedge \dots \wedge \mathfrak{J}^*_{\mathfrak{p}_k^*}) . (\eta_1, ..., \eta_d)$$

$$= \mathfrak{Q}^* . (\eta_1, ..., \eta_d).$$

So far, we have only used the fact that $x^{(1)}, ..., x^{(k)}$ are simple on V^*. Since U is simple on V, we know, from Theorem I, that if K' denotes the algebraic closure of K in $K(\xi_1, ..., \xi_n)$, then $K' \subseteq \mathfrak{J}_{\mathfrak{p}}$. Hence [Th. VI]

$$\mathfrak{J}_{\mathfrak{p}} . \mathfrak{p} = \mathfrak{J}_{\mathfrak{p}} \wedge (\mathfrak{Q}^* . \mathfrak{p}) = \mathfrak{J}_{\mathfrak{p}} \wedge (\mathfrak{Q}^* . (\eta_1, ..., \eta_d)) = \mathfrak{J}_{\mathfrak{p}} . (\eta_1, ..., \eta_d).$$

Hence we have:

THEOREM VII. *If U is simple on V, the ideal of non-units of $\mathfrak{J}_{\mathfrak{p}}$ has a basis consisting of d elements.*

We now consider the converse of Theorem VII. Let $\mathfrak{p}$ be a maximal ideal in $\mathfrak{J}$ such that $\mathfrak{J}_{\mathfrak{p}}$ has a basis consisting of d elements, say $\zeta_1, ..., \zeta_d$. The ideal $\mathfrak{p}$ defines a variety U consisting of a set of conjugate points $x^{(1)}, ..., x^{(r)}$ over K, and we consider, as usual, a normal extension K^* of K which is a splitting extension for V and contains the non-homogeneous coordinates of the points $x^{(1)}, ..., x^{(r)}$. V^* denotes the absolutely irreducible component of V whose generic point ξ over K^* is the given generic point ξ of V over K, and $x^{(1)}, ..., x^{(k)}$ are the points of U on V^*. To prove that U is simple on V, we have to show

(*a*) that one of the points $x^{(h)}$ $(h \leq k)$ is simple on V^*, and

(*b*) that K', the algebraic closure of K in $K(\xi_1, ..., \xi_n)$, is contained in $\mathfrak{J}_{\mathfrak{p}}$. The second condition implies that U is not contained in the locus of points common to more than one component of V over K^*, and hence, since U is irreducible over K and of dimension zero, that none of its points is in this locus. This, taken with the first condition, implies that $x^{(h)}$ is simple on V, and therefore that U is simple on V.

We are given that

$$\mathfrak{J}_{\mathfrak{p}} . (\zeta_1, ..., \zeta_d) = \mathfrak{J}_{\mathfrak{p}} . \mathfrak{p}.$$

From the definition of $\mathfrak{Q}^*$, we have $\mathfrak{J}_{\mathfrak{p}} \subseteq \mathfrak{Q}^*$, and, by Theorem III, Corollary I, $\mathfrak{Q}^* \subseteq \mathfrak{J}^*_{\mathfrak{p}^*_h}$. Hence $\mathfrak{J}^*_{\mathfrak{p}^*_h}$ is an extension of $\mathfrak{J}_{\mathfrak{p}}$. We therefore have

$$\begin{aligned}\mathfrak{J}^*_{\mathfrak{p}^*_h} \cdot (\zeta_1, \ldots, \zeta_d) = \mathfrak{J}^*_{\mathfrak{p}^*_h} \cdot (\mathfrak{J}_{\mathfrak{p}} \cdot (\zeta_1, \ldots, \zeta_d)) &= \mathfrak{J}^*_{\mathfrak{p}^*_h} \cdot (\mathfrak{J}_{\mathfrak{p}} \cdot \mathfrak{p}) \\ &= \mathfrak{J}^*_{\mathfrak{p}^*_h} \cdot \mathfrak{p} \\ &= \mathfrak{J}^*_{\mathfrak{p}^*_h} \cdot \mathfrak{p}^*_h.\end{aligned}$$

Hence the ideal of non-units of $\mathfrak{J}^*_{\mathfrak{p}^*_h}$ has a basis consisting of d elements. Therefore $x^{(h)}$ is a simple point of V^* [§ 3, Th. IV]. We note also that $\zeta_1, \ldots, \zeta_d$ are uniformising parameters at $x^{(h)}$.

In order to prove that $K' \subseteq \mathfrak{J}_{\mathfrak{p}}$, we have first to prove some lemmas:

Lemma I. $$\mathfrak{J}_{\mathfrak{p}} \wedge (\mathfrak{J}^*_{\mathfrak{p}^*_h} \cdot \mathfrak{p}^{*\rho}_h) = \mathfrak{J}_{\mathfrak{p}} \cdot \mathfrak{p}^\rho,$$
for any integer ρ.

There is nothing to prove if $\rho = 0$. If $\rho = 1$, we have

$$\mathfrak{J}_{\mathfrak{p}} \cdot \mathfrak{p} \subseteq \mathfrak{J}^*_{\mathfrak{p}^*_h} \cdot \mathfrak{p} = \mathfrak{J}^*_{\mathfrak{p}^*_h} \cdot \mathfrak{p}^*_h.$$

Hence $$\mathfrak{J}_{\mathfrak{p}} \cdot \mathfrak{p} \subseteq \mathfrak{J}_{\mathfrak{p}} \wedge (\mathfrak{J}^*_{\mathfrak{p}^*_h} \cdot \mathfrak{p}^*_h).$$

Since $\mathfrak{J}_{\mathfrak{p}} \cdot \mathfrak{p}$ is a maximal ideal in $\mathfrak{J}_{\mathfrak{p}}$, we must have equality, or else

$$\mathfrak{J}_{\mathfrak{p}} \wedge (\mathfrak{J}^*_{\mathfrak{p}^*_h} \cdot \mathfrak{p}^*_h) = \mathfrak{J}_{\mathfrak{p}}.$$

This, however, would imply that $\mathfrak{J}^*_{\mathfrak{p}^*_h} \cdot \mathfrak{p}^*_h$ contains unity, and is therefore the unit ideal, which is absurd. Hence the lemma is proved for $\rho = 1$.

For any value of ρ we have

$$\mathfrak{J}_{\mathfrak{p}} \cdot \mathfrak{p}^\rho \subseteq \mathfrak{J}^*_{\mathfrak{p}^*_h} \cdot (\mathfrak{p}^*_1 \ldots \mathfrak{p}^*_k)^\rho = \mathfrak{J}^*_{\mathfrak{p}^*_h} \cdot \mathfrak{p}^{*\rho}_h;$$

hence $$\mathfrak{J}_{\mathfrak{p}} \cdot \mathfrak{p}^\rho \subseteq \mathfrak{J}_{\mathfrak{p}} \wedge (\mathfrak{J}^*_{\mathfrak{p}^*_h} \cdot \mathfrak{p}^{*\rho}_h).$$

If equality does not hold, there must exist some element α in $\mathfrak{J}_{\mathfrak{p}} \cdot \mathfrak{p}^\sigma$ but not in $\mathfrak{J}_{\mathfrak{p}} \cdot \mathfrak{p}^{\sigma+1}$, for some value of σ less than ρ, which lies in $\mathfrak{J}^*_{\mathfrak{p}^*_h} \cdot \mathfrak{p}^{*\rho}_h \subseteq \mathfrak{J}^*_{\mathfrak{p}^*_h} \cdot \mathfrak{p}^{*\sigma+1}_h$. We show that this is impossible. If α is in $\mathfrak{J}_{\mathfrak{p}} \cdot \mathfrak{p}^\sigma$ and not in $\mathfrak{J}_{\mathfrak{p}} \cdot \mathfrak{p}^{\sigma+1}$, we can write

$$\alpha = \Sigma A_{i_1 \ldots i_d} \zeta_1^{i_1} \ldots \zeta_d^{i_d} \quad (i_1 + \ldots + i_d = \sigma),$$

where $A_{i_1 \ldots i_d}$ is in $\mathfrak{J}_{\mathfrak{p}}$, and at least one of these coefficients is not in $\mathfrak{J}_{\mathfrak{p}} \cdot \mathfrak{p}$, and hence is not in $\mathfrak{J}^*_{\mathfrak{p}^*_h} \cdot \mathfrak{p}^*_h$ (case $\rho = 1$). If $a^*_{i_1 \ldots i_d}$ is the element of K^* such that

$$A_{i_1 \ldots i_d} = a^*_{i_1 \ldots i_d} \; (\mathfrak{J}^*_{\mathfrak{p}^*_h} \cdot \mathfrak{p}^*_h),$$

(for short, the *residue* of $A_{i_1\dots i_d}$ at $x^{(h)}$), not all the $a^*_{i_1\dots i_d}$ are zero. Hence
$$\alpha = \Sigma a^*_{i_1\dots i_d}\zeta_1^{i_1}\dots\zeta_d^{i_d}\ (\mathfrak{J}^*_{\mathfrak{p}^*_h}.\mathfrak{p}^{*\sigma+1}_h).$$

If α is in $\mathfrak{J}^*_{\mathfrak{p}^*_h}.\mathfrak{p}^{*\sigma+1}_h$, we have, therefore,
$$\Sigma a^*_{i_1\dots i_d}\zeta_1^{i_1}\dots\zeta_d^{i_d} = 0\ (\mathfrak{J}^*_{\mathfrak{p}^*_h}.\mathfrak{p}^{*\sigma+1}_h),$$

and, as in the proof of Theorem V of § 3, we deduce that each $a^*_{i_1\dots i_d}$ is zero. Thus we have a contradiction. Hence α is not in $\mathfrak{J}^*_{\mathfrak{p}^*_h}.\mathfrak{p}^{*\sigma+1}_h$, and the truth of the lemma follows.

Lemma II. If α is any element of $\mathfrak{J}_\mathfrak{p}$, and
$$\alpha = \psi_0+\psi_1+\dots$$
is the expansion of α in a power series in $\zeta_1, \dots, \zeta_d$ at $x^{(h)}$, the coefficients of the homogeneous forms ψ_ρ of degree ρ in $\zeta_1, \dots, \zeta_d$ are all residues of elements of $\mathfrak{J}_\mathfrak{p}$ at $x^{(h)}$.

It is clear that $\alpha = \psi_0\ (\mathfrak{J}^*_{\mathfrak{p}^*_h}.\mathfrak{p}^*_h)$; hence the lemma is true for ψ_0, whatever element α in $\mathfrak{J}_\mathfrak{p}$ is considered. We first show that if ψ_ρ is the first non-zero term in the expansion, the lemma is true for ψ_ρ. For, in this case, we see that $\alpha\in\mathfrak{J}_\mathfrak{p}\wedge(\mathfrak{J}^*_{\mathfrak{p}^*_h}.\mathfrak{p}^{*\rho}_h) = \mathfrak{J}_\mathfrak{p}.\mathfrak{p}^\rho$ [Lemma I]. Hence
$$\alpha = \Sigma A_{i_1\dots i_d}\zeta_1^{i_1}\dots\zeta_d^{i_d}\quad (i_1+\dots+i_d = \rho),$$
where $A_{i_1\dots i_d}\in\mathfrak{J}_\mathfrak{p}$. If $a^*_{i_1\dots i_d}$ is the residue of $A_{i_1\dots i_d}$ at $x^{(h)}$, it follows that
$$\alpha = \Sigma a^*_{i_1\dots i_d}\zeta_1^{i_1}\dots\zeta_d^{i_d}\ (\mathfrak{J}^*_{\mathfrak{p}^*_h}.\mathfrak{p}^{*\rho+1}_h).$$
Hence
$$\psi_\rho = \Sigma a^*_{i_1\dots i_d}\zeta_1^{i_1}\dots\zeta_d^{i_d},$$
and so the coefficients of ψ_ρ are residues.

We may now proceed by induction, assuming as hypothesis of induction that the coefficients of $\psi_0, \psi_1, \dots, \psi_{m-1}$ are residues (whatever element α of $\mathfrak{J}_\mathfrak{p}$ is being considered). We have to show that the coefficients of ψ_m are residues. Let ρ be any positive integer less than m. If
$$\psi_\rho = \Sigma a^*_{i_1\dots i_d}\zeta_1^{i_1}\dots\zeta_d^{i_d}\quad (i_1+\dots+i_d = \rho),$$
we know, from the hypothesis of induction, that $a^*_{i_1\dots i_d}$ is the residue at $x^{(h)}$ of some element, say $A_{i_1\dots i_d}$, of $\mathfrak{J}_\mathfrak{p}$. Let
$$\beta_\rho = \Sigma A_{i_1\dots i_d}\zeta_1^{i_1}\dots\zeta_d^{i_d}.$$
We apply the hypothesis of induction to the coefficients $A_{i_1\dots i_d}$ in β_ρ. Then
$$\beta_\rho = \psi_\rho+\psi^{(0)}_{\rho,\rho+1}+\psi^{(0)}_{\rho,\rho+2}+\dots,$$

where the coefficients of $\psi_{\rho,i}$ are residues if $i \leqslant \rho+m-1$. Similarly, if

$$-\psi^{(0)}_{\rho,\rho+1} = \Sigma b^*_{i_1\ldots i_d}\zeta_1^{i_1}\ldots\zeta_d^{i_d} \quad (i_1+\ldots+i_d = \rho+1),$$

$b^*_{i_1\ldots i_d}$ is the residue of some element $B^{(0)}_{i_1\ldots i_d}$ of $\mathfrak{J}_\mathfrak{p}$. We then obtain the expansion of

$$\beta_{\rho,1} = \Sigma B^{(0)}_{i_1\ldots i_d}\zeta_1^{i_1}\ldots\zeta_d^{i_d}$$

in the form

$$\beta_{\rho,1} = \psi^{(1)}_{\rho,\rho+1}+\psi^{(1)}_{\rho,\rho+2}+\ldots \quad (\psi^{(0)}_{\rho,\rho+1}+\psi^{(1)}_{\rho,\rho+1} = 0).$$

In this way we construct elements $\beta_{\rho,1}, \beta_{\rho,2}, \ldots, \beta_{\rho,m-2}$ of $\mathfrak{J}_\mathfrak{p}$ such that

$$\beta_{\rho,j} = \psi^{(j)}_{\rho,\rho+j}+\psi^{(j)}_{\rho,\rho+j+1}+\ldots,$$

where $\psi_{\rho,\rho+j}+\psi^{(1)}_{\rho,\rho+j}+\ldots+\psi^{(j)}_{\rho,\rho+j} = 0 \quad (j = 1, \ldots, m-2),$

and the coefficients of $\psi^{(j)}_{\rho,i}$ are residues if $i \leqslant \rho+j+m-1$. Let

$$\gamma_\rho = \beta_\rho+\beta_{\rho,1}+\ldots+\beta_{\rho,m-2}.$$

From the manner in which the elements $\beta_{\rho,i}$ have been constructed, we see that

$$\gamma_\rho = \psi_\rho+\phi_{\rho+m-1}+\text{terms of higher degree},$$

where $\phi_{\rho+m-1}$ is a form of degree $\rho+m-1$ whose coefficients are residues at $x^{(h)}$.

Now let

$$\omega = \alpha-\gamma_1-\gamma_2-\ldots-\gamma_{m-1}.$$

Then

$$\omega = \psi_0+\psi_m-\phi_m+\text{terms of higher degree}.$$

ψ_0 is in K^*, and hence satisfies an irreducible equation $f(z) = 0$ over K. We may suppose that

$$f(z) = (z-\psi_0)(z-\psi_0^{(2)})\ldots(z-\psi_0^{(g)}).$$

Then, clearly,

$$f(\omega) = (\psi_m-\phi_m)f'(\psi_0) \quad (\mathfrak{J}^*_{\mathfrak{p}^*_h}.\mathfrak{p}_h^{*m+1}).$$

$f(\omega)$ is an element of $\mathfrak{J}_\mathfrak{p}$, and when it is expanded as a power series in $\zeta_1, \ldots, \zeta_d$, its first term is clearly $(\psi_m-\phi_m)f'(\psi_0)$. By a special case of our lemma proved above, the coefficients in this are residues. But $f'(\psi_0)$ is the residue of $f'(\omega)$, and is not in $\mathfrak{J}_\mathfrak{p}.\mathfrak{p}$. Hence the coefficients of $\psi_m-\phi_m$ are residues of elements of $\mathfrak{J}_\mathfrak{p}$. Since, by construction, the coefficients of ϕ_m are residues, it follows that the coefficients of ψ_m are residues. The induction is thus completed and the lemma is proved.

Now let θ be any element of K'. Then there exist elements α and β of $\mathfrak{J}$ such that $\alpha-\theta\beta = 0$. Let the expansion of β at $x^{(h)}$ be

$$\beta = \Sigma b^*_{i_1\ldots i_d}\zeta_1^{i_1}\ldots\zeta_d^{i_d}+\text{terms of higher degree},$$

where $i_1 + \ldots + i_d = \rho$, and at least one coefficient, say $b^*_{j_1\ldots j_d}$, is not zero. By the last lemma, $b^*_{j_1\ldots j_d}$ is the residue of some element B of $\mathfrak{J}_\mathfrak{p}$, which is not in $\mathfrak{J}_\mathfrak{p}.\mathfrak{p}$, since $b^*_{j_1\ldots j_d}$ is not zero. Since $\alpha = \theta\beta$ and $\theta \in K^*$, the expansion of α must be of the form

$$\alpha = \Sigma a^*_{i_1\ldots i_d} \zeta_1^{i_1} \ldots \zeta_d^{i_d} + \text{terms of degree greater than } \rho,$$

where $a^*_{i_1\ldots i_d} = \theta b^*_{i_1\ldots i_d}$. The element $a^*_{j_1\ldots j_d}$ is the residue of some element A of $\mathfrak{J}_\mathfrak{p}$. Hence $\theta = a^*_{j_1\ldots j_d}/b^*_{j_1\ldots j_d}$ is the residue of $\gamma_1 = A/B$ of $\mathfrak{J}_\mathfrak{p}$.

Let
$$\gamma_1 = \theta + \psi_1^{(1)} + \ldots$$

be the expansion of γ_1 at $x^{(h)}$, $\psi_j^{(1)}$ denoting a form of degree j in $\zeta_1, \ldots, \zeta_d$. If
$$\psi_1^{(1)} = \Sigma c_i^* \zeta_i,$$

c_i^* is the residue at $x^{(h)}$ of some element C_i of $\mathfrak{J}_\mathfrak{p}$. If

$$\gamma_2 = \gamma_1 - \Sigma C_i \zeta_i,$$

γ_2 is an element of $\mathfrak{J}_\mathfrak{p}$ whose expansion at $x^{(h)}$ is of the form

$$\gamma_2 = \theta + \psi_2^{(2)} + \ldots.$$

Proceeding in this way, we can find, for any integer n, an element γ_n of $\mathfrak{J}_\mathfrak{p}$ such that
$$\gamma_n = \theta + \psi_n^{(n)} + \ldots.$$

Hence
$$\alpha - \gamma_n\beta = 0 \ (\mathfrak{J}^*_{\mathfrak{p}_h^*}.\mathfrak{p}_h^{*n}).$$

But $\alpha - \gamma_n\beta$ is in $\mathfrak{J}_\mathfrak{p}$; hence it is in $\mathfrak{J}_\mathfrak{p} \wedge (\mathfrak{J}^*_{\mathfrak{p}_h^*}.\mathfrak{p}_h^{*n}) = \mathfrak{J}_\mathfrak{p}.\mathfrak{p}^n$ [Lemma I]. Therefore $\alpha \in \mathfrak{J}_\mathfrak{p}.(\beta, \mathfrak{p}^n)$, for $n = 1, 2, \ldots$. Since $\mathfrak{p}$ is a maximal ideal of $\mathfrak{J}_\mathfrak{p}$, we can apply Theorem X of § 2, Chapter XV, and deduce that α belongs to $\mathfrak{J}_\mathfrak{p}(\beta)$. Hence there exists an element γ of $\mathfrak{J}_\mathfrak{p}$ such that $\alpha = \gamma\beta$. We conclude that $\theta = \gamma$ belongs to $\mathfrak{J}_\mathfrak{p}$. Since θ is any element of K', we thus have $K' \subseteq \mathfrak{J}_\mathfrak{p}$.

Summing up these results, we have

THEOREM VIII. *A necessary and sufficient condition that the variety U defined by the maximal ideal $\mathfrak{p}$ of $\mathfrak{J}$ be simple on V is that the ideal of non-units of $\mathfrak{J}_\mathfrak{p}$ have a basis of d elements.*

It is now a simple matter to extend the foregoing results to irreducible simple subvarieties of V of any dimension. If $\mathfrak{p}$ is any prime ideal of V, and U is the irreducible variety, of dimension s, say, defined by $\mathfrak{p}$ on V, U is simple on V if it contains a point which is a simple point of V. It is clear that U is simple on V if and only if a generic point of U is simple on V, and also that U is simple on V if and only if it contains an algebraic point which is simple on V. This last condition is the same as saying that U contains an irreducible variety of dimension zero which is simple. Our object is to determine necessary and sufficient conditions that U be simple on V in terms of the quotient ring $Q(U) = \mathfrak{J}_{\mathfrak{p}}$.

Let

$$f_i(x_1, \ldots, x_n) = 0 \quad (i = 1, \ldots, r)$$

be a basis for the equations of V, and let $\xi = (\xi_1, \ldots, \xi_n)$ be a generic point of V. We may assume that the coordinate system is chosen sufficiently generally so that $\xi_1, \ldots, \xi_s$ are algebraically independent modulo $\mathfrak{p}$. Since U is of dimension s, $\xi_{s+1}, \ldots, \xi_n$ are then algebraically dependent on $\xi_1, \ldots, \xi_s$ modulo $\mathfrak{p}$. If U is simple, the matrix $(\partial f_i/\partial x_j)$ is of rank $n-d$ at some point of U, and without loss of generality we may assume that

$$\alpha(x) = \begin{vmatrix} \dfrac{\partial f_1}{\partial x_{d+1}} & \cdot & \cdot & \dfrac{\partial f_1}{\partial x_n} \\ \cdot & \cdot & \cdot & \cdot \\ \cdot & \cdot & \cdot & \cdot \\ \dfrac{\partial f_{n-d}}{\partial x_{d+1}} & \cdot & \cdot & \dfrac{\partial f_{n-d}}{\partial x_n} \end{vmatrix} \tag{2}$$

does not vanish on U. Then $\alpha(\xi) \neq 0$ $(\mathfrak{p})$. Conversely, if $\alpha(\xi) \neq 0$ $(\mathfrak{p})$, $\alpha(x)$ does not vanish on U; hence $(\partial f_i/\partial x_j)$ is of rank $n-d$ at some point of U. Therefore U is simple on V.

Now let $\Omega = K(\xi_1, \ldots, \xi_s)$, and consider the variety $\bar{V}$ in the affine space $(x_{s+1}, \ldots, x_n)$ over Ω defined by the equations

$$f_i(\xi_1, \ldots, \xi_s, x_{s+1}, \ldots, x_n) = 0 \quad (i = 1, \ldots, r).$$

This contains the point $(\xi_{s+1}, \ldots, \xi_n)$. Let $\phi^*(x_{s+1}, \ldots, x_n)$ be any polynomial over Ω which vanishes at $(\xi_{s+1}, \ldots, \xi_n)$. We can write

$$\phi^*(x_{s+1}, \ldots, x_n) = \frac{\phi(\xi_1, \ldots, \xi_s, x_{s+1}, \ldots, x_n)}{\psi(\xi_1, \ldots, \xi_s)},$$

where $\phi(x_1, \ldots, x_n)$ and $\psi(x_1, \ldots, x_s)$ are polynomials over K. Since

$$\phi^*(\xi_{s+1}, \ldots, \xi_n) = 0,$$

we have

$$\phi(\xi_1, \ldots, \xi_n) = 0,$$

and hence

$$\phi(x_1, \ldots, x_n) = 0$$

on V. Therefore there exist polynomials $a_i(x_1, \ldots, x_n)$ over K such that

$$\phi(x_1, \ldots, x_n) = \Sigma a_i(x_1, \ldots, x_n) f_i(x_1, \ldots, x_n),$$

and hence we have

$$\phi^*(x_{s+1}, \ldots, x_n) = \Sigma \frac{a_i(\xi_1, \ldots, \xi_s, x_{s+1}, \ldots, x_n)}{\psi(\xi_1, \ldots, \xi_s)} f_i(\xi_1, \ldots, \xi_s, x_{s+1}, \ldots, x_n).$$

Hence $\phi^*(x_{s+1}, \ldots, x_n)$ vanishes on $\overline{V}$. The point $(\xi_{s+1}, \ldots, \xi_n)$ is therefore a generic point of $\overline{V}$. Since $K(\xi_1, \ldots, \xi_n)$ is of dimension d over K, it follows that $\Omega(\xi_{s+1}, \ldots, \xi_n) = K(\xi_1, \ldots, \xi_n)$ is of dimension $d-s$ over Ω.

The integral domain of $\overline{V}$ in A_{n-s} is $\overline{\mathfrak{J}} = \Omega[\xi_{s+1}, \ldots, \xi_n]$. It consists of elements α/β, where $\alpha \in \mathfrak{J}$, and $\beta \in K[\xi_1, \ldots, \xi_s]$. The integral domain $\overline{\mathfrak{J}}$ is an extension of $\mathfrak{J}$. Let us consider the ideal $\overline{\mathfrak{J}}.\mathfrak{p}$. It consists of elements α/β, where $\alpha \in \mathfrak{p}$ and $\beta \in K[\xi_1, \ldots, \xi_s]$. If α/β and α'/β' are two such elements whose product belongs to $\overline{\mathfrak{J}}.\mathfrak{p}$,

$$\frac{\alpha\alpha'}{\beta\beta'} = \frac{\gamma}{\delta},$$

where $\gamma \in \mathfrak{p}$ and $\delta \in K[\xi_1, \ldots, \xi_s]$. Then

$$\alpha\alpha'\delta = \gamma\beta\beta' = 0 \quad (\mathfrak{p}).$$

Since $\delta \neq 0 \ (\mathfrak{p})$, $$\alpha\alpha' = 0 \quad (\mathfrak{p}),$$

and therefore α or α' belongs to $\mathfrak{p}$. Hence α/β or α'/β' belongs to $\overline{\mathfrak{J}}.\mathfrak{p}$, which is therefore prime.

Since $\xi_{s+1}, \ldots, \xi_n$ are algebraically dependent on $\xi_1, \ldots, \xi_s$, modulo $\mathfrak{p}$, it follows that they are algebraic over Ω. Hence $\overline{\mathfrak{p}} = \overline{\mathfrak{J}}.\mathfrak{p}$ defines an irreducible variety $\overline{U}$ of dimension zero on $\overline{V}$. We show that $\overline{U}$ is simple on $\overline{V}$ if and only if U is simple on V.

If the coordinate system is sufficiently general, U is simple on V if and only if $\alpha(\xi)$, defined by equation (2), is not in $\mathfrak{p}$. Similarly, $\overline{U}$ is simple on $\overline{V}$ if and only if $\alpha(\xi)$ is not in $\overline{\mathfrak{p}}$. Since $\mathfrak{p} \subseteq \overline{\mathfrak{p}}$, it is clear that if $\alpha(\xi)$ is not in $\overline{\mathfrak{p}}$ it is not in $\mathfrak{p}$. Hence if $\overline{U}$ is simple on $\overline{V}$, U is simple on V. If $\alpha(\xi) \in \overline{\mathfrak{p}}$, then $\alpha(\xi) = \beta/\gamma$, where $\beta \in \mathfrak{p}$ and

$$\gamma \in K[\xi_1, \ldots, \xi_s].$$

Hence $$\gamma\alpha(\xi) = \beta = 0 \quad (\mathfrak{p}),$$

and since $\gamma \neq 0 \ (\mathfrak{p})$, it follows that $\alpha(\xi) = 0 \ (\mathfrak{p})$. Hence if $\alpha(\xi) \neq 0 \ (\mathfrak{p})$, then $\alpha(\xi) \neq 0 \ (\overline{\mathfrak{p}})$. Hence if U is simple on V, $\overline{U}$ is simple on $\overline{V}$.

It should be noted that the geometrical significance of $\bar{U}$ and $\bar{V}$ is that they are the sections of U and V by the $(n-s)$-space

$$x_i = \xi_i \quad (i = 1, \ldots, s),$$

where $\xi_1, \ldots, \xi_s$ are independent indeterminates over K. The implication of the results we have proved is then clear.

It is clear that $\mathfrak{J}_{\mathfrak{p}} \subseteq \bar{\mathfrak{J}}_{\bar{\mathfrak{p}}}$. Let $\bar{\eta}$ be an element of $\bar{\mathfrak{J}}_{\bar{\mathfrak{p}}}$. We can write $\bar{\eta} = \bar{\alpha}/\bar{\beta}$, where $\bar{\alpha}$ and $\bar{\beta}$ belong to $\bar{\mathfrak{J}}$, and $\bar{\beta}$ does not belong to $\bar{\mathfrak{p}}$. Therefore we can write

$$\bar{\alpha} = \frac{\lambda}{\mu}, \quad \bar{\beta} = \frac{\omega}{\nu},$$

where λ, ω belong to $\mathfrak{J}$ and μ, ν to $K[\xi_1, \ldots, \xi_s]$. Moreover, ω, μ, ν do not belong to $\mathfrak{p}$. Then

$$\bar{\eta} = \frac{\lambda\nu}{\mu\omega},$$

and so $\bar{\eta}$ is contained in $\mathfrak{J}_{\mathfrak{p}}$. Therefore $\bar{\mathfrak{J}}_{\bar{\mathfrak{p}}} \subseteq \mathfrak{J}_{\mathfrak{p}}$. Hence we have

$$\mathfrak{J}_{\mathfrak{p}} = \bar{\mathfrak{J}}_{\bar{\mathfrak{p}}}.$$

Now $\bar{U}$ is a simple variety (of dimension zero) on $\bar{V}$ if and only if the quotient ring $\bar{\mathfrak{J}}_{\bar{\mathfrak{p}}}$ has the property that its ideal of non-units $\bar{\mathfrak{J}}_{\bar{\mathfrak{p}}}.\bar{\mathfrak{p}}$ has a basis consisting of $d-s$ elements. Since U is simple on V if and only if $\bar{U}$ is simple on $\bar{V}$, and $\mathfrak{J}_{\mathfrak{p}} = \bar{\mathfrak{J}}_{\bar{\mathfrak{p}}}$, we obtain

THEOREM IX. *If U is an irreducible variety of dimension s on the irreducible variety V of dimension d, U is simple on V if and only if the ideal of non-units of the quotient ring of U on V has a basis consisting of $d-s$ elements.*

A basis for the ideal of non-units of $Q(U)$ is called a set of *uniformising parameters* for V at U.

We conclude this section with a generalisation of Theorem I of § 3.

THEOREM X. *The quotient ring of any simple irreducible variety U of dimension s on V is integrally closed in its quotient field.*

The proof is similar to that of Theorem I, § 3, and we use the same notation. If the coordinate axes are sufficiently general, we may assume that $\xi_1, \ldots, \xi_d$ are algebraically independent over K, that $\xi_{d+1}, \ldots, \xi_n$ are integrally dependent on $K[\xi_1, \ldots, \xi_d]$, and that ξ_{d+1} is a primitive element of $K(\xi_1, \ldots, \xi_n)$ over $K(\xi_1, \ldots, \xi_d)$. Any element of the integral closure $\mathfrak{J}^*$ of $\mathfrak{J}$ in $K(\xi_1, \ldots, \xi_n)$ is of the form

$$\eta = \frac{r_0 + r_1\xi_{d+1} + \ldots + r_{g-1}\xi_{d+1}^{g-1}}{D^2(\xi_1, \ldots, \xi_d)},$$

where $r_i \in K[\xi_1, \ldots, \xi_d]$. If U contains a simple point (possibly algebraic) we may assume the axes sufficiently general to ensure that $D^2(x_1, \ldots, x_d)$ does not vanish at this point. Hence $D^2(\xi_1, \ldots, \xi_d)$ is not in $\mathfrak{p}$, and therefore $\eta \in \mathfrak{J}_\mathfrak{p}$. Hence $\mathfrak{J}^* \subseteq \mathfrak{J}_\mathfrak{p}$, and therefore [XV, § 8, Th. IV] $\mathfrak{J}_\mathfrak{p}$ is integrally closed.

5. Normal varieties in affine space. Let V be a variety in affine space A_n which is irreducible and of dimension d over the ground field K, and let $(\xi_1, \ldots, \xi_n)$ be the non-homogeneous coordinates of a generic point of V. The variety V is said to be *normal* in A_n (*affine-normal*) if its integral domain $\mathfrak{J} = K[\xi_1, \ldots, \xi_n]$ is integrally closed in its quotient field. By XV, § 8, Th. V, $\mathfrak{J}$ is integrally closed if and only if the quotient rings of all its prime ideals are integrally closed, and for this it is sufficient that the quotient rings of the maximal prime ideals of $\mathfrak{J}$ be integrally closed. Hence we have

THEOREM I. *The irreducible variety V in A_n is affine-normal if and only if the quotient rings of its irreducible subvarieties are integrally closed, and for this it is sufficient that the quotient rings of its irreducible subvarieties of dimension zero be integrally closed.*

Experience shows that normal varieties play a fundamental role in the geometry of varieties, so much so, indeed, that the more advanced parts of the birational theory of varieties are almost entirely stated in terms of them. In Chapter XVIII, where we shall consider the theory of birational correspondences, the importance of the concept of normal varieties will be manifest, but we can give here some immediate results which will help the reader to form some notion of the importance of the idea.

We first observe that, since $\mathfrak{J}$ is integrally closed in its quotient field, we can apply the multiplicative theory of ideals [XV, § 7] to ideals in $\mathfrak{J}$. If $\mathfrak{p}$ is a prime ideal in $\mathfrak{J}$, we know [XV, § 7, Th. VI] that $\mathfrak{p}$ is quasi-equal to the unit ideal, $\mathfrak{p} \sim \mathbf{1}$, if and only if there exists a prime ideal $\mathfrak{p}'$ in $\mathfrak{J}$, not the zero ideal, which is a proper multiple of $\mathfrak{p}$. If U is the irreducible variety defined by $\mathfrak{p}$, there exists a prime ideal $\mathfrak{p}'$ properly contained in $\mathfrak{p}$ if and only if there exists an irreducible subvariety U' of V which contains U as a proper subvariety. If $U \subset U'$, $\dim U < \dim U' < d$; hence $\dim U \leqslant d-2$. Conversely, if $\dim U \leqslant d-2$, we consider a non-zero element of $\mathfrak{p}$, say $v(\xi)$. Then $v(x) = 0$ meets V in an unmixed variety of dimension $d-1$ containing U; hence there exists an irreducible component U' of the intersection which contains U as a proper subvariety. Hence $\mathfrak{p} \sim \mathbf{1}$.

Let $\mathfrak{i}$ be any ideal in $\mathfrak{I}$, and let $[\mathfrak{q}_1, \ldots, \mathfrak{q}_k, \ldots, \mathfrak{q}_l]$ be an uncontractible representation of $\mathfrak{i}$ as an intersection of primary ideals. Let $\mathfrak{p}_i$ be the radical of $\mathfrak{q}_i$, σ_i the index of $\mathfrak{q}_i$, and suppose that $\mathfrak{p}_1, \ldots, \mathfrak{p}_k$ are $(d-1)$-dimensional, while $\mathfrak{p}_{k+1}, \ldots, \mathfrak{p}_l$ are of dimension less than $d-1$. Then [XV, § 7, Th. VIII] the maximal ideal quasi-equivalent to $\mathfrak{i}$ is

$$\mathfrak{i}^* = [\mathfrak{q}_1, \ldots, \mathfrak{q}_k] = [\mathfrak{p}_1^{(\sigma_1)}, \ldots, \mathfrak{p}_k^{(\sigma_k)}],$$

where $\mathfrak{p}_j^{(\sigma_j)}$ is the symbolic σ_jth power of $\mathfrak{p}_j$. Thus, to pass from $\mathfrak{i}$ to $\mathfrak{i}^*$, we merely omit the components of $\mathfrak{i}$ of dimension less than $d-1$. The integer σ_j $(j \leqslant k)$ is called *the multiplicity* of $\mathfrak{p}_j$ in $\mathfrak{i}^*$, or the multiplicity of the corresponding variety U_j of $d-1$ dimensions. The multiplicative theory of ideals in $\mathfrak{I}$ is thus equivalent to a multiplicative theory of varieties of dimension $d-1$ on V.

In Chapter XI we developed a multiplicative theory of varieties on V, and, for consistency, we have to show that the two theories agree where they overlap. We shall shortly prove that every $(d-1)$-dimensional variety on an affine-normal variety is simple. Hence we can apply Theorem V of § 2. Let

$$\boldsymbol{U} = \sigma_1 \boldsymbol{U}_1 + \ldots + \sigma_k \boldsymbol{U}_k$$

be a multiplicative $(d-1)$-dimensional algebraic variety on V. Let $\mathfrak{p}_i$ be the prime ideal in $\mathfrak{I}$ defined by $\boldsymbol{U}_i$, and let $F_i(u_0, \ldots, u_{d-1})$ be the Cayley form of U_i. Then the Cayley form of the multiplicative variety $\boldsymbol{U}$ is

$$\prod_{i=1}^{k} F_i^{\sigma_i}(u_0, \ldots, u_{d-1}).$$

If $S^0, \ldots, S^{d-1}$ are $(n+1) \times (n+1)$ skew-symmetric matrices whose elements are independent indeterminates over K, the coefficients of the distinct power-products of the s^i_{jk} in

$$\prod_{i=1}^{k} F_i^{\sigma_i}(S^0\xi, \ldots, S^{d-1}\xi)$$

define an ideal $\mathfrak{i}$. But [§ 2, Th. V] the coefficients of the power-products in $F_i(S^0\xi, \ldots, S^{d-1}\xi)$ generate an ideal of the form $[\mathfrak{p}_i, \mathfrak{q}_{i1}, \mathfrak{q}_{i2}, \ldots]$, where $\mathfrak{q}_{ij}$ is embedded in $\mathfrak{p}_i$. Hence

$$[\mathfrak{p}_i, \mathfrak{q}_{i1}, \mathfrak{q}_{i2}, \ldots] \sim \mathfrak{p}_i,$$

and therefore
$$\mathfrak{i} \sim \mathfrak{p}_1^{(\sigma_1)} \ldots \mathfrak{p}_k^{(\sigma_k)}$$
$$\sim [\mathfrak{p}_1^{(\sigma_1)}, \ldots, \mathfrak{p}_k^{(\sigma_k)}].$$

It may appear, at first, that the multiplicative theory of varieties on V given by ideal theory is less powerful than that previously considered, since our present theory deals only with $(d-1)$-dimensional varieties, whereas our earlier theory dealt with varieties of any dimension. Nevertheless, we have, in fact, gained much, for we have defined multiplicities of $(d-1)$-dimensional varieties intrinsically, in terms of properties of a ring of the function field of V, and this will be found to be of immense value in later developments.

We should also note that when $\mathfrak{J}$ is integrally closed a principal ideal in $\mathfrak{J}$ has no embedded components [XV, § 7, Th. X].

The concept of multiplicity which we have developed for ideals in $\mathfrak{J}$ can be extended to give an example of a *valuation* of the function field $K(\xi_1, \ldots, \xi_n)$ of V, a concept to be discussed at length in the next chapter. Let U be an irreducible variety of dimension $d-1$ on V, in A_n, and let $\mathfrak{p}$ be the corresponding prime ideal in $\mathfrak{J}$. Let ζ be any non-zero element of the field $K(\xi_1, \ldots, \xi_n)$, and let α and β be elements of $\mathfrak{J}$ such that $\zeta = \alpha/\beta$. Then we can factorise $\mathfrak{J}.(\alpha)$ and $\mathfrak{J}.(\beta)$ in $\mathfrak{J}$:
$$\mathfrak{J}.(\alpha) \sim \mathfrak{p}^{\rho}\mathfrak{p}_1^{\rho_1} \ldots \mathfrak{p}_k^{\rho_k},$$
$$\mathfrak{J}.(\beta) \sim \mathfrak{p}^{\sigma}\mathfrak{p}_1^{\sigma_1} \ldots \mathfrak{p}_k^{\sigma_k},$$
where $\mathfrak{p}_1, \ldots, \mathfrak{p}_k$ are $(d-1)$-dimensional prime ideals of $\mathfrak{J}$ and $\rho, \rho_1, \ldots, \rho_k$, $\sigma, \sigma_1, \ldots, \sigma_k$ are non-negative integers. Let us define $v(\zeta) = \rho - \sigma$ to be the *value* of ζ at U. Clearly this does not depend on the particular choice of α and β. For if α', β' are any other elements of $\mathfrak{J}$ such that $\zeta = \alpha'/\beta'$, we have
$$\mathfrak{J}.(\alpha') \sim \mathfrak{p}^{\rho'}\mathfrak{p}_1'^{\rho_1'} \ldots,$$
$$\mathfrak{J}.(\beta') \sim \mathfrak{p}^{\sigma'}\mathfrak{p}_1'^{\sigma_1'} \ldots,$$
and since $\alpha\beta' = \alpha'\beta$, we have $\rho + \sigma' = \rho' + \sigma$, by the uniqueness theorem of quasi-factorisation [XV, § 7, Th. V], that is,
$$\rho' - \sigma' = \rho - \sigma = v(\zeta).$$

It is clear from our definition that
$$v(\zeta\eta) = v(\zeta) + v(\eta). \tag{1}$$

Now let η, ζ be two elements of $K(\xi_1, \ldots, \xi_n)$, and suppose that $v(\eta) \geqslant v(\zeta)$. We may write $\eta = \alpha/\gamma$, $\zeta = \beta/\gamma$, where α, β, γ are in $\mathfrak{J}$. If

$$\mathfrak{J}.(\alpha) \sim \mathfrak{p}^{\rho}\mathfrak{p}_1^{\rho_1}\ldots,$$

$$\mathfrak{J}.(\beta) \sim \mathfrak{p}^{\sigma}\mathfrak{p}_1^{\sigma_1}\ldots,$$

$$\mathfrak{J}.(\gamma) \sim \mathfrak{p}^{\tau}\mathfrak{p}_1^{\tau_1}\ldots,$$

$v(\eta) \geqslant v(\zeta)$ implies $\rho \geqslant \sigma$. Hence $\mathfrak{J}.(\alpha+\beta) \geqslant \mathfrak{p}^{\sigma}$, and therefore

$$\mathfrak{J}.(\alpha+\beta) \sim \mathfrak{p}^{\sigma'}\mathfrak{p}_1^{\sigma_1'}\ldots,$$

where $\sigma' \geqslant \sigma$. Hence

$$v(\eta+\zeta) = \sigma' - \tau \geqslant \sigma - \tau = v(\zeta),$$

that is,

$$v(\eta+\zeta) \geqslant \min[v(\eta), v(\zeta)]. \tag{2}$$

Finally, if ζ is any element of $\mathfrak{J}$ in $\mathfrak{p}$,

$$v(\zeta) > 0;$$

and if ζ is in K, $\mathfrak{J}.(\zeta) \sim \mathbf{1}$, and therefore $v(\zeta) = 0$.

Thus having selected an irreducible variety U of dimension $d-1$, we have constructed a mapping of the elements ζ of $K(\xi_1, \ldots, \xi_n)$ on the additive group of integers $v(\gamma)$, which satisfies the relations (1) and (2). This is what is called a *valuation* of the field; the fact that not all the elements of the field map on the zero integer implies that the valuation is not *trivial*, and the property $v(\zeta) = 0$ whenever ζ is a non-zero element of K shows that the valuation satisfies an important restriction which we have to impose on the valuations we consider in the next chapter.

We now prove a result assumed earlier in this section.

THEOREM II. *Every irreducible variety of dimension $d-1$ on an affine normal variety V of dimension d is simple on V.*

Let U be an irreducible variety of dimension $d-1$ on V, and let $\mathfrak{p}$ be the prime ideal of U in $\mathfrak{J}$. Let $\omega_i = \omega_i(\xi_1, \ldots, \xi_n)$ $(i = 1, \ldots, r)$ be a basis for $\mathfrak{p}$, and let $\omega = \omega(\xi_1, \ldots, \xi_n)$ be an element of $\mathfrak{p}$ which is not in $\mathfrak{p}^{(2)}$, the symbolic square of $\mathfrak{p}$. If $\mathfrak{i} = \mathfrak{J}.(\omega)$, $\mathfrak{i} \subseteq \mathfrak{p}$. Hence $\mathfrak{i} \geqslant \mathfrak{p}$, and therefore [XV, §7, Th. III] there exist prime ideals $\mathfrak{p}_1, \ldots, \mathfrak{p}_k$, all different from $\mathfrak{p}$, such that

$$\mathfrak{i} \sim \mathfrak{p}\mathfrak{p}_1\ldots\mathfrak{p}_k, \quad \text{where} \quad \mathfrak{p}\mathfrak{p}_1\ldots\mathfrak{p}_k = 0\ (\mathfrak{i}).$$

Let $\phi = \phi(\xi_1, \ldots, \xi_n)$ be an element of $\mathfrak{p}_1 \ldots \mathfrak{p}_k$ which is not in $\mathfrak{p}$. Then $\omega_j \phi \in \mathfrak{p}\mathfrak{p}_1 \ldots \mathfrak{p}_k \subseteq \mathfrak{i}$, and hence

$$\omega_j \phi = a_j \omega \quad (j = 1, \ldots, r),$$

where $a_j = a_j(\xi_1, \ldots, \xi_n) \in \mathfrak{J}$.

Let

$$\omega_i(x_1, \ldots, x_n) = 0 \quad (i = r+1, \ldots, s)$$

be a basis for the equations of V, and let $a_i(x_1, \ldots, x_n)$ be zero for $i = r+1, \ldots, s$. Then define

$$\nu_i(x_1, \ldots, x_n) = \omega_i(x)\, \phi(x) - a_i(x)\, \omega(x) \quad (i = 1, \ldots, s).$$

The equations $\qquad \omega_i(x) = 0 \quad (i = 1, \ldots, s)$

form a basis for the equations of U, and hence, since $\omega(x)$ vanishes on U, there exist polynomials $b_i(x_1, \ldots, x_n)$ $(i = 1, \ldots, s)$ such that

$$\omega(x) = \sum_1^s b_i(x)\, \omega_i(x).$$

Hence $\quad \nu_i(x) = \sum_{j=1}^{s} [\phi(x)\, \delta_{ij} - a_i(x)\, b_j(x)]\, \omega_j(x) \quad (i = 1, \ldots, s).$

Since, by construction, $\nu_i(\xi)$ is zero for each value of i, we have s equations satisfied by V. If we can prove that the matrix $(\partial \nu_i / \partial x_j)$ is of rank $n-d$ at a generic point η of U, it will follow that U is simple on V [§ 4, p. 137].

Since η is a generic point of U, and the equations $\omega_i(x) = 0$ $(i = 1, \ldots, s)$ form a basis for the equations of U, the matrix $(\partial \omega_i / \partial \eta_j)$ is of rank $n - d + 1$. Again, since

$$\omega_j(\eta) = 0 \quad (j = 1, \ldots, s),$$

it follows that $\quad \dfrac{\partial \nu_i}{\partial \eta_k} = \sum_{j=1}^{s} [\phi(\eta)\, \delta_{ij} - a_i(\eta)\, b_j(\eta)] \dfrac{\partial \omega_j}{\partial \eta_k}.$

Since $\phi(\xi) \neq 0$ $(\mathfrak{p})$, $\phi(x)$ does not vanish on U; hence $\phi(\eta) \neq 0$.

By direct calculation the determinant of the matrix

$$(\phi(\eta)\, \delta_{ij} - a_i(\eta)\, b_j(\eta))$$

is

$$[\phi(\eta)]^s - [\phi(\eta)]^{s-1} \sum_{i=1}^{s} a_i(\eta)\, b_i(\eta).$$

The rank of this matrix is therefore $\rho = s-1$ or s, according as the relation

$$\phi(\eta) = \sum_{i=1}^{s} a_i(\eta)\, b_i(\eta)$$

is or is not satisfied.

It follows from II, § 6, Th. I that the rank of $(\partial v_i/\partial \eta_k)$ is at least $\rho+n-d+1-s \geqslant n-d$. But the rank cannot exceed $n-d$, hence it is exactly $n-d$ and $\rho = s-1$. Since the rank of $(\partial v_i/\partial \eta_k)$ is $n-d$, U is simple on V and Theorem II is proved.

Now let V be any irreducible variety in A_n, of dimension d, not necessarily normal, and let $(\xi_1, \ldots, \xi_n)$ be a generic point. From Theorems II and VII of XV, § 8, we know that there exists a finite set of elements $\eta_1, \ldots, \eta_s$ in the quotient field of $\mathfrak{J} = K[\xi_1, \ldots, \xi_n]$ such that $\mathfrak{J}^* = \mathfrak{J}[\eta_1, \ldots, \eta_s] = K[\xi_1, \ldots, \xi_n, \eta_1, \ldots, \eta_s]$ is the integral closure of $\mathfrak{J}$. Let $\zeta_1, \ldots, \zeta_t$ be any finite set of elements in $\mathfrak{J}^*$ such that $\mathfrak{J}^* = K[\zeta_1, \ldots, \zeta_t]$. Consider an affine space A_t of t dimensions over K, and let V^* be the variety in it whose generic point is $(\zeta_1, \ldots, \zeta_t)$. The integral domain of V^* is $\mathfrak{J}^*$; hence V^* is affine normal, and its quotient field is the quotient field of $\mathfrak{J}$. Hence V^* is birationally equivalent to V. Any variety V^* obtained from V in this way is called a normal variety *associated with* V. We give some properties of the relation between V and V^*.

Let U^* be an irreducible variety on V^*, of dimension s, and let $\mathfrak{p}^*$ be its prime ideal in $\mathfrak{J}^*$. Then [XV, § 4, Th. III] $\mathfrak{p} = \mathfrak{J} \wedge \mathfrak{p}^*$ is a prime ideal. It determines an irreducible variety U of V. We show that U is of dimension s. Let $\zeta_1, \ldots, \zeta_{s+1}$ be any $s+1$ elements of $\mathfrak{J}$ (hence of $\mathfrak{J}^*$). They are algebraically dependent in $\mathfrak{J}^*$ modulo $\mathfrak{p}^*$. Hence there exists a non-zero polynomial $f(z_1, \ldots, z_{s+1})$ over K such that $f(\zeta_1, \ldots, \zeta_{s+1}) \in \mathfrak{p}^*$. But since $\zeta_1, \ldots, \zeta_{s+1}$ are in $\mathfrak{J}$, $f(\zeta_1, \ldots, \zeta_{s+1}) \in \mathfrak{J}$, and therefore $f(\zeta_1, \ldots, \zeta_{s+1})$ belongs to $\mathfrak{J} \wedge \mathfrak{p}^* = \mathfrak{p}$. Hence any $s+1$ elements of $\mathfrak{J}$ are algebraically dependent modulo $\mathfrak{p}$. Therefore U is of dimension s, at most. Suppose that U is of dimension s' $(s' \leqslant s)$. We can find s' elements $\tau_1, \ldots, \tau_{s'}$ in $\mathfrak{J}$ such that any element of $\mathfrak{J}$ is integrally dependent on them, modulo $\mathfrak{p}$. For $\mathfrak{J}/\mathfrak{p}$ is the integral domain of U, and since U is of dimension s' we know that we can find s' elements in its integral domain on which the integral domain is integrally dependent. We then take $\tau_1, \ldots, \tau_{s'}$ to be elements of $\mathfrak{J}$ which correspond to this integral base of $\mathfrak{J}/\mathfrak{p}$. Now any element of $\mathfrak{J}^*$ is integrally dependent on $\mathfrak{J}$, and hence on $\tau_1, \ldots, \tau_{s'}$ modulo $\mathfrak{p}^*$. Hence U^* is of dimension s' at most, that is, $s' \geqslant s$. It follows that $s = s'$. Thus each irreducible variety U^* of dimension s on V^* defines a unique variety of dimension s on V.

Next let U be an irreducible variety of dimension s on V, and let $\mathfrak{p}$ be its prime ideal in $\mathfrak{J}$. Let $\mathfrak{J}^*.\mathfrak{p} = [\mathfrak{q}_1^*, \ldots, \mathfrak{q}_k^*]$, where $\mathfrak{q}_i^*$ is a $\mathfrak{p}_i^*$-primary ideal in $\mathfrak{J}^*$. Now $\mathfrak{p} = \mathfrak{J}_\wedge \mathfrak{p}_i^*$, for at least one value of i [XV, § 4, Th. V, Cor.]; and $\mathfrak{p} \subseteq \mathfrak{J}_\wedge \mathfrak{q}_i^* \subseteq \mathfrak{J}_\wedge \mathfrak{p}_i^*$ for all i. Let U_i^* be the variety on V^* defined by $\mathfrak{p}_i^*$, and let U_i be the variety defined by $\mathfrak{J}_\wedge \mathfrak{p}_i^*$. Then it follows that $U_i \subseteq U$, and, for at least one value of i, $U_i = U$. Suppose that $U_1, \ldots, U_r$ coincide with U, and that $U_{r+1}, \ldots, U_k$ are subvarieties of U. Since $\dim U_i^* = \dim U_i$, $U_1^*, \ldots, U_r^*$ are of dimension s, and $U_{r+1}^*, \ldots, U_k^*$ are of dimension less than s. We say that $U_1^*, \ldots, U_r^*$ are the varieties of V^* which correspond to U on V.

If U^* is any variety of V^* which determines the variety U of V, we have $\mathfrak{J}_\wedge \mathfrak{p}^* = \mathfrak{p}$, and therefore $\mathfrak{J}^*.\mathfrak{p} = 0 \ (\mathfrak{p}^*)$ [XV, § 4, Th. VI]. Since $\mathfrak{J}^*.\mathfrak{p}$ is of dimension s, and U^* is of dimension s, it follows that U^* is one of the varieties $U_1^*, \ldots, U_r^*$. In particular, if $\mathfrak{J}_\mathfrak{p}$ is integrally closed, Theorem VIII of XV, § 8 tells us that $r = 1$, that $\mathfrak{q}_1^* = \mathfrak{p}_1^*$, and that $\mathfrak{J}^*_{\mathfrak{p}_1^*} = \mathfrak{J}_\mathfrak{p}$. Hence if the quotient ring of U is integrally closed there is a unique variety U^* on V^* corresponding to it, and $Q(U) = Q(U^*)$. Thus we have:

THEOREM III. *A correspondence exists between the irreducible varieties of dimension s on an irreducible variety V in affine space, and the irreducible varieties of dimension s on an affine-normal variety V^* associated with V. To each variety U^* of dimension s on V^* there corresponds a unique variety U of dimension s on V, and to each variety U of dimension s on V there corresponds a finite number of varieties on V^*. When the quotient ring of U is integrally closed, the variety U^* on V^* which corresponds to it is unique, and $Q(U) = Q(U^*)$.*

The relation between two normal varieties in affine space associated with the same variety V can be obtained at once, since each can be regarded as a normal variety associated with the other. The correspondence between the irreducible subvarieties is one-to-one.

The reader will observe that we have spoken of correspondences between V and V^*, but have not shown that this concept agrees with the definition of Chapter XI. Here, however, we have not considered corresponding *points* of V and V^*, but corresponding irreducible *varieties*. The ideas here considered do, however, bear a close relationship to those of Chapter XI. In fact, it is easy to set up a correspondence between V and V^* of the type considered in Chapter XI, such that to each point of V there corresponds a finite number of points on V^*, and to each point of V^* there corresponds a unique point of V. If U is any irreducible variety of V, given by the prime ideal $\mathfrak{p}$ of $\mathfrak{I}$, and $\mathfrak{I}^*.\mathfrak{p}$ defines $U^* = U_1^* + \ldots + U_k^*$, the aggregate of points of V^* which correspond to the points of U constitute the whole variety U^*; and if U^* is any irreducible subvariety of V^*, given by the prime ideal $\mathfrak{p}^*$ of $\mathfrak{I}^*$, the locus of the points P of V which correspond to the points P^* on U^*, as P^* describes U^*, is given by the ideal $\mathfrak{I}_\wedge \mathfrak{p}^*$ of $\mathfrak{I}$. We do not, however, prove these results here, for they are contained in more general results which will be proved for birational correspondences in Chapter XVIII.

6. Projectively normal varieties. Most of the results in this chapter have been proved for varieties in affine space, although those which have been expressed in terms of the quotient ring of an irreducible subvariety of the given irreducible variety can be interpreted as projective properties. While the results proved for varieties in an affine space have considerable importance in their own right, we are mainly concerned with the use which can be made of them to derive properties of a variety in projective space. We are now in a position to use the results already obtained in this chapter to introduce and examine the important concept of *projectively normal varieties*.

Let V be an irreducible variety of dimension d over the ground field K, lying in the projective space S_n. As usual, we shall assume that V does not lie in the prime $x_0 = 0$. Let $\xi = (\xi_0, \ldots, \xi_n)$ be a generic point of V, and assume, as we may, that its coordinates have been normalised so that $\xi_0 = 1$. The most general form of the coordinates of this point is

$$(\tau_0, \tau_1, \ldots, \tau_n) = (\tau_0, \tau_0\xi_1, \ldots, \tau_0\xi_n),$$

where τ_0 lies in some extension of K and is not zero. We shall find it convenient to suppose that τ_0 is an indeterminate over the function field $\Sigma = K(\xi_1, \ldots, \xi_n)$ of V. Then, since ξ_i is in Σ, $\tau_i = \tau_0\xi_i$ is an indeterminate over Σ for $i = 1, \ldots, n$. In this section we shall always assume that the homogeneous coordinates of the generic point $(\tau_0, \ldots, \tau_n)$ of V_d are such that

(i) $$\tau_i = \tau_0 \xi_i;$$

(ii) $$\Sigma^* = K(\tau_0, \ldots, \tau_n) = K(\tau_0, \tau_0\xi_1, \ldots, \tau_0\xi_n) = \Sigma(\tau_0)$$

is a pure transcendental extension of Σ.

If we regard $(x_0, \ldots, x_n)$ as the non-homogeneous coordinates of a point in A_{n+1}, $(\tau_0, \ldots, \tau_n)$ is the generic point of a variety V^* of dimension $d+1$, whose function field is Σ^*. To each point $(t'_0, \ldots, t'_n)$ of V^*, different from $(0, \ldots, 0)$, there corresponds a unique point with homogeneous coordinates $(t'_0, \ldots, t'_n)$ of V, while to each point $(x'_0, \ldots, x'_n)$ of V there corresponds the set of points of V^* whose coordinates are of the form $(t'x'_0, \ldots, t'x'_n)$. These constitute a line through $(0, \ldots, 0)$. Thus V^* is a 'cone' with vertex at $(0, \ldots, 0)$, and the section of V^* by the prime at infinity of A_{n+1} is V_n. We shall not, however, use this representation.

If t is any non-zero element of K, $(t\tau_0, \ldots, t\tau_n)$ is also a generic point of V, and $K(t\tau_0, \ldots, t\tau_n) = \Sigma(\tau_0) = \Sigma^*$. Let ζ be any element of Σ^*. Then we may write

$$\zeta = \frac{a_0 + a_1 + \ldots + a_\rho}{b_0 + b_1 + \ldots + b_\sigma},$$

where a_i and b_i are forms in $\tau_0, \ldots, \tau_n$ over K of degree i. The mapping in Σ^* defined by $\tau_i \to t\tau_i$ takes ζ into $\zeta^{(t)}$, where

$$\zeta^{(t)} = \frac{a_0 + ta_1 + \ldots + t^\rho a_\rho}{b_0 + tb_1 + \ldots + t^\sigma b_\sigma}.$$

ζ is said to be *homogeneous of degree* λ if $\zeta^{(t)} = t^\lambda \zeta$ for every choice of t in K. Suppose that ζ is homogeneous of degree λ, where $\lambda \geqslant 0$. Then

$$a_0 + ta_1 + \ldots + t^\rho a_\rho = \zeta t^\lambda (b_0 + tb_1 + \ldots + t^\sigma b_\sigma).$$

Since this relation holds for $t = 0, 1, 2, \ldots$ we deduce that

$$a_\mu = 0 \quad (\mu < \lambda), \qquad a_{\lambda+\mu} = \zeta b_\mu \quad (\mu = 0, 1, \ldots),$$

and since for one value of μ at least, say $\mu = \nu$, $b_\nu \neq 0$, we have $\zeta = a_{\lambda+\nu}/b_\nu$, that is, ζ is the ratio of a homogeneous polynomial in $\tau_0, \ldots, \tau_n$ over K of degree $\lambda+\nu$, to a homogeneous polynomial of degree ν. A similar argument holds if $\lambda < 0$. Hence any element of

Σ^* which is homogeneous of degree λ is of the form $\alpha\tau_0^\lambda$, where $\alpha \in \Sigma$; and conversely, if

$$\alpha = \beta(\xi_1, \ldots, \xi_n)/\gamma(\xi_1, \ldots, \xi_n)$$

and $$\zeta = \alpha\tau_0^\lambda = \tau_0^{\lambda+\nu}\beta(\tau_1/\tau_0, \ldots, \tau_n/\tau_0)/\tau_0^\nu\gamma(\tau_1/\tau_0, \ldots, \tau_n/\tau_0),$$

ζ is the ratio of a homogeneous polynomial of degree $\lambda+\nu$ to a homogeneous polynomial of degree ν, and it is therefore homogeneous of degree λ. In particular, an element of Σ^* which is homogeneous of degree zero belongs to Σ, and conversely any element of Σ is homogeneous of degree zero.

We shall denote the ring $K[\tau_0, \ldots, \tau_n]$ by $\mathfrak{J}^*$. Its quotient field is, of course, Σ^*. We shall say that the variety V is *projectively normal* if every homogeneous element of Σ^* of degree of homogeneity greater than zero which is integrally dependent on $\mathfrak{J}^*$ lies in $\mathfrak{J}^*$. The existence of a variety $\bar{V}$ with the same function field as V which is projectively normal cannot be proved simply by finding a $\mathfrak{J}^*$-basis $\tau_{n+1}, \ldots, \tau_{n+r}$ for the integral closure of $\mathfrak{J}^*$ and then considering the variety in S_{n+r} whose generic point is $(\tau_0, \ldots, \tau_{n+r})$; for $\tau_{n+1}, \ldots, \tau_{n+r}$ may not all be of degree of homogeneity one. If they are not, it is easy to see that

$$K(\tau_0/\tau_0, \ldots, \tau_n/\tau_0, \tau_{n+1}/\tau_0, \ldots, \tau_{n+r}/\tau_0)$$

is of dimension $d+1$ over K, so that the variety with the generic point $(\tau_0, \ldots, \tau_{n+r})$ is of dimension greater than that of V. We must therefore prove the following theorem.

THEOREM I. *If V is an irreducible variety of dimension d over K, there exists a birational transform of V which is projectively normal.*

In proving this theorem we shall obtain a well-defined class of birational transforms of V which are projectively normal. These varieties will be called the *derived normal varieties* of V_d.

The first stage of our proof is to show that every element of Σ^* which is integrally dependent on $\mathfrak{J}^*$ is a sum of homogeneous elements of Σ^*, all of which are of non-negative degrees of homogeneity, and each of which is integrally dependent on $\mathfrak{J}^*$. Let ζ be any element of Σ^* which is integrally dependent on $\mathfrak{J}^*$. Then we have an equation

$$\zeta^m + \alpha_1\zeta^{m-1} + \ldots + \alpha_m = 0,$$

where $\alpha_1, \ldots, \alpha_m$ are in $\mathfrak{J}^*$. Since $\Sigma^* = \Sigma(\tau_0)$ and $\mathfrak{J}^* \subseteq \Sigma[\tau_0]$, this implies that ζ is an element of $\Sigma(\tau_0)$ integrally dependent on $\Sigma[\tau_0]$.

But [XV, § 8, Th. VI] $\Sigma[\tau_0]$ is integrally closed in its quotient field. Hence $\zeta \in \Sigma[\tau_0]$. Therefore we can write

$$\zeta = \zeta_0 + \zeta_1 + \ldots + \zeta_\rho,$$

where ζ_i is homogeneous of degree i. If t is any non-zero element of K, the mapping $\tau_i \to t\tau_i$ transforms $\mathfrak{I}^*$ into itself; hence it transforms any element of Σ^* which is integrally dependent on $\mathfrak{I}^*$ into another element of Σ^* which is integrally dependent on $\mathfrak{I}^*$. In particular,

$$\zeta^{(t)} = \zeta_0 + t\zeta_1 + \ldots + t^\rho \zeta_\rho$$

is integrally dependent on $\mathfrak{I}^*$. Take $t = 0, 1, \ldots, \rho$. Then $\zeta^{(0)}, \ldots, \zeta^{(\rho)}$ are integrally dependent on $\mathfrak{I}^*$, and we have $\rho + 1$ equations to determine $\zeta_0, \ldots, \zeta_\rho$ in terms of $\zeta^{(0)}, \ldots, \zeta^{(\rho)}$. The determinant of the coefficients of these equations is a non-zero element of K, hence a unit of $\mathfrak{I}^*$, and therefore $\zeta_0, \ldots, \zeta_\rho$ are expressible as sums of elements integrally dependent on $\mathfrak{I}^*$, and so are themselves integrally dependent on $\mathfrak{I}^*$. This is what we had to prove.

The integral closure $\bar{\mathfrak{I}}$ of $\mathfrak{I}^*$ in Σ^* has a finite $\mathfrak{I}^*$-basis [XV, § 8, Th. VII]. Any member of this basis is a sum of homogeneous elements each of which is integrally dependent on $\mathfrak{I}^*$; hence we may assume that the $\mathfrak{I}^*$-basis of $\bar{\mathfrak{I}}$ consists of homogeneous elements. Since the ring $\mathfrak{I}^*$ is generated over K by the homogeneous elements $\tau_0, \ldots, \tau_n$, it follows from XV, § 8, Th. II that $\bar{\mathfrak{I}}$ can be generated over K by a finite number of homogeneous elements of non-negative degree:

$$\bar{\mathfrak{I}} = K[\eta_0, \ldots, \eta_r],$$

where $\eta_0, \ldots, \eta_r$ are homogeneous elements of Σ^* of non-negative degree. Let us now consider the elements of $\bar{\mathfrak{I}}$ which are homogeneous of degree zero. If ζ is homogeneous of degree zero and satisfies the equation

$$\zeta^m + \alpha_1 \zeta^{m-1} + \ldots + \alpha_m = 0,$$

where $\alpha_i \in \mathfrak{I}^*$, we write

$$\alpha_i = \sum_{j \geqslant 0} \alpha_{ij},$$

where α_{ij} is of degree of homogeneity j, and $\alpha_{i0} \in K$. Then, considering the automorphisms $\tau_i \to t\tau_i$ of Σ^*, we have the equation

$$\zeta^m + \sum_{j \geqslant 0} \alpha_{1j} t^j \zeta^{m-1} + \ldots + \sum_{j \geqslant 0} \alpha_{mj} t^j = 0,$$

and as this holds for all t in K, we obtain the equation

$$\zeta^m + \alpha_{10}\zeta^{m-1} + \ldots + \alpha_{m0} = 0.$$

Since ζ is an element of Σ^* of degree of homogeneity zero, it belongs to Σ, and by the above equation it is algebraic over K. Hence ζ belongs to the algebraic closure K' of K in Σ. Conversely, if ζ is any element of K', it satisfies an equation

$$\zeta^m + a_1\zeta^{m-1} + \ldots + a_m = 0,$$

where $a_i \in K$. Hence ζ is integrally dependent on $\mathfrak{J}^*$, and therefore lies in $\bar{\mathfrak{J}}$.

Let σ be the degree of K' over K. Then in $\bar{\mathfrak{J}}$ there are σ independent elements over K of degree of homogeneity zero, and without loss of generality we may assume that in $\bar{\mathfrak{J}} = K[\eta_0, \ldots, \eta_r]$, $\eta_0, \ldots, \eta_\rho$ are of positive degree of homogeneity, while $\eta_{\rho+1}, \ldots, \eta_r$ are of degree of homogeneity zero, where $r - \rho = \sigma$. Let η_i $(i \leqslant \rho)$ be of degree of homogeneity d_i. Any element of $\bar{\mathfrak{J}}$ of degree of homogeneity k can be expressed as a linear combination, with coefficients in K, of products

$$\eta_0^{i_0} \ldots \eta_\rho^{i_\rho}\zeta \quad (i_j \geqslant 0),$$

where

$$i_0 d_0 + \ldots + i_\rho d_\rho = k, \tag{1}$$

and ζ is of degree of homogeneity zero and lies in $\bar{\mathfrak{J}}$. Since ζ is a linear combination, with coefficients in K, of $\eta_{\rho+1}, \ldots, \eta_r$, it follows that the products $\eta_0^{i_0} \ldots \eta_\rho^{i_\rho}\eta_{\rho+j}$, where $i_0, \ldots, i_\rho$ satisfy (1), and $j = 1, \ldots, r-\rho$, form a basis for the K-module formed by elements of $\bar{\mathfrak{J}}$ of degree of homogeneity k. We shall denote an independent K-basis of this module by $\zeta_0^{(k)}, \ldots, \zeta_{n_k}^{(k)}$.

An integer δ $(\delta > 0)$ will be called a *character of homogeneity* of V if it has the property that a K-basis for the module of elements of $\bar{\mathfrak{J}}$ of degree of homogeneity $k\delta$, where k is any positive integer, can be constructed from the power-products of $\zeta_0^{(\delta)}, \ldots, \zeta_{n_\delta}^{(\delta)}$ of degree k, that is, if

$$\zeta_i^{(k\delta)} = f_i(\zeta_0^{(\delta)}, \ldots, \zeta_{n_\delta}^{(\delta)}) \quad (i = 0, \ldots, n_{k\delta}),$$

where $f_i(z_0, \ldots, z_{n_\delta})$ is a homogeneous polynomial over K. We shall prove shortly that characters of homogeneity of V exist, but as yet little is known of the possible characters of homogeneity of a variety. However, it is easily seen that if δ is any character of homogeneity of V, so is $t\delta$, where t is any positive integer. For, since δ is a character of homogeneity of V, $\zeta_i^{(kt\delta)}$ is equal to a homogeneous polynomial over K of degree kt in $\zeta_0^{(\delta)}, \ldots, \zeta_{n_\delta}^{(\delta)}$, and hence is equal to a homogeneous polynomial of degree k in the power-products of $\zeta_0^{(\delta)}, \ldots, \zeta_{n_\delta}^{(\delta)}$ of degree t. These power-products are, in turn, expressible as linear combinations of $\zeta_0^{(t\delta)}, \ldots, \zeta_{n_{t\delta}}^{(t\delta)}$, and from this the result follows. If V is projectively normal, $\bar{\mathfrak{J}} = \mathfrak{J}^*$. Any element of $\mathfrak{J}^*$ of degree of homogeneity k is a homogeneous polynomial in $\tau_0, \ldots, \tau_n$ of degree k. Hence, if V is projectively normal, the integer one is a character of homogeneity.

We establish the existence of characters of homogeneity of V by showing that if d is the least common multiple of $d_0, \ldots, d_\rho$ (where d_i is the degree of homogeneity of η_i), ρd is a character of homogeneity of V. Let k be any integer greater than ρ. To find the elements of $\bar{\mathfrak{J}}$ of degree of homogeneity kd, we consider the power-products $\eta_0^{i_0} \ldots \eta_\rho^{i_\rho} \eta_{\rho+i}$, where

$$i_0 d_0 + \ldots + i_\rho d_\rho = kd. \tag{2}$$

If $i_j < d/d_j$ $(i = 0, \ldots, \rho)$,

$$i_0 d_0 + \ldots + i_\rho d_\rho < (\rho + 1) d \leqslant kd.$$

Hence at least one i_j is greater than or equal to d/d_j. Suppose, for instance, that $i_0 \geqslant d/d_0$. Then

$$\eta_0^{i_0} \ldots \eta_\rho^{i_\rho} \eta_{\rho+j} = \eta_0^{d/d_0} \eta_0^{i_0'} \eta_1^{i_1} \ldots \eta_\rho^{i_\rho} \eta_{\rho+j},$$

where $i_0' \geqslant 0$ and

$$i_0' d_0 + i_1 d_1 + \ldots + i_\rho d_\rho = (k-1) d.$$

Hence if $k > \rho$, any power-product of $\eta_0, \ldots, \eta_r$ of degree of homogeneity kd is the product of an element of $\bar{\mathfrak{J}}$ of degree of homogeneity d and a power-product of degree of homogeneity $(k-1)d$. If $k - 1 > \rho$, we treat this power-product in the same way, and continue thus. If $k = m + \rho$, we see that $\eta_0^{i_0} \ldots \eta_\rho^{i_\rho} \eta_{\rho+j}$ is equal to the product of m elements of degree of homogeneity d and a power-product of degree of homogeneity ρd. If $m = t\rho$, it follows that any power-product $\eta_0^{i_0} \ldots \eta_\rho^{i_\rho} \eta_{\rho+j}$ of degree of homogeneity $(t+1)\rho d$ is equal to the product of $t+1$ elements of degree of homogeneity ρd. Hence any element of $\bar{\mathfrak{J}}$ of degree of homogeneity $(t+1)\rho d$ is equal to a homogeneous form over K of elements of degree of homogeneity ρd. This shows that ρd is a character of homogeneity of V.

Now let δ be any character of homogeneity of V, and let $V^{(\delta)}$ be the variety in S_{n_δ} which has $(\zeta_0^{(\delta)}, \ldots, \zeta_{n_\delta}^{(\delta)})$ as a generic point over the field K. We establish Theorem I by proving that $V^{(\delta)}$ is birationally equivalent to V, and that it is projectively normal. Since $(\tau_0)^\delta$ is an element of $\bar{\mathfrak{J}}$ of degree of homogeneity δ, it is equal to a linear combination of $\zeta_0^{(\delta)}, \ldots, \zeta_{n_\delta}^{(\delta)}$ with coefficients in K; without any loss of generality we may suppose that $\zeta_0^{(\delta)} = (\tau_0)^\delta$. Then, for any value of i, $\tau_i(\tau_0)^{\delta-1}$ is an element of $\bar{\mathfrak{J}}$ of degree of homogeneity δ; hence

$$\tau_i(\tau_0)^{\delta-1} = \sum_{j=0}^{n_\delta} a_{ij}\zeta_j^{(\delta)}.$$

Therefore
$$\tau_i/\tau_0 = \sum_j a_{ij}\zeta_j^{(\delta)}/\zeta_0^{(\delta)},$$

and hence
$$K(\tau_1/\tau_0, \ldots, \tau_n/\tau_0) \subseteq K(\zeta_1^{(\delta)}/\zeta_0^{(\delta)}, \ldots, \zeta_{n_\delta}^{(\delta)}/\zeta_0^{(\delta)}).$$

On the other hand, $\zeta_i^{(\delta)}/\zeta_0^{(\delta)}$ is an element of Σ^* of degree of homogeneity zero; hence it is contained in Σ. Therefore

$$K(\zeta_1^{(\delta)}/\zeta_0^{(\delta)}, \ldots, \zeta_{n_\delta}^{(\delta)}/\zeta_0^{(\delta)}) \subseteq K(\tau_1/\tau_0, \ldots, \tau_n/\tau_0) = \Sigma,$$

and we conclude that

$$K(\zeta_1^{(\delta)}/\zeta_0^{(\delta)}, \ldots, \zeta_{n_\delta}^{(\delta)}/\zeta_0^{(\delta)}) = \Sigma.$$

But $K(\zeta_1^{(\delta)}/\zeta_0^{(\delta)}, \ldots, \zeta_{n_\delta}^{(\delta)}/\zeta_0^{(\delta)})$ is the function field of $V^{(\delta)}$. It therefore follows that V and $V^{(\delta)}$ have the same function field; that is, they are birationally equivalent.

The field $\Sigma^{(\delta)} = K(\zeta_0^{(\delta)}, \ldots, \zeta_{n_\delta}^{(\delta)})$, which is equal to $\Sigma(\zeta_0^{(\delta)}) = \Sigma(\tau_0^\delta)$, is a pure transcendental extension of the function field Σ of $V^{(\delta)}$. In considering the degree of homogeneity of elements of $\Sigma^{(\delta)}$, the extended field associated with $V^{(\delta)}$, we must count $\zeta_0^{(\delta)}$ as of degree of homogeneity one. If ζ is any element of $\Sigma^{(\delta)}$ of degree of homogeneity ρ $(\rho > 0)$, then, since $\Sigma^{(\delta)} \subseteq \Sigma^* = \Sigma(\tau_0)$, ζ can be regarded as an element of Σ^*, and then its degree of homogeneity is $\rho\delta$. If ζ is integrally dependent on $K[\zeta_0^{(\delta)}, \ldots, \zeta_{n_\delta}^{(\delta)}]$, it is integrally dependent on $\mathfrak{J}^* = K[\tau_0, \ldots, \tau_n]$, since $K[\zeta_0^{(\delta)}, \ldots, \zeta_{n_\delta}^{(\delta)}] \subseteq \bar{\mathfrak{J}}$, which is integrally dependent on $\mathfrak{J}^*$. Hence, $\zeta \in \bar{\mathfrak{J}}$. But, since δ is a character of homogeneity of V, and ζ is of degree of homogeneity $\rho\delta$, regarded as an element of Σ^*, ζ is equal to a homogeneous polynomial in $\zeta_0^{(\delta)}, \ldots, \zeta_{n_\delta}^{(\delta)}$, and hence lies in $K[\zeta_0^{(\delta)}, \ldots, \zeta_{n_\delta}^{(\delta)}]$. Now any element η of $\Sigma^{(\delta)}$ which is integrally dependent on $K[\zeta_0^{(\delta)}, \ldots, \zeta_{n_\delta}^{(\delta)}]$ is the sum of homogeneous elements, each of which is integrally dependent on this ring; hence $\eta \in K[\zeta_0^{(\delta)}, \ldots, \zeta_{n_\delta}^{(\delta)}]$. This ring is therefore integrally closed, and hence $V^{(\delta)}$ is projectively normal.

The variety $V^{(\delta)}$ is determined by V and the character of homogeneity δ, to within a projective transformation. For $V^{(\delta)}$ is determined by the K-module of elements of $\bar{\mathfrak{J}}$ of degree of homogeneity δ. The set $\zeta_0^{(\delta)}, \ldots, \zeta_{n_\delta}^{(\delta)}$ is an independent basis of this module. If $\eta_0^{(\delta)}, \ldots, \eta_{n'}^{(\delta)}$ is any other independent basis for the module, $n' = n_\delta$, and

$$\zeta_i^{(\delta)} = \sum_{j=0}^{n_\delta} a_{ij} \eta_j^{(\delta)} \quad (i = 0, \ldots, n_\delta),$$

$$\eta_i^{(\delta)} = \sum_{j=0}^{n_\delta} b_{ij} \zeta_j^{(\delta)} \quad (i = 0, \ldots, n_\delta),$$

where a_{ij} and b_{ij} are in K. Then

$$\zeta_i^{(\delta)} = \sum_j \sum_k a_{ij} b_{jk} \zeta_k^{(\delta)},$$

and since $\zeta_0^{(\delta)}, \ldots, \zeta_{n_\delta}^{(\delta)}$ are linearly independent over K, $\sum_{j=0}^{n_\delta} a_{ij} b_{jk} = \delta_{ik}$. Hence (a_{ij}) and (b_{ij}) are non-singular matrices. If $'V^{(\delta)}$ is the variety in S_{n_δ} having $(\eta_0^{(\delta)}, \ldots, \eta_{n_\delta}^{(\delta)})$ as generic point, $'V^{(\delta)}$ is obtained from $V^{(\delta)}$ by the projective transformation

$$y_i = \sum_j b_{ij} x_j \quad (i = 0, \ldots, n_\delta).$$

We should also note that $V^{(\delta)}$ does not lie in any linear subspace of S_{n_δ}. For if it lay in

$$c_0 z_0 + c_1 z_1 + \ldots + c_{n_\delta} z_{n_\delta} = 0,$$

we should have

$$c_0 \zeta_0^{(\delta)} + \ldots + c_{n_\delta} \zeta_{n_\delta}^{(\delta)} = 0,$$

whereas, by arrangement, $\zeta_0^{(\delta)}, \ldots, \zeta_{n_\delta}^{(\delta)}$ are linearly independent over K.

The varieties $V^{(\delta_1)}, V^{(\delta_2)}, \ldots$ defined from V and corresponding to the characters of homogeneity $\delta_1, \delta_2, \ldots$, are called the *derived normal varieties* corresponding to V.

We now consider the relation between two derived normal varieties $V^{(\delta)}$ and $V^{(\delta')}$ derived from V. We first introduce a concept which can be defined for any irreducible variety U of d dimensions in S_n, whether it is projectively normal or not. We again denote the normalised generic point of U by $(1, \xi_1, \ldots, \xi_n)$, and let τ_0 be an indeterminate over $K(\xi_1, \ldots, \xi_n)$, and write $\tau_i = \tau_0 \xi_i$. Then $(\tau_0, \ldots, \tau_n)$ is a generic point of U. The elements of $K[\tau_0, \ldots, \tau_n]$ of degree of homogeneity k, where k is any positive integer, form a finite K-module. Let $\zeta_0, \ldots, \zeta_{n'}$ be an independent K-basis for this module.

Without loss of generality we may suppose that $\zeta_0 = \tau_0^k$, and that each ζ_i is a power-product of τ's. Let U' be the variety in $S_{n'}$ whose generic point is $(\zeta_0, \ldots, \zeta_{n'})$. Then, by an argument given above, we show that

$$K(\tau_1/\tau_0, \ldots, \tau_n/\tau_0) = K(\zeta_1/\zeta_0, \ldots, \zeta_{n'}/\zeta_0);$$

that is, U' is birationally equivalent to U. We consider the correspondence C between S_n and $S_{n'}$ given by the generic point (τ, ζ) in $S_{n,n'}$. The varieties U and U' are the object- and image-varieties of the correspondence. We show that to each point of U there corresponds, in this correspondence, a unique point of U', and to each point of U' there corresponds a unique point of U. The varieties U and U' are then said to be in *one-to-one correspondence, without exception.*

Since U and U' are the object- and image-varieties of the correspondence, to each point of one there corresponds at least one point of the other. Let $\zeta_i = f_i(\tau_0, \ldots, \tau_n)$, where $f_i(x_0, \ldots, x_n)$ is homogeneous of degree k. We first show that for no proper specialisation of τ can the corresponding specialisations of all $f_i(\tau_0, \ldots, \tau_n)$ vanish. Indeed, for $i = 0, \ldots, n$, we can write

$$\tau_i^k = \sum_{j=0}^{n'} a_{ij} f_j(\tau) \quad (a_{ij} \in K),$$

and since for any specialisation t' of τ at least one t'_ρ is different from zero, for this value of ρ,

$$(t'_\rho)^k = \sum_{j=0}^{n'} a_{\rho j} f_j(t') \neq 0,$$

and the result follows.

Again, if $(x_0, \ldots, x_n)$ and $(z_0, \ldots, z_{n'})$ are the coordinate systems in S_n and $S_{n'}$, respectively, we have, among the equations of C, the equations

$$z_i f_j(x) - z_j f_i(x) = 0 \quad (i, j = 0, \ldots, n'),$$

and since for no point x' of U is

$$f_i(x') = 0 \quad (i = 0, \ldots, n'),$$

these equations show that to each point x' of U there corresponds a unique point of U'.

To show that to each point of U' there corresponds a unique point of U, we consider any specialisation $z' = (z'_0, \ldots, z'_{n'})$ of ζ. One z'_i at least, say z'_ρ, is different from zero. Let

$$\zeta_\rho = f_\rho(\tau_0, \ldots, \tau_n) = \tau_0^{k_0} \ldots \tau_n^{k_n},$$

and let j be chosen so that $k_j > 0$. Since $\tau_0^{k_0} \ldots \tau_j^{k_j-1} \ldots \tau_n^{k_n}\tau_i$ is homogeneous of degree k, we must have

$$\tau_0^{k_0} \ldots \tau_j^{k_j-1} \ldots \tau_n^{k_n}\tau_i = \sum_{\lambda=0}^{n'} b_{i\lambda}\zeta_\lambda \quad (i = 0, \ldots, n),$$

where $b_{i\lambda} \in K$, and $b_{j\lambda} = 0$ for $\lambda \neq \rho$, while $b_{j\rho} = 1$. Then the correspondence C satisfies the equations

$$x_i z_\rho - x_j \sum_{\lambda=0}^{n'} b_{i\lambda} z_\lambda = 0 \quad (i = 0, \ldots, n).$$

From these equations it follows that $(x'_0, \ldots, x'_n)$, where

$$x'_i = \sum_{\lambda=0}^{n'} b_{i\lambda} z'_\lambda,$$

is the only point of U which can correspond to z'.

Hence the correspondence between U and U' is one-to-one without exception. The variety U' obtained from U is called *the transform of U by primals of order k*.

It is clear that if δ is any character of homogeneity of V, and k is any positive integer, $V^{(k\delta)}$ is the transform of $V^{(\delta)}$ by primals of order k. If $V^{(\delta)}$ and $V^{(\delta')}$ are two derived normal varieties, and Δ is the least common multiple of δ and δ', where

$$\Delta = e\delta = e'\delta',$$

$V^{(\Delta)}$ is the transform of $V^{(\delta)}$ by primals of order e, and $V^{(\Delta)}$ is also the transform of $V^{(\delta')}$ by primals of order e', and it follows that the correspondence between $V^{(\delta)}$ and $V^{(\delta')}$ is one-to-one without exception. Hence we have

THEOREM II. *If $V^{(\delta)}$ and $V^{(\delta')}$ are derived normal varieties of V corresponding to the characters of homogeneity δ and δ', and if $\Delta = e\delta = e'\delta'$ be the least common multiple of δ and δ', the transform of $V^{(\delta)}$ by primals of order e is projectively equivalent to the transform of $V^{(\delta')}$ by primals of order e', and there is a one-to-one correspondence (without exception) between the points of $V^{(\delta)}$ and the points of $V^{(\delta')}$.*

Now suppose that V is projectively normal in S_n, and A_n is any affine space derived from S_n (with the proviso that V is not entirely at infinity in A_n). Then we show that V is affine-normal in A_n.

If the prime at infinity is

$$c_0 x_0 + \ldots + c_n x_n = 0,$$

then the integral domain of V in A_n is

$$\mathfrak{J} = K\left[\frac{\tau_0}{\Sigma c_i \tau_i}, \frac{\tau_1}{\Sigma c_i \tau_i}, \ldots, \frac{\tau_n}{\Sigma c_i \tau_i}\right].$$

Let η be any element of the quotient field of $\mathfrak{J}$ which is integrally dependent on $\mathfrak{J}$; then there exists an integer ρ $(\rho > 0)$ such that $\zeta = \eta(\Sigma c_i \tau_i)^\rho$ is integrally dependent on $\mathfrak{J}^* = K[\tau_0, \ldots, \tau_n]$. Since V is projectively normal, ζ is equal to a form $f(\tau_0, \ldots, \tau_n)$ over K of degree ρ. Hence

$$\eta = f\left(\frac{\tau_0}{\Sigma c_i \tau_i}, \ldots, \frac{\tau_n}{\Sigma c_i \tau_i}\right) \in \mathfrak{J}.$$

Therefore V is affine-normal in A_n. Thus we have

THEOREM III. *If V is a projectively normal variety in S_n, and A_n is any affine space defined from S_n by taking as prime at infinity any prime not containing V, then V is affine-normal in A_n.*

We can now apply Theorem II of § 5, and deduce

THEOREM IV. *Every irreducible variety of dimension $d-1$ on a projectively normal variety V of dimension d is simple on V;*

and from Theorems I and IV we deduce, as a particular case,

THEOREM V. *Any irreducible variety of dimension one can be birationally transformed into one on which every point is simple.*

The reader may be tempted to assume that the converse of Theorem III is true, namely, that if V is a variety in S_n such that whenever we pass from S_n to an affine space A_n, we find that V is affine-normal in A_n, then V is projectively normal in S_n; but this is not true, as the following example will show. In S_3 consider the variety V having the generic point $(1, t, t^3, t^4)$, where t is an indeterminate over the ground field K. The reader will be able to verify that every point of this variety is simple; hence the quotient ring of every zero-dimensional variety is integrally closed. If we pass to any affine space A_3, the integral domain of V in A_3 is the intersection of the quotient rings of the zero-dimensional varieties on it [XV, § 5, Th. X], and hence is integrally closed. V is therefore affine-normal in A_3. We show, however, that V is not projectively normal in S_3. Let τ_0 be an indeterminate over the function field $K(t)$ of V, and write $\tau_1 = \tau_0 t$, $\tau_2 = \tau_0 t^3$, $\tau_3 = \tau_0 t^4$. V is projectively normal if and only if $K[\tau_0, \ldots, \tau_3]$ is integrally closed in its quotient field $K(t, \tau_0)$. Now if $\zeta = \tau_0 \tau_2/\tau_1$, $\zeta^2 = (\tau_0 t^2)^2 = \tau_0 \tau_3$. Hence ζ is integrally dependent on $K[\tau_0, \ldots, \tau_3]$. But ζ is not in this ring. It follows that V is not projectively normal.

The truth is that the fact that a variety is affine-normal, however we choose the prime at infinity, is essentially a *local* property of the variety; that is, it can be expressed in terms of properties of subvarieties (as, in fact, we have expressed it in the foregoing example), whereas the property of being projectively normal is a property of the variety as a whole.

The next theorem will be used to obtain a new characterisation of a projectively normal variety, but it is of some importance in its own right. Let V be an algebraic variety in S_n of dimension d and order g, and let $(\tau_0, \ldots, \tau_n)$ be a generic point of V, where τ_0 is transcendental over the function field $\Sigma = K(\tau_1/\tau_0, \ldots, \tau_n/\tau_0)$ of V. Let V^* be an irreducible variety, in S_m, birationally equivalent to V. We can choose the coordinates $(\tau_0^*, \ldots, \tau_m^*)$ of a generic point of V^* to be elements of $\Sigma(\tau_0)$ of degree of homogeneity one. Then if

$$\tau_i = \sum_{j=0}^{m} a_{ij}\tau_j^* \quad (i = 0, \ldots, n),$$

where a_{ij} is in K, we say that V is a *projection* of V^*.

If the maximum number of linearly independent relations

$$\sum_0^n a_i\tau_i = 0$$

satisfied by $(\tau_0, \ldots, \tau_n)$ is r, V lies in a space of $(n-r)$ dimensions and not in a lower space. We say that *the system of prime sections of* V *is complete* if any variety V^* *of order* g of which V is the projection also lies in a space of $n-r$ dimensions. This nomenclature is borrowed from the theory of linear systems, which plays a fundamental role in the theory of the birational invariants of algebraic varieties, but is beyond the scope of this work. We now prove

THEOREM VI. *The system of prime sections of V is complete if and only if every element of $\Sigma(\tau_0)$ of degree of homogeneity one which is integrally dependent on $K[\tau_0, \ldots, \tau_n]$ lies in this ring.*

We first make a preliminary transformation of coordinates in S_n so that V lies in

$$x_{n'+1} = 0, \ldots, x_n = 0$$

(where $n' = n-r$) and does not lie in any linear subspace of this. Then we can regard V as a variety in $S_{n'}$, with generic point

$(\tau_0, \ldots, \tau_{n'})$; and we know that $\tau_0, \ldots, \tau_{n'}$ are linearly independent over K. We may further suppose that if $\xi_i = \tau_i/\tau_0$, then $\xi_1, \ldots, \xi_d$ are algebraically independent over K, and $\xi_{d+1}, \ldots, \xi_{n'}$ are integrally dependent on $K[\xi_1, \ldots, \xi_d]$, while ξ_{d+i} is a primitive element of $K(\xi_1, \ldots, \xi_{n'})$ over $K[\xi_1, \ldots, \xi_d]$ $(i = 1, \ldots, n'-d)$. The irreducible equation connecting $\xi_1, \ldots, \xi_d, \xi_{d+i}$ is of order g and contains ξ_{d+i}^g as a term. We can express these results in terms of the homogeneous coordinates of the generic point τ of V. The elements τ_i $(i = 0, \ldots, n')$ are integrally dependent on $K[\tau_0, \ldots, \tau_d]$, and τ_{d+i} is a primitive element of $K(\tau_0, \ldots, \tau_{n'})$ over $K(\tau_0, \ldots, \tau_d)$, while the irreducible homogeneous equation

$$f(t_0, \ldots, t_d, t_{d+i}) = 0$$

satisfied by $\tau_0, \ldots, \tau_d, \tau_{d+i}$ is of order g in t_{d+i}, and is of total order g. Using Theorem I of Chapter X, § 9, we see that the irreducible equation relating $\tau_0, \ldots, \tau_d$ and $\tau = \sum_0^{n'} a_i \tau_i$ is homogeneous of total degree g at most, and there are elements τ for which the total degree is exactly g.

If $(\tau_0^*, \ldots, \tau_m^*)$ is a generic point of a variety V^* of which V is a projection, we have

$$\tau_i = \sum_{j=0}^{m} a_{ij} \tau_j^* \quad (i = 0, \ldots, n'),$$

and since $\tau_0, \ldots, \tau_{n'}$ are linearly independent over K, the matrix (a_{ij}) must be of rank $n'+1$. Hence we can choose the coordinate system in S_m so that $\tau_i = \tau_i^*$ $(i = 0, \ldots, n')$, and write the generic point of V^* as $(\tau_0, \ldots, \tau_m)$. We have to show

(i) that if V^* is of order g, and if every element of $K(\tau_0, \ldots, \tau_{n'})$ which is of degree of homogeneity one and which is integrally dependent on $K[\tau_0, \ldots, \tau_{n'}]$ lies in this ring, then

$$\tau_{n'+i} = \sum_{j=0}^{n'} b_{ij} \tau_j \quad (i = 1, \ldots, m-n');$$

(ii) that if there exist elements of $K(\tau_0, \ldots, \tau_{n'})$ of degree of homogeneity one which are integrally dependent on $K[\tau_0, \ldots, \tau_{n'}]$, but are not in this ring, then we can find a variety V^* of order g of which V is the projection, and V^* is not contained in an n'-space.

(i) Since τ_{d+1} is a primitive element of $K(\tau_0, \ldots, \tau_{n'})$ over $K(\tau_0, \ldots, \tau_d)$, and $\tau_{n'+i}$ lies in the extended field, $\tau = a\tau_{d+1} + b\tau_{n'+i}$ is a primitive element of the extension, save for a finite number of values of the ratio a/b in K. We choose values of a and b in K for which $b \neq 0$, and so that the element τ is primitive. Hence the irreducible equation satisfied by $\tau_0, \ldots, \tau_d, \tau$ is of degree g exactly in τ. Moreover, since V^* is of order g, the irreducible relation satisfied by $\tau_0, \ldots, \tau_d, \tau$ is of total degree g at most. From these two facts we deduce that the relation is of the form

$$\tau^g + B_1(\tau_0, \ldots, \tau_d)\,\tau^{g-1} + \ldots + B_g(\tau_0, \ldots, \tau_d) = 0,$$

where $B_i(\tau) \in K[\tau_0, \ldots, \tau_{n'}]$. Now, τ is an element of $\Sigma(\tau_0)$ of degree of homogeneity one, and the equation above shows that τ is integrally dependent on $K[\tau_0, \ldots, \tau_{n'}]$. Since, by hypothesis, any element of $\Sigma(\tau_0)$ of degree of homogeneity one which is integrally dependent on this ring lies in it, we have

$$\tau \in K[\tau_0, \ldots, \tau_{n'}],$$

and since τ is of degree of homogeneity one, we must have

$$\tau = a\tau_{d+1} + b\tau_{n'+i} = \sum_{j=0}^{n'} c_{ij}\tau_j.$$

Hence, since $b \neq 0$,
$$\tau_{n'+i} = \sum_{j=0}^{n'} b_{ij}\tau_j,$$

and (i) is proved.

(ii) If there exists an element τ of $\Sigma(\tau_0)$, of degree of homogeneity one, which is integrally dependent on $K[\tau_0, \ldots, \tau_{n'}]$ but is not in it, $\tau_0, \ldots, \tau_{n'}, \tau$ are elements of $\Sigma(\tau_0)$ which are of degree of homogeneity one, and are linearly independent over K. Hence the variety V^* in $S_{n'+1}$ whose generic point is $(\tau_0, \ldots, \tau_{n'}, \tau)$ lies in $S_{n'+1}$ but not in any linear subspace of it, and V is a projection of V^*. To prove (ii) we have only to show that V^* is of order g.

Let $\bar{\tau} = a\tau + \sum_{i=0}^{n'} a_i\tau_i$, where $a, a_0, \ldots, a_{n'}$ are in K. Then since τ is integrally dependent on $K[\tau_0, \ldots, \tau_{n'}]$, $\bar{\tau}$ is integrally dependent on $K[\tau_0, \ldots, \tau_{n'}]$. But, by hypothesis, $\tau_{d+1}, \ldots, \tau_{n'}$ are integrally dependent on $K[\tau_0, \ldots, \tau_d]$. Hence $\bar{\tau}$ is integrally dependent on $K[\tau_0, \ldots, \tau_d]$. Thus there exists a homogeneous equation

$$F(\tau_0, \ldots, \tau_d, \bar{\tau}) \equiv \bar{\tau}^\rho + B_1(\tau_0, \ldots, \tau_d)\,\bar{\tau}^{\rho-1} + \ldots + B_\rho(\tau_0, \ldots, \tau_d) = 0,$$

connecting $\tau_0, \ldots, \tau_d, \bar{\tau}$. But, since $\bar{\tau} \in \Sigma(\tau_0)$, which is an extension of degree g of $K(\tau_0, \ldots, \tau_d)$, the irreducible equation

$$f(t_0, \ldots, t_d, \bar{t}) = 0$$

satisfied by $\tau_0, \ldots, \tau_d, \bar{\tau}$ must be of order g at most in $\bar{t}$. Now $F(t_0, \ldots, t_d, \bar{t})$ contains $f(t_0, \ldots, t_d, \bar{t})$ as a factor, and this clearly implies that the coefficient of the highest power of $\bar{t}$ in $f(t_0, \ldots, t_d, \bar{t})$ lies in K. Hence we may assume that $F(t_0, \ldots, t_d, \bar{t})$ is irreducible and that $\rho \leqslant g$. Also for some $\bar{\tau}$, e.g. when $a = a_i = 0$ $(i \neq d+1)$, ρ must be g exactly. It follows from the form of Theorem I of Chapter X, § 9 stated above that V^* is of order g. Thus Theorem VI is proved.

We use this result to prove

THEOREM VII. *A necessary and sufficient condition that a variety V in S_n be projectively normal is that the transform of V by primals of order h should have a complete system of prime sections for each value of h.*

Suppose, first, that V is projectively normal. Then each positive integer h is a character of homogeneity of V. Therefore the transform $V^{(h)}$ of V by primals of order h is projectively normal. Hence, by Theorem VI, the system of prime sections of $V^{(h)}$ is complete.

Conversely, suppose that V is not projectively normal. We can then find an element τ of $K(\tau_0, \ldots, \tau_n)$ of degree of homogeneity h $(h > 0)$ which is integrally dependent on $K[\tau_0, \ldots, \tau_n]$ but is not in this ring. Construct the transform $V^{(h)}$ of V by primals of order h, and let $(\tau_0^{(h)}, \ldots, \tau_{n_h}^{(h)})$ be a generic point of it. Without loss of generality we may assume that $\tau_0^{(h)} = (\tau_0)^h$. When we consider $V^{(h)}$, we must assume that $\tau_0^{(h)}$ is of degree of homogeneity one. Since $\tau/\tau_0^{(h)} = \tau/(\tau_0)^h$ is in the function field of V and $V^{(h)}$, it is of degree of homogeneity zero with respect both to V and $V^{(h)}$, and hence τ is of degree of homogeneity one with respect to $V^{(h)}$. Since τ is integrally dependent on $K[\tau_0, \ldots, \tau_n]$, it satisfies an equation

$$\tau^\rho + a_1(\tau_0, \ldots, \tau_n)\,\tau^{\rho-1} + \ldots + a_\rho(\tau_0, \ldots, \tau_n) = 0 \quad (a_i(\tau) \in K[\tau_0, \ldots, \tau_n]),$$

and by equating terms of like degree of homogeneity, we see that we may assume that $a_i(\tau)$ is a form of degree ih. Hence we find that τ is integrally dependent on $K[\tau_0^{(h)}, \ldots, \tau_{n_h}^{(h)}]$.

Since
$$K[\tau_0^{(h)}, \ldots, \tau_{n_h}^{(h)}] \subseteq K[\tau_0, \ldots, \tau_n],$$

and τ is not in $K[\tau_0, \ldots, \tau_n]$, it follows that there are elements of $K(\tau_0^{(h)}, \ldots, \tau_{n_h}^{(h)})$ of degree of homogeneity one (with respect to $V^{(h)}$) which are integrally dependent on $K[\tau_0^{(h)}, \ldots, \tau_{n_h}^{(h)}]$, but are not in this ring. Hence, by Theorem VI, the system of prime sections of $V^{(h)}$ is not complete.

The results proved in this section will give the reader some idea of the special properties possessed by normal varieties. Experience gained in further investigations will reveal that in the study of the birational invariants of algebraic varieties, projectively normal varieties play a fundamental part. In short, we find that the best procedure always is to pass from a variety V to a derived normal variety $V^{(\delta)}$, and, having examined the relation between V and $V^{(\delta)}$, to carry out all subsequent investigations on projectively normal varieties. The properties of projectively normal varieties which make them peculiarly suited to further investigations are of two kinds: (i) the local properties, summed up in the fact that the quotient ring of every irreducible subvariety is integrally closed (for this, affine-normality is sufficient, provided the subvariety under discussion is not at infinity), and (ii) the property in the large, which was proved in Theorem VII. In future, we shall always construct varieties which are projectively normal, remembering that, by Theorem III, however we choose the prime at infinity the variety is affine-normal in the derived affine space, and hence we shall usually use the word normal to imply that the variety is projectively normal.

CHAPTER XVII

VALUATION THEORY

In Chapter XVI, § 5, we introduced, as an example, the idea of a valuation of a function field. This idea, which generalises the idea of a branch of a curve in classical algebraic geometry, is of fundamental importance in the theory of birational transformations, and we proceed in this chapter to develop the general properties of valuations, and then, in Chapter XVIII, pass on to some applications. We shall begin by considering the main properties of valuations of any field, and pass to the particular study of the valuations of algebraic function fields. In this, we shall take the ground field K to be without characteristic, but we shall not assume that it is algebraically closed.

1. Ordered Abelian groups. In Chapter I, § 1, we defined an Abelian group as a group in which the law of combination has the commutative property, and we mentioned that in the case of an Abelian group the law of combination is often written as an addition, the group then being called an additive group. In this section we consider an additive group Γ, with elements $a, b, c, \ldots$, and suppose that the elements of Γ, *other than zero*, are divided into two mutually exclusive sets Γ_p and Γ_n. If a is in Γ_p we write $a > 0$, and if it is in Γ_n we write $a < 0$; thus every element x of Γ satisfies one and only one of the relations

$$x > 0, \quad x = 0, \quad x < 0.$$

The elements in Γ_p are said to be *positive*, those in Γ_n *negative*. The division of the non-zero elements of Γ into the two sets Γ_p, Γ_n is called an *ordering* of Γ if the following properties hold:

(i) if $a < 0$, then $-a > 0$;

(ii) if $a > 0$ and $b > 0$, then $a + b > 0$.

We are concerned in this section with the elementary properties of ordered additive groups.

(1) If $a > 0$, we prove that $-a < 0$. Clearly, $-a$ is not equal to zero. If $-a > 0$, we have, from property (ii), $0 = a + (-a) > 0$, that is, 0 is in Γ_p. This contradicts the definition of Γ_p. Hence $-a$ is not zero, and is not in Γ_p; therefore it is in Γ_n.

(2) If $a-b>0$, we shall write $a>b$. We show that this is a transitive relation. Suppose that $a>b$, and that $b>c$. Then

$$a-b>0 \quad \text{and} \quad b-c>0.$$

Hence, by (ii), $$a-c=(a-b)+(b-c)>0,$$

that is, $$a>c.$$

Similarly, if $a-b<0$, we write $a<b$, and show that this relation is also transitive, and is equivalent to $b>a$.

(3) Suppose that $a>b$, and that c is any element of Γ. Then

$$(a+c)-(b+c)=a-b>0,$$

that is, $$a+c>b+c.$$

(4) Let a be any element of Γ different from zero, and let na, where n is an integer, denote the sum of n elements of Γ each equal to a. If $ma=na$ $(m>n)$, then $(m-n)a=0$, and $(m-n)(-a)=0$. Suppose that $a>0$, then, by (ii),

$$2a = a+a>0,$$

$$3a = 2a+a>0, \ldots$$

$$(m-n)a>0,$$

and we have a contradiction. Similarly, if $a<0$, and hence $-a>0$, we obtain the relation $(m-n)(-a)>0$. Thus, whether $a>0$ or $a<0$, we obtain a contradiction, and we conclude that the elements $0, \pm a, \pm 2a, \ldots$ are all distinct. Thus, in particular, any ordered group, other than one consisting of the zero element only, has an infinite number of elements.

A few examples of ordered additive groups may be given.

(*a*) Any subgroup of the additive group of real numbers, with the usual definition of positive and negative numbers, is ordered. In particular, the additive group of integers is ordered.

(*b*) Let Γ and Γ' be any two ordered additive groups; $a, b, \ldots$ will denote elements of Γ, and $a', b', \ldots$ will denote elements of Γ'. Consider the pairs (a, a') consisting of any element a of Γ and any element a' of Γ'. We form a group Γ^* consisting of these pairs, by means of the addition law

$$(a, a')+(b, b')=(a+b, a'+b').$$

It can be verified at once that Γ^* is an additive group; we call it the *direct sum* of Γ and Γ'. The zero of Γ^* is easily seen to be $(0, 0)$.

We now say that $(a, a') > 0$ if and only if either $a > 0$, or $a = 0$, $a' > 0$. We prove that this defines an ordering of Γ^*. If (a, a') is any non-zero element of Γ^*, either $(a, a') > 0$, or $a < 0$, or $a = 0$, $a' < 0$. Hence either $(a, a') > 0$ or $-(a, a') = (-a, -a') > 0$. Thus we have

(i) if $(a, a') < 0$, then $-(a, a') > 0$.

(ii) Suppose that $(a, a') > 0$ and $(b, b') > 0$. If either $a > 0$ or $b > 0$, $a + b > 0$; hence $(a, a') + (b, b') = (a+b, a'+b') > 0$. If $a = 0$ and $b = 0$, then $a' > 0$ and $b' > 0$. Hence $(a, a') + (b, b') = (0, a'+b') > 0$. Hence Γ^* satisfies the conditions for an ordered group. Apart from the ordering, Γ^* is determined symmetrically by Γ and Γ', but in the ordering Γ has a privileged position. Interchanging the roles of Γ and Γ', we obtain a different ordering of Γ^*.

Given any finite set $\Gamma_1, \ldots, \Gamma_r$ of additive groups, we can, by a simple induction, define the direct sum Γ^* of the r groups. If $\Gamma_1, \ldots, \Gamma_r$ are each ordered, we can, corresponding to a given arrangement of these r groups, define, as above, the ordering of Γ^*. We thus obtain $r!$ different orderings of Γ^*.

We now consider any ordered additive group Γ. If for every pair of positive elements x and y of Γ we can find integers m and n (depending on x and y) such that

$$mx > y, \quad ny > x,$$

then Γ is said to be *Archimedean-ordered*. Clearly, any subgroup of the ordered additive group of real numbers is Archimedean-ordered. On the other hand, if Γ^* is the direct sum of two ordered additive groups Γ and Γ', as in example (*b*) above, Γ^* is not Archimedean-ordered. For if $x = (a, a')$, where $a > 0$, and $y = (0, b')$, where $b' > 0$, then x and y are positive, and

$$(a, a') = x > ny = (0, nb')$$

for every integer n.

THEOREM I. *Every Archimedean-ordered additive group Γ is isomorphic to a subgroup of the additive group of real numbers.*

If Γ consists of the zero element only, the theorem is trivial. If Γ contains a non-zero element, it necessarily contains a positive element. Let a be a positive element of Γ, and let b be any other positive element of Γ. We consider the set R_b of ordered pairs (m, n) of positive integers such that

$$ma \geqslant nb.$$

If (m, n) is any pair of positive numbers, and r is any positive integer,

$$ma \geqslant nb \quad \text{or} \quad ma < nb,$$

according as

$$rma \geqslant rnb \quad \text{or} \quad rma < rnb.$$

Let λ be any positive rational number. If $\lambda = m/n = p/q$, where m, n, p, q are positive integers, the above result enables us to prove at once that (m, n) belongs to R_b if and only if (p, q) belongs to R_b. If (m, n) belongs to R_b, we say that λ belongs to $\mathscr{R}_b$. We prove that $\mathscr{R}_b$ defines a Dedekind section, and hence a real number, which we will denote by $\bar{b}$. In the first place, not all positive rational numbers belong to $\mathscr{R}_b$. For, since Γ is Archimedean-ordered, there exists a positive integer n such that $a < nb$. Hence $1/n$ is not in $\mathscr{R}_b$.

Next, if $\lambda = m/n$ is in $\mathscr{R}_b$, and $\mu = r/s > \lambda$, we show that μ is in $\mathscr{R}_b$. Since λ is in $\mathscr{R}_b$, we have

$$ma \geqslant nb.$$

Hence

$$n(ra-sb) = (nr-ms)a+s(ma-nb) > 0,$$

since

$$(nr-ms)a > 0,$$

and therefore

$$ra > sb,$$

and so μ is in $\mathscr{R}_b$.

These two results show that $\mathscr{R}_b$ is a Dedekind section defining a positive real number $\bar{b}$. Keeping a fixed, each positive element x of Γ thus determines a positive real number $\bar{x}$. Clearly, $\bar{a} = 1$. We define a real number $\bar{y}$ corresponding to any negative element y of Γ by the formula

$$\bar{y} = -\overline{(-y)},$$

and, further, we associate zero with the zero element of Γ. We thus map each element of Γ on a real number. We show that the real numbers obtained as the maps of elements of Γ form a subgroup Δ of the additive group of real numbers; and that if Δ is ordered in the usual way, the mapping of Γ on Δ is an isomorphism, which preserves order.

We first show that if x and y are any two elements of Γ, and $\bar{x}$, $\bar{y}$ are the corresponding real numbers, then the real number corresponding to $x+y$ is $\bar{x}+\bar{y}$. This is clear if either x or y is zero; so we need only consider the three cases (i) $x>0$, $y>0$; (ii) $x<0$, $y<0$; (iii) x and y of different signs.

(i) If $x > 0$ and $y > 0$, then $x+y > 0$; also, by definition, $\bar{x} > 0$, $\bar{y} > 0$, $\overline{x+y} > 0$. Suppose that

$$\bar{x}+\bar{y} > \overline{(x+y)}.$$

Then we can find rational numbers $\lambda = p/q$, $\mu = r/s$, such that

$$\lambda < \bar{x}, \quad \mu < \bar{y}, \quad \lambda + \mu > \overline{(x+y)}.$$

The first two of these relations imply that

$$pa < qx, \quad ra < sy.$$

Hence
$$(ps+qr)\,a < qs(x+y),$$

that is
$$\lambda + \mu < \overline{(x+y)}.$$

Thus we obtain a contradiction. Similarly, if we assume that

$$\bar{x} + \bar{y} < \overline{(x+y)},$$

we obtain a contradiction. Hence we must have

$$\bar{x} + \bar{y} = \overline{(x+y)}.$$

(ii) If $x < 0$, $y < 0$, we have

$$\overline{(x+y)} = -\overline{(-x-y)} = -\overline{(-x)} - \overline{(-y)} = \bar{x} + \bar{y}.$$

(iii) Suppose, say, that $x > 0$, $y < 0$. If $x + y \geqslant 0$, we have

$$x + y + (-y) = x;$$

hence, by (i),
$$\overline{x+y} + \overline{(-y)} = \bar{x},$$

and therefore
$$\overline{x+y} = \bar{x} - \overline{(-y)} = \bar{x} + \bar{y}.$$

On the other hand, if $x + y < 0$, we have

$$-(x+y) + x = -y,$$

and therefore
$$-\overline{(x+y)} + \bar{x} = -\bar{y},$$

that is,
$$\overline{x+y} = \bar{x} + \bar{y}.$$

Thus Δ is a subgroup of the additive group of real numbers. If the homomorphism of Γ on Δ were not an isomorphism, there would exist non-zero elements of Γ which map on the zero of Δ. But no such elements exist, since if $x > 0$, then $\bar{x} > 0$, and if $x < 0$, then $\bar{x} < 0$. Finally, since positive elements of Γ correspond to positive elements of Δ, and conversely, the isomorphism of Γ on Δ preserves the ordering.

The isomorphism which we have obtained of Γ on a subgroup of the additive group of real numbers depends on the choice of the element a of Γ which maps on 1. If we choose a different element, b, say, to be the element which maps on 1, we obtain a different subgroup Δ' of the group of real numbers. If, in the mapping of Γ on Δ', a maps on the real number ρ, then the map of any element x of Γ on Δ' is $\rho\bar{x}$, where $\bar{x}$ is the map of x on Δ. If Γ is mapped on the subgroup of the group of real numbers in the manner described above so that the element a is mapped on 1, Γ is said to be *normalised* with respect to this element.

In our use of ordered additive groups, we may always replace a group by any isomorphic group similarly ordered. In particular, we shall usually represent an Archimedean-ordered group by a subgroup of the additive group of real numbers.

Next, let us consider an additive group Γ whose ordering is not Archimedean. There must exist in Γ two positive elements ξ and η such that $\xi > n\eta$, for all integers n. For this element ξ we consider the set of all positive elements η in Γ such that $\xi > n\eta$ for all integers n. If η_1 and η_2 are two elements of this set, and $\eta_1 \geqslant \eta_2$, say,

$$\xi > 2n\eta_1 \geqslant n(\eta_1 + \eta_2),$$

for all integers n; hence $\eta_1 + \eta_2$ belongs to the set. Thus the set is closed under addition. Similarly,

$$\xi > n(\eta_1 + \eta_2) > n(\eta_1 - \eta_2) > n(-\eta_1 - \eta_2);$$

hence the subgroup Γ' of Γ generated by elements of the set has the property that $\xi > n\eta$, when n is any positive integer and η is any element of Γ'. The subgroup Γ' has the property that if η is any positive element of Γ' and η' is any positive element of Γ less than η, then η' belongs to Γ'. For $\xi > n\eta > n\eta'$, for any integer n.

A subgroup Δ of an ordered additive group Γ is said to be *isolated* if it has the property that if η and η' are elements of Γ and $\eta > \eta' > 0$, and η is in Δ, then η' necessarily belongs to Δ. From what has been said above, it is clear that if the ordering of Γ is not Archimedean, it contains an isolated subgroup. On the other hand, if Γ is Archimedean-ordered, we can regard it as a subgroup of the additive group of real numbers. Let Δ be an isolated subgroup of Γ, not consisting solely of the zero element. If η is any positive element of Δ, and ζ any positive element of Γ, there clearly exists an integer n such that $\zeta < n\eta$. Hence, since Δ is an isolated subgroup containing $n\eta$, it must contain ζ. It must also contain $-\zeta$; and so $\Delta = \Gamma$. Hence we have

THEOREM II. *A necessary and sufficient condition that the ordering of an additive group be non-Archimedean is that it contains proper subgroups which are isolated.*

In the example (*b*) above, of an ordered additive group Γ^* which is the direct sum of two ordered additive groups Γ and Γ', it is easy to see that the elements of Γ^* of the form $(0, a')$ form an isolated subgroup.

We now consider some properties of the isolated subgroups of an ordered additive group Γ. If Δ is any subgroup of Γ, we can use it to divide the elements of Γ into classes of equivalent elements. Two elements a and a' of Γ are said to be equivalent with respect to Δ if $a-a'$ is in Δ. In particular, the elements equivalent to zero are the elements of Δ. This is a true equivalence relation, being reflexive, symmetric and transitive. Let (a) denote the set of elements of Γ equivalent to a; then $(a) = (a')$ if and only if a is equivalent to a'. If a is equivalent to a', and if b is equivalent to b',

$$[a+b]-[a'+b'] = [a-a']+[b-b']$$

is in Δ, since $a-a'$ and $b-b'$ are in Δ, and Δ is a group. Hence $(a+b)$ is determined uniquely by (a) and (b), and we may therefore define the addition of equivalent sets by the formula

$$(a)+(b) = (a+b).$$

The addition of sets is obviously associative and commutative. If (a) and (b) are any two sets, and $(c) = (b-a)$, then the set (c) is a solution of the equation

$$(a)+x = (b).$$

Hence, with this law of addition, the sets of equivalent elements of Γ with respect to Δ form an additive group, which we call the *factor group* of Γ by Δ, and denote by Γ/Δ. The zero of Γ/Δ is the set (0), that is, the set formed by the elements of Δ.

If Γ is ordered, and Δ is an isolated subgroup of Γ, we can deduce an ordering for Γ/Δ. Consider any non-zero element (a) of Γ/Δ. Suppose that this contains a positive element a. If η is any positive element of Δ, $a > \eta$, otherwise, by the characteristic property of an isolated subgroup, we could deduce that a belonged to Δ, and hence that $(a) = 0$. Therefore a is greater than any element of Δ. Any other element of (a) is of the form $a \pm \eta$, where η is a non-negative element of Δ. But since a is greater than any element of Δ, $a \pm \eta > 0$, and it follows that every element of (a) is positive, and greater than every element of Δ. We then write $(a) > 0$. Similarly, if the set $(b) \neq 0$ contains a negative element, we show that every element of (b) is negative and less than every element of Δ. We then write $(b) < 0$. We thus divide the non-zero elements of Γ/Δ into mutually exclusive sets, positive and negative. Clearly, we see that

(i) if $\quad (a) < 0, \quad -(a) = (-a) > 0;$

(ii) if $\quad (a) > 0, \quad (b) > 0, \quad (a+b) = (a)+(b) > 0.$

Hence we have an ordering of Γ/Δ, and

THEOREM III. *If Δ is an isolated subgroup of Γ, then Γ/Δ is an ordered additive group.*

We next prove

THEOREM IV. *If Δ_1 and Δ_2 are two isolated subgroups of Γ, then either Δ_1 contains Δ_2, or Δ_2 contains Δ_1. If $\Delta_1 \supset \Delta_2$, Δ_1/Δ_2 is an isolated subgroup of Γ/Δ_2.*

The theorem is trivial if Δ_1 coincides with Δ_2; hence we may assume that one of them, say Δ_1, contains an element a_1 which is not in Δ_2. Now a_1 is not zero, since zero belongs to Δ_2, and $-a_1$ is also an element in Δ_1, but is not in Δ_2. Hence, without loss of generality, we may assume that $a_1 > 0$. Let a_2 be any positive element of Δ_2. If $a_2 \geqslant a_1$, it follows from the fact that Δ_2 is an isolated subgroup that a_1 is also in Δ_2, and we have a contradiction. Hence $a_2 < a_1$. But since Δ_1 is also an isolated subgroup, it follows that a_2 is in Δ_1. Hence every positive element of Δ_2 is in Δ_1. Since Δ_1 and Δ_2 are groups, it follows that every element of Δ_2 is in Δ_1, that is, $\Delta_2 \subset \Delta_1$. From the definition of an isolated subgroup, it follows that Δ_2, regarded as a subgroup of Δ_1, is an isolated subgroup.

Now consider the elements of the factor group Γ/Δ_2 which arise from elements of Γ which lie in Δ_1. If a lies in Δ_1, and $a'-a$ lies in Δ_2, then $a' = a'-a+a$ lies in Δ_1, since $\Delta_2 \subset \Delta_1$. Hence if a lies in Δ_1, every element equivalent to a with respect to Δ_2 lies in Δ_1. Thus the elements of Δ_1 are divided into equivalence classes with respect to Δ_2, each class being a complete class of equivalent elements of Γ with respect to Δ_2. These classes form a subgroup of Γ/Δ_2, which is clearly isomorphic with Δ_1/Δ_2, and may be identified with it. If (a) is any positive element of Δ_1/Δ_2, and (b) is any positive element of Γ/Δ_2 such that $(a) > (b)$, then $a > b > 0$, and a is in Δ_1. Therefore b is in Δ_1, and hence (b) is in Δ_1/Δ_2. Thus Δ_1/Δ_2 is an isolated subgroup of Γ/Δ_2.

Now let Δ_2 be any isolated subgroup of Γ, and let Γ^* be an isolated subgroup of Γ/Δ_2. We consider the set of elements of Γ which correspond to elements of Γ^* in the mapping of Γ on Γ/Δ_2. If a and b are two such elements, $a+b$ and $-a$ map on $(a)+(b)$ and $-(a)$, which are in Γ^*. Hence the elements of Γ which map on the elements of Γ^* form a subgroup Δ_1 of Γ. Since any element of Δ_2 maps on an element of Γ^*, $\Delta_2 \subset \Delta_1$. Let a be any positive element of Δ_1, and let b be any positive element of Γ such that $a > b$. Then $(a) \geqslant (b) \geqslant 0$, and (a) is in Γ^*. Since Γ^* is isolated, it follows that (b) is in Γ^*, and hence b is in Δ_1. Therefore Δ_1 is an isolated subgroup of Γ.

We have thus established a one-to-one correspondence between the isolated subgroups of Γ which contain a given isolated subgroup Δ and the isolated subgroups of Γ/Δ. In this correspondence it is easily seen that if Δ' is an isolated subgroup of Γ containing Δ, then $(\Gamma/\Delta)/(\Delta'/\Delta) \cong \Gamma/\Delta'$. The only case with which we shall be concerned is that in which Γ contains only a finite number of distinct isolated subgroups. By Theorem IV, these can be arranged in order so that each subgroup contains the succeeding one. Suppose that

$$\Gamma \supset \Gamma_1 \supset \Gamma_2 \supset \dots \supset \Gamma_k = 0$$

are all the isolated subgroups of Γ. Then Γ is said to *be of rank k*. The group Γ_{i+1} is an isolated subgroup of Γ_i. If Γ_i/Γ_{i+1} were not Archimedean-ordered, there would exist a proper subgroup of it which was isolated. This, in turn, would define a proper subgroup Γ' of Γ_i, properly containing Γ_{i+1}, which was isolated in Γ_i. We show that Γ' is an isolated subgroup of Γ. It is clear that Γ' is a subgroup of Γ. If a is any positive element of Γ', and b is any positive element of Γ such that $a > b$, it follows from the fact that a is in Γ_i and Γ_i is isolated in Γ that b is in Γ_i. Then, since Γ' is isolated in Γ_i, it follows that b is in Γ'. Thus Γ' is isolated in Γ. But the sequence $\Gamma_1, \Gamma_2, \dots$ contains all the isolated subgroups of Γ. Hence $\Gamma' = \Gamma_{i+1}$ or Γ_i, which is inconsistent with the fact that Γ' is a proper subgroup of Γ_i and properly contains Γ_{i+1}. Hence our assumption that Γ_i/Γ_{i+1} was not Archimedean-ordered is false. Thus we have

THEOREM V. *If Γ is of rank k, the isolated subgroups can be arranged in order $\Gamma \supset \Gamma_1 \supset \dots \supset \Gamma_k = 0$, and Γ_i/Γ_{i+1} is Archimedean-ordered.*

2. Valuations of a field. Let Σ be any field, and Γ any ordered additive group. A mapping $\xi \to v(\xi)$ of the *non-zero* elements of Σ, on the elements of Γ, is called a *valuation* of Σ, if it has the properties

$$\text{(i)} \qquad v(\xi\eta) = v(\xi) + v(\eta),$$

$$\text{(ii)} \qquad v(\xi+\eta) \geq \min[v(\xi), v(\eta)], \quad \text{if} \quad \xi+\eta \neq 0.$$

If $v(\xi) = 0$ for every non-zero element of Σ, the valuation is said to be *trivial*. As an example of a valuation, consider the field Σ of functions of the complex variable z which are meromorphic in the neighbourhood of $z = 0$. Any non-zero function ζ of the field can then be expanded in a series

$$\zeta = z^\sigma(a_0 + a_1 z + \ldots) \quad (a_0 \neq 0),$$

where σ is an integer. Let Γ be the additive group of integers, and define $v(\zeta)$ to be σ. This defines a mapping of Σ on Γ which is a valuation. For if

$$\xi = z^\alpha(a_0 + a_1 z + \ldots) \qquad (a_0 \neq 0),$$

and

$$\eta = z^\beta(b_0 + b_1 z + \ldots) \qquad (b_0 \neq 0),$$

then

$$\xi\eta = z^{\alpha+\beta}[a_0 b_0 + (a_0 b_1 + a_1 b_0) z + \ldots] \qquad (a_0 b_0 \neq 0),$$

and

$$\xi + \eta = z^\sigma[c_0 + \ldots] \qquad (c_0 \neq 0),$$

where $\sigma \geq \min[\alpha, \beta]$, and is finite unless $\xi + \eta = 0$. Hence

$$v(\xi) = \alpha, \quad v(\eta) = \beta,$$

$$v(\xi\eta) = \alpha + \beta, \quad v(\xi+\eta) = \sigma \geq \min[\alpha, \beta].$$

The theorems proved in this section will hold for valuations of arbitrary fields, but the applications to algebraic geometry which come later will deal with valuations of the function field of an algebraic variety defined over a ground field K (of characteristic zero), and in this case it will be necessary to impose a third condition on the mapping of Σ on Γ:

$$\text{(iii)} \qquad \text{if } a \in K, \ a \neq 0, \quad \text{then } v(a) = 0.$$

To avoid needless repetition, we shall state in the course of each theorem any modification which is necessary when Σ is a function field and the condition (iii) is to be satisfied.

We begin with a few general results on valuations of a field Σ.

(1) If ξ is any non-zero element of Σ, the equation

$$1\xi = \xi$$

implies, by (i), $$v(1)+v(\xi) = v(\xi),$$

and hence $$v(1) = 0.$$

Since $(-1)^2 = 1$, $$v(-1)+v(-1) = v(1) = 0,$$

and hence $$v(-1) = 0.$$

Therefore $$v(-\xi) = v(-1)+v(\xi) = v(\xi)$$

for every non-zero element ξ of Σ.

By (ii),

$$\begin{aligned} v(2) &\geqslant \min\,[v(1), v(1)] \\ &\geqslant 0, \end{aligned}$$

and, by elementary induction, we have

$$v(m) \geqslant 0,$$

where m is any integer not a multiple of the characteristic of Σ.

We show that if Σ is of finite characteristic p, and m is any integer not a multiple of p, then $v(m) = 0$. For if m is not a multiple of p, there exists another integer n such that $mn = 1 \ (p)$. Then

$$0 = v(1) = v(m)+v(n),$$

and since $v(m) \geqslant 0$, $v(n) \geqslant 0$, this is only possible if $v(m) = v(n) = 0$.

Now consider the case of a field without characteristic. In this case there may exist a positive integer m such that $v(m) > 0$. Let q be the smallest positive integer with this property. We show that q is prime. Suppose that it is composite: $q = ab$, where $a > 1$, $b > 1$. Then, since $a < q$, $b < q$, $v(a) = v(b) = 0$. Hence

$$v(q) = v(a)+v(b) = 0,$$

and we have a contradiction. Hence q must be prime.

Let m be any integer greater than q such that $v(m) > 0$. If $m = aq+b$, where $a \geqslant 1$, $1 \leqslant b < q$, then

$$\begin{aligned} 0 = v(b) &= v(m-aq) \\ &\geqslant \min\,[v(m), v(aq)] > 0, \end{aligned}$$

since, by hypothesis, $v(m) > 0$, and $v(aq) = v(a) + v(q) > 0$. Hence we have a contradiction. Therefore m must be a multiple of q. If $m = q^a p_1^{a_1} \dots p_k^{a_k}$, where $q, p_1, \dots, p_k$ are distinct primes, it follows that $v(p_i) = 0$, and that

$$v(m) = av(q) + a_1 v(p_1) + \dots + a_k v(p_k) = av(q).$$

We consider, in particular, the case in which Σ is the field of rational numbers, and ζ is any element of Σ. Let

$$\zeta = q^\lambda p_1^{\lambda_1} \dots p_k^{\lambda_k},$$

where $\lambda, \lambda_1, \dots, \lambda_k$ are positive or negative integers, and $q, p_1, \dots, p_k$ are distinct primes. If $\mu, \mu_1 \dots, \mu_k$ are positive integers such that $\lambda + \mu, \lambda_1 + \mu_1, \dots, \lambda_k + \mu_k$ are positive,

$$\begin{aligned}(\lambda + \mu)\, v(q) &= v(q^{\lambda+\mu} p_1^{\lambda_1+\mu_1} \dots p_k^{\lambda_k+\mu_k}) \\ &= v(\zeta q^\mu p_1^{\mu_1} \dots p_k^{\mu_k}) \\ &= v(\zeta) + \mu v(q).\end{aligned}$$

Hence $v(\zeta) = \lambda v(q)$. The values of elements of Σ are therefore all the multiples of $v(q)$. Hence they form an Archimedean-ordered additive group, and if this is normalised with respect to $v(q)$, the value group Γ is the additive group of integers. Thus the non-trivial valuations of the field of rational numbers are obtained by selecting any prime p and writing $v(p^\lambda p_1^{\lambda_1} \dots p_k^{\lambda_k}) = \lambda$.

(2) If ξ is any non-zero element of Σ,

$$0 = v(1) = v(\xi\xi^{-1}) = v(\xi) + v(\xi^{-1}).$$

Hence
$$v(\xi^{-1}) = -v(\xi).$$

In the mapping $\xi \to v(\xi)$ of Σ on Γ it may happen that not all elements of Γ appear as the images of elements of Σ. If a and b are two elements of Γ which are images of elements ξ and η of Σ, respectively:

$$v(\xi) = a, \quad v(\eta) = b,$$

then
$$v(\xi\eta) = a + b, \quad v(\xi^{-1}) = -v(\xi) = -a;$$

hence the elements of Γ which are images of elements of Σ form a subgroup Γ^* of Γ; and we can define the same valuation of Σ by using Γ^* instead of Γ. It is usually convenient to do this, and in future we shall assume that every element of Γ is the image of some element of Σ. Γ is then called the *valuation group* of the valuation, and $v(\xi)$ is called the *value* of ξ.

(3) The second condition for a mapping of Σ on Γ to be a valuation is an inequality:

$$v(\xi+\eta) \geqslant \min[v(\xi), v(\eta)].$$

We now show that this inequality can be replaced by equality in the case in which $v(\xi)$ and $v(\eta)$ are unequal. Let us suppose that $v(\xi) > v(\eta)$. If $\zeta = \xi+\eta$,

$$v(\zeta) \geqslant \min[v(\xi), v(\eta)] = v(\eta).$$

But we have

$$\begin{aligned} v(\eta) &= v(\zeta-\xi) \\ &\geqslant \min[v(\zeta), v(-\xi)] \\ &= \min[v(\zeta), v(\xi)]. \end{aligned}$$

Since

$$v(\eta) < v(\xi),$$

we must have

$$\min[v(\zeta), v(\xi)] = v(\zeta);$$

hence

$$v(\eta) \geqslant v(\zeta).$$

Thus $v(\eta) = v(\zeta)$, and it follows that

$$v(\xi+\eta) = v(\zeta) = v(\eta).$$

Thus if $v(\xi) \neq v(\eta)$, we have

$$v(\xi+\eta) = \min[v(\xi), v(\eta)].$$

(4) In defining a valuation, we have excluded the element zero of Σ. Sometimes we shall find it convenient to define formally a value of zero, which is not an element of Γ, and to denote it by ∞. If we assume that the conditions of (i) and (ii) defining a valuation hold when we include the value of zero, (i) tells us that

$$\begin{aligned} \infty = v(0) = v(0\times\xi) &= v(0)+v(\xi) \\ &= \infty + v(\xi); \end{aligned}$$

and (ii) tells us that

$$\infty = v(0) = v(\xi-\xi) \geqslant \min[v(\xi), v(-\xi)] = v(\xi).$$

Thus ∞ has two of the properties of 'infinity' as it is used in analysis. As a general rule ∞ behaves in the same way as the infinity of analysis, but the reader is warned not to assume any property of it which has not been proved.

A valuation is thus sometimes regarded as a mapping of the whole field Σ on (Γ, ∞), rather than as a mapping of the non-zero elements of Σ on Γ.

Having established the foregoing elementary results, we now pass to a more detailed study of a non-trivial valuation $\mathfrak{B}$ of a field Σ to a valuation group Γ. Let us consider the set of elements of Σ whose values are non-negative, together with the zero of Σ. If ξ and η are two elements of this set, we have, from (i),

$$v(\xi\eta) = v(\xi)+v(\eta) \geqslant 0,$$

and $$v(\xi+\eta) \geqslant \min[v(\xi), v(\eta)] \geqslant 0.$$

Hence the set of elements in question form a ring $\mathfrak{R}$, which we call the *valuation ring* of $\mathfrak{B}$. Since $\mathfrak{B}$ is not a trivial valuation, there exists an element ζ of Σ whose value in Γ is not zero. Then either $v(\zeta)$ or $v(\zeta^{-1}) = -v(\zeta)$ is negative, and hence either ζ or ζ^{-1} is not in $\mathfrak{R}$. Therefore $\mathfrak{R}$ is a proper subring of Σ, and since it contains unity and has no divisors of zero, it is an integral domain. If ζ is any element of Σ, either ζ is in $\mathfrak{R}$, or $v(\zeta) < 0$,in which case $v(\zeta^{-1}) > 0$, and therefore ζ^{-1} is in $\mathfrak{R}$. It follows that Σ is the quotient field of $\mathfrak{R}$.

Next, let us consider the elements of Σ (including zero) whose values are positive. If ξ and η are two such elements, we have, from (ii),

$$v(\xi-\eta) \geqslant \min[v(\xi), v(\eta)] > 0,$$

and if ζ is any element of $\mathfrak{R}$, $v(\zeta) \geqslant 0$, and hence, from (i),

$$v(\zeta\xi) = v(\zeta)+v(\xi) > 0.$$

Hence the elements considered form an ideal $\mathfrak{p}$ in $\mathfrak{R}$. This ideal $\mathfrak{p}$ is clearly a prime ideal in $\mathfrak{R}$. For if ξ, η are two elements of $\mathfrak{R}$ which do not lie in $\mathfrak{p}$,

$$v(\xi) = 0, \quad v(\eta) = 0,$$

and hence, by (i), $$v(\xi\eta) = v(\xi)+v(\eta) = 0,$$

and therefore $\xi\eta$ is not in $\mathfrak{p}$.

If ζ is any element of $\mathfrak{R}$, its inverse ζ^{-1} lies in $\mathfrak{R}$ if and only if $v(\zeta^{-1}) \geqslant 0$, that is, $v(\zeta) \leqslant 0$. But since ζ is in $\mathfrak{R}$, $v(\zeta) \geqslant 0$. Hence ζ has an inverse in $\mathfrak{R}$ if and only if it does not lie in $\mathfrak{p}$. *Hence* $\mathfrak{p}$ *is the ideal of non-units of* $\mathfrak{R}$. It is clearly maximal in $\mathfrak{R}$. For if $\mathfrak{i}$ is any proper factor of $\mathfrak{p}$, it contains an element ζ of $\mathfrak{R}$ not in $\mathfrak{p}$; hence $v(\zeta) = 0$. Therefore ζ^{-1} lies in $\mathfrak{R}$, and so $\zeta^{-1}\zeta = 1$ lies in $\mathfrak{i}$; that is, $\mathfrak{i}$ is the unit ideal.

$\mathfrak{p}$ is usually called the *ideal of the valuation* $\mathfrak{B}$.

Now consider any proper ideal $\mathfrak{i}$ of $\mathfrak{R}$. If ζ is any element of $\mathfrak{i}$, $v(\zeta) > 0$, for otherwise $\mathfrak{R}$ would contain ζ^{-1}, and so $\mathfrak{i}$ would contain $\zeta^{-1}\zeta = 1$, and be the unit ideal. Hence

$$\mathfrak{i} = 0 \quad (\mathfrak{p}).$$

If ξ is any element of $\mathfrak{i}$, and if η is any element of $\mathfrak{R}$ such that $v(\eta) \geqslant v(\xi)$, then $v(\eta/\xi) \geqslant 0$, and therefore η/ξ is in $\mathfrak{R}$. Hence $\eta = (\eta/\xi)\xi$ is in $\mathfrak{i}$. Thus the elements of Σ which lie in $\mathfrak{i}$ are just those elements whose values in (Γ, ∞) lie in a set Ω with the following properties:

(i) the elements of Ω are non-negative;

(ii) if a is in Ω and b is any element of Γ (or ∞) such that $b > a$, then b is in Ω.

Any set in (Γ, ∞) with these two properties is called an *over-class* of (Γ, ∞). Conversely, given any over-class Ω of (Γ, ∞), the elements of Σ, including zero, whose values are in Ω form an ideal of $\mathfrak{R}$. To see this, let Ω be any over-class of (Γ, ∞), and let S be the set of elements of Σ whose values are in Ω. Since Ω contains infinity, S contains the zero of Σ, and since each element of Ω is non-negative, $S \subseteq \mathfrak{R}$. Again, if $\xi, \eta \in S$, and, say, $v(\xi) \geqslant v(\eta)$,

$$v(\xi - \eta) \geqslant \min[v(\xi), v(\eta)] = v(\eta),$$

and since $v(\eta) \in \Omega$, it follows that $v(\xi - \eta) \in \Omega$, that is, $\xi - \eta \in S$. If ζ is any element of $\mathfrak{R}$, $v(\zeta) \geqslant 0$. Hence

$$v(\zeta\xi) = v(\zeta) + v(\xi) \geqslant v(\xi).$$

Therefore $\zeta\xi \in S$. Hence S is an ideal in $\mathfrak{R}$.

If $\mathfrak{i}$ is an ideal of $\mathfrak{R}$, and Ω is the corresponding over-class in (Γ, ∞), Ω may or may not have a smallest member. Suppose that it has a smallest member a, and let ζ be an element of $\mathfrak{i}$ such that $v(\zeta) = a$. If ξ is any element of $\mathfrak{i}$, $v(\xi) \geqslant a = v(\zeta)$, and therefore $v(\xi/\zeta) \geqslant 0$. Hence $\eta = \xi/\zeta$ is in $\mathfrak{R}$. Therefore any element of $\mathfrak{i}$ can be written in the form $\eta\zeta$, where η is some member of $\mathfrak{R}$. Thus $\mathfrak{i}$ is a principal ideal, having a basis consisting of the element ζ. Suppose, conversely, that $\mathfrak{i}$ is an ideal in $\mathfrak{R}$ having a finite basis $\omega_1, \ldots, \omega_r$. We may suppose that this is arranged so that $v(\omega_1) \leqslant v(\omega_2) \leqslant \ldots \leqslant v(\omega_r)$. Then, just as above, we show that $\omega_i = \rho_i \omega_1$ $(i = 2, \ldots, r)$, where ρ_i is in $\mathfrak{R}$. Hence the basis for $\mathfrak{i}$ can be reduced to the single element ω_1, so that $\mathfrak{i}$ is a principal ideal. $v(\omega_1)$ is then in the corresponding over-class Ω. If ζ is any element of $\mathfrak{i}$, $\zeta = \rho\omega_1$, where ρ is in $\mathfrak{R}$, and $v(\zeta) = v(\rho) + v(\omega_1) \geqslant v(\omega_1)$. Hence Ω has a first member. In some valuation groups of a field, every over-class has a first member, and hence every ideal in the ring of the valuation is a principal ideal. But examples can be given of valuation groups in which there are over-classes which have no first members. The corresponding ideals of the valuation ring cannot have a finite basis.

If $\mathfrak{i}_1$ and $\mathfrak{i}_2$ are ideals in the ring $\mathfrak{R}$ of a valuation $\mathfrak{B}$ of Σ, and if $\mathfrak{i}_1$ contains an element ζ not in $\mathfrak{i}_2$, the value $v(\eta)$ of any element η of $\mathfrak{i}_2$ is greater than $v(\zeta)$. Since $v(\eta) > v(\zeta)$, $(\eta\zeta^{-1})\,\zeta = \eta$ must lie in $\mathfrak{i}_1$, and so $\mathfrak{i}_2 \subset \mathfrak{i}_1$. Hence, given any two ideals in $\mathfrak{R}$, one of them must be a factor of the other.

Let us now find the prime ideals of $\mathfrak{R}$, by considering the corresponding over-classes of (Γ, ∞). If $\mathfrak{i}$ is a prime ideal of $\mathfrak{R}$, Ω the corresponding over-class, let us consider two non-negative elements a and b of Γ which do not lie in Ω. Then there exist two elements ξ and η of $\mathfrak{R}$ which do not lie in $\mathfrak{i}$ such that

$$v(\xi) = a, \quad v(\eta) = b.$$

Since $\mathfrak{i}$ is prime, $\xi\eta$ is not in $\mathfrak{i}$; hence $v(\xi\eta) = a+b$ is not in Ω. Hence the non-negative elements of Γ which are not in Ω form a set closed under addition. We show that this implies that the non-negative elements of Γ not in Ω, and their negatives, form an isolated subgroup Δ of Γ. We first see that they form a subgroup. If a and b are either non-negative and not in Ω or else are the negatives of such elements, we prove that $a+b$ and $-a$ have the same properties. We have already seen that if $a \geqslant 0$ and $b \geqslant 0$, $a+b$ is not in Ω. If $-a > 0$, $-b > 0$ and are not in Ω, $-(a+b) > 0$ and is not in Ω, hence $(a+b)$ is the negative of some positive element of Γ which is not in Ω. Finally, if say, $a \geqslant 0$, $b < 0$, we consider first the case in which $a+b \geqslant 0$. Then $a > a+b \geqslant 0$ is not in Ω, and hence $a+b$, which is non-negative, is not in Ω. If $a+b < 0$, $-a-b > 0$ and $-b > -a-b > 0$; hence $-a-b$ is positive and not in Ω, since $-b$ is not in Ω. Thus the elements in question form a subgroup Δ. If a is any positive element of Δ, and b is any positive element of Γ such that $a > b$, b is not in Ω, hence it is in Δ. Therefore Δ is isolated.

Conversely, if Δ is an isolated subgroup of Γ, it is clear from the definition of an isolated subgroup that the positive elements of (Γ, ∞) which are not in Δ form an over-class Ω, and that the ideal $\mathfrak{i}$ corresponding to Ω is prime. The ideal $\mathfrak{p}$ of non-units of $\mathfrak{R}$ clearly corresponds to the isolated subgroup of Γ consisting of zero only. Hence we have:

THEOREM I. *If $\mathfrak{V}$ is a valuation of the field Σ to (Γ, ∞), the elements of Σ whose values are non-negative form a ring $\mathfrak{R}$. If k is the rank of Γ, and $\Gamma_1, \ldots, \Gamma_k$ are the isolated subgroups of Γ:*

$$\Gamma \supset \Gamma_1 \supset \ldots \supset \Gamma_k = 0,$$

there are k prime ideals $\mathfrak{p}_1, \ldots, \mathfrak{p}_k$ in $\mathfrak{R}$. The elements of $\mathfrak{p}_i$ are the elements of $\mathfrak{R}$ whose values are not in Γ_i. Also $\mathfrak{p}_i = 0\ (\mathfrak{p}_{i+1})$, and $\mathfrak{p} = \mathfrak{p}_k$ is the ideal formed by the non-units of $\mathfrak{R}$, and is a maximal ideal of $\mathfrak{R}$.

The purpose of the next theorem is to show that a valuation is uniquely determined by its valuation ring.

THEOREM II. *If $\mathfrak{V}_1$ and $\mathfrak{V}_2$ are two valuations of Σ, to (Γ_1, ∞) and (Γ_2, ∞) respectively, which have the same valuation ring $\mathfrak{R}$, then Γ_1 is isomorphic to Γ_2, and if these are identified, $v_1(\xi) = v_2(\xi)$ for every element ξ of Σ.*

We first show that if ξ and η are two non-zero elements of Σ such that $v_1(\xi) = v_1(\eta)$, then $v_2(\xi) = v_2(\eta)$. If $v_1(\xi) = v_1(\eta)$, $v_1(\xi/\eta) = 0$, and therefore ξ/η is a unit of $\mathfrak{R}$. Hence $v_2(\xi/\eta) = 0$, that is, $v_2(\xi) = v_2(\eta)$. Similarly, if $v_2(\xi) = v_2(\eta)$, then $v_1(\xi) = v_1(\eta)$.

Let a_1 be any element of Γ_1, and let $\xi, \eta, \ldots$ be elements of Σ such that

$$v_1(\xi) = v_1(\eta) = \ldots = a_1.$$

By the results we have just proved,

$$v_2(\xi) = v_2(\eta) = \ldots,$$

and if this common value in Γ_2 is a_2, a_1 determines a_2 uniquely. Similarly, any element a_2 of Γ_2 determines a unique element of Γ_1, and a one-to-one correspondence is established between the two groups. We have to show that this correspondence is an isomorphism, in which the ordering is preserved. Let a_1 and b_1 be two elements of Γ_1, a_2 and b_2 the corresponding elements of Γ_2, and let ξ and η be elements of Σ such that

$$v_1(\xi) = a_1, \quad v_1(\eta) = b_1, \quad v_2(\xi) = a_2, \quad v_2(\eta) = b_2.$$

Then

$$v_1(\xi\eta) = a_1 + b_1, \quad v_2(\xi\eta) = a_2 + b_2;$$

hence $a_1 + b_1$ corresponds to $a_2 + b_2$. The correspondence is thus an isomorphism. If $a_1 > 0$, and $v_1(\xi) = a_1$, ξ is a non-unit of $\mathfrak{R}$, and hence

$a_2 = v_2(\xi)$ is positive. Hence ordering is preserved, and the theorem is proved.

Given a field Σ and a ring $\mathfrak{R}$ in it, it is important to be able to tell whether $\mathfrak{R}$ is the ring of some valuation. In the first place $\mathfrak{R}$ must be an integral domain, and Σ must be its quotient field. Moreover, if Σ is a function field of an algebraic variety defined over the ground field K, and if the valuation satisfies condition (iii), p. 172, any element of K other than zero has value zero, and is therefore in $\mathfrak{R}$. Hence $K \subseteq \mathfrak{R}$. In addition to these elementary conditions, which we shall assume are satisfied in what follows, $\mathfrak{R}$ must satisfy further conditions in order that it be a valuation ring.

THEOREM III. *An integral domain $\mathfrak{R}$ is a valuation ring of its quotient field Σ if and only if it has the property that if ξ is any element of Σ, then either ξ or ξ^{-1} belongs to $\mathfrak{R}$.*

The necessity of the condition has already been proved. We therefore consider the sufficiency, and suppose that $\mathfrak{R}$ is an integral domain satisfying the conditions of the theorem. If $\mathfrak{R} = \Sigma$, it is the valuation ring of Σ for the trivial valuation, and the theorem is proved. Suppose, now, that $\mathfrak{R}$ does not coincide with Σ. The elements of Σ, other than zero, form a commutative group G under multiplication. The elements of $\mathfrak{R}$ whose inverses in Σ also lie in $\mathfrak{R}$ likewise form a commutative group G_1 under multiplication in Σ, which is a subgroup of G. The fact that these groups are written as multiplicative groups and not as additive groups is purely a question of notation, since they are commutative, and we can define the factor group G/G_1 as in § 1. This factor group is also a commutative group, and can be written as an additive group Γ. Every element ζ of Σ, other than zero, thus determines a unique element of Γ, which we denote by $v(\zeta)$. The elements which map on the zero of Γ are just those which lie in G_1. Clearly $v(1) = 0$. Since the mapping of the multiplicative group Σ (zero excepted) on the additive group Γ is a homomorphism, we have

$$v(\xi\eta) = v(\xi) + v(\eta). \tag{1}$$

Let a, b be any two elements of Γ, and let ξ, ξ', ... be elements of Σ such that

$$v(\xi) = v(\xi') = \dots = a,$$

while $\eta, \eta', \dots$ are elements of Σ such that

$$v(\eta) = v(\eta') = \dots = b.$$

By hypothesis, ξ/η or η/ξ (or both) lie in $\mathfrak{R}$. Suppose that ξ/η lies in $\mathfrak{R}$. Since ξ and ξ' both map on the element a of Γ, ξ'/ξ is in G_1, and hence in $\mathfrak{R}$. Similarly, η/η' lies in $\mathfrak{R}$. Therefore

$$\xi'/\eta' = (\xi'/\xi)(\eta/\eta')(\xi/\eta)$$

lies in $\mathfrak{R}$. Thus $\xi/\eta, \xi'/\eta', \ldots$ all lie in $\mathfrak{R}$. We then write $a \geqslant b$, and $b \leqslant a$. If, also, $b \geqslant a$, we must have ξ/η in G_1, hence ξ and η must map on the same element of Γ, that is, $a = b$. Conversely, $a = b$ implies that ξ/η lies in G_1; hence $a \geqslant b$ and $b \geqslant a$. If $a \geqslant b$ and $a \neq b$ we write $a > b$. We see that $a > 0$ implies that $\xi = \xi/1$ is in $\mathfrak{R}$ while ξ^{-1} is not in $\mathfrak{R}$.

(i) If $v(\xi) < 0$, ξ is not in $\mathfrak{R}$. Hence ξ^{-1} is in $\mathfrak{R}$, and its inverse is not. Therefore $v(\xi^{-1}) > 0$. But $0 = v(1) = v(\xi . \xi^{-1}) = v(\xi) + v(\xi^{-1})$. Hence

$$-v(\xi) = v(\xi^{-1}) > 0.$$

(ii) If $v(\xi) > 0$, $v(\eta) > 0$, ξ and η are in $\mathfrak{R}$. Hence $\xi\eta$ is in $\mathfrak{R}$, that is, $v(\xi\eta) \geqslant 0$. If $v(\xi\eta) = 0$, $\xi\eta$ is in G_1, and hence it has an inverse ζ in $\mathfrak{R}$. Then $\xi\eta\zeta = 1$, so that $\eta\zeta$ is the inverse of ξ and lies in $\mathfrak{R}$. Hence $v(\xi) = 0$, and we have a contradiction. Therefore

$$v(\xi) + v(\eta) = v(\xi\eta) > 0.$$

Thus we have an ordering of Γ.

By (1), the mapping $\xi \rightarrow v(\xi)$ of Σ on Γ satisfies the first condition for a valuation. To prove the second condition, let ξ and η be non-zero elements of Σ, and suppose that $v(\xi) \geqslant v(\eta)$. Then ξ/η is in $\mathfrak{R}$. Since unity is also in $\mathfrak{R}$, $1 + \xi/\eta$ is in $\mathfrak{R}$. Therefore

$$v(1 + \xi/\eta) = v(\xi + \eta) - v(\eta) \geqslant 0,$$

and hence $$v(\xi + \eta) \geqslant v(\eta) = \min[v(\xi), v(\eta)].$$

Thus the mapping $\xi \rightarrow v(\xi)$ of the non-zero elements of Σ on Γ is a valuation, and $\mathfrak{R}$ is therefore a valuation ring.

If the quotient field Σ of $\mathfrak{R}$ is the function field of an algebraic variety over K, we must have $K \subseteq \mathfrak{R}$. Then if a is any non-zero element of K, a^{-1} is also in K, and hence a and a^{-1} are in G_1. Therefore $v(a) = 0$, and the third condition is satisfied.

Before considering a second characterisation of a valuation ring it is convenient to prove a lemma.

Lemma. If the integral domain $\mathfrak{R}$ has the property that any proper extension of it which lies in the quotient field Σ of $\mathfrak{R}$ contains the inverse of a non-unit of $\mathfrak{R}$, then $\mathfrak{R}$ is integrally closed in Σ.

Let $\mathfrak{R}^*$ be the integral closure of $\mathfrak{R}$ in Σ. If $\mathfrak{R}$ is not integrally closed, $\mathfrak{R}^* \supset \mathfrak{R}$, and hence, by hypothesis, there exists a non-unit ξ of $\mathfrak{R}$ whose inverse ξ^{-1} lies in $\mathfrak{R}^*$. Since ξ^{-1} is in $\mathfrak{R}^*$, it is integrally dependent on $\mathfrak{R}$; hence we have a relation

$$\xi^{-n} + a_1 \xi^{-n+1} + \ldots + a_n = 0 \quad (a_i \in \mathfrak{R}),$$

and therefore

$$1 + a_1 \xi + \ldots + a_n \xi^n = 0.$$

If

$$\eta = -(a_1 + a_2 \xi + \ldots + a_n \xi^{n-1}),$$

η lies in $\mathfrak{R}$, and is the inverse of ξ, which contradicts the fact that ξ is a non-unit of $\mathfrak{R}$. Hence the assumption that $\mathfrak{R}$ is not integrally closed in Σ is false.

From this lemma we prove

THEOREM IV. *The ring $\mathfrak{R}$ of a valuation $\mathfrak{B}$ of Σ is integrally closed in Σ.*

Since $\mathfrak{R}$ consists of all those elements of Σ whose values in $\mathfrak{B}$ are non-negative, any proper extension of $\mathfrak{R}$ in Σ will contain an element ζ such that $v(\zeta) < 0$. Then $v(\zeta^{-1}) > 0$, that is, ζ^{-1} is a non-unit of $\mathfrak{R}$. Hence $\mathfrak{R}$ has the property that any proper extension of it in Σ contains the inverse of a non-unit of $\mathfrak{R}$. Thus, from the Lemma, $\mathfrak{R}$ is integrally closed in Σ.

We now come to our second characterisation of a valuation ring. We have already established the following properties of a valuation ring $\mathfrak{R}$: (i) the non-units of $\mathfrak{R}$ form an ideal in $\mathfrak{R}$; (ii) any extension of $\mathfrak{R}$ in Σ contains the inverse of a non-unit of $\mathfrak{R}$. These two properties are sufficient to characterise a valuation ring; precisely, we have to prove

THEOREM V. *An integral domain $\mathfrak{R}$ is the ring of some valuation of its quotient field Σ if and only if it satisfies the two conditions* (i) *the non-units of $\mathfrak{R}$ form an ideal in $\mathfrak{R}$, and* (ii) *any proper extension of $\mathfrak{R}$ in Σ contains the inverse of a non-unit of $\mathfrak{R}$.*

We note that if Σ is the function field of a variety defined over the ground field K, and the valuation is required to satisfy the condition that $v(a) = 0$ for any non-zero element a of K, we must add to Theorem V the condition that K is contained in $\mathfrak{R}$.

The proof consists in showing that a ring $\mathfrak{R}$ which satisfies the given conditions satisfies the condition of Theorem III for a valuation ring. We have to show that if $\mathfrak{R}$ satisfies the conditions of Theorem V, and if ξ is any non-zero element of Σ, then either ξ or

ξ^{-1} must lie in $\mathfrak{R}$. Suppose that, for a given ξ, ξ^{-1} does not lie in $\mathfrak{R}$. Then $\mathfrak{R}[\xi^{-1}]$ is a proper extension of $\mathfrak{R}$. By hypothesis, $\mathfrak{R}[\xi^{-1}]$ contains the inverse of some non-unit η of $\mathfrak{R}$, so that

$$\eta^{-1} = a_0 + a_1\xi^{-1} + \ldots + a_r\xi^{-r} \quad (a_i \in \mathfrak{R}).$$

Hence
$$(1 - a_0\eta)\,\xi^r - a_1\eta\xi^{r-1} - \ldots - a_r\eta = 0.$$

Now $a_0\eta$ is a non-unit of $\mathfrak{R}$, while 1 is, of course, a unit. Since the non-units of $\mathfrak{R}$ form an ideal, it follows that $1 - a_0\eta$ must be a unit of $\mathfrak{R}$. Hence it has an inverse ζ in $\mathfrak{R}$. Therefore

$$\xi^r - a_1\eta\zeta\xi^{r-1} - \ldots - a_r\eta\zeta = 0;$$

that is, ξ is integrally dependent on $\mathfrak{R}$. But, by the Lemma proved above, the conditions of Theorem V imply that $\mathfrak{R}$ is integrally closed in Σ; hence ξ is in $\mathfrak{R}$. Thus we see that if ξ is any non-zero element of Σ, either ξ or ξ^{-1} lies in $\mathfrak{R}$, and therefore, by Theorem III, we conclude that $\mathfrak{R}$ is a valuation ring.

Theorem IV tells us that the ring of any valuation is integrally closed in its quotient field. Suppose that we have any set of valuations $\mathfrak{B}_1, \mathfrak{B}_2, \ldots$ of the field Σ, and let $\mathfrak{R}_1, \mathfrak{R}_2, \ldots$ be their valuation rings. Let $\mathfrak{R}$ be the set of elements of Σ common to $\mathfrak{R}_1, \mathfrak{R}_2, \ldots$. If α and β are any two elements of $\mathfrak{R}$, $\alpha + \beta$ and $\alpha\beta$ must lie in each of the rings $\mathfrak{R}_1, \mathfrak{R}_2, \ldots$, and hence must lie in $\mathfrak{R}$. Hence $\mathfrak{R}$ is a ring. If ζ is any element of Σ which is integrally dependent on $\mathfrak{R}$, ζ is integrally dependent on each of the rings $\mathfrak{R}_1, \mathfrak{R}_2, \ldots$. But these rings, being valuation rings, are integrally closed. Hence ζ lies in $\mathfrak{R}_1, \mathfrak{R}_2, \ldots$, and therefore ζ lies in $\mathfrak{R}$. Thus $\mathfrak{R}$ is integrally closed in Σ, and therefore in its quotient field, which is contained in Σ. The following converse of this result is important.

THEOREM VI. *If $\mathfrak{R}$ is an integral domain which is integrally closed in the field Σ, $\mathfrak{R}$ is the intersection of the valuation rings of Σ which contain it.*

We shall at present only prove this theorem when Σ is the quotient field of $\mathfrak{R}$. The extension to more general fields Σ (containing $\mathfrak{R}$ and hence its quotient field) will be a simple consequence of Theorem VII. In applying the result to algebraic geometry it is usually desirable to impose the condition that $\mathfrak{R}$ contains the ground field K; then all the valuations whose rings contain $\mathfrak{R}$ have the property $v(a) = 0$, where a is any non-zero element of K.

To prove Theorem VI, our first step is to show that if ξ is any element of Σ which is not in $\mathfrak{R}$, there exists a ring $\mathfrak{R}_\xi$ in Σ which contains $\mathfrak{R}$ but not ξ, and which is a valuation ring of its quotient field. Since ξ is not in $\mathfrak{R}$, and since $\mathfrak{R}$ is integrally closed in Σ, ξ cannot be contained in $\mathfrak{R}[\xi^{-1}]$; for if

$$\xi = a_0 + a_1\xi^{-1} + \ldots + a_r\xi^{-r} \quad (a_i \in \mathfrak{R}),$$

$$\xi^{r+1} - a_0\xi^r - \ldots - a_r = 0,$$

and it would follow from the fact that $\mathfrak{R}$ is integrally closed in Σ that ξ was in $\mathfrak{R}$.

Now let us well-order the elements of Σ which are not in $R[\xi^{-1}]$, and then, by transfinite induction, construct a function $\phi(x)$ of these elements satisfying the conditions:

(i) $\phi(x) = 1$, if the ring obtained by adjoining to $R[\xi^{-1}]$ the element x and the elements $\phi(y)\,y$, for all elements y which precede x, does not contain ξ;

(ii) $\phi(x) = 0$, otherwise.

Then the set of elements of Σ consisting of $\mathfrak{R}[\xi^{-1}]$ and the elements $\phi(x)\,x$ as x ranges over $\Sigma - \mathfrak{R}[\xi^{-1}]$ is clearly a ring, which we denote by $\mathfrak{R}_\xi$. If Σ_ξ is its quotient field, $\Sigma \supseteq \Sigma_\xi \supseteq$ (the quotient field of $\mathfrak{R}$). We prove that $\mathfrak{R}_\xi$ is a valuation ring of Σ_ξ. In the first place, any extension of $\mathfrak{R}_\xi$ in Σ_ξ is an extension of $\mathfrak{R}_\xi$ in Σ. But any extension of $\mathfrak{R}_\xi$ in Σ must contain some element x of Σ for which $\phi(x) = 0$. By the definition of $\phi(x)$, it follows that this extension contains ξ, which is the inverse of ξ^{-1}, a non-unit of $\mathfrak{R}_\xi$. Hence any extension of $\mathfrak{R}_\xi$ in Σ_ξ contains the inverse of a non-unit of $\mathfrak{R}_\xi$. It follows from Theorem V that to prove that $\mathfrak{R}_\xi$ is a valuation ring we have only to prove that the non-units of $\mathfrak{R}_\xi$ form an ideal in $\mathfrak{R}_\xi$. If the non-units of $\mathfrak{R}_\xi$ do not form an ideal, there must exist two elements ζ' and η' of $\mathfrak{R}_\xi$ which are non-units, such that $\alpha = \zeta' + \eta'$ is a unit. Then $\zeta = \alpha^{-1}\zeta'$, $\eta = \alpha^{-1}\eta'$ are non-units such that $\zeta + \eta = 1$. We show that this is impossible.

Since ζ and η are non-units of $\mathfrak{R}_\xi$, ζ^{-1} and η^{-1} are not in $\mathfrak{R}_\xi$. Hence $\mathfrak{R}_\xi[\zeta^{-1}]$ and $\mathfrak{R}_\xi[\eta^{-1}]$ are proper extensions of $\mathfrak{R}_\xi$ and therefore contain ξ. Hence

$$\xi = a_0 + a_1\zeta^{-1} + \ldots + a_r\zeta^{-r} \quad (a_i \in \mathfrak{R}_\xi),$$

$$\xi = b_0 + b_1\eta^{-1} + \ldots + b_s\eta^{-s} \quad (b_i \in \mathfrak{R}_\xi);$$

therefore, if $t = \max[r, s]$,

$$\xi = \alpha/\zeta^t = \beta/\eta^t,$$

where α and β are in $\mathfrak{R}_\xi$. Let p and q be any two non-negative integers such that $p+q = t$. Then

$$(\eta^p\zeta^q\xi)^t = (\zeta^t\xi)^q(\eta^t\xi)^p = \alpha^q\beta^p.$$

Hence $\eta^p\zeta^q\xi$ is integrally dependent on $\mathfrak{R}_\xi$. But, by our lemma, R_ξ is integrally closed. Hence $\eta^p\zeta^q\xi = c_{pq}$, where c_{pq} is in $\mathfrak{R}_\xi$. Now

$$\begin{aligned}\xi &= (\eta+\zeta)^t\xi \\ &= \sum_{p=0}^{t}\binom{t}{p}\eta^p\zeta^{t-p}\xi = \sum_{p=0}^{t}\binom{t}{p}c_{p\,t-p},\end{aligned}$$

and hence ξ lies in $\mathfrak{R}_\xi$. This is a contradiction, and from it we conclude that the non-units of $\mathfrak{R}_\xi$ form an ideal. Hence [Theorem V] $\mathfrak{R}_\xi$ is a valuation ring of Σ_ξ.

If Σ is the quotient field of $\mathfrak{R}$, $\Sigma_\xi = \Sigma$, and hence $\mathfrak{R}_\xi$ is a valuation ring of Σ. Now $\mathfrak{R}_\xi$ is a valuation ring of Σ which excludes ξ, but contains $\mathfrak{R}$. Hence as ξ ranges over the elements of Σ which are not in $\mathfrak{R}$, we obtain a set of valuation rings, each of which contains $\mathfrak{R}$, but whose intersection does not contain any element ξ of Σ which is not in $\mathfrak{R}$. $\mathfrak{R}$ is therefore the intersection of these valuation rings, and Theorem VI is proved in the case in which Σ is the quotient field of $\mathfrak{R}$.

If Σ is an extension of the quotient field of $\mathfrak{R}$, the same argument will hold if we can prove that Σ_ξ coincides with Σ. In Theorem VII we shall show that if Σ_ξ is a proper subfield of Σ, there exists a valuation ring $\mathfrak{R}^*_\xi$ of Σ such that $\mathfrak{R}^*_\xi \wedge \Sigma_\xi = \mathfrak{R}_\xi$. Now $\mathfrak{R}^*_\xi$ is a proper extension of $\mathfrak{R}_\xi$ (since its quotient field Σ is a proper extension of Σ_ξ). Hence, by the characteristic property of $\mathfrak{R}_\xi$, $\mathfrak{R}^*_\xi$ contains ξ. Since Σ_ξ also contains ξ, $\mathfrak{R}^*_\xi \wedge \Sigma_\xi$ must contain ξ, and we have a contradiction. It follows that $\Sigma_\xi = \Sigma$, and the proof of Theorem VI is complete.

Let Σ_1 and Σ_2 be two fields, and suppose that $\Sigma_1 \supset \Sigma_2$. Consider any valuation $\mathfrak{B}_1$ of Σ_1, and let Γ_1 be its valuation group, $\mathfrak{R}_1$ the ring of the valuation. The mapping of Σ_1 on Γ_1 induces a mapping of Σ_2 on Γ_1, or on a subset Γ_2 of Γ_1, which is at once seen to be a subgroup of Γ_1. Γ_2 is an ordered additive group, and it is clear that the mapping of Σ_2 on Γ_2 induced by $\mathfrak{B}_1$ is a valuation $\mathfrak{B}_2$ of Σ_2. The elements of the ring $\mathfrak{R}_2$ of $\mathfrak{B}_2$ are those elements of Σ_2 which map on non-negative elements of Γ_2, that is, those elements of Σ_1 which lie in Σ_2 and map on non-negative elements of Γ_1. Hence

$$\mathfrak{R}_2 = \mathfrak{R}_1 \wedge \Sigma_2.$$

Thus any valuation of a field Σ_1 induces in this way a unique valuation of any subfield Σ_2 of Σ_1 (which may, of course, be a trivial valuation). The valuation $\mathfrak{B}_2$ is called the *induced valuation* of Σ_2 determined by $\mathfrak{B}_1$, and $\mathfrak{B}_1$ is called an *extended* valuation of $\mathfrak{B}_2$. The question then arises whether, given an arbitrary valuation $\mathfrak{B}_2$ of Σ_2, there exists in a given extension Σ_1 of Σ_2 a valuation $\mathfrak{B}_1$ which is an extended valuation of $\mathfrak{B}_2$. This is answered by

THEOREM VII. *If Σ^* is any extension of the field Σ, and $\mathfrak{B}$ is any valuation of Σ, there exists at least one valuation of Σ^* which is an extended valuation of $\mathfrak{B}$.*

If $\mathfrak{B}$ is a trivial valuation of Σ, the trivial valuation of Σ^* satisfies the requirements of the theorem. We may therefore confine ourselves to the case in which the valuation group Γ has non-zero elements. By convention, we have agreed to consider only valuations to groups of finite rank. If k is the rank of Γ, we denote its isolated subgroups by $\Gamma_1, \ldots, \Gamma_k$, where

$$\Gamma \supset \Gamma_1 \supset \ldots \supset \Gamma_k = 0.$$

$\Gamma_{k-1}/\Gamma_k \cong \Gamma_{k-1}$ is Archimedean-ordered [§ 1, Th. V]. Let a be a positive element of Γ_{k-1}, and let b be any positive element of Γ. If b is not in Γ_{k-1}, $b > a$, and, if b is in Γ_{k-1}, there exists an integer n such that $nb > a$, since Γ_{k-1} is Archimedean-ordered. We choose some element ζ of Σ such that $v(\zeta) = a$. It follows that if ξ is any element of Σ whose value is positive, then there exists an integer n such that $v(\xi^n) > v(\zeta)$.

Let $\mathfrak{R}$ be the valuation ring of $\mathfrak{B}$. Since $v(\zeta) > 0$, ζ is in $\mathfrak{R}$, and is a non-unit of this ring. We construct the ring $\mathfrak{R}^* = \mathfrak{R}_{\zeta^{-1}}$, in Σ^*, as in the proof of Theorem VI. This is a valuation ring. To prove Theorem VII, we have to show (i) that $\mathfrak{R}^* \wedge \Sigma = \mathfrak{R}$; and (ii) that the quotient field of $\mathfrak{R}^*$ is Σ^*.

(i) $\mathfrak{R}^* \supseteq \mathfrak{R}$ and $\Sigma \supseteq \mathfrak{R}$; hence $\overline{\mathfrak{R}} = \mathfrak{R}^* \wedge \Sigma \supseteq \mathfrak{R}$. Suppose that $\overline{\mathfrak{R}}$ is a proper extension of $\mathfrak{R}$. Then it contains the inverse of a non-unit, say ξ, of $\mathfrak{R}$, since $\mathfrak{R}$ is a valuation ring for Σ. We have seen above that there exists an integer n such that $v(\xi^n) > v(\zeta)$. Hence ξ^n/ζ is contained in $\mathfrak{R}$. Now $\overline{\mathfrak{R}} \supset \mathfrak{R}$ and $\xi^{-1} \in \overline{\mathfrak{R}}$; hence $\overline{\mathfrak{R}} \supseteq \mathfrak{R}[\xi^{-1}]$. It follows that $(\xi^n/\zeta)\,\xi^{-n} = \zeta^{-1}$ lies in $\bar{\mathfrak{R}}$. Since $\bar{\mathfrak{R}} \subseteq \mathfrak{R}^*$, $\zeta^{-1} \in \mathfrak{R}^* = \mathfrak{R}_{\zeta^{-1}}$, which is not the case, since by construction $\mathfrak{R}^*$ does not contain ζ^{-1}. Hence our assumption that $\bar{\mathfrak{R}}$ is a proper extension of $\mathfrak{R}$ is false.

(ii) We first show that $\mathfrak{R}^*$ is a proper extension of $\mathfrak{R}$. Suppose that this is not the case. Then any extension of $\mathfrak{R}$ in Σ^* contains ζ^{-1}. In particular, if ξ is any element of Σ^* not in Σ, $\mathfrak{R}[\xi]$ and $\mathfrak{R}[\xi^{-1}]$ both contain ζ^{-1}. Hence we have equations

$$a_n + a_{n-1}\xi + \dots + a_0\xi^n = \zeta^{-1} \quad (a_i \in \mathfrak{R}),$$

$$c_0 + c_1\xi^{-1} + \dots + c_m\xi^{-m} = \zeta^{-1} \quad (c_i \in \mathfrak{R}).$$

Hence
$$b_0\xi^n + b_1\xi^{n-1} + \dots + b_n = 0,$$
and
$$d_0\xi^m + d_1\xi^{m-1} + \dots + d_m = 0,$$
where
$$b_i = a_i\zeta = 0 \ (\mathfrak{p}) \quad (i = 0, \dots, n-1),$$
$$b_n = a_n\zeta - 1 \neq 0 \ (\mathfrak{p}),$$
and
$$d_i = c_i\zeta = 0 \ (\mathfrak{p}) \quad (i = 1, \dots, m),$$
$$d_0 = c_0\zeta - 1 \neq 0 \ (\mathfrak{p}),$$
where $\mathfrak{p}$ is the ideal of non-units of $\mathfrak{R}$. The equations

$$b_0x^n + \dots + b_n = 0$$
and
$$d_0x^m + \dots + d_m = 0$$
have a common root; hence their resultant D vanishes. But [IV, § 3]

$$D = \begin{vmatrix} d_0 & d_1 & . & . & d_m & . & . & 0 \\ 0 & d_0 & d_1 & . & . & d_m & . & 0 \\ . & . & . & . & . & . & . & . \\ . & . & . & d_0 & d_1 & . & . & d_m \\ b_0 & b_1 & . & . & b_n & . & . & . \\ 0 & b_0 & . & . & . & b_n & . & . \\ . & . & . & . & . & . & . & . \\ . & . & . & b_0 & b_1 & . & . & b_n \end{vmatrix}$$
$$= d_0^n b_n^m \ (\mathfrak{p}).$$

Hence $d_0^n b_n^m = 0 \ (\mathfrak{p})$. This, however, is impossible, since d_0 and b_n are not in $\mathfrak{p}$, and $\mathfrak{p}$ is prime. Our assumption is therefore false, and $\mathfrak{R}^*$ is a proper extension of $\mathfrak{R}$. Since $\mathfrak{R}^* \wedge \Sigma = \mathfrak{R}$, there must be elements of R^* which are not in Σ, and it follows that the quotient field Σ' of $\mathfrak{R}^*$ is a proper extension of Σ:

$$\Sigma^* \supseteq \Sigma' \supset \Sigma.$$

If Σ' is not equal to Σ^*, we can begin again with $\mathfrak{R}^*$ in place of $\mathfrak{R}$, and construct a ring $\mathfrak{R}^{**}$ which is a proper extension of $\mathfrak{R}^*$, and has the property $\mathfrak{R}^{**}{}_\wedge\Sigma' = \mathfrak{R}^*$. Now $\zeta \in \Sigma \subseteq \Sigma'$; hence ζ^{-1} is in Σ'. Since $\mathfrak{R}^{**}$ is a proper extension of $\mathfrak{R}^*$ and lies in Σ^*, it must, by the method of construction of $\mathfrak{R}^*$, contain ζ^{-1}. Hence $\mathfrak{R}^*$ contains ζ^{-1}, which again is a contradiction. Hence $\Sigma' = \Sigma^*$, that is, the quotient field of $\mathfrak{R}^*$ is Σ^*.

There exist a number of theorems concerning the possible valuations of Σ^* which are extensions of a given valuation of Σ, when Σ^* is derived from Σ by some standard process, such as algebraic extension or the adjunction of indeterminates. We shall prove later such theorems of this type as we shall require for the purposes of algebraic geometry, but as a preliminary we must prove some results concerning sets of valuations of a given field Σ.

THEOREM VIII. *If $\mathfrak{R}_1$ and $\mathfrak{R}_2$ are two valuation rings of a field Σ, and $\mathfrak{p}_1$ and $\mathfrak{p}_2$ are the ideals of the corresponding valuations (that is, the ideals of the non-units of $\mathfrak{R}_1$ and $\mathfrak{R}_2$ respectively),*

(i) *if $\mathfrak{R}_1 \subset \mathfrak{R}_2$, then $\mathfrak{p}_2 \subset \mathfrak{p}_1$; and the rank of $\mathfrak{R}_1 >$ rank of $\mathfrak{R}_2$;*

(ii) *if $\mathfrak{R}_1 \nsubseteq \mathfrak{R}_2$, then there exists an element ξ in Σ which is a unit of $\mathfrak{R}_1$ and a non-unit of $\mathfrak{R}_2$.*

The rank of a valuation is the rank of its valuation group.

We can omit the case $\mathfrak{R}_1 = \mathfrak{R}_2$, since in this case the valuations are the same [Theorem II].

(i) If ξ is a non-unit of $\mathfrak{R}_2$, ξ^{-1} is not in $\mathfrak{R}_2$, and hence is not in $\mathfrak{R}_1 \subset \mathfrak{R}_2$. Therefore $\xi = (\xi^{-1})^{-1}$ is a non-unit of $\mathfrak{R}_1$, and therefore we have $\mathfrak{p}_2 \subseteq \mathfrak{p}_1$. Let η be any element of $\mathfrak{R}_2$ which is not in $\mathfrak{R}_1$. Then $\eta+1$ is in $\mathfrak{R}_2$ and not in $\mathfrak{R}_1$. The elements η and $\eta+1$ cannot both be non-units of $\mathfrak{R}_2$, since the non-units of $\mathfrak{R}_2$ form an ideal $\mathfrak{p}_2$ which does not contain $(\eta+1)-\eta = 1$. Hence we can find a unit τ (equal to η or $\eta+1$) of $\mathfrak{R}_2$ which is not in $\mathfrak{R}_1$. Then $\zeta = \tau^{-1}$ is a unit of $\mathfrak{R}_2$ which is a non-unit of $\mathfrak{R}_1$. It follows that there are elements of $\mathfrak{p}_1$ which are not in $\mathfrak{p}_2$. Therefore

$$\mathfrak{p}_2 \subset \mathfrak{p}_1.$$

Now let $\mathfrak{P}_2$ be any ideal of $\mathfrak{R}_2$, other than the unit ideal.

$$\mathfrak{P}_2 \subseteq \mathfrak{p}_2 \subset \mathfrak{p}_1 \subset \mathfrak{R}_1;$$

hence $\mathfrak{P}_2$ is contained in $\mathfrak{R}_1$. To see that $\mathfrak{P}_2$ is an ideal of $\mathfrak{R}_1$, it is sufficient to observe, first, that if α and β are two elements of $\mathfrak{P}_2$, $\alpha-\beta$ is in $\mathfrak{P}_2$, since $\mathfrak{P}_2$ is an ideal of $\mathfrak{R}_2$, and, secondly, that if α is in $\mathfrak{P}_2$ and ρ is in $\mathfrak{R}_1$, ρ is in $\mathfrak{R}_2$, since $\mathfrak{R}_1 \subset \mathfrak{R}_2$, and therefore $\rho\alpha$ is in $\mathfrak{P}_2$. If $\mathfrak{P}_2$ is a prime ideal of $\mathfrak{R}_2$, let γ and δ be two elements of $\mathfrak{R}_1$ such that $\gamma\delta \in \mathfrak{P}_2$. Then γ and δ belong to $\mathfrak{R}_2$, and hence γ or δ is in $\mathfrak{P}_2$. Hence $\mathfrak{P}_2$ is a prime ideal of $\mathfrak{R}_1$.

Thus any proper prime ideal of $\mathfrak{R}_2$ is a proper prime ideal of $\mathfrak{R}_1$, and in addition $\mathfrak{p}_1$ is a proper prime ideal of $\mathfrak{R}_1$. Hence $\mathfrak{R}_1$ has more prime ideals than $\mathfrak{R}_2$. But the number of prime ideals of a valuation ring is equal to its rank [Theorem I]. Hence the rank of $\mathfrak{R}_1$ is greater than the rank of $\mathfrak{R}_2$.

(ii) If $\mathfrak{R}_1 \not\subseteq \mathfrak{R}_2$, there exists an element η in $\mathfrak{R}_1$ which is not contained in $\mathfrak{R}_2$. Replacing η, if necessary, by $\eta+1$ (as in (i)), we may suppose that η is a unit of $\mathfrak{R}_1$ which is not in $\mathfrak{R}_2$. Then $\xi = \eta^{-1}$ is a unit of $\mathfrak{R}_1$ and a non-unit of $\mathfrak{R}_2$.

The following generalisation of Theorem VIII (ii) is sometimes useful. It concerns any finite number of valuation rings $\mathfrak{R}_1, \ldots, \mathfrak{R}_k$. These rings all have zero and unity in common, but we shall make the assumption that they have in common at least two elements α and β which are units of $\mathfrak{R}_i$ (for $i = 1, \ldots, k$) and are such that $\alpha-\beta$ is also a unit of each of these rings. This assumption is always satisfied in the case of the valuations of an algebraic function field, since the rings $\mathfrak{R}_1, \ldots, \mathfrak{R}_k$ all contain the ground field K, of characteristic zero, and any non-zero of K is a unit of $\mathfrak{R}_i$.

THEOREM IX. *If $\mathfrak{R}_1, \ldots, \mathfrak{R}_k$ are valuation rings of a field Σ having the properties (a) $\mathfrak{R}_i \not\subseteq \mathfrak{R}_j$ $(i = 1, \ldots, k;\ j = 1, \ldots, i-1, i+1, \ldots, k)$; (b) there exist elements α and β in $\mathfrak{R}_1 \wedge \mathfrak{R}_2 \wedge \ldots \wedge \mathfrak{R}_k$ such that $\alpha, \beta, \alpha-\beta$ are units of $\mathfrak{R}_i$ $(i = 1, \ldots, k)$; then there exist elements $\xi_1, \ldots, \xi_k$ in $\mathfrak{R}_1 \wedge \mathfrak{R}_2 \wedge \ldots \wedge \mathfrak{R}_k$ such that*

$$\xi_i = 0 \ (\mathfrak{p}_j) \quad (j \neq i), \qquad \xi_i \neq 0 \ (\mathfrak{p}_i).$$

The case $k = 2$ has already been proved in Theorem VIII (ii). Suppose that it has been proved for $k-1$ rings $(k \geqslant 3)$. Then there exists an element η_{ij} of Σ lying in $\mathfrak{R}_1, \ldots, \mathfrak{R}_{i-1}, \mathfrak{R}_{i+1}, \ldots, \mathfrak{R}_k$, which is a unit of $\mathfrak{R}_j$ and a non-unit of $\mathfrak{R}_t$ (for all t different from i and j). If η_{ij} is in $\mathfrak{R}_i$, we write $\zeta_{ij} = \eta_{ij}$. If η_{ij} is not in $\mathfrak{R}_i$, η_{ij}^{-1} is a unit of $\mathfrak{R}_j$, and is a non-unit of $\mathfrak{R}_i$, but does not lie in $\mathfrak{R}_t$ $(t \neq i, j)$. $\eta_{ij}^{-1}-\alpha$ and $\eta_{ij}^{-1}-\beta$ are both in $\mathfrak{R}_i$ and in $\mathfrak{R}_j$, but are not in $\mathfrak{R}_t$ $(t \neq i, j)$. They are both units of $\mathfrak{R}_i$ (since α and β are units), and one at least is a unit of $\mathfrak{R}_j$. For, if they were both non-units, $(\eta_{ij}^{-1}-\alpha)-(\eta_{ij}^{-1}-\beta) = \beta-\alpha$ would be a non-unit, contrary to hypothesis. Suppose that $\eta_{ij}^{-1}-\alpha$ is a unit of $\mathfrak{R}_j$. Then if $\zeta_{ij} = (\eta_{ij}^{-1}-\alpha)^{-1}$, ζ_{ij} is in $\mathfrak{R}_i$; it is a unit of $\mathfrak{R}_j$, and a non-unit of $\mathfrak{R}_t$ $(t \neq i, j)$.

Let

$$\xi_j = \prod_{i \neq j} \zeta_{ij}.$$

Since $\zeta_{ij} \neq 0$ $(\mathfrak{p}_j)$ for $i = 1, \ldots, j-1, j+1, \ldots, k$, and $\mathfrak{p}_j$ is prime, $\xi_j \neq 0$ $(\mathfrak{p}_j)$. If $t \neq j$, we can choose i different from j and t (since $k \geqslant 3$), and for this value of j

$$\zeta_{ij} = 0 \quad (\mathfrak{p}_t).$$

Hence
$$\xi_j = 0 \quad (\mathfrak{p}_t).$$

Thus $\xi_1, \ldots, \xi_k$ have the required properties.

We conclude this section by proving a preliminary result on the extended valuations of a valuation $\mathfrak{B}$ of a field Σ, when Σ^* is an algebraic extension of Σ, of finite degree r. More precise results will be obtained in the next section. Let Γ be the valuation group of $\mathfrak{B}$, Γ^* that of an extended valuation $\mathfrak{B}^*$ of Σ^*. Let σ be any non-zero element of Σ^*, and let

$$a_0\sigma^n + \ldots + a_n = 0$$

be the algebraic equation with coefficients in Σ which it satisfies $(1 \leqslant n \leqslant r)$. There are at least two non-zero terms in this equation. If all the non-zero terms of this equation had different values, it would follow from the preliminary result (3) of this section that

$$v^*(a_0\sigma^n + \ldots + a_n) = \min[v^*(a_0\sigma^n), \ldots, v^*(a_n)],$$

and is therefore finite, whereas $v^*(0) = \infty$. Thus there must exist two terms $a_i\sigma^{n-i}$ and $a_j\sigma^{n-j}$ which have the same (finite) values in $\mathfrak{B}^*$. If $j > i$, we obtain $(j-i)v^*(\sigma) = v^*(\sigma^{j-i}) = v^*(a_j/a_i) = v(a_j/a_i)$. Since $0 \leqslant i < j \leqslant r$, for every element σ of Σ^*, $j-i$ is a factor of $r!$, and hence

$$r!\,v^*(\sigma) = v([a_j/a_i]^{r!/(j-i)})$$

is equal to the value in $\mathfrak{B}$ of some element of Σ.

Now Γ is a subgroup of Γ^*, and it has the property that $r!$ times any element of Γ^* is in Γ.

Let Γ^* be of rank k, and let

$$\Gamma^* \supset \Gamma_1^* \supset \ldots \supset \Gamma_k^* = 0$$

be its isolated subgroups. Then $\Gamma_i^* \wedge \Gamma \supseteq \Gamma_{i+1}^* \wedge \Gamma$, and it follows at once from the definition of an isolated subgroup that $\Gamma_i^* \wedge \Gamma$ is an isolated subgroup of Γ. Further, it can be verified at once that $\Gamma_i^* \wedge \Gamma / \Gamma_{i+1}^* \wedge \Gamma$ is isomorphic to a subgroup of the additive group of real numbers; hence it is Archimedean-ordered. It follows that all the isolated subgroups of Γ are in the sequence

$$\Gamma \supseteq \Gamma_1^* \wedge \Gamma \supseteq \ldots \supseteq \Gamma_k^* \wedge \Gamma = 0.$$

To show that Γ is of rank k, we have only to show that

$$\Gamma_i^* \wedge \Gamma \supset \Gamma_{i+1}^* \wedge \Gamma.$$

Let α be any positive element of Γ_i^* which is not in Γ_{i+1}^*. Then $r!\alpha$ is contained in Γ_i^* and in Γ; that is, $r!\alpha \in \Gamma_i^* \wedge \Gamma$. If $r!\alpha \in \Gamma_{i+1}^* \wedge \Gamma$, $r!\alpha \in \Gamma_{i+1}^*$, and since $r!\alpha > \alpha > 0$, it follows from the definition of an isolated subgroup that $\alpha \in \Gamma_{i+1}^*$, and we have a contradiction. Hence there are elements of $\Gamma_i^* \wedge \Gamma$ which are not in $\Gamma_{i+1}^* \wedge \Gamma$. Hence $\Gamma_i^* \wedge \Gamma \supset \Gamma_{i+1}^* \wedge \Gamma$. Thus Γ and Γ^* are of the same rank. Again, if Γ is *discrete*, that is, if it is isomorphic with the additive group of integers, it follows that Γ^* is also discrete. For Γ^* is of rank one and has a first positive element α ($\geqslant 1/r!$). If β is any other positive element of Γ^*, and $n\alpha \leqslant \beta < (n+1)\alpha$, then $\beta - n\alpha \geqslant 0$ is an element of Γ^*, and is therefore either zero or $\geqslant \alpha$. Since $\beta - n\alpha < \alpha$, it follows that $\beta = n\alpha$. Since Γ^* contains $n\alpha$ for all values of n, the fact that Γ^* is discrete follows immediately. We thus have

THEOREM X. *If Σ^* is an algebraic extension of Σ of finite degree, any extension to Σ^* of a rank k valuation (a discrete valuation) of Σ is of rank k (discrete).*

We can use this result and Theorem VIII (i) to prove that if $\mathfrak{B}_1^*$ and $\mathfrak{B}_2^*$ are two distinct extensions of a valuation $\mathfrak{B}$ of Σ, where Σ^* is an algebraic extension of Σ of finite degree, then the valuation ring of one cannot contain the valuation ring of the other. Using this result we can then apply Theorem IX to any finite number of distinct extensions of $\mathfrak{B}$ to Σ^*.

Indeed, by Theorem VIII (i), if the valuation ring $\mathfrak{R}_1^*$ of $\mathfrak{B}_1^*$ is contained in the valuation ring $\mathfrak{R}_2^*$ of $\mathfrak{B}_2^*$, the rank of $\mathfrak{B}_1^*$ is greater than the rank of $\mathfrak{B}_2^*$. But, by Theorem X, the ranks of $\mathfrak{B}_1^*$ and of $\mathfrak{B}_2^*$ are equal to the rank of $\mathfrak{B}$. Hence we have a contradiction. Thus we must have $\mathfrak{R}_1^* \nsubseteq \mathfrak{R}_2^*$, $\mathfrak{R}_2^* \nsubseteq \mathfrak{R}_1^*$.

3. **Residue fields.** Let Σ be any field, $\mathfrak{B}$ a valuation of Σ to the valuation group Γ, and, as usual, denote the ring of the valuation by $\mathfrak{R}$, and the ideal of non-units of $\mathfrak{R}$ by $\mathfrak{p}$. Then $\mathfrak{p}$ is a maximal ideal of $\mathfrak{R}$ [§ 2, Th. I]; hence [XV, § 3, Th. III] the remainder-class ring $\mathfrak{R}/\mathfrak{p}$ is a field. We call this the *residue field* of $\mathfrak{B}$. The image of any element ξ of $\mathfrak{R}$ in $\mathfrak{R}/\mathfrak{p}$ is called the *residue* of ξ. If $\mathfrak{R}$ contains a field K, every element of K, other than zero, has an inverse in K, and hence in $\mathfrak{R}$, and is therefore a unit of $\mathfrak{R}$. Thus every non-zero element of K maps on a non-zero element of $\mathfrak{R}/\mathfrak{p}$, which therefore contains a field $\bar{K}$ not consisting of zero only, on which K is mapped homomorphically. By Theorem I of Chapter XV, § 3, $\bar{K}$ is isomorphic with K. Thus when Σ is a function field over a ground field K, and $\mathfrak{B}$ is such that $v(a) = 0$ for every non-zero element of K, the residue field of $\mathfrak{B}$ contains a field $\bar{K}$ isomorphic with K. The field $\mathfrak{R}/\mathfrak{p}$ is, of course, only defined to within an isomorphism, but it is usual in this case to identify $\bar{K}$ with K, and thus, in the case of a function field over K, the residue field of any valuation is an extension of K.

Suppose that Γ is of rank k, and let

$$\Gamma \supset \Gamma_1 \supset \ldots \supset \Gamma_k = 0$$

be its isolated subgroups. To each of the subgroups $\Gamma_1, \ldots, \Gamma_k$ there corresponds a prime ideal, $\mathfrak{p}_1, \ldots, \mathfrak{p}_k$ of $\mathfrak{R}$. Let Δ be any isolated subgroup of Γ; if $\Delta = \Gamma_i$, it is clear from Theorem V of § 1 that Γ/Γ_i is of rank i, and Γ_i is of rank $k-i$. We shall denote the rank of Δ by k', and the prime ideal of $\mathfrak{R}$ corresponding to Δ by $\mathfrak{P}$. If ξ is any non-zero element of Σ, it is mapped by the valuation $\mathfrak{B}$ on an element $v(\xi)$ of Γ, and in turn the mapping of Γ on the factor group Γ/Δ leads to a mapping of ξ on an element $v'(\xi)$ of Γ/Δ. We prove that this mapping $\xi \to v'(\xi)$ is a valuation of Σ. In the first place, we know that Γ/Δ is an ordered additive group. Next, if ξ, η are any two elements of Σ, $\xi\eta \to v(\xi\eta) = v(\xi)+v(\eta)$ of Γ, and this maps on $v'(\xi)+v'(\eta)$ of Γ/Δ, by the known properties of the mapping of a group on its factor group. Finally, since we know that if α and β are elements of Γ such that $\alpha \geqslant \beta$, then α and β map on elements α' and β' of Γ/Δ such that $\alpha' \geqslant \beta'$, we have, from the relations

$$\xi+\eta \to v(\xi+\eta) \geqslant \min[v(\xi), v(\eta)] = \beta,$$

that $$\xi+\eta \to v'(\xi+\eta) \geqslant \min[v'(\xi), v'(\eta)] = \beta'.$$

Thus the mapping on Γ/Δ is a valuation $\mathfrak{B}'$, say. If α' is any element of Γ/Δ, α any element of Γ which corresponds to α' in the mapping of Γ on Γ/Δ, there exists an element ξ of Σ such that $v(\xi) = \alpha$. Then $v'(\xi) = \alpha'$; and it follows that the valuation group of $\mathfrak{B}'$ is Γ/Δ. It is then clear that $\mathfrak{B}'$ is a trivial valuation if and only if $\Delta = \Gamma$. In the general case, the rank of $\mathfrak{B}'$ is $k-k'$.

Let us now determine the ring of the valuation $\mathfrak{B}'$. Let ξ be any element of Σ such that $v'(\xi) \geqslant 0$. Then there exists some element α of Δ such that $v(\xi) \geqslant \alpha$. We define an element η of Σ as follows: if $\alpha \geqslant 0$, $\eta = 1$, and if $\alpha < 0$, η is some element of $\mathfrak{R}$ such that $v(\eta) = -\alpha$. Then if $\zeta = \xi\eta$, we have, in both cases, $v(\zeta) \geqslant 0$, that is, ζ is in $\mathfrak{R}$. Since $v(\eta)$ is in Δ, η is not in the prime ideal $\mathfrak{P}$ of $\mathfrak{R}$ which corresponds to Δ. Hence $\xi = \zeta/\eta$ is in the quotient ring $\mathfrak{R}_{\mathfrak{P}}$ of $\mathfrak{P}$. Conversely, if ζ and η are two elements of $\mathfrak{R}$ such that η is not in $\mathfrak{P}$, and $\xi = \zeta/\eta$, $v'(\zeta) \geqslant 0$, $v'(\eta) = 0$, and so $v'(\xi) \geqslant 0$. Hence the valuation ring of $\mathfrak{B}'$ is $\mathfrak{R}_{\mathfrak{P}}$.

The ideal of the valuation $\mathfrak{B}'$ is the ideal of non-units of $\mathfrak{R}_{\mathfrak{P}}$, that is, it is the ideal $\mathfrak{R}_{\mathfrak{P}} . \mathfrak{P}$. The residue field of $\mathfrak{B}'$ is, therefore, $\mathfrak{R}_{\mathfrak{P}}/\mathfrak{R}_{\mathfrak{P}} . \mathfrak{P}$, which is isomorphic with the quotient field of the remainder-class ring $\mathfrak{R}/\mathfrak{P}$ [XV, § 5, Th. XI, Corollary]. It is also clear that if Σ is an algebraic function field over K such that $v(a) = 0$ for every non-zero element of K, then $v'(a)$ is also zero for every non-zero a in K. The results proved are summed up in

THEOREM I. *If $\mathfrak{B}$ is a valuation of Σ of rank k to the valuation group Γ, and if Δ is an isolated subgroup of Γ of rank k', then there exists a valuation $\mathfrak{B}'$ of Σ to Γ/Δ, of rank $k-k'$. If $\mathfrak{R}$ is the valuation ring of $\mathfrak{B}$, and $\mathfrak{P}$ the prime ideal of $\mathfrak{R}$ which corresponds to Δ, then the valuation ring of $\mathfrak{B}'$ is $\mathfrak{R}_{\mathfrak{P}}$, and the residue field of $\mathfrak{B}'$ is isomorphic with the quotient field of $\mathfrak{R}/\mathfrak{P}$.*

In the case $\mathfrak{P} = \mathfrak{p}$ (i.e. $\Delta = 0$), it follows that $\mathfrak{R}_{\mathfrak{P}} = \mathfrak{R}$.

Preserving the same notation as in Theorem I, let Σ' be the residue field of $\mathfrak{B}'$. Let ξ' be any non-zero element of Σ', and let $\xi, \eta, \ldots$ be elements of $\mathfrak{R}_{\mathfrak{P}}$ which map on ξ' in the mapping of $\mathfrak{R}_{\mathfrak{P}}$ on $\mathfrak{R}_{\mathfrak{P}}/\mathfrak{R}_{\mathfrak{P}} . \mathfrak{P}$. Since ξ' is not zero, $\xi, \eta, \ldots$ do not lie in $\mathfrak{R}_{\mathfrak{P}} . \mathfrak{P}$, and hence

$$v'(\xi) = v'(\eta) = \ldots = 0,$$

and $$v(\xi),\ v(\eta),\ \ldots \text{ lie in } \Delta.$$

Since $\xi-\eta$ maps on the zero of Σ', it lies in $\mathfrak{R}_{\mathfrak{P}}\,.\,\mathfrak{P}$, and hence $v(\xi-\eta)$ is greater than any element of Δ. Hence

$$v(\xi-\eta)>v(\eta),$$

and therefore

$$\begin{aligned} v(\xi) &= v(\xi-\eta+\eta) \\ &= \min\,[v(\xi-\eta), v(\eta)] \\ &= v(\eta). \end{aligned}$$

Thus $\xi, \eta, \ldots$ all have the same value in $\mathfrak{B}$, and this value lies in Δ. This element of Δ is thus uniquely determined by ξ', and we denote it by $v_1(\xi')$. We thus determine a unique element of Δ corresponding to each non-zero element of Σ'. Any element of Δ arises in this way from at least one element of Σ'. For if α is any element of Δ, there exists an element ξ of Σ such that $v(\xi) = \alpha$. Since $v(\xi)\in\Delta$, $v'(\xi) = 0$, and hence ξ determines a non-zero element ξ' of Σ'. For this element we have $v_1(\xi') = \alpha$.

Let ξ'_1, ξ'_2 be two non-zero elements of Σ', and let ξ_1, ξ_2 be elements of Σ which correspond to them. Then

$$v_1(\xi'_1\xi'_2) = v(\xi_1\xi_2) = v(\xi_1)+v(\xi_2) = v_1(\xi'_1)+v_1(\xi'_2),$$

and

$$v_1(\xi'_1+\xi'_2) = v(\xi_1+\xi_2) \geqslant \min\,[v(\xi_1), v(\xi_2)] = \min\,[v_1(\xi'_1), v_1(\xi'_2)].$$

Hence the mapping $\xi'\rightarrow v_1(\xi')$ of Σ' on Δ is a valuation of Σ' to the valuation group Δ. We shall denote it by $\mathfrak{B}_1$.

The valuation ring of $\mathfrak{B}_1$ is easily determined. For $v_1(\xi')\geqslant 0$ if and only if $v(\xi)\geqslant 0$. Two elements ξ, η of $\mathfrak{R}$ determine the same element of Σ' if and only if $\xi-\eta$ lies in $\mathfrak{P}$. Hence the valuation ring of $\mathfrak{B}_1$ is isomorphic with $\mathfrak{R}/\mathfrak{P}$. The ideal of non-units of this ring is $\mathfrak{p}/\mathfrak{P}$. Thus we have

THEOREM II. *In the notation of Theorem I, the valuation $\mathfrak{B}$ of Σ determines a valuation $\mathfrak{B}_1$ of the residue field Σ' of $\mathfrak{B}'$ to the valuation group Δ. The ring of this valuation $\mathfrak{B}_1$ is isomorphic with $\mathfrak{R}/\mathfrak{P}$, and the ideal of the valuation is $\mathfrak{p}/\mathfrak{P}$.*

The effect of Theorems I and II can be described as a breaking up of a valuation $\mathfrak{B}$ of Σ of rank k $(k>1)$ into a valuation $\mathfrak{B}'$ of Σ of rank $k-k'$, and a valuation $\mathfrak{B}_1$ of the residue field Σ' of $\mathfrak{B}'$ of rank k'. We cannot, however, determine the value of any element ξ of Σ in $\mathfrak{B}$ directly from a knowledge of $\mathfrak{B}'$ and $\mathfrak{B}_1$; since if ξ is an element of Σ which is not a unit of the valuation ring of $\mathfrak{B}'$, its image ξ' in Σ' is either zero or infinity, according as $v'(\xi)>0$ or $v'(\xi)<0$, and hence $v_1(\xi')$ does not properly exist. Nevertheless, $\mathfrak{B}$ is uniquely determined by $\mathfrak{B}'$ and $\mathfrak{B}_1$. For it is determined if we know the valuation ring $\mathfrak{R}$, and this is found as follows: ξ belongs to $\mathfrak{R}$ if either

$$v'(\xi) > 0,$$

or
$$v'(\xi) = 0 \quad \text{and} \quad v_1(\xi') \geqslant 0,$$

ξ' being the residue of ξ in $\mathfrak{B}'$.

The following theorem is a converse of Theorems I and II:

THEOREM III. *Let $\mathfrak{B}'$ be a valuation of Σ to the valuation group Γ', and let Σ' be the residue field of $\mathfrak{B}'$. Let $\mathfrak{B}_1$ be a valuation of Σ' to the valuation group Δ. Then there exists a valuation $\mathfrak{B}$ of Σ which can be decomposed, as in Theorems I and II, into $\mathfrak{B}'$ and $\mathfrak{B}_1$.*

When Σ is a function field over the ground field K, and $v'(a) = 0$, $v_1(a) = 0$ for all non-zero elements a of K, then $v(a) = 0$ for all non-zero elements a of K.

It is clear from the remarks made above that if $\mathfrak{B}$ exists its valuation ring must consist of those elements ξ of Σ such that either $v'(\xi) > 0$ or $v'(\xi) = 0$ and $v_1(\xi') \geqslant 0$, where ξ' is the residue of ξ in $\mathfrak{B}'$. Let $\mathfrak{R}$ be the set of such elements ξ of Σ. We first show that $\mathfrak{R}$ is a ring. Let ξ, η be two elements of $\mathfrak{R}$.

(i) If $v'(\xi) > 0$, $v'(\eta) > 0$, then

$$v'(\xi\eta) = v'(\xi) + v'(\eta) > 0 \quad \text{and} \quad v'(\xi+\eta) \geqslant \min[v'(\xi), v'(\eta)] > 0.$$

(ii) If $v'(\xi) > 0$, $v'(\eta) = 0$, $v_1(\eta') \geqslant 0$, then

$$v'(\xi\eta) = v'(\xi) + v'(\eta) > 0 \quad \text{and} \quad v'(\xi+\eta) = v'(\eta) = 0.$$

Moreover, since ξ is a non-unit of the valuation ring of $\mathfrak{B}'$, its residue ξ' in $\mathfrak{B}'$ is zero; hence $v_1(\xi'+\eta') = v_1(\eta') \geqslant 0$.

(iii) If $v'(\xi) = 0$, $v_1(\xi') \geqslant 0$ and $v'(\eta) > 0$, the argument given in (ii) tells us that

$$v'(\xi\eta) > 0, \quad v'(\xi+\eta) = 0, \quad v_1(\xi'+\eta') \geqslant 0.$$

(iv) If $v'(\xi) = 0$, $v_1(\xi') \geqslant 0$ and $v'(\eta) = 0$, $v_1(\eta') \geqslant 0$, then

$$v'(\xi\eta) = v'(\xi) + v'(\eta) = 0, \quad v_1(\xi'\eta') = v_1(\xi') + v_1(\eta') \geqslant 0,$$

and
$$v'(\xi+\eta) \geqslant \min[v'(\xi), v'(\eta)] = 0.$$

If $v'(\xi+\eta) = 0$, $\xi'+\eta' \neq 0$, and

$$v_1(\xi'+\eta') \geqslant \min[v_1(\xi'), v_1(\eta')] \geqslant 0.$$

Hence in all cases, if ξ and η lie in $\mathfrak{R}$, so do $\xi\eta$ and $\xi+\eta$; hence $\mathfrak{R}$ is a ring.

To prove that $\mathfrak{R}$ is a valuation ring, we use Theorem III of § 2. Let ζ be any element of Σ. We have to show that either ζ or ζ^{-1} lies in $\mathfrak{R}$. Suppose that ζ does not lie in $\mathfrak{R}$. Then either $v'(\zeta) < 0$, in which case $v'(\zeta^{-1}) > 0$, or $v'(\zeta) = 0$ and $v_1(\zeta') < 0$, where ζ' is the residue of ζ in $\mathfrak{B}'$, and in this case we have $v'(\zeta^{-1}) = 0$, $v_1((\zeta')^{-1}) > 0$. Hence in either case, ζ^{-1} lies in $\mathfrak{R}$.

Let $\mathfrak{B}$ be the valuation of Σ determined by $\mathfrak{R}$, and Γ its valuation group. The elements of $\mathfrak{R}$ for which $v'(\xi) > 0$ form an ideal $\mathfrak{P}$ in $\mathfrak{R}$. For if ξ and η are two elements of $\mathfrak{R}$ such that $v'(\xi) > 0$ and $v'(\eta) > 0$, and if ρ is any element of $\mathfrak{R}$, then

$$v'(\xi - \eta) \geqslant \min[v'(\xi), v'(\eta)] > 0,$$

and
$$v'(\rho\xi) = v'(\rho) + v'(\xi) > 0,$$

since $v'(\rho) \geqslant 0$. If σ and τ are two elements of $\mathfrak{R}$ not in $\mathfrak{P}$, then

$$v'(\sigma\tau) = v'(\sigma) + v'(\tau) = 0,$$

and hence $\sigma\tau$ is not in $\mathfrak{P}$. Hence $\mathfrak{P}$ is a prime ideal of $\mathfrak{R}$.

$\mathfrak{P}$ therefore determines an isolated subgroup Γ^* of Γ. We can then decompose $\mathfrak{B}$ into a valuation B^* of Σ to the valuation group Γ/Γ^* and a valuation $\mathfrak{B}_1^*$ of the residue field of $\mathfrak{B}^*$ to Γ^*. The valuation ring of $\mathfrak{B}^*$ is the quotient ring $\mathfrak{R}_{\mathfrak{P}}$; hence it consists of elements $\zeta = \xi/\eta$, where ξ and η belong to $\mathfrak{R}$, and $v(\xi) \geqslant 0$ and $v(\eta)$ is in Γ^*. Hence it consists of elements ζ such that $v'(\zeta) \geqslant 0$. The valuation ring of $\mathfrak{B}^*$ therefore coincides with that of $\mathfrak{B}'$, and $\mathfrak{B}'$ and $\mathfrak{B}^*$ may therefore be identified. Then $\Gamma' = \Gamma/\Gamma^*$. The residue field of $\mathfrak{B}^*$ is therefore Σ'. The valuation ring of $\mathfrak{B}_1^*$ is the set of elements of Σ' which are residues of elements of Σ in $\mathfrak{R}$, and hence it coincides with the valuation ring of $\mathfrak{B}_1$. Thus we may identify Γ^* with Δ. The valuation $\mathfrak{B}$ can therefore be decomposed into the valuation $\mathfrak{B}'$ of Σ and the valuation $\mathfrak{B}_1$ of the residue field of $\mathfrak{B}'$.

The valuation $\mathfrak{B}$ of Σ is said to be *compounded* of $\mathfrak{B}'$ and $\mathfrak{B}_1$.

A particular case of some interest arises when the valuation $\mathfrak{B}'$ is discrete, that is, when the valuation group of $\mathfrak{B}'$ is the additive group of integers. We can then find an element ζ of Σ such that $v'(\zeta) = 1$. Now let ξ be any non-zero element of Σ, and let $v'(\xi) = n$ (n an integer, positive, zero, or negative). Let $\bar{\xi} = \xi\zeta^{-n}$. Then $v'(\bar{\xi}) = 0$, and therefore the residue $\bar{\xi}'$ of $\bar{\xi}$ in $\mathfrak{B}'$ is not zero. Let $v_1(\bar{\xi}') = \alpha$. We write $w(\xi) = (n, \alpha)$. Similarly, if η is any other element of Σ we can define $w(\eta) = (m, \beta)$, where $v'(\eta) = m$, $\bar{\eta} = \eta\zeta^{-m}$,

$v_1(\bar{\eta}') = \beta$. Since $v'(\xi\eta) = m+n$ and $\xi\eta = \bar{\xi}\bar{\eta}\zeta^{n+m}$ and $v_1(\bar{\xi}'\bar{\eta}') = \alpha+\beta$, we have $w(\xi\eta) = (m+n, \alpha+\beta)$. Also $v'(\xi+\eta) \geqslant \min[m,n]$, and if $v'(\xi+\eta) = \min[m,n] = m$, say, $\xi+\eta = (\bar{\xi}\zeta^{n-m}+\bar{\eta})\zeta^m$,

$$v_1([\bar{\xi}\zeta^{n-m}+\bar{\eta}]') = v_1(\bar{\eta}') \quad \text{if} \quad n>m,$$

and
$$v_1(\bar{\xi}'+\bar{\eta}') \geqslant \min[v_1(\bar{\xi}'), v_1(\bar{\eta}')] \quad \text{if} \quad n=m.$$

Hence if we interpret the pairs (n, α) as the elements of the direct sum of Γ' and Δ, ordered as in § 1, we have a mapping which is a valuation of the non-zero elements of Σ on this direct sum. The ring of this valuation consists of those elements ξ of Σ such that $v'(\xi) \geqslant 0$ or $v'(\xi) = 0$, $v_1(\xi') \geqslant 0$; hence it coincides with $\mathfrak{R}$, and the valuation coincides with $\mathfrak{B}$. It follows that in this case Γ is the direct sum of Γ' and Δ. Hence we have

THEOREM IV. *If $\mathfrak{B}'$ is a valuation of Σ to the discrete group Γ', and $\mathfrak{B}_1$ is a valuation of the residue field of $\mathfrak{B}'$ to Δ, the group of the valuation $\mathfrak{B}$ of Σ obtained by compounding $\mathfrak{B}'$ and $\mathfrak{B}$ is the direct sum of Γ' and Δ.*

We conclude this section with a theorem on the possible extensions of a valuation $\mathfrak{B}$ of Σ to valuations of an algebraic extension Σ^*, of finite degree, of Σ.

THEOREM V. *If $\mathfrak{B}$ is a valuation of the field Σ, and Σ^* is an algebraic extension of Σ of finite degree m, then there are at most m valuations of Σ^* which are extensions of $\mathfrak{B}$.*

We assume that the residue field of $\mathfrak{B}$ contains an infinite number of elements.

We first prove that if $\mathfrak{B}^*$ is any valuation of Σ^* which is an extension of $\mathfrak{B}$, the residue field of $\mathfrak{B}^*$ is isomorphic with an algebraic extension, of degree m at most, of the residue field of $\mathfrak{B}$. Clearly, any element of Σ which is a unit of the valuation ring of $\mathfrak{B}$ is a unit of the valuation ring of $\mathfrak{B}^*$. Hence the residue field of $\mathfrak{B}^*$ contains a field isomorphic with the residue field of $\mathfrak{B}$. We therefore regard it as an extension of the residue field of $\mathfrak{B}$. We shall denote the residue of any element σ^* of Σ^* in $\mathfrak{B}^*$ by $\bar{\sigma}^*$. If $a^* = a \in \Sigma$, $\bar{a}^* = \bar{a}$. Now let $\bar{\sigma}^*$ be any non-zero element of the residue field of $\mathfrak{B}^*$, σ^* an element of Σ^* which corresponds to it. σ^* satisfies an equation

$$a_0\sigma^{*n} + a_1\sigma^{*n-1} + \ldots + a_n = 0 \quad (a_i \in \Sigma)$$

over Σ, of degree n $(n \leqslant m)$. By multiplying through by a suitable element of Σ, we may suppose that $\min[v(a_0), \ldots, v(a_n)] = 0$. Now since σ^* has a non-zero residue, $v^*(\sigma^*) = 0$. From this we can show that $v(a_i) = 0$ for some i less than n. For if $v(a_n) < v(a_i)$ $(i = 1, \ldots, n-1)$, $v^*(a_n) < v^*(a_i \sigma^{*n-i})$, and hence

$$v^*[a_0 \sigma^{*n} + \ldots + a_n] = v^*(a_n) = 0,$$

which conflicts with the fact that $v^*(0) = \infty$. Hence when we pass to the residue field, we have

$$\bar{a}_0 \bar{\sigma}^{*n} + \ldots + \bar{a}_n = 0,$$

where at least one of the powers of $\bar{\sigma}^*$ has a non-zero coefficient. Hence σ^* satisfies an equation of degree n $(n \leqslant m)$ at most, with coefficients in the residue field of $\mathfrak{B}$.

Now suppose that $\mathfrak{B}$ has at least $m+1$ extensions $\mathfrak{B}_1^*, \ldots, \mathfrak{B}_{m+1}^{*'}$. We saw in § 2 (p. 192) that $\mathfrak{R}_i^* \not\subseteq \mathfrak{R}_j^*$ $(i \neq j)$, and hence [§ 2, Th. IX] we can find elements $\xi_1^*, \ldots, \xi_{m+1}^*$ in the rings of the valuations $\mathfrak{B}_i^*$ such that the residue $\bar{\xi}_i^*$ of ξ_i in $\mathfrak{B}_i^*$ is not zero, while the residue of ξ_i^* in $\mathfrak{B}_j^*$ $(i \neq j)$ is zero. $\bar{\xi}_i^*$ is algebraic over the residue field of $\mathfrak{B}$; let its conjugates be $\bar{\xi}_i^{*(\sigma)}$ $(\sigma = 1, \ldots, n_i)$. Since the residue field of $\mathfrak{B}$ has an infinite number of elements, we can find $\alpha_1, \ldots, \alpha_{m+1}$ in Σ such that

$$\bar{\alpha}_i \bar{\xi}_i^{*(\sigma)} \neq \bar{\alpha}_j \bar{\xi}_j^{*(\rho)},$$

for $\sigma = 1, \ldots, n_i$, $\rho = 1, \ldots, n_j$, and all distinct pairs i, j. Now let $\sigma^* = \Sigma \alpha_i \xi_i^*$. The residue of this in $\mathfrak{B}_i^*$ is $\bar{\alpha}_i \bar{\xi}_i^*$. Hence the $m+1$ residues of σ^* all lie in some extension of the residue field of Σ, and are all distinct. But σ^* satisfies an equation over Σ of degree less than or equal to m; hence its residues are roots of an equation of degree not exceeding m whose coefficients are in the residue field of $\mathfrak{B}$. Therefore at most m of them are distinct, and we have a contradiction. Thus the theorem is proved.

4. Valuations of algebraic function fields. In §§ 2 and 3 we have considered general properties of valuations of any field Σ, but have added comments to indicate how the results obtained apply when Σ is a field of algebraic functions over a ground field K. Our ultimate object is to apply our theory to valuations of the function field of an algebraic variety defined over a ground field K of characteristic zero. In this and the next section we consider this case more particularly. However, there is one danger which we must avoid. Suppose that Σ is the function field of an algebraic variety V of dimension d over K. Then we can find d elements $\zeta_1, \ldots, \zeta_d$ of Σ which are algebraically independent over K. Then Σ is a simple algebraic extension of finite degree of $K(\zeta_1, \ldots, \zeta_d)$. If we consider a valuation $\mathfrak{B}$ of Σ (satisfying, always, the condition (iii) of § 2, namely, that if a is any non-zero value of K, $v(a) = 0$), we shall prove that the residue field Σ' of the valuation is of degree of transcendency d', where $d' < d$; that is, that there exist d' elements of Σ' which are algebraically independent over K, but such that any $d'+1$ elements are algebraically dependent over K. But it does not necessarily follow that Σ' is the function field of an algebraic variety, since if $\zeta_1, \ldots, \zeta_{d'}$ are d' algebraically independent members of Σ', Σ' is not necessarily a *finite* extension of $K(\zeta'_1, \ldots, \zeta'_{d'})$. For some purposes we shall find it necessary to consider valuations of fields such as Σ', which, while of finite degree of transcendency over K, are not the function fields of algebraic varieties. It should be noted that the results stated in §§ 2 and 3 for algebraic function fields apply to all fields of a finite degree of transcendency over K, whether they are function fields of algebraic varieties or not. But to avoid any confusion in what follows, we shall say that Σ is *the function field of an algebraic variety of dimension* d when the theorem in question is restricted to the case in which Σ is a finite (simple) algebraic extension of a pure transcendental extension of K of dimension d, and that Σ is *an extension of* K *of finite degree of transcendency* d, when the theorem is valid whether Σ is the function field of an algebraic variety or not.

Let Σ be an extension of K of degree of transcendency d, and let $\mathfrak{B}$ be a valuation of it. As usual, let $\mathfrak{R}$ denote the valuation ring of $\mathfrak{B}$, $\mathfrak{p}$ the ideal of non-units of $\mathfrak{R}$. The residue field of $\mathfrak{B}$ is isomorphic with $\mathfrak{R}/\mathfrak{p}$ and contains a subfield isomorphic with K [§ 3]; this we identify with K, so that the residue field of $\mathfrak{B}$ is an extension of K. We define the *dimension of* $\mathfrak{B}$ to be the degree of transcendency of the residue field over K.

THEOREM I. *The dimension of a non-trivial valuation of an extension Σ of K of degree of transcendency d is less than d.*

Since any $d+1$ elements of $\mathfrak{R}$ are algebraically dependent over K, they are, of course, algebraically dependent modulo $\mathfrak{p}$. Hence the dimension of the valuation cannot exceed d. Suppose that it is exactly d. Then we can find d elements $\xi_1, \ldots, \xi_d$ of $\mathfrak{R}$ which are algebraically independent modulo $\mathfrak{p}$. Let η be any non-zero element of $\mathfrak{p}$. The elements $\xi_1, \ldots, \xi_d, \eta$ are algebraically dependent over K; hence we have an irreducible equation

$$a_0(\xi_1, \ldots, \xi_d)\,\eta^r + a_1(\xi_1, \ldots, \xi_d)\,\eta^{r-1} + \ldots + a_r(\xi_1, \ldots, \xi_d) = 0,$$

where $a_r(\xi_1, \ldots, \xi_d)$ is not zero. Since $a_i(\xi)$ is in $\mathfrak{R}$, and η is in $\mathfrak{p}$, we have

$$a_r(\xi_1, \ldots, \xi_d) = 0 \ \ (\mathfrak{p}),$$

which contradicts our assumption that $\xi_1, \ldots, \xi_d$ are algebraically independent modulo $\mathfrak{p}$. Hence $\mathfrak{R}/\mathfrak{p}$ is of dimension less than d over K; that is, the dimension of the residue field of the valuation is less than d.

As an immediate consequence of this we have the

Corollary. If Σ is of degree of transcendency zero, every valuation of Σ is trivial.

THEOREM II. *If Σ is an extension of K of degree of transcendency d, the rank of a valuation of Σ cannot exceed d. More generally, the rank plus the dimension of a valuation is not greater than d.*

Let the rank of the valuation $\mathfrak{B}$ of Σ be k, and let $\Gamma_1, \Gamma_2, \ldots$ be the isolated subgroups of the valuation group Γ:

$$\Gamma \supset \Gamma_1 \supset \ldots \supset \Gamma_k = 0.$$

By Theorems I and II of § 3, we can decompose $\mathfrak{B}$ into a valuation $\mathfrak{B}'$ of Σ to Γ/Γ_1, and a valuation $\mathfrak{B}_1$ of the residue field Σ'_0 of $\mathfrak{B}'$ to Γ_1. Similarly, we can decompose the valuation $\mathfrak{B}_1$ into a valuation $\mathfrak{B}'_1$ to Γ_1/Γ_2, and a valuation $\mathfrak{B}_2$ of the residue field Σ'_1 of $\mathfrak{B}'_1$ to Γ_2. Proceeding in this way we construct residue fields $\Sigma'_0, \Sigma'_1, \ldots, \Sigma'_{k-1}$, and for each of these fields a valuation $\mathfrak{B}_i$ of Σ'_{i-1} to Γ_i. Let d'_i be the dimension of Σ'_i. Then, by Theorem I, $d'_0 < d$, $d'_i < d'_{i-1}$. Therefore $d'_i < d - i$. Since the valuation of Σ'_{k-2} to Γ_{k-1} is not trivial, $d'_{k-2} > 0$. Hence

$$d - k + 2 > d'_{k-2} > 0,$$

that is,

$$d - k + 2 \geqslant 2,$$

and therefore

$$k \leqslant d.$$

The proof of the last part of the theorem is implicit in what precedes. In particular, it follows that a non-trivial valuation of a field of degree of transcendency one over K is of rank one, and is therefore Archimedean-ordered.

The remaining theorems of this section deal with the special case in which Σ is the function field of an irreducible algebraic variety over K of dimension d. Our first results deal with the possible valuations of Σ of dimension $d-1$.

We first prove a lemma on the possible valuations of Σ when it is a pure transcendental extension $K(x)$ of K of dimension one.

Lemma. Any non-trivial valuation of $K(x)$ can either be obtained by selecting an irreducible polynomial $f(x)$ in $K[x]$ of degree greater than zero and defining the value of a rational function $R(x)$ as a, when $R(x)$ is written in the form

$$R(x) = [f(x)]^a [f_1(x)]^{a_1} \dots [f_k(x)]^{a_k},$$

where $a, a_1, \dots, a_k$ are integers, positive, negative or zero, and $f(x)$, $f_1(x), \dots, f_k(x)$ are distinct irreducible polynomials; or else by writing the elements of $K(x)$ as rational functions $R(x^{-1})$ of x^{-1}, defining the value of $R(x^{-1})$ as before with $f(x^{-1}) = x^{-1}$.

It is obvious that if $f(x)$ is any irreducible polynomial, or if $f(x) = x^{-1}$, then the mapping $R(x) \to a$ gives a valuation of $K(x)$. In order to prove that any valuation of $K(x)$ is obtained in this way, we first observe that not all elements of $K[x]$ can have value zero. For otherwise, since any element of $K(x)$ is of the form $a(x)/b(x)$, where $a(x)$ and $b(x)$ are in $K[x]$, we should have

$$v(a/b) = v(a) - v(b) = 0,$$

and hence the valuation would be trivial.

Case (i). If $v(x) \geqslant 0$, $v(f(x)) \geqslant 0$ for any polynomial $f(x)$ in $K[x]$. For if

$$f(x) = a_0 + a_1 x + \dots + a_n x^n,$$

$$v(a_i x^i) = v(a_i) + iv(x) = iv(x) \geqslant 0,$$

and therefore $\quad v(f(x)) \geqslant \min [v(a_i x^i)] \geqslant 0.$

Hence in this case $K[x] \subseteq \mathfrak{R}$, the ring of the valuation. If $\mathfrak{p}$ is the ideal of non-units of $\mathfrak{R}$, and $\mathfrak{P} = K[x]_\wedge \mathfrak{p}$, $\mathfrak{P}$ is a prime ideal of $K[x]$ [XV, § 4]. Also, $\mathfrak{P}$ is not the zero ideal, since we have seen that there are elements of $K[x]$ whose values are non-zero, and therefore positive, since $K[x] \subseteq \mathfrak{R}$. Since $\mathfrak{P}$ does not contain any element of K other than zero, it is not the unit ideal. Hence it is a proper prime ideal of $K[x]$. But $K[x]$ is a principal ideal ring [XV, § 1]. Hence $\mathfrak{P}$ has a basis consisting of a single element $f(x)$, which must be irreducible, since $\mathfrak{P}$ is prime. If $g(x)$ is any irreducible polynomial, not a constant multiple of $f(x)$, we can find polynomials $a(x)$ and $b(x)$ such that

$$a(x)\,f(x)+b(x)\,g(x) = 1.$$

Hence

$$\begin{aligned} 0 = v(1) &\geqslant \min\,[v(af), v(bg)] = \min\,[v(a)+v(f), v(b)+v(g)] \\ &\geqslant \min\,[v(f), v(g)], \end{aligned}$$

and since $v(f) > 0$, $v(g) \geqslant 0$, it follows that $v(g) = 0$. If, therefore, any rational function $R(x)$ is written in the form

$$R(x) = [f(x)]^a\,[f_1(x)]^{a_1} \ldots [f_k(x)]^{a_k},$$

as in the enunciation of the lemma,

$$\begin{aligned} v(R) &= av(f)+a_1 v(f_1)+\ldots+a_k v(f_k) \\ &= av(f), \end{aligned}$$

since $v(f_i) = 0$. Thus the valuation group consists of integral multiples of $v(f)$; if we normalise the valuation group so that $v(f) = 1$, Γ is the additive group of integers. Thus the lemma is proved in the case $v(x) \geqslant 0$.

Case (ii). If $v(x) < 0$, we then write the elements $K(x)$ as rational functions of x^{-1}. Then $K[x^{-1}] \subseteq \mathfrak{R}$, and $\mathfrak{P}' = K[x^{-1}] \wedge \mathfrak{p}$ is a proper prime ideal of $K[x^{-1}]$. Since $x^{-1} \in K[x^{-1}] \wedge \mathfrak{p}$ and is irreducible, it is clear that x^{-1} is a basis for $\mathfrak{P}'$, and the remainder of the lemma follows as in case (i).

This lemma, taken with Theorem X of § 2, leads at once to the

Corollary. If Σ is the function field of an algebraic variety of dimension one, any non-trivial valuation of Σ is discrete.

THEOREM III. *If Σ is the function field of an algebraic variety of dimension d over K, and $\mathfrak{B}$ is a non-trivial valuation of Σ of dimension $d-1$, then $\mathfrak{B}$ is a discrete valuation, and the residue field Σ' of $\mathfrak{B}$ is the function field of an algebraic variety of dimension $d-1$ over K.*

We prove the last part of the theorem first. Since $\mathfrak{B}$ is of dimension $d-1$, there exist $d-1$ elements $\xi_1, \ldots, \xi_{d-1}$ of the valuation ring $\mathfrak{R}$ which are algebraically independent modulo $\mathfrak{p}$, the ideal of non-units of $\mathfrak{R}$. Let η be any non-zero element of $\mathfrak{p}$. If there exists an algebraic equation connecting $\xi_1, \ldots, \xi_{d-1}, \eta$ over K,

$$a_0(\xi_1, \ldots, \xi_{d-1})+a_1(\xi_1, \ldots, \xi_{d-1})\,\eta+\ldots+a_r(\xi_1, \ldots, \xi_{d-1})\,\eta^r = 0,$$

we may assume $a_0(\xi_1, \ldots, \xi_{d-1}) \neq 0$. It follows that

$$a_0(\xi_1, \ldots, \xi_{d-1}) = 0 \ (\mathfrak{p}),$$

which contradicts the fact that $\xi_1, \ldots, \xi_{d-1}$ are independent modulo $\mathfrak{p}$. Hence $\xi_1, \ldots, \xi_{d-1}, \eta$ are independent over K. Since Σ is the function field of a variety of dimension d over K, it is a finite algebraic extension of $K(\xi_1, \ldots, \xi_{d-1}, \eta)$, of degree m, say. Let ζ be any element of this extension, and let

$$\sum_{i=0}^{s} \sum_{j=0}^{m'} a_{ij}(\xi_1, \ldots, \xi_{d-1}) \eta^i \zeta^j = 0 \quad (m' \leqslant m) \tag{1}$$

be the irreducible equation connecting $\xi_1, \ldots, \xi_{d-1}, \eta, \zeta$ over K. Passing to the residue field Σ', we have, since $\eta \in \mathfrak{p}$,

$$\sum_j a_{0j}(\xi'_1, \ldots, \xi'_{d-1}) \zeta'^j = 0.$$

Not all the coefficients $a_{0j}(\xi'_1, \ldots, \xi'_{d-1})$ can be zero, for if they were, $a_{0j}(\xi_1, \ldots, \xi_{d-1})$ would be zero for each value of j, since $\xi_1, \ldots, \xi_{d-1}$ are algebraically independent modulo $\mathfrak{p}$. Then (1) would be reducible, contrary to hypothesis. Hence not all $a_{0j}(\xi'_1, \ldots, \xi'_{d-1})$ are zero, and ζ' is algebraic over $K(\xi'_1, \ldots, \xi'_{d-1})$, and satisfies an equation of degree m at most.

It follows that Σ' is an extension of $K(\xi'_1, \ldots, \xi'_{d-1})$ of degree m at most, and is therefore the function field of a variety of dimension $d-1$.

Since $\xi_1, \ldots, \xi_{d-1}$ are algebraically independent over K, and $\xi'_1, \ldots, \xi'_{d-1}$ are algebraically independent over K,

$$K(\xi_1, \ldots, \xi_{d-1}) \cong K(\xi'_1, \ldots, \xi'_{d-1}).$$

Without confusion we may identify ξ_i and ξ'_i. Σ can then be regarded as the function field of a variety of dimension one over

$$\Omega = K(\xi_1, \ldots, \xi_{d-1}).$$

The valuation $\mathfrak{B}$ has the property that $v(\tau) = 0$ for any non-zero element τ in Ω. Hence we can apply the corollary to the lemma proved above. It follows that $\mathfrak{B}$ is discrete.

THEOREM IV. *If Σ is the function field of an algebraic variety of dimension d over K, any valuation $\mathfrak{B}$ of rank d is to a valuation group which is the direct sum of d discrete groups.*

If $d = 1$, the theorem is contained in Theorem III. If $d > 1$, we consider the largest isolated proper subgroup Γ_1 of the valuation group Γ of $\mathfrak{B}$. Then Γ_1 is of rank $d-1$. The valuation $\mathfrak{B}$ can be decomposed into a valuation $\mathfrak{B}'$ of Σ to Γ/Γ_1, and a valuation $\mathfrak{B}_1$ of the residue field Σ' of $\mathfrak{B}'$ to Γ_1. By Theorem III, Σ' is the function field of an algebraic variety. Since Γ_1 is of rank $d-1$, and the dimension of Σ' is an integer less than d, it follows from Theorem II that Σ' is of dimension $d-1$ exactly. Hence, as hypothesis of induction, we may assume that Γ_1 is the direct sum of $d-1$ discrete groups. Moreover, since the residue field of $\mathfrak{B}'$ is of dimension $d-1$ over K, $\mathfrak{B}'$ is of dimension $d-1$, and hence, by Theorem III, the valuation group of $\mathfrak{B}'$ is discrete. Now apply Theorem IV of § 3. From this it follows that Γ is the direct sum of Γ/Γ_1 and Γ_1, and this proves our theorem.

This result means that if $\mathfrak{B}$ is of rank d, the value of any element of Σ can be written in the form $(m_1, \ldots, m_k)$, where $m_1, \ldots, m_k$ are integers. We have

$$(m_1, \ldots, m_k) + (n_1, \ldots, n_k) = (m_1 + n_1, \ldots, m_k + n_k),$$

and

$$(m_1, \ldots, m_k) > (n_1, \ldots, n_k)$$

if

$$m_1 = n_1, \quad \ldots, \quad m_{i-1} = n_{i-1}, \quad m_i > n_i,$$

for some i $(1 \leqslant i \leqslant k)$. Given any set of integers $m_1, \ldots, m_k$, there exists an element of Σ whose value is $(m_1, \ldots, m_k)$.

Next, let us consider valuations of the function field Σ of an algebraic variety of dimension d over K which are of rank one. If $\mathfrak{B}$ is such a valuation of Σ, we may take the value group Γ to be a subgroup of the additive group of real numbers [§ 1, Th. I]. A finite set k of real numbers $\alpha_1, \ldots, \alpha_k$ is said to be *rationally dependent* if there exist integers $p_1, \ldots, p_k$, not all zero, such that

$$p_1\alpha_1 + \ldots + p_k\alpha_k = 0;$$

otherwise, they are said to be *rationally independent*. If a set of real numbers contains k rationally independent numbers, while every subset of $k+1$ numbers is rationally dependent, the set is said to be of *rational rank* k. If the elements of Γ form a set of rational rank k, $\mathfrak{B}$ is said to be of rational rank k.

THEOREM V. *The rational rank of a valuation $\mathfrak{B}$ of rank one of the function field Σ of an algebraic variety of dimension d cannot exceed d.*

Suppose that the rational rank of $\mathfrak{B}$ is k. (If the rational rank were infinite, it would be sufficient to consider a subgroup of Γ of rational rank $k = d+1$.) Let $\alpha_1, \ldots, \alpha_k$ be k rationally independent elements of Γ, and let $\zeta_1, \ldots, \zeta_k$ be elements of Σ such that $v(\zeta_i) = \alpha_i$. We prove that $\zeta_1, \ldots, \zeta_k$ are algebraically independent over K. Let

$$f(\zeta_1, \ldots, \zeta_k) = \Sigma a_{i_1 \ldots i_k} \zeta_1^{i_1} \ldots \zeta_k^{i_k}$$

be any polynomial in $\zeta_1, \ldots, \zeta_k$, in which the coefficients $a_{i_1 \ldots i_k}$ are in K and are not all zero. If $a_{i_1 \ldots i_k}$ is not zero,

$$\begin{aligned} v(a_{i_1 \ldots i_k} \zeta_1^{i_1} \ldots \zeta_k^{i_k}) &= v(a_{i_1 \ldots i_k}) + i_1 v(\zeta_1) + \ldots + i_k v(\zeta_k) \\ &= i_1 \alpha_1 + \ldots + i_k \alpha_k. \end{aligned}$$

No two non-zero terms of $f(\zeta_1, \ldots, \zeta_k)$ can have the same value, for, if

$$i_1 \alpha_1 + \ldots + i_k \alpha_k = j_1 \alpha_1 + \ldots + j_k \alpha_k,$$

then $i_1 = j_1, \ldots, i_k = j_k$, since $\alpha_1, \ldots, \alpha_k$ are rationally independent. Hence

$$v(f) = p_1 \alpha_1 + \ldots + p_k \alpha_k,$$

where $\zeta_1^{p_1} \ldots \zeta_k^{p_k}$ is the power-product of least value actually present in $f(\zeta_1, \ldots, \zeta_k)$. It follows that the value of $f(\zeta_1, \ldots, \zeta_k)$ is finite unless

$$f(x_1, \ldots, x_k) \equiv 0.$$

Therefore $f(\zeta_1, \ldots, \zeta_k)$ is not zero unless $f(x_1, \ldots, x_k)$ is identically zero. Hence $\zeta_1, \ldots, \zeta_k$ are algebraically independent over K. Since Σ is of degree of transcendency d, we deduce that $k \leqslant d$, which proves the theorem.

Let us now suppose that the rational rank k of $\mathfrak{B}$ has its maximum value, $k = d$. Let $\alpha_1, \ldots, \alpha_d$ be rationally independent elements of Γ, and let $\zeta_1, \ldots, \zeta_d$ be elements of Σ such that $v(\zeta_i) = \alpha_i$. Then $\zeta_1, \ldots, \zeta_d$ are algebraically independent over K, so that Σ is an algebraic extension of $K(\zeta_1, \ldots, \zeta_d)$ of finite degree, m say. As in the proof of Theorem V, we show that if $f(x_1, \ldots, x_d) \in K[x_1, \ldots, x_d]$, $v[f(\zeta_1, \ldots, \zeta_d)]$ is of the form $r_1 \alpha_1 + \ldots + r_d \alpha_d$, where $r_1, \ldots, r_d$ are integers. Let ζ be any element of Σ, and let

$$f(x_1, \ldots, x_d, x) \equiv \sum_{i=0}^{m'} a_i(x_1, \ldots, x_d)\, x^i = 0 \quad (m' \leqslant m)$$

be the irreducible equation connecting $\zeta_1, \ldots, \zeta_d, \zeta$. Following the argument of Theorem X of § 2, we see that $m!\,v(\zeta)$ is equal to the value of an element of $K(\zeta_1, \ldots, \zeta_d)$. Hence the value of any element ζ of Σ is of the form

$$r_1 \frac{\alpha_1}{m!} + \ldots + r_k \frac{\alpha_k}{m!},$$

where $r_1, \ldots, r_k$ are integers.

The values of elements of Σ are thus seen to form a submodule $\mathfrak{M}'$ of the $\mathfrak{J}$-module $\mathfrak{J}.\left(\frac{\alpha_1}{m!}, \ldots, \frac{\alpha_k}{m!}\right)$, where $\mathfrak{J}$ is the ring of integers. Since $\mathfrak{J}$ is a principal ideal ring, it follows from XV, § 6, Th. I (Corollary) that $\mathfrak{M}'$ is an $\mathfrak{J}$-module with a basis consisting of d' elements, where $d' \leqslant d$. Since $\mathfrak{M}'$ contains d rationally independent real numbers, $d' = d$. Hence the valuation group of Γ consists of the real numbers which can be written in the form

$$s_1\beta_1 + \ldots + s_k\beta_k,$$

where $s_1, \ldots, s_k$ are integers, and $\beta_1, \ldots, \beta_k$ are rationally independent real numbers. The numbers $\beta_1, \ldots, \beta_k$ are said to form an *integral basis* for Γ, and our result can be stated as

THEOREM VI. *If Σ is the function field of an algebraic variety of dimension d, and $\mathfrak{V}$ is a valuation of Σ of rank one and rational rank d, the valuation group of $\mathfrak{V}$ has an integral basis.*

THEOREM VII. *If Σ is the function field of an algebraic variety of dimension d, and $\mathfrak{V}$ is a valuation of Σ of rank one and rational rank k, the dimension of $\mathfrak{V}$ cannot exceed $d-k$.*

Let $\alpha_1, \ldots, \alpha_k$ be rationally independent elements of the valuation group Γ of $\mathfrak{V}$. Without loss of generality, we may suppose $\alpha_i > 0$ $(i = 1, \ldots, k)$ (since if $\alpha_j < 0$ we can replace it by $-\alpha_j$). Let $\zeta_1, \ldots, \zeta_k$ be elements of Σ such that $v(\zeta_i) = \alpha_i$ $(i = 1, \ldots, k)$. As in Theorem V, we can show that $\zeta_1, \ldots, \zeta_k$ are algebraically independent over K, and since $v(\zeta_i) > 0$, $\zeta_1, \ldots, \zeta_k$ are in $\mathfrak{p}$, the ideal of the valuation $\mathfrak{V}$. Let $\zeta_{k+1}, \ldots, \zeta_{d+1}$ be any $d-k+1$ elements of the valuation ring $\mathfrak{R}$. $\zeta_1, \ldots, \zeta_{d+1}$ are algebraically dependent over K; hence there is a non-zero irreducible polynomial over K, $f(x_1, \ldots, x_{d+1})$, such that

$$f(\zeta_1, \ldots, \zeta_{d+1}) = 0.$$

Since $\zeta_i \in \mathfrak{p}$ $(i \leqslant k)$, $\zeta_{k+1}, \ldots, \zeta_{d+1}$ are algebraically dependent modulo $\mathfrak{p}$, and hence $\mathfrak{V}$ is of dimension less than $d-k+1$.

We conclude this section with some remarks on the construction of zero-dimensional valuations of Σ of rank one and rational rank k. We shall not actually require these results in the sequel, and as the proofs are in some cases difficult, we shall omit them, referring the reader to the original memoirs mentioned in the Bibliographical Notes (p. 332). Let $\xi_1, \ldots, \xi_d$ be d algebraically independent elements of Σ, and let $\alpha_1, \ldots, \alpha_d$ be d rationally independent real numbers. We can define a valuation of $K(\xi_1, \ldots, \xi_d)$ which is of rank one and rational rank d as follows. Let t be a new indeterminate, and let $a_1, \ldots, a_d$ be any elements of K. If $f(\xi_1, \ldots, \xi_d)$ is any element of $K[\xi_1, \ldots, \xi_d]$, we can write

$$f(\xi_1, \ldots, \xi_d) = g(\xi_1 - a_1, \ldots, \xi_d - a_d)$$
$$= \Sigma a_{i_1 \ldots i_d} (\xi_1 - a_1)^{i_1} \ldots (\xi_d - a_d)^{i_d}.$$

In this identity, replace $\xi_i - a_i$ by t^{α_i}; then

$$f(a_1 + t^{\alpha_1}, \ldots, a_d + t^{\alpha_d}) = \Sigma a_{i_1 \ldots i_d} t^{\Sigma i_j \alpha_j}.$$

Since $\Sigma i_j \alpha_j \neq \Sigma k_j \alpha_j$ unless $i_j = k_j$ $(j = 1, \ldots, d)$, this series in t is only zero if $f(x_1, \ldots, x_d) \equiv 0$. We define the value of $f(\xi_1, \ldots, \xi_d)$ to be the value of the least exponent of t in this series. If $R(\xi_1, \ldots, \xi_d)$ is any rational function of $\xi_1, \ldots, \xi_d$ we can, in the same way, obtain a series in t for it, and define the value of $R(\xi_1, \ldots, \xi_d)$ as the lowest index which occurs in this series. It is easily verified that this method of defining the value of an element in $K(\xi_1, \ldots, \xi_d)$ is a valuation of this field to the value group Γ consisting of the real numbers of the form

$$p_1 \alpha_1 + \ldots + p_d \alpha_d,$$

where $p_1, \ldots, p_d$ are any integers. We can then extend this valuation to a valuation of Σ, and the methods used to establish Theorem X of § 2 show that this valuation is of rank one and rational rank d. It follows from Theorem VII that the valuation is of dimension zero.

This method of obtaining a valuation of Σ of rank one and rational rank d illustrates a general theorem on zero-dimensional valuations of rank one of the function field of an algebraic variety. Let $\xi_1, \ldots, \xi_n$ be a finite set of elements of Σ such that $K(\xi_1, \ldots, \xi_n) = \Sigma$ (for instance, $\xi_1, \ldots, \xi_n$ may be the coordinates of a generic point of the variety considered). If $\mathfrak{B}$ is any zero-dimensional valuation of Σ of rank one to the subgroup Γ of the additive group of real numbers, there corresponds to each ξ_i a series

$$z_i(t) = t^{\sigma_i}(a_{0i} + a_{1i} t^{\mu_1} + a_{2i} t^{\mu_2} + \ldots) \quad (\mu_1 > 0, \mu_2 > \mu_1, \ldots)$$

in t, where the coefficients a_{ji} belong to some algebraic extension Ω of K, and the exponents σ_i, μ_j belong to Γ. These series have the property that if $f = f(\xi_1, \ldots, \xi_n)$ is any non-zero element of $K(\xi_1, \ldots, \xi_n)$, and

$$f(z_1(t), \ldots, z_n(t)) = t^{\sigma}(b_0 + b_1 t^{\mu_1} + \ldots),$$

where $b_0 \neq 0$ and $\mu_1, \mu_2, \ldots$ are positive, then $v(f) = \sigma$.

The reader who is familiar with the theory of the branches of a curve in classical algebraic geometry will recognise that our method of obtaining a valuation of dimension zero and rank one bears a resemblance to the method of defining the order of a rational function f on a curve on a branch of the curve. In some respects the whole of valuation theory can be regarded as a generalisation of the classical theory of branches, but to push the analogy too far would be confusing rather than helpful. The 'branches' which occur in valuation theory, given by the expansions $\xi_i = z_i(t)$ are not, in general, algebraic. This is obvious if the exponents which appear in the power series are not rational. But even when the valuation group is the group of integers, so that the series are ordinary power series, the 'branch' which arises from a valuation of dimension zero cannot be algebraic, unless $d = 1$. Indeed, if $d > 1$ and the branch $\xi_i = z_i(t)$ were algebraic, there would exist elements $f(\xi_1, \ldots, \xi_n)$ of Σ which were not zero, but were such that $f(z_1(t), \ldots, z_n(t)) \equiv 0$, and so our method of determining the value of any element of Σ would fail. An algebraic branch can sometimes be used to determine a valuation, but it is of rank greater than one.

5. The centre of a valuation. In §4 we considered some properties of the valuations of the function field of an algebraic variety, but these properties concerned the field in an abstract way, having no reference to the geometry of the variety from which the field was derived. We now consider an irreducible variety V_d, in S_n, of dimension d over the ground field K, and examine the relation between a valuation $\mathfrak{B}$ of the function field Σ for V_d and the geometry of V_d. The results of this section form the basis on which the application of valuation theory to geometry rests.

Let $(\zeta_0, \ldots, \zeta_n)$ be the coordinates of a generic point of V_d. The ratios ζ_i/ζ_j lie in the function field Σ, but the ζ_i may be multiplied by any element lying in an extension of Σ. Let us, however, impose the condition that $\zeta_0, \ldots, \zeta_n$ all lie in Σ. Let $\mathfrak{B}$ be any valuation of Σ, and let k be an integer such that

$$v(\zeta_k) \leqslant v(\zeta_i) \quad (i = 0, \ldots, n).$$

Consider the set of homogeneous polynomials $f(x_0, \ldots, x_n)$ over K such that

$$v[f(\zeta_0, \ldots, \zeta_n)] > mv(\zeta_k), \tag{1}$$

where m is the degree of $f(x_0, \ldots, x_n)$. If $f(x_0, \ldots, x_n)$ and $g(x_0, \ldots, x_n)$ are two polynomials of the same degree m with this property,

$$v[f(\zeta_0, \ldots, \zeta_n) - g(\zeta_0, \ldots, \zeta_n)] \geq \min\{v[f(\zeta_0, \ldots, \zeta_n)], v[g(\zeta_0, \ldots, \zeta_n)]\} > mv(\zeta_k);$$

and if $a(x_0, \ldots, x_n)$ is any homogeneous polynomial of degree r, the value of $a(\zeta_0, \ldots, \zeta_n)$ is not less than the value of its term of least value, and since $v(\zeta_k) \leq v(\zeta_i)$ it follows that

$$v[a(\zeta_0, \ldots, \zeta_n)] \geq rv(\zeta_k),$$

and therefore

$$v[a(\zeta_0, \ldots, \zeta_n)\, f(\zeta_0, \ldots, \zeta_n)] > (m+r)\, v(\zeta_k).$$

Hence the homogeneous polynomials $f(x_0, \ldots, x_n)$ which satisfy (1) form a homogeneous ideal $\mathfrak{i}$ in $K[x_0, \ldots, x_n]$. If $f(x_0, \ldots, x_n)$ vanishes on V_d, $f(\zeta_0, \ldots, \zeta_n) = 0$, and hence $f(x_0, \ldots, x_n)$ belongs to $\mathfrak{i}$; that is, $\mathfrak{i}$ is a factor of the prime ideal which defines V_d. Further, if $f(x_0, \ldots, x_n)$ and $g(x_0, \ldots, x_n)$ are homogeneous polynomials, of degrees m and r respectively, which do not belong to $\mathfrak{i}$, then

$$v[f(\zeta_0, \ldots, \zeta_n)] = mv(\zeta_k), \quad v[g(\zeta_0, \ldots, \zeta_n)] = rv(\zeta_k),$$

and hence

$$v[f(\zeta_0, \ldots, \zeta_n)\, g(\zeta_0, \ldots, \zeta_n)] = (m+r)\, v(\zeta_k).$$

It follows that $f(x_0, \ldots, x_n)\, g(x_0, \ldots, x_n)$ is not in $\mathfrak{i}$. The ideal $\mathfrak{i}$ must therefore be a prime ideal. The ideal $\mathfrak{i}$ therefore defines an irreducible variety C contained in V_d. To prove that if the valuation $\mathfrak{B}$ is not trivial, C is a proper subvariety of V_d we note, in the first place, that every element of Σ can be written as $f(\zeta_0, \ldots, \zeta_n)/g(\zeta_0, \ldots, \zeta_n)$, where $f(x)$ and $g(x)$ are forms of like degree. If every non-zero form $f(x)$ of degree m is such that $f(\zeta_0, \ldots, \zeta_n)$ has value $mv(\zeta_k)$, it would follow that every non-zero element of Σ has value zero, and the valuation would be trivial. Hence there must exist a polynomial $f(x_0, \ldots, x_n)$ which does not vanish on V_d and yet belongs to $\mathfrak{i}$, unless $\mathfrak{B}$ is trivial. Again, since x_k does not belong to $\mathfrak{i}$, this ideal is not the unit ideal. Hence if $\mathfrak{B}$ is not trivial C is not vacuous, and C is a proper subvariety of V_d.

C is called the *centre* of $\mathfrak{B}$ on V_d. Thus we have

THEOREM I. *If $\mathfrak{B}$ is any non-trivial valuation of the function field of the irreducible variety V_d, it defines a proper irreducible subvariety C of V_d, called the centre of $\mathfrak{B}$ on V_d.*

We have defined the centre of a valuation in terms of homogeneous coordinates, since it is a projective concept, but in fact it is more convenient to use non-homogeneous coordinates. If V_d does not lie in $x_0 = 0$, and we normalise the coordinates with respect to x_0, that is, if we take x_0 as the prime at infinity, C lies at a finite distance if and only if $v(\zeta_0) = v(\zeta_k)$. This is the case if and only if $v(\zeta_i/\zeta_0) \geqslant 0$ for $i = 1, \ldots, n$, that is, if and only if $K[\zeta_1/\zeta_0, \ldots, \zeta_n/\zeta_0]$ is contained in the ring of the valuation. When we are dealing with a finite number of given valuations we can always choose the coordinates so that their centres do not lie in $x_0 = 0$, and this we generally do. We then regard V_d as a variety in affine space, $x_0 = 0$ being the prime at infinity. The centres of the varieties are then all at a finite distance. If a new valuation is introduced, its centre is at a finite distance if and only if the integral domain of the variety is contained in the ring of the valuation.

We now consider a variety V_d in the affine space A_n, and let $(\xi_1, \ldots, \xi_n)$ be the non-homogeneous coordinates of a generic point. Consider a valuation $\mathfrak{B}$ of the function field $\Sigma = K(\xi_1, \ldots, \xi_n)$, whose centre on V_d is at a finite distance. If $\mathfrak{J} = K[\xi_1, \ldots, \xi_n]$ is the integral domain of V_d in A_n, and $\mathfrak{R}$ is the valuation ring of $\mathfrak{B}$, $\mathfrak{J} \subseteq \mathfrak{R}$. Let $\mathfrak{p}$ be the ideal of non-units of $\mathfrak{R}$. The centre C of $\mathfrak{B}$ is defined by the ideal in $\mathfrak{J}$ formed by the elements whose value is positive; hence the prime ideal $\mathfrak{P}$ of C is given by the relation

$$\mathfrak{P} = \mathfrak{J} \wedge \mathfrak{p}.$$

The dimension e of C is equal to the maximum number of elements which can be found in $\mathfrak{J}$ which are algebraically independent modulo $\mathfrak{P}$. Let $\eta_1, \ldots, \eta_e$ be elements of $\mathfrak{J}$ which are algebraically independent modulo $\mathfrak{P}$. Then if $f(x_1, \ldots, x_e)$ is any polynomial over K such that $f(\eta_1, \ldots, \eta_e) \in \mathfrak{P}$, $f(x_1, \ldots, x_e) \equiv 0$. If $g(x_1, \ldots, x_e)$ is any polynomial over K such that $g(\eta_1, \ldots, \eta_e) \in \mathfrak{p}$, $g(\eta_1, \ldots, \eta_e)$ is both in $\mathfrak{J}$ and in $\mathfrak{p}$; hence it is in $\mathfrak{P} = \mathfrak{J} \wedge \mathfrak{p}$. Therefore $g(x_1, \ldots, x_e) \equiv 0$, and so $\eta_1, \ldots, \eta_e$ are algebraically independent elements of $\mathfrak{R}$ modulo $\mathfrak{p}$. It follows that the dimension of $\mathfrak{B}$ is not less than e. Hence:

THEOREM II. *The dimension of the centre of a valuation of the function field of an algebraic variety is not greater than the dimension of the valuation.*

We can show by an example that the dimension of the centre may actually be less than the dimension of the valuation. We take V_d

to be the projective plane over the field K. The ring $\mathfrak{J}$ is then the ring of polynomials in two independent indeterminates, x, y, say: $\mathfrak{J} = K[x, y]$, $\Sigma = K(x, y)$. We define a valuation $\mathfrak{V}$ of Σ as follows. Consider any non-zero element ζ of Σ. We can write

$$\zeta = a(x, y)/b(x, y),$$

where $a(x, y)$ and $b(x, y)$ are polynomials. Let

$$a(x, y) = a_r(x, y) + a_{r+1}(x, y) + \dots$$

and $$b(x, y) = b_s(x, y) + b_{s+1}(x, y) + \dots,$$

where $a_i(x, y)$, $b_i(x, y)$ are homogeneous of degree i, and $a_r(x, y)$ and $b_s(x, y)$ are not zero. We write $v(\zeta) = r - s$. This value depends only on ζ, and not on the representation of ζ in the form $a(x, y)/b(x, y)$. For if $\zeta = c(x, y)/d(x, y)$, where

$$c(x, y) = c_p(x, y) + c_{p+1}(x, y) + \dots$$

and $$d(x, y) = d_q(x, y) + d_{q+1}(x, y) + \dots,$$

where $c_p(x, y)$ and $d_q(x, y)$ are not zero, then the equation

$$a(x, y)\, d(x, y) = b(x, y)\, c(x, y)$$

gives the relation $$r + q = s + p,$$

and since $r - s = p - q$, $v(\zeta)$ is determined uniquely by ζ. The reader will verify at once that the mapping $\zeta \rightarrow v(\zeta)$ of the non-zero elements of Σ on the ordered additive group of integers is a non-trivial valuation $\mathfrak{V}$.

Since Σ is of dimension two, the dimension of $\mathfrak{V}$ is zero or one. Now $v(y/x) = 0$; hence y/x is in the valuation ring $\mathfrak{R}$, and it is not algebraically dependent on K modulo the ideal $\mathfrak{p}$ of non-units of $\mathfrak{R}$. Hence the dimension of $\mathfrak{V}$ is one. The centre C of $\mathfrak{V}$ is defined by the prime ideal $\mathfrak{P} = \mathfrak{J} \wedge \mathfrak{p} = \mathfrak{J} \cdot (x, y)$ and is therefore given by $x = 0$, $y = 0$; that is, it is of dimension zero.

We return now to a valuation $\mathfrak{V}$ of the function field of an arbitrary variety V_d in A_n. Let $\mathfrak{V}$ be of dimension f, and suppose that the centre C of $\mathfrak{V}$ is at a finite distance and is of dimension e, where $e < f$. Let $\xi_{i_1}, \dots, \xi_{i_e}$ be elements of $\mathfrak{J}$ which are algebraically independent modulo $\mathfrak{P}$. Then they are algebraically independent elements of the valuation ring $\mathfrak{R}$ of $\mathfrak{V}$, modulo $\mathfrak{p}$ [cf. Theorem II]. We can find in $\mathfrak{R}$ a further $f - e$ elements $\eta_1, \dots, \eta_{f-e}$ such that $\xi_{i_1}, \dots, \xi_{i_e}, \eta_1, \dots, \eta_{f-e}$ are algebraically independent modulo $\mathfrak{p}$. In the space A_{n+f-e} consider the variety V'_d whose generic point is $(\xi_1, \ldots, \xi_n, \eta_1, \ldots, \eta_{f-e})$. This variety has the same function field Σ as V_d; hence it is a birational transform of V_d. Since $\mathfrak{J} \subseteq \mathfrak{R}$, and $\eta_i \in \mathfrak{R}$, the integral domain $\mathfrak{J}' = K[\xi_1, \ldots, \xi_n, \eta_1, \ldots, \eta_{f-e}]$ of V'_d is contained in $\mathfrak{R}$, and therefore the centre C' of $\mathfrak{B}$ on V'_d is at a finite distance. Since $\xi_{i_1}, \ldots, \xi_{i_e}, \eta_1, \ldots, \eta_{f-e}$ are algebraically independent elements of $\mathfrak{R}$ modulo $\mathfrak{p}$ and lie in $\mathfrak{J}'$, they are algebraically independent elements of $\mathfrak{J}'$ modulo $\mathfrak{P}'$, where $\mathfrak{P}' = \mathfrak{J}' \wedge \mathfrak{p}$. It follows that C' is of dimension f at least. But, by Theorem II, the dimension of C' cannot exceed f; hence it must be exactly f.

Now $\mathfrak{J} \subseteq \mathfrak{J}'$, and $\mathfrak{P} = \mathfrak{J} \wedge \mathfrak{p} = \mathfrak{J} \wedge (\mathfrak{J}' \wedge \mathfrak{p}) = \mathfrak{J} \wedge \mathfrak{P}'$. Any element of the quotient ring [p. 99] $Q(C) = \mathfrak{J}_{\mathfrak{P}}$ is of the form α/β, where α and β are in $\mathfrak{J}$, and β is not in $\mathfrak{P}$. Then α and β are in $\mathfrak{J}'$, and β is not in $\mathfrak{P}'$, since, otherwise, it would lie in $\mathfrak{J} \wedge \mathfrak{P}' = \mathfrak{P}$. Hence $\alpha/\beta \in \mathfrak{J}'_{\mathfrak{P}} = Q(C')$. Thus we have $Q(C) \subseteq Q(C')$. Hence we have

THEOREM III. *If the centre C on V_d of a valuation $\mathfrak{B}$ of the function field of the variety V_d is of dimension less than the dimension f of the valuation, there exists a birational transform V'_d of V_d on which the centre C' of $\mathfrak{B}$ is of dimension f, and which is such that $Q(C) \subseteq Q(C')$.*

The significance of the last part of the theorem will become apparent in Chapter XVIII.

We now seek to characterise the centre C of a valuation $\mathfrak{B}$ of the function field of V_d in terms of properties of its quotient ring. We may, as usual, assume that C is at a finite distance in A_n. Since the integral domain $\mathfrak{J}$ of V_d is contained in the ring $\mathfrak{R}$ of $\mathfrak{B}$, and the prime ideal $\mathfrak{P}$ of C in $\mathfrak{J}$ is equal to $\mathfrak{J} \wedge \mathfrak{p}$, where $\mathfrak{p}$ is the prime ideal of non-units of $\mathfrak{R}$, the quotient ring $Q(C) = \mathfrak{J}_{\mathfrak{P}} \subseteq \mathfrak{R}_{\mathfrak{p}} = \mathfrak{R}$. We can, however, go further. Let ζ be any element of $Q(C)$, so that $\zeta = \alpha/\beta$, where α and β are in $\mathfrak{J}$, and β is not in $\mathfrak{P}$. Since α and β are in $\mathfrak{J}$, and therefore in $\mathfrak{R}$, $v(\alpha) \geqslant 0$, and $v(\beta) \geqslant 0$. Further, since β is not in $\mathfrak{P}$, and hence not in $\mathfrak{p}$, $v(\beta) = 0$. The element ζ is a non-unit of $Q(C)$ if and only if α is in $\mathfrak{P}$, and hence in $\mathfrak{p}$, that is, if and only if $v(\alpha) > 0$. Then $v(\zeta) = v(\alpha) - v(\beta) > 0$ if and only if ζ is a non-unit of $Q(C)$. Hence if ζ is a non-unit of $Q(C)$ it is a non-unit of $\mathfrak{R}$.

Now consider any irreducible variety D on V_d such that $Q(D) \subseteq \mathfrak{R}$. Without loss of generality we may suppose that the prime at infinity is chosen so that it does not contain D or the centre C of $\mathfrak{B}$. If α is any element of $\mathfrak{J}$ such that $v(\alpha) > 0$, that is, if α is any element of $\mathfrak{P}$, $v(\alpha^{-1}) < 0$, and hence α^{-1} is not in $\mathfrak{R}$ and therefore not in $Q(D)$.

Since α is in $\mathfrak{J}$ it is in $Q(D)$; hence it must be a non-unit of $Q(D)$. Thus every element of $\mathfrak{J}$ in $\mathfrak{P}$ is a non-unit of $Q(D)$ which lies in $\mathfrak{J}$, and hence it lies in the prime ideal $\mathfrak{P}'$ of D. Thus $\mathfrak{P} \subseteq \mathfrak{P}'$, and hence $Q(D) \subseteq Q(C)$. Hence [XVI, § 2, Th. IV], $D \subseteq C$. Conversely, if $D \subseteq C$, $Q(D) \subseteq Q(C) \subseteq \mathfrak{R}$. Thus $Q(D) \subseteq \mathfrak{R}$ is a necessary and sufficient condition that D be contained in the centre of $\mathfrak{B}$.

If D is a proper subvariety of C, these exists a polynomial $f(x_1, \ldots, x_n)$ which vanishes on D but not on C. Then $f(\xi_1, \ldots, \xi_n)$ is a unit of $Q(C)$ and hence of $\mathfrak{R}$, but is not a unit of $Q(D)$. Hence there are non-units of $Q(D)$ which are units of $\mathfrak{R}$, whereas we have seen that this is not the case if $D = C$. We therefore have the following theorem,

THEOREM IV. *A necessary and sufficient condition that an irreducible variety D be contained in the centre of a valuation $\mathfrak{B}$ of the function field of a variety V_d is that the quotient ring of D be contained in the valuation ring of $\mathfrak{B}$. The variety D is the centre of $\mathfrak{B}$ on V_d if and only if one of the two following additional and equivalent conditions is satisfied*: (i) *D is not a proper subvariety of any irreducible variety on V_d whose quotient ring is contained in the valuation ring*; (ii) *every non-unit of the quotient ring of D is a non-unit of the valuation ring.*

From this theorem, we can deduce a relation between the centre of a valuation $\mathfrak{B}$ of Σ of rank greater than one and the centre of the valuation $\mathfrak{B}'$, where $\mathfrak{B}$ is decomposed into the valuation $\mathfrak{B}'$ of Σ followed by the valuation $\mathfrak{B}_1$ of the residue field of $\mathfrak{B}'$ [§ 3, Th. II]. The valuation ring $\mathfrak{R}'$ of $\mathfrak{B}'$ is the quotient ring of $\mathfrak{R}$ with respect to a prime ideal of $\mathfrak{R}$. Hence $\mathfrak{R} \subseteq \mathfrak{R}'$. If C is the centre of $\mathfrak{R}$, C' that of $\mathfrak{R}'$, we have $Q(C) \subseteq \mathfrak{R} \subseteq \mathfrak{R}'$; hence, by Theorem IV, $C \subseteq C'$. Thus we have

THEOREM V. *If a valuation $\mathfrak{B}$ is decomposed into a valuation $\mathfrak{B}'$ followed by a valuation of the residue field of $\mathfrak{B}'$, the centre of $\mathfrak{B}$ is contained in the centre of $\mathfrak{B}'$.*

If the variety V_d is defined over a ground field K which is not algebraically closed, we frequently have to extend the ground field K. If $\Sigma = K(\xi_1, \ldots, \xi_n)$ and we extend K to K^*, $\Sigma^* = K^*(\xi_1, \ldots, \xi_n)$ is an extension of Σ, and any valuation $\mathfrak{B}$ of Σ extends to one or more valuations $\mathfrak{B}^*$ of Σ^*. Over K^*, $(\xi_1, \ldots, \xi_n)$ is the generic point of a variety V_d^*, which is either V_d itself, or is a component of V_d [X, § 12]. It is necessary to consider the relation between the centre C of $\mathfrak{B}$ on V_d, and the centre C^* of $\mathfrak{B}^*$ on V_d^*.

We begin with the case in which V_d is not absolutely irreducible. Then there exist elements of Σ which are algebraically dependent on K but are not in K [X, § 11]. These elements define a field K', the algebraic closure of K in Σ. We first consider the case $K^* = K'$. Over the field K', $(\xi_1, \ldots, \xi_n)$ is the generic point of a variety V'_d which is one of the absolutely irreducible components of V_d [X, § 10]. The integral domain $\mathfrak{J}' = K'[\xi_1, \ldots, \xi_n]$ is, in general, a proper extension of $\mathfrak{J} = K[\xi_1, \ldots, \xi_n]$, but

$$\Sigma' = K'(\xi_1, \ldots, \xi_n) = K(\xi_1, \ldots, \xi_n) = \Sigma,$$

since $K' \subseteq \Sigma$. Thus $\mathfrak{B}$ is a valuation of Σ'. Before, however, we speak of its centre, we have to show that $v(a') = 0$ for every non-zero element of K'. This follows from the Corollary to Theorem I of § 4, since K' is of degree of transcendency zero over K. If $\mathfrak{R}$ is the valuation ring of $\mathfrak{B}$, it follows that $K' \subseteq \mathfrak{R}$. If we assume, as usual, that C is at a finite distance, we have $\xi_i \in \mathfrak{R}$ $(i = 1, \ldots, n)$. Hence $\mathfrak{J}' \subseteq \mathfrak{R}$, and therefore the centre of $\mathfrak{B}$ on V' is at a finite distance. If $\mathfrak{p}$ is the ideal of non-units of $\mathfrak{R}$, $\mathfrak{P} = \mathfrak{J}_\wedge \mathfrak{p}$ is the ideal of C in $\mathfrak{J}$, and $\mathfrak{P}' = \mathfrak{J}'_\wedge \mathfrak{p}$ is the ideal of C' in $\mathfrak{J}'$. Thus, $\mathfrak{P} = \mathfrak{J}_\wedge \mathfrak{P}'$. Hence

$$\mathfrak{J}' . \mathfrak{P} = 0 \ (\mathfrak{P}').$$

Let
$$\phi_i(x_1, \ldots, x_n) = 0 \quad (i = 1, \ldots, r) \tag{2}$$

be a basis for the equations of C over K. Then

$$\phi_i(\xi_1, \ldots, \xi_n) \quad (i = 1, \ldots, r)$$

is a basis for $\mathfrak{P}$ in $\mathfrak{J}$, and hence for $\mathfrak{J}' . \mathfrak{P}$ in $\mathfrak{J}'$. Since $\mathfrak{J}' . \mathfrak{P} \subseteq \mathfrak{P}'$ it follows that C' satisfies all the equations (2), and hence $C' \subseteq C$.

Since K' is algebraic over K, the dimension of the residue field of $\mathfrak{B}$ over K' is the same as its dimension over K. Hence the dimension of $\mathfrak{B}$ is the same whether K or K' is taken as the ground field. Next, if C is of dimension f, there exist f elements of $\mathfrak{J}$ which are algebraically independent over K modulo $\mathfrak{P}$. These are algebraically independent modulo $\mathfrak{p}$, and hence, since they lie in $\mathfrak{J}'$, there are f elements of $\mathfrak{J}'$ which are algebraically independent over K' modulo $\mathfrak{P}'$. Hence the dimension of C' is at least f, and since $C' \subseteq C$ it follows that the dimension of C' is exactly f. It is then clear that C' is one of the components into which C breaks up when K is extended to K'.

The only other case with which we shall be concerned is that in which K^* is an algebraic extension of K which contains K'. We can pass from K to K^* by two stages, first from K to K', and then from K' to K^*. We have already dealt with the first case, in which we pass from V_d to the variety V'_d, which is absolutely irreducible. Thus we need only consider the case in which V_d is absolutely irreducible over K, and pass directly from K to K^*. In this case, if K^* is a proper extension of K, $\Sigma^* = K^*(\xi_1, \ldots, \xi_n)$ is a proper extension of Σ, but V_d^*, as a set of points, coincides with V_d, and may continue to be denoted by this symbol.

Since Σ^* is a proper extension of Σ, we have to use Theorem VII of § 2 to conclude that there are valuations of Σ which are extensions of $\mathfrak{B}$. If K^* is an algebraic extension of K, Σ^* is an algebraic extension of Σ, and if K^* is an extension of K of finite degree m, Σ^* is of finite degree m over Σ. Then, by Theorem V of § 3, there are at most m valuations of Σ^* which are extensions of $\mathfrak{B}$. There is, in any case, one, and we select one, say $\mathfrak{B}^*$. Since $\mathfrak{B}^*$ induces the valuation $\mathfrak{B}$ of Σ, the value $v^*(a)$ of any non-zero element a of K in $\mathfrak{B}^*$ is zero. $\mathfrak{B}^*$ also induces a valuation of K^*. In this valuation every element of K has value zero, and therefore, since K^* is algebraic over K, this valuation of K^* is trivial. Hence $v^*(a) = 0$ for every non-zero element a of K^*. Hence we can speak of the centre of $\mathfrak{B}^*$, and since $\mathfrak{J}^* = K^*[\xi_1, \ldots, \xi_n] \subseteq \mathfrak{R}^*$, the centre of $\mathfrak{B}^*$ is at a finite distance on V_d. Suppose that the centre C^* of $\mathfrak{B}^*$ is not contained in C. Then C^* and its conjugates over K constitute a variety over K which is not contained in C. Then there exists a polynomial $f(x_1, \ldots, x_n)$ over K which vanishes on C but not on C^*. The element

$$\zeta = f(\xi_1, \ldots, \xi_n)$$

is therefore an element of $\mathfrak{J}$ which belongs to the prime ideal $\mathfrak{P}$ of C in $\mathfrak{J}$, but does not belong to the prime ideal $\mathfrak{P}^*$ of C^* in $\mathfrak{R}^*$. This, however, is impossible. For since $\zeta \in \mathfrak{P}$, $\zeta \in \mathfrak{p}$, and hence $v(\zeta) > 0$. Therefore $v^*(\zeta) > 0$, and therefore $\zeta \in \mathfrak{P}^*$. Hence $C^* \subseteq C$.

A proof exactly similar to that used in the case $K^* = K'$ above shows that $\mathfrak{B}^*$ is of the same dimension as $\mathfrak{B}$, and that C^* is of the same dimension as C. C^* is therefore one of the components into which C decomposes when K is extended to K^*. The converse of this result is obvious. If C^* is the centre of any valuation $\mathfrak{B}^*$ of Σ^*, given by the prime ideal $\mathfrak{P}^*$ of $\mathfrak{J}^*$, the centre C of the induced valuation of Σ is given by the prime ideal $\mathfrak{P} = \mathfrak{J} \wedge \mathfrak{P}^*$ of $\mathfrak{J}$. The centre C is therefore the irreducible variety over K of which C^* is a component when the ground field is extended to K^*.

We sum up our results in

THEOREM VI. *If $\mathfrak{B}$ is any valuation of the function field Σ of the irreducible variety V_d over K, and K^* is any algebraic extension of K which contains the algebraic closure of K in Σ, and if V_d^* is the variety over K^* whose generic point is a generic point of V_d over K, there is an extension $\mathfrak{B}^*$ of $\mathfrak{B}$ to the function field of V_d^*. The centre C^* of $\mathfrak{B}^*$ on V_d^* is one of the components into which C breaks up when K is extended to K^*. The dimension of $\mathfrak{B}^*$ is equal to the dimension of $\mathfrak{B}$.*

We now turn to some problems concerning the construction of valuations with a given centre. We first show that the problem of finding the valuations of the function field Σ of a variety V_d whose centre C is at a finite distance in A_n can be reduced to the case in which V_d is affine-normal.

We consider any irreducible variety V_d whose integral domain is $\mathfrak{J} = K[\xi_1, \ldots, \xi_n]$, and let C be any irreducible variety on V_d, at a finite distance, given by the prime ideal $\mathfrak{P}$ of $\mathfrak{J}$. Let V_d^* be an affine-normal variety whose integral domain $\mathfrak{J}^*$ is the integral closure of $\mathfrak{J}$ in $\Sigma = K(\xi_1, \ldots, \xi_n)$. Consider the ideal $\mathfrak{J}^*.\mathfrak{P}$ in $\mathfrak{J}^*$. If $\mathfrak{J}^*.\mathfrak{P} = [\mathfrak{Q}_1^*, \ldots, \mathfrak{Q}_k^*]$, and if $\mathfrak{P}_i^*$ is the radical of $\mathfrak{Q}_i^*$, $\mathfrak{J}_\wedge \mathfrak{P}_i^*$ is a prime ideal of $\mathfrak{J}$ which is a factor of $\mathfrak{P}$, and, for at least one value of i, $\mathfrak{J}_\wedge \mathfrak{P}_i^* = \mathfrak{P}$ [XV, § 8, Th. VIII]. Moreover, if $\mathfrak{P}^*$ is any ideal of $\mathfrak{J}^*$ such that $\mathfrak{J}_\wedge \mathfrak{P}^* = \mathfrak{P}$, then $\mathfrak{P}^* = \mathfrak{P}_i^*$, for some value of i. Let $\mathfrak{Q}_1^*, \ldots, \mathfrak{Q}_k^*$ be arranged so that $\mathfrak{J}_\wedge \mathfrak{P}_i^* = \mathfrak{P}$ $(i = 1, \ldots, r)$, while $\mathfrak{J}_\wedge \mathfrak{P}_i^*$ is a proper factor of $\mathfrak{P}$ if $i > r$. Let C_i^* be the irreducible variety on V_d^* defined by $\mathfrak{P}_i^*$.

Let $\mathfrak{B}$ be any valuation of Σ whose centre on V_d is C, and let $\mathfrak{R}$ be its valuation ring, $\mathfrak{p}$ the ideal of non-units of $\mathfrak{R}$. Then $\mathfrak{J} \subseteq \mathfrak{R}$, since C is at a finite distance. Any element of $\mathfrak{J}^*$ is integrally dependent on $\mathfrak{J}$, and hence on $\mathfrak{R}$. But $\mathfrak{R}$ is integrally closed [§ 2, Th. IV]. Hence $\mathfrak{J}^* \subseteq \mathfrak{R}$. Hence the centre C^* of $\mathfrak{B}$ on V_d^* is at a finite distance. If C^* is given by the prime ideal $\mathfrak{P}^*$ of $\mathfrak{J}^*$, then $\mathfrak{P}^* = \mathfrak{J}^*{}_\wedge \mathfrak{p}$, and so

$$\mathfrak{J}_\wedge \mathfrak{P}^* = \mathfrak{J}_\wedge (\mathfrak{J}^*{}_\wedge \mathfrak{p}) = \mathfrak{J}_\wedge \mathfrak{p} = \mathfrak{P}.$$

Hence C^* must be one of the varieties $C_1^*, \ldots, C_r^*$. Conversely, if $\mathfrak{B}$ is any valuation of Σ whose centre is C_i^*, for any i $(1 \leqslant i \leqslant r)$, the centre of $\mathfrak{B}$ on V_d is given by the prime ideal

$$\mathfrak{J}_\wedge \mathfrak{p} = \mathfrak{J}_\wedge (\mathfrak{J}^*{}_\wedge \mathfrak{p}) = \mathfrak{J}_\wedge \mathfrak{P}_i^* = \mathfrak{P};$$

hence its centre on V_d is C.

Thus to find the valuations of Σ whose centres on V_d are C, we have merely to find the valuations whose centres on V_d^* are any of the varieties C_i^* $(i = 1, \ldots, r)$. The principal advantage of working on an affine-normal variety is that the quotient ring of any subvariety is integrally closed. In fact, in our present problem of finding the valuations of Σ with a given centre C, it is sufficient to consider the case when the quotient ring of C is integrally closed, but there are slight advantages in assuming that V_d is normal.

Now suppose that V_d is affine-normal, and that C is an irreducible variety of dimension $d-1$ on it, given by the prime ideal $\mathfrak{P}$. It will be recalled that in § 5 of Chapter XVI we used the multiplicative theory of ideals in the integral domain of an affine-normal variety to give an example of a valuation, and we showed that the prime ideal of an irreducible subvariety of dimension $d-1$ serves to determine a valuation. In particular, by the method described, we can use the prime ideal $\mathfrak{P}$ of C to define a valuation $\mathfrak{V}$, and it is clear that C is the centre of $\mathfrak{V}$ on V_d. We now prove that this is the only valuation of Σ whose centre is C. We use the results of § 7 of Chapter XV on the multiplicative theory of ideals in an integrally closed ring.

Let η and ζ be any two non-zero elements of the integral domain $\mathfrak{I}$ of V_d. Express the ideals $\mathfrak{I}.(\eta)$, $\mathfrak{I}.(\zeta)$ in quasi-factors:

$$\mathfrak{i} = \mathfrak{I}.(\eta) \sim \mathfrak{P}^a \mathfrak{P}_1^{a_1}, \ldots, \mathfrak{P}_r^{a_r},$$
$$\mathfrak{j} = \mathfrak{I}.(\zeta) \sim \mathfrak{P}^b \mathfrak{P}_1^{b_1}, \ldots, \mathfrak{P}_r^{b_r},$$

where $\mathfrak{P}, \mathfrak{P}_1, \ldots, \mathfrak{P}_r$ are all different prime ideals of dimension $d-1$, and $a, a_1, \ldots, a_r, b, b_1, \ldots, b_r$ are positive integers or zero. We show that if $a = b$ there exist two elements η', ζ' of $\mathfrak{I}$, such that neither lies in $\mathfrak{P}$, which satisfy the equation $\eta/\zeta = \eta'/\zeta'$. We can find an element of $\mathfrak{I}$ which is in $\mathfrak{P}_1, \ldots, \mathfrak{P}_r$ but is not in $\mathfrak{P}$. If ζ' is a sufficiently high power of this element,

$$\mathfrak{k} = \mathfrak{I}.(\eta\zeta') \sim \mathfrak{P}^a \mathfrak{P}_1^{c_1} \ldots \mathfrak{P}_r^{c_r},$$

where $c_i \geqslant b_i$ $(i = 1, \ldots, r)$. Then $\mathfrak{k} \geqslant \mathfrak{j}$, and hence $\mathfrak{k}\mathfrak{j}^{-1} \subseteq \mathfrak{I}$. But

$$\mathfrak{k}\mathfrak{j}^{-1} = \mathfrak{I}.(\eta\zeta'/\zeta).$$

Hence $\eta' = \eta\zeta'/\zeta$ belongs to $\mathfrak{I}$. Also

$$\mathfrak{I}.(\eta') \sim \mathfrak{P}_1^{c_1-b_1} \ldots \mathfrak{P}_r^{c_r-b_r},$$

so that η' is not in $\mathfrak{P}$.

Let ξ be any element of $\mathfrak{P}$ which is not in the symbolic square, $\mathfrak{P}^{(2)}$, of $\mathfrak{P}$. Then
$$\mathfrak{J}.(\xi) \sim \mathfrak{P}\mathfrak{P}_1^{d_1}\mathfrak{P}_2^{d_2}\ldots.$$
If α/β is any element of Σ (α and β being in $\mathfrak{J}$), we can find an integer λ (positive, zero, or negative) such that $\alpha/\beta\xi^\lambda = \alpha'/\beta'$, where
$$\mathfrak{J}.(\alpha') \sim \mathfrak{P}^e\mathfrak{P}_1^{e_1}\ldots,$$
$$\mathfrak{J}.(\beta') \sim \mathfrak{P}^e\mathfrak{P}_1^{f_1}\ldots.$$
Hence $\alpha/\beta\xi^\lambda = \alpha''/\beta''$, where α'' and β'' are in $\mathfrak{J}$, but not in $\mathfrak{P}$. If $\mathfrak{B}$ is any valuation of Σ whose centre is C, $v(\alpha'') = 0$ and $v(\beta'') = 0$, since α'' and $\beta'' \in \mathfrak{J} \subseteq \mathfrak{R}$ but are not in $\mathfrak{P}$. Hence
$$v(\alpha/\beta) = v(\xi^\lambda\alpha''/\beta'') = \lambda v(\xi).$$
It follows that the valuation group of $\mathfrak{B}$ consists of the integral multiples of $v(\xi)$. If we normalise the valuation group so that $v(\xi) = 1$, the value group is the group of integers. (This verifies Theorem III of § 4.) It is then clear that if
$$\mathfrak{J}.(\alpha) \sim \mathfrak{P}^\rho\mathfrak{P}_1^{\rho_1}\ldots\mathfrak{P}_s^{\rho_s},$$
$$\mathfrak{J}.(\beta) \sim \mathfrak{P}^\sigma\mathfrak{P}_1^{\sigma_1}\ldots\mathfrak{P}_s^{\sigma_s},$$
then $v(\alpha/\beta) = \rho - \sigma$. Thus the valuation is necessarily the one discussed in § 5 of Chapter XVI. Moreover, if $v(\alpha/\beta) \geqslant 0$, $\rho \geqslant \sigma$, and hence
$$\frac{\alpha}{\beta} = \frac{\alpha''\xi^{\rho-\sigma}}{\beta''} = \frac{\alpha_1}{\beta''},$$
where $\alpha_1 \in \mathfrak{J}$, and β'' belongs to $\mathfrak{J}$ but not to $\mathfrak{P}$. Hence any element of the valuation ring $\mathfrak{R}$ of $\mathfrak{B}$ lies in $Q(C)$; that is, $\mathfrak{R} \subseteq Q(C)$. But we already know that $Q(C) \subseteq \mathfrak{R}$; hence $\mathfrak{R} = Q(C)$. Thus we have

THEOREM VII. *If the variety V_d is affine-normal, and C is any irreducible variety of dimension $d-1$ on V_d, there is a unique valuation of the function field of V_d whose centre on V_d is C, and the ring of this valuation is the quotient ring of C.*

If V_d is not affine-normal there are, corresponding to C, a finite number of irreducible varieties of dimension $d-1$ on the derived affine-normal variety V_d^*. The valuations of the function field Σ of V_d whose centres on V_d^* are at one of these varieties are the only valuations whose centres on V_d are C; hence there are only a finite number of valuations which have C as centre. If $Q(C)$ is integrally closed, there corresponds only one variety on V_d^*, and the valuation whose centre on V_d is C is then unique.

If V_d is affine-normal, and C is of dimension e, less than $d-1$, there are infinitely many valuations whose centres are C. Since V_d is affine-normal, the quotient ring is integrally closed; hence [§ 2, Th. VI] it is the intersection of the valuation rings which contain it. These valuations may be of any dimension r, where $e \leqslant r \leqslant d-1$, and their centres contain C but need not coincide with it. It is convenient to have a more powerful result, to the effect that $Q(C)$ is the intersection of the rings of the e-dimensional valuations which contain it. Since the centre of such a valuation contains C and is of dimension e at most, it must coincide with C. The first step towards this result is to prove

THEOREM VIII. *If C is a variety of dimension e on V_d, and C is the centre of a valuation $\mathfrak{B}$ of dimension f of the function field Σ of V_d, $(e<f)$, we can compound $\mathfrak{B}$ with a valuation of the residue field of $\mathfrak{B}$ to give a valuation of Σ of dimension e whose centre on V_d is C.*

Let Σ' be the residue field of $\mathfrak{B}$. Since $Q(C)$ is contained in the valuation ring $\mathfrak{R}$ of $\mathfrak{B}$, every element of it has an image in Σ'. An element of $Q(C)$ has image zero in Σ' if and only if it is a non-unit of $\mathfrak{R}$, and hence a non-unit of $Q(C)$ [Th. IV]. Hence the image F' of $Q(C)$ in Σ' is isomorphic with the remainder-class ring of $Q(C)$ with respect to its ideal of non-units; that is, F' is isomorphic with the function field of C [XV, § 5, Th. XI, Cor.]. In this isomorphism elements of the ground field K are self-corresponding. Hence F' is a field which is an extension of dimension e of K. If F' is not integrally closed in Σ', its integral closure F'' is an algebraic extension of F', and is still of dimension e over K. The field F'' is then a proper subfield of Σ', which is of dimension f over K, and since it is the intersection of the valuation rings of Σ' which contain it, there exists a valuation $\overline{\mathfrak{B}}$ of Σ' whose valuation ring contains F', but does not contain an assigned element ζ' of Σ' which is not in F''. The valuation $\overline{\mathfrak{B}}$ is thus a non-trivial valuation of Σ' whose valuation ring contains F'. If we construct the valuation $\mathfrak{B}^*$ of Σ obtained by compounding $\mathfrak{B}$ and $\overline{\mathfrak{B}}$, the valuation ring $\mathfrak{R}^*$ of $\mathfrak{B}^*$ contains every element of $\mathfrak{R}$ whose image in Σ' lies in the valuation ring of $\overline{\mathfrak{B}}$. Hence $\mathfrak{R}^*$ contains $Q(C)$.

By Theorem V, the centre of $\mathfrak{B}^*$ is contained in C, the centre of $\mathfrak{B}$, and it also contains C, since its valuation ring contains $Q(C)$. Hence the centre of $\mathfrak{B}^*$ is C. The residue field of $\mathfrak{B}^*$ is isomorphic with the residue field of the valuation $\overline{\mathfrak{B}}$ of Σ'; hence it is of dimension less than f. Thus the dimension f^* of the valuation $\mathfrak{B}^*$ satisfies the inequalities $e \leqslant f^* < f$. If $f^* > e$, we repeat our construction, using $\mathfrak{B}^*$ in place of $\mathfrak{B}$, and obtain a new valuation $\mathfrak{B}^{**}$ of dimension f^{**}, where $e \leqslant f^{**} < f^*$, and whose centre is C. After a finite number of steps, we must arrive at a valuation $\mathfrak{B}'$ of the function field of V_d which is of dimension e and whose centre on V_d is C. The valuation group Γ' of $\mathfrak{B}'$ contains an isolated subgroup Δ such that the factor group Γ'/Δ is isomorphic to Γ, where Γ is the valuation group of $\mathfrak{B}$. The valuation ring of $\mathfrak{B}'$ is a subring of the valuation ring of $\mathfrak{B}$, and, of course, contains $Q(C)$, and $\mathfrak{B}'$ is compounded of the valuation $\mathfrak{B}$ of Σ and a valuation to Δ of the residue field of $\mathfrak{B}$.

THEOREM IX. *If $\mathfrak{B}$ is a valuation of Σ whose centre on V_d is C, and D is any irreducible subvariety of C, we can compound $\mathfrak{B}$ with a valuation of its residue field to obtain a valuation of Σ whose centre is D.*

The proof is similar to that of Theorem VIII. If the dimension of $\mathfrak{B}$ is greater than e, the dimension of C, we can compound $\mathfrak{B}$ with a valuation $\mathfrak{B}_1$ of its residue field to give a valuation $\mathfrak{B}'$ of Σ of dimension e whose centre is C. We then consider the residue field Σ'' of $\mathfrak{B}'$, and in it the ring $\mathfrak{S}''$ formed by the residues of the elements of $Q(D)$. Since D is a proper subvariety of the centre of $\mathfrak{B}'$, there are non-units of $Q(D)$ which are units of the valuation ring of $\mathfrak{B}'$ [Theorem IV]. The images of these elements in Σ'' are non-units of $\mathfrak{S}''$ which are different from zero. Hence $\mathfrak{S}''$ contains non-zero elements which are not units of $\mathfrak{S}''$; therefore it is not a field. Since $\mathfrak{S}''$ is not a field its integral closure $\mathfrak{S}'''$ in Σ'' is not a field (since it cannot contain the inverse of a non-unit of $\mathfrak{S}''$). Hence, just as in Theorem VIII, we can find a non-trivial valuation of Σ'' whose valuation ring contains $\mathfrak{S}''$. Compounding $\mathfrak{B}'$ with this, we obtain a valuation $\mathfrak{B}_1$ of Σ whose valuation ring contains $Q(D)$. The dimension of $\mathfrak{B}_1$ is less than e, the dimension of $\mathfrak{B}'$. Hence the centre of $\mathfrak{B}_1$ on V_d is a variety C_1 such that $D \subseteq C_1 \subset C$. If $C_1 \neq D$, we just repeat the argument, using C_1 in place of C, and after a finite number of steps we obtain a valuation of Σ, which is obtained by compounding $\mathfrak{B}$ with some valuation of its residue field, whose centre on V_d is D.

From Theorems VIII and IX we can quickly deduce

THEOREM X. *If C is an irreducible variety on V_d of dimension e whose quotient ring $Q(C)$ is integrally closed, then $Q(C)$ is the inter-*

section of the valuation rings of the e-dimensional valuations of the function field of V_d which contain it (and hence have C as their centre).

Let $\mathfrak{B}$ be any valuation of the function field Σ whose valuation ring $\mathfrak{R}$ contains $Q(C)$. The centre E of $\mathfrak{B}$ contains C. By Theorem VIII we can compound $\mathfrak{B}$ with a valuation of its residue field to obtain a valuation of Σ whose centre on V_d is C, and by further compounding this we obtain a valuation $\mathfrak{B}^*$ of Σ of dimension e whose centre is C. Since $\mathfrak{B}^*$ is obtained by compounding $\mathfrak{B}$ with some valuation of its residue field, the valuation ring $\mathfrak{R}^*$ of $\mathfrak{B}^*$ is contained in $\mathfrak{R}$. Hence

$$\mathfrak{R} \supseteq \mathfrak{R}^* \supseteq Q(C).$$

Hence in constructing the intersection of the valuation rings which contain $Q(C)$ we may omit $\mathfrak{R}$, since any element of it in the intersection must lie in $\mathfrak{R}^*$. Thus in forming the intersection we may omit all valuations $\mathfrak{B}$ of dimension greater than e. The theorem then follows.

The theorems we have proved in the preceding pages on valuations with a given centre are sufficient for most applications. More complete results have been obtained by Zariski, and we conclude by quoting, without proof, his most comprehensive result.

THEOREM XI. *Let $C_0, C_1, \ldots, C_{r-1}$ be irreducible varieties on V_d such that*

$$C_0 \supseteq C_1 \supseteq \ldots \supseteq C_{r-1},$$

and let $\rho_0, \ldots, \rho_{r-1}$ be integers such that

$$d > \rho_0 > \rho_1 > \ldots > \rho_{r-1},$$

and

$$\rho_i \geqslant \dim C_i, \quad r \leqslant d.$$

Then there exists a sequence of valuations $\mathfrak{B}_0, \mathfrak{B}_1, \ldots, \mathfrak{B}_{r-1}$ of the function field of V_d with the properties

(i) *$\mathfrak{B}_i$ is compounded from $\mathfrak{B}_{i-1}$;*

(ii) *$\mathfrak{B}_i$ is of dimension ρ_i and rank $i+1$;*

(iii) *the centre of $\mathfrak{B}_i$ on V_d is C_i.*

CHAPTER XVIII

BIRATIONAL TRANSFORMATIONS

1. Birational correspondences. Let S_m and S'_n be two projective spaces defined over the ground field K, and let $(x_0, \ldots, x_m)$ be a coordinate system in S_m, $(y_0, \ldots, y_n)$ a coordinate system in S'_n. Let U be an irreducible variety of d dimensions in S_m, and let $(\xi_0, \ldots, \xi_m)$ be a generic point of U; similarly, let U' be an irreducible variety in S'_n, and $(\eta_0, \ldots, \eta_n)$ a generic point of it. We may suppose that U does not lie in $x_h = 0$, and that U' is not in $y_k = 0$. The varieties U and U' are birationally equivalent if there exists an isomorphism between their function fields $K(\xi_0/\xi_h, \ldots, \xi_m/\xi_h)$ and $K(\eta_0/\eta_k, \ldots, \eta_n/\eta_k)$ in which elements of K are self-corresponding. It may happen that when U and U' are birationally equivalent there is more than one such isomorphism. Further investigation shows that this is the exception rather than the rule, but we must take account of the possibility. In any case, if U and U' are birationally equivalent, we select one isomorphism between their function fields in which elements of K are self-corresponding, and identify corresponding elements. If the function fields of the given varieties U and U' coincide originally, as happens, for instance, when U' is a normal variety derived from U, the isomorphism which we naturally consider is that in which each element of the common function field corresponds to itself. When the function fields of U and U' have been identified, we have

$$\eta_i/\eta_k \in K(\xi_0/\xi_h, \ldots, \xi_m/\xi_h) \quad (i = 0, \ldots, n).$$

Hence there exist $n+1$ homogeneous polynomials $\phi_i(x_0, \ldots, x_m)$ $(i = 0, \ldots, n)$ of the same degree, with coefficients in K, where $\phi_k(\xi_0, \ldots, \xi_m) \neq 0$, such that

$$\frac{\eta_i}{\eta_k} = \frac{\phi_i(\xi_0, \ldots, \xi_m)}{\phi_k(\xi_0, \ldots, \xi_m)} \quad (i = 0, \ldots, n);$$

and similarly there exist $m+1$ homogeneous polynomials

$$\psi_i(y_0, \ldots, y_n) \quad (i = 0, \ldots, m)$$

of the same degree, where $\psi_h(\eta_0, \ldots, \eta_n) \neq 0$, such that

$$\frac{\xi_i}{\xi_h} = \frac{\psi_i(\eta_0, \ldots, \eta_n)}{\psi_h(\eta_0, \ldots, \eta_n)}.$$

From these relations we deduce that

$$\xi_i \psi_j[\phi_0(\xi), \ldots, \phi_n(\xi)] - \xi_j \psi_i[\phi_0(\xi), \ldots, \phi_n(\xi)] = 0;$$

hence $$x_i \psi_j[\phi_0(x), \ldots, \phi_n(x)] - x_j \psi_i[\phi_0(x), \ldots, \phi_n(x)]$$

vanishes on U; similarly

$$y_i \phi_j[\psi_0(y), \ldots, \psi_m(y)] - y_j \phi_i[\psi_0(y), \ldots, \psi_m(y)]$$

vanishes on U'.

The variety W in the two-way space $S_{m,n}$ whose generic point is (ξ, η) defines an irreducible correspondence Γ between S_m and S'_n, in which the object-variety, in S_m, has generic point ξ, and is therefore the variety U, and the image-variety, in S'_n, has generic point η, and is therefore the variety U'. Since

$$\eta_i \phi_j(\xi) - \eta_j \phi_i(\xi) = 0 \quad (i = 0, \ldots, n; j = 0, \ldots, n),$$

it follows that if x' is any point of U, the points of U' which correspond to it in Γ satisfy the equations

$$y_i \phi_j(x') - y_j \phi_i(x') = 0.$$

Hence if there exists a value of i such that $\phi_i(x')$ is not zero, there is a unique point y' of U' corresponding to x'. If $\phi_i(x')$ is zero for all values of i, there corresponds to x' at least one point of U', but there may be more. Similarly, if y' is any point of U', there corresponds to y' at least one point x' of U, and if there exists a value of j such that $\psi_j(y')$ is not zero, x' is uniquely determined by y'. In general terms, the correspondence Γ between U and U' may be said to be one-to-one, with exceptions, the exceptional points of U, to which there correspond more than one point of U', lying in some sublocus of U; and similarly the exceptional points of U' lie in some sublocus of U'.

Conversely, suppose that between the two varieties U and U' there exists an irreducible correspondence such that to a generic point of each variety there corresponds a unique point of the other. Let ξ be a normalised generic point of U. Then if the equations of the correspondence are

$$f_i(x, y) = 0 \quad (i = 1, \ldots, r),$$

the equations $$f_i(\xi, y) = 0 \quad (i = 1, \ldots, r)$$

have a unique solution η, and hence if the coordinates of η are normalised,

$$\eta_i \in K(\xi_1, \ldots, \xi_m).$$

But [XI, § 3] η is a generic point of U', regarded as a variety over K, and we similarly find that, since ξ is the unique point of U which corresponds to η,

$$\xi_i \in K(\eta_1, \ldots, \eta_n).$$

Thus $K(\xi_1, \ldots, \xi_m) = K(\eta_1, \ldots, \eta_n)$, and therefore U and U' are birationally equivalent. The correspondence, moreover, establishes an isomorphism between the function fields of U and U'.

An irreducible correspondence between two varieties U and U' in which to a generic point of each there corresponds a unique point of the other is called a *birational correspondence*. The mapping of the points of U on the corresponding points of U' is called a *birational transformation* of U into U'. The results established show that if U and U' are birationally equivalent, and we select an isomorphism between the function fields of U and U' in which elements of K are self-corresponding, we can deduce a birational correspondence between U and U'; and conversely, if there exists a birational correspondence between two varieties U and U', then U and U' are birationally equivalent, and the birational correspondence defines an isomorphism between their function fields. It is clear that if the function field of a variety U possesses an automorphism which leaves the elements of the ground field invariant, and which is different from the identical automorphism, we can, corresponding to such an automorphism, define a birational transformation of U into itself.

Our object in this and the following sections is to lay the foundations of a general theory of birational correspondences, and to consider some of the more important birational transformations which arise in practice. We shall use the notation introduced above. U and U' will denote two irreducible varieties in S_m and S'_n which are birationally equivalent, and we suppose that an isomorphism between their function fields has been selected. We then identify corresponding elements of the function fields, and denote the common function field by Σ. We denote by Γ the birational correspondence between U and U' determined by the isomorphism, as above. But instead of using the variety W in $S_{m,n}$ to investigate the birational correspondence Γ, we find it more convenient to represent $S_{m,n}$ by a Segre variety in S_{mn+m+n}, and to consider the variety U^* which represents W on this Segre variety. In a suitable coordinate system $(z_{00}, \ldots, z_{mn})$ in S_{mn+m+n} we may suppose U^*

given by the generic point $(\zeta_{00}, \ldots, \zeta_{mn})$, where $\zeta_{ij} = \xi_i \eta_j$. The function field of U^* is

$$K(\zeta_{00}/\zeta_{hk}, \ldots, \zeta_{mn}/\zeta_{hk}) = K(\xi_0/\xi_h, \ldots, \xi_m/\xi_h, \eta_0/\eta_k, \ldots, \eta_n/\eta_k)$$
$$= K(\xi_0/\xi_h, \ldots, \xi_m/\xi_h) = K(\eta_0/\eta_k, \ldots, \eta_n/\eta_k) = \Sigma.$$

Hence U^* is birationally equivalent to U and to U', and since the function field of U^* is not merely isomorphic with that of U (or U'), but actually coincides with it, a birational correspondence is established between U^* and U (or U'). The variety U^* is called the *join* of U and U'.

If U is not in $x_h = 0$, $\xi_h \neq 0$, and if U' is not in $y_k = 0$, $\eta_k \neq 0$; hence $\zeta_{hk} \neq 0$, so that U^* is not in $z_{hk} = 0$. If we pass to affine spaces by taking $x_h = 0$, $y_k = 0$, and $z_{hk} = 0$ as the prime at infinity in S_m, S'_n, and S_{mn+m+n}, the integral domains of U, U', U^* are, respectively,

$$\mathfrak{J}_h = K[\xi_0/\xi_h, \ldots, \xi_m/\xi_h], \quad \mathfrak{J}'_k = K[\eta_0/\eta_k, \ldots, \eta_n/\eta_k],$$

$$\mathfrak{J}^*_{hk} = K[\zeta_{00}/\zeta_{hk}, \ldots, \zeta_{mn}/\zeta_{hk}] = K[\xi_0/\xi_h, \ldots, \xi_m/\xi_h, \eta_0/\eta_k, \ldots, \eta_n/\eta_k].$$

Hence $$\mathfrak{J}_h \subseteq \mathfrak{J}^*_{hk}, \quad \mathfrak{J}'_k \subseteq \mathfrak{J}^*_{hk},$$

and $\mathfrak{J}^*_{hk}$ is the smallest ring in Σ which contains both $\mathfrak{J}_h$ and $\mathfrak{J}'_k$; that is, it is the *join* of $\mathfrak{J}_h$ and $\mathfrak{J}'_k$.

Let V be an irreducible subvariety of U, and let $\bar{\xi}$ be a generic point of it. Let $\bar{\eta}$ be a point of U' corresponding to $\bar{\xi}$. Then if $\bar{\zeta}_{ij} = \bar{\xi}_i \bar{\eta}_j$, $\bar{\zeta}$ is a point of U^* which corresponds to $\bar{\xi}$ in the birational correspondence between U and U^*, and corresponds to $\bar{\eta}$ in the correspondence between U' and U^*. Let V' be the subvariety of U' whose generic point is $\bar{\eta}$, and let V^* be the subvariety of U^* whose generic point is $\bar{\zeta}$. If we apply the theory of Chapter XI, we see that, since $\bar{\zeta}_{ij} = \bar{\xi}_i \bar{\eta}_j$, V^* defines an irreducible correspondence Δ between S_m and S'_n in which the object and image-varieties are V and V'. Since $V^* \subseteq U^*$, any pair of points of V and V' which correspond in Δ correspond in Γ. Conversely, let V and V' be irreducible subvarieties of U and U' between which there exists an irreducible correspondence Δ such that every pair of points which correspond in Δ correspond in Γ. To a generic point of V there corresponds in Γ at least one (generic) point of V'. For if $\bar{\xi}$ is a generic point of V, its image in Δ is an irreducible variety on V', and a generic point $\bar{\eta}$ of this is a generic point, over K, of V', and since $\bar{\xi}$ and $\bar{\eta}$ correspond in Δ they correspond in Γ. Two irreducible varieties V and V' contained in U and U' respectively are said to correspond in the birational correspondence Γ if and only if there exists an irreducible correspondence Δ having V and V' as object-variety and image-variety such that every pair of points which correspond in Δ also correspond in Γ.

If V and V' are irreducible subvarieties of U and U' respectively which correspond in Γ, the correspondence Δ between V and V' is defined by a variety V^* contained in U^*. Suppose that V^* is not in $z_{hk} = 0$. Then the generic point $\bar{\zeta}$ of V^* is of the form $\bar{\zeta}_{ij} = \bar{\xi}_i \bar{\eta}_j$, where $\bar{\xi}$ is a generic point of V and $\bar{\eta}$ is a generic point of V', and $\bar{\xi}_h \neq 0$, $\bar{\eta}_k \neq 0$. We then pass from the projective spaces to affine spaces, and consider the prime ideals $\mathfrak{P}$, $\mathfrak{P}'$, $\mathfrak{P}^*$ of $\mathfrak{I}_h$, $\mathfrak{I}'_k$, $\mathfrak{I}^*_{hk}$ which define V, V', V^* respectively. Any element σ of $\mathfrak{I}_h$ is of the form

$$f(\xi_0/\xi_h, \ldots, \xi_m/\xi_h) = f(\zeta_{0k}/\zeta_{hk}, \ldots, \zeta_{mk}/\zeta_{hk}),$$

where $f(x_0, \ldots, x_m)$ is a homogeneous polynomial over K, and σ belongs to $\mathfrak{P}$ if and only if $f(x_0, \ldots, x_m)$ vanishes on V; that is, if and only if $f(z_{0k}, \ldots, z_{mk})$ vanishes on V^*. Hence σ belongs to $\mathfrak{P}$ if and only if it belongs to $\mathfrak{P}^*$. Since $\mathfrak{I}_h \subseteq \mathfrak{I}^*_{hk}$, it follows that $\mathfrak{P} = \mathfrak{I}_h \wedge \mathfrak{P}^*$. Similarly, $\mathfrak{P}' = \mathfrak{I}'_k \wedge \mathfrak{P}^*$. Hence if V is an irreducible variety contained in U and not lying in $x_h = 0$, given by the prime ideal $\mathfrak{P}$ in $\mathfrak{I}_h$, and V' is an irreducible variety contained in U' and not lying in $y_k = 0$, given by the prime ideal $\mathfrak{P}'$ in $\mathfrak{I}'_k$, and if V' corresponds to V in Γ, then there exists a prime ideal $\mathfrak{P}^*$ in $\mathfrak{I}^*_{hk}$ such that $\mathfrak{P} = \mathfrak{I}_h \wedge \mathfrak{P}^*$ and $\mathfrak{P}' = \mathfrak{I}'_k \wedge \mathfrak{P}^*$.

Conversely, let V be an irreducible variety in U given by the prime ideal $\mathfrak{P}$ of $\mathfrak{I}_h$, and let V' be an irreducible variety in U' given by the prime ideal $\mathfrak{P}'$ of $\mathfrak{I}'_k$, and suppose that there exists a prime ideal $\mathfrak{P}^*$ of $\mathfrak{I}^*_{hk}$ such that $\mathfrak{P} = \mathfrak{I}_h \wedge \mathfrak{P}^*$ and $\mathfrak{P}' = \mathfrak{I}'_k \wedge \mathfrak{P}^*$. The ideal $\mathfrak{P}^*$ defines an irreducible variety V^* contained in U^*, which defines a correspondence Δ between corresponding subvarieties of U and U'. A form $f(x_0, \ldots, x_m)$ vanishes on the object-variety of Δ if and only if $f(z_{0k}, \ldots, z_{mk})$ vanishes on V^*. Hence

$$\sigma = f(\xi_0/\xi_h, \ldots, \xi_m/\xi_h) = f(\zeta_{0k}/\zeta_{hk}, \ldots, \zeta_{mk}/\zeta_{hk})$$

belongs to the prime ideal of the object-variety in $\mathfrak{I}_h$ if and only if σ belongs to $\mathfrak{P}^*$. Hence the object-variety of Δ is given by the prime ideal $\mathfrak{I}_h \wedge \mathfrak{P}^* = \mathfrak{P}$, and it is therefore V. Similarly the image-variety is V'. Hence V and V' correspond in Γ.

Let V be an irreducible subvariety of U given by the prime ideal $\mathfrak{P}$ of $\mathfrak{I}_h$, and let V' be an irreducible subvariety of U' given by the prime ideal $\mathfrak{P}'$ of $\mathfrak{J}'_k$. Suppose, in the first place, that these varieties correspond in the birational correspondence Γ. Then, by what has been proved above, there exists a prime ideal $\mathfrak{P}^*$ of $\mathfrak{J}_{hk}$ such that $\mathfrak{P} = \mathfrak{J}_h \wedge \mathfrak{P}^*$, and $\mathfrak{P}' = \mathfrak{J}'_k \wedge \mathfrak{P}^*$. Let V^* be the irreducible subvariety of U^* defined by the prime ideal $\mathfrak{P}^*$ of $\mathfrak{J}_{hk}$. There exists a valuation $\mathfrak{B}$ of Σ whose centre on U^* is V^* [XVII, § 5, Ths. VII, IX]. Let $\mathfrak{R}$ be the valuation ring of $\mathfrak{B}$, and $\mathfrak{p}$ the ideal of non-units of $\mathfrak{R}$. Then $\mathfrak{J}^*_{hk} \subseteq \mathfrak{R}$, and $\mathfrak{P}^* = \mathfrak{J}^*_{hk} \wedge \mathfrak{p}$. Since $\mathfrak{J}_h \subseteq \mathfrak{J}_{hk}$, $\mathfrak{J}_h \subseteq \mathfrak{R}$; hence the centre of $\mathfrak{B}$ on U is not in $x_h = 0$. It is therefore given by the prime ideal $\mathfrak{J}_h \wedge \mathfrak{p}$ of $\mathfrak{J}_h$. But, since $\mathfrak{J}_h \subseteq \mathfrak{J}^*_{hk}$,

$$\mathfrak{J}_h \wedge \mathfrak{p} = \mathfrak{J}_h \wedge \mathfrak{J}^*_{hk} \wedge \mathfrak{p} = \mathfrak{J}_h \wedge \mathfrak{P}^* = \mathfrak{P},$$

and hence the centre of $\mathfrak{B}$ on U is V. Similarly, the centre of $\mathfrak{B}$ on U' is V'.

Next, let us suppose that there exists a valuation $\mathfrak{B}$ of Σ whose centre on U is V, and whose centre on U' is V'. Then $\mathfrak{J}_h \subseteq \mathfrak{R}$, and $\mathfrak{J}'_k \subseteq \mathfrak{R}$, where $\mathfrak{R}$ is the valuation ring of $\mathfrak{B}$. Since $\mathfrak{J}^*_{hk}$ is the smallest ring of Σ which contains $\mathfrak{J}_h$ and $\mathfrak{J}'_k$, it follows that $\mathfrak{J}^*_{hk} \subseteq \mathfrak{R}$. Let $\mathfrak{p}$ be the ideal of non-units of $\mathfrak{R}$, and let $\mathfrak{P}^* = \mathfrak{J}^*_{hk} \wedge \mathfrak{p}$. Since $\mathfrak{P} = \mathfrak{J}_h \wedge \mathfrak{p}$, we have

$$\mathfrak{P} = \mathfrak{J}_h \wedge \mathfrak{J}^*_{hk} \wedge \mathfrak{p} = \mathfrak{J}_h \wedge \mathfrak{P}^*,$$

and, similarly,

$$\mathfrak{P}' = \mathfrak{J}'_k \wedge \mathfrak{P}^*.$$

Hence there exists a prime ideal $\mathfrak{P}^*$ of $\mathfrak{J}^*_{hk}$ such that $\mathfrak{P} = \mathfrak{J}_h \wedge \mathfrak{P}^*$, and $\mathfrak{P}' = \mathfrak{J}'_k \wedge \mathfrak{P}^*$. Therefore, by a result proved above, V and V' correspond in Γ. Thus we have:

THEOREM I. *A necessary and sufficient condition that two irreducible subvarieties V and V' of U and U' respectively should correspond in the birational correspondence which exists between U and U' is that there should exist a valuation of the common function field Σ of U and U' whose centre on U is V, and whose centre on U' is V'.*

This property of corresponding varieties in a birational correspondence is sometimes taken as a definition of corresponding varieties. In what follows we shall use Theorem I to determine whether varieties correspond. An example of its usefulness is provided by

THEOREM II. *If U and U' are in birational correspondence, to any irreducible subvariety V of U there corresponds at least one variety V' on U'.*

A direct proof of this from our definition of corresponding varieties is implicit in what has been proved above, but we give a proof using valuation theory. There exists at least one valuation $\mathfrak{B}$ of the common function field Σ of U and U', whose centre on U is V. This valuation has centre at some irreducible variety V' on U'. By Theorem I, V' corresponds to V.

It is easy to verify that the special cases of one-to-one correspondences between algebraic varieties which we have met in previous chapters are birational correspondences, and that varieties which correspond in them satisfy Theorem I. For instance, a projective correspondence between n-spaces is a birational correspondence, and varieties which correspond in the projective correspondence correspond according to Theorem I. Again, if U is an irreducible variety in A_m, and U' is a derived normal variety, we saw in XVII, § 5, how to associate with each irreducible variety of U' a unique variety of U, and with each irreducible variety of U at least one, and at most a finite number of varieties of U'. If V and V' are two irreducible subvarieties of U and U', they are associated in this way if and only if there is a valuation of the common function field of U and U' which has V as its centre on U, and V' as its centre on U' [p. 216]. Hence V and V' are associated if and only if they correspond in the birational transformation between U and U', and the use of the term 'corresponding varieties' in XVII, § 5, is therefore justified.

We now consider some elementary properties of the birational correspondence Γ between U and U'.

THEOREM III. *If V and V' are corresponding irreducible subvarieties of U and U', and W is an irreducible subvariety of V, then there exists a variety W' on U' corresponding to W, which is contained in V'.*

Since V and V' correspond in Γ, there is a valuation $\mathfrak{B}$ of the common function field Σ of U and U' whose centre on U is V and whose centre on U' is V'. By Th. IX, p. 220, we can compound $\mathfrak{B}$ with a valuation of its residue field to obtain a valuation $\mathfrak{B}^*$ of Σ whose centre on U is W. Since $\mathfrak{B}^*$ is compounded from $\mathfrak{B}$, its centre W' on U' is contained in the centre V' of $\mathfrak{B}$ [XVII, § 5, Th. V]. W' is then a subvariety of U' which corresponds to W and fulfils the requirements of the theorem.

THEOREM IV. *If V is an irreducible subvariety of U, and if either*

(i) *there exists a subvariety V' of U' whose quotient ring $Q(V')$ is contained in the quotient ring $Q(V)$ of V, and every non-unit of $Q(V')$ is a non-unit of $Q(V)$;*

or (ii) *there exists an integer k such that $\eta_i/\eta_k \in Q(V)$ $(i = 0, \ldots, n)$, where $(\eta_0, \ldots, \eta_n)$ is a generic point of U';*
then there is a unique variety V' of U' which corresponds to V in Γ, and $\dim V' \leqslant \dim V$.

(i) To find the varieties on U' which correspond to V, we have to find the centres on U' of the valuations of the function field Σ whose centres on U are V. Let $\mathfrak{B}$ be any such valuation. Then [XVII, §5, Th. IV] $Q(V)$ is contained in the valuation ring $\mathfrak{R}$ of $\mathfrak{B}$, and every non-unit of $Q(V)$ is a non-unit of $\mathfrak{R}$. Then

$$Q(V') \subseteq Q(V) \subseteq \mathfrak{R}.$$

If α is any non-unit of $Q(V')$, then, by hypothesis, α is a non-unit of $Q(V)$, and hence it is a non-unit of $\mathfrak{R}$. By the converse of the result already quoted, it follows that V' is the centre of $\mathfrak{B}$ on U'. Hence V' is the *unique* variety on U' which corresponds to V.

(ii) Since $\eta_i/\eta_k \in Q(V)$ $(i = 0, \ldots, n)$,

$$\mathfrak{J}'_k = K[\eta_0/\eta_k, \ldots, \eta_n/\eta_k] \subseteq Q(V).$$

Let $\mathfrak{P}$ be the ideal of non-units of $Q(V)$. Then $\mathfrak{P}' = \mathfrak{J}'_k \wedge \mathfrak{P}$ is the prime ideal of an irreducible variety V' of U' not in $y_k = 0$. We consider the quotient ring $Q(V')$ of V'. Any element of $Q(V')$ can be written in the form α/β, where α and β are in $\mathfrak{J}'_k$, and β is not in $\mathfrak{P}'$. Since $\mathfrak{J}'_k \subseteq Q(V)$, α and β are in $Q(V)$, and since β is not in $\mathfrak{P}'$, it is not in $\mathfrak{P}$, and hence it is a unit of $Q(V)$. Hence $\alpha/\beta \in Q(V)$; that is, $Q(V') \subseteq Q(V)$. Moreover, α/β is a non-unit of $Q(V')$ if and only if α lies in $\mathfrak{P}'$, and therefore in $\mathfrak{P}$. Hence if α/β is a non-unit of $Q(V')$, $\alpha/\beta \in \mathfrak{P}$, and is therefore a non-unit of $Q(V)$. Hence the non-units of $Q(V')$ are non-units of $Q(V)$. By (i), V' is the unique variety of U' which corresponds to V in Γ.

The function field of V' is the remainder-class ring of $Q(V')$ modulo the ideal of non-units of this ring [XV, §5, Th. XI, Cor.]. Hence the dimension of V' is equal to the maximum number of elements of $Q(V')$ which are algebraically independent, over K, modulo the ideal of non-units of $Q(V')$. If the dimension of V' is e', there exist e' elements of $Q(V')$, say $\zeta_1, \ldots, \zeta_{e'}$, such that $f(\zeta_1, \ldots, \zeta_{e'})$ is a unit of $Q(V')$ for all non-zero polynomials $f(z_1, \ldots, z_{e'})$ over K. Then since $Q(V') \subseteq Q(V)$, $f(\zeta_1, \ldots, \zeta_{e'})$ is a unit of $Q(V)$ for all non-zero polynomials $f(z_1, \ldots, z_{e'})$. Hence $\zeta_1, \ldots, \zeta_{e'}$ are independent elements of $Q(V)$ over K, modulo $\mathfrak{P}$, and therefore the dimension of V is not less than e'. Hence $\dim V' \leqslant \dim V$.

Corollary. If V is an irreducible variety on U, and there exists an irreducible variety V' on U' such that $Q(V') = Q(V)$, then

$$dim\ V = dim\ V',$$

and V' is the unique variety corresponding to V, and V is the unique variety corresponding to V'.

THEOREM V. *If there exist integers h and k such that $\mathfrak{I}_h \subseteq \mathfrak{I}'_k$, then to each irreducible variety V', contained in U', which does not lie in $y_k = 0$, there corresponds a unique variety V of U; also $Q(V) \subseteq Q(V')$, and $dim\ V \leqslant dim\ V'$.*

Since V' is not in $y_k = 0$, it is defined by a prime ideal $\mathfrak{P}'$ of $\mathfrak{I}'_k$. Let $\mathfrak{B}$ be any valuation of Σ whose centre on U' is V', and let $\mathfrak{R}$ be the valuation ring of $\mathfrak{B}$, and $\mathfrak{p}$ the ideal of non-units of $\mathfrak{R}$. Then $\mathfrak{P}' = \mathfrak{I}'_k \wedge \mathfrak{p}$. Since $\mathfrak{I}_h \subseteq \mathfrak{I}'_k \subseteq \mathfrak{R}$, the centre V of $\mathfrak{B}$ on U is not in $x_h = 0$, and is given by the prime ideal

$$\mathfrak{P} = \mathfrak{I}_h \wedge \mathfrak{p} = \mathfrak{I}_h \wedge \mathfrak{I}'_k \wedge \mathfrak{p} = \mathfrak{I}_h \wedge \mathfrak{P}'.$$

The variety V is thus uniquely determined by V'.

Since $\mathfrak{I}_h \subseteq \mathfrak{I}'_k$ and $\mathfrak{P} = \mathfrak{I}_h \wedge \mathfrak{P}'$, it follows that $Q(V) \subseteq Q(V')$, and the argument used in Theorem IV shows that dim $V \leqslant$ dim V'.

An immediate application of this result is to the case in which U and U' are varieties in affine space, and the integral domain of U' is the integral closure of the integral domain of U. In this way we can obtain new proofs of results already obtained. A second application is to varieties in affine space which are integrally equivalent, that is, which have the same integral domain. Then it follows that to each irreducible subvariety of either there corresponds a unique subvariety of the other, of the same dimension. We also note that if $\mathfrak{I}_h \subseteq \mathfrak{I}'_k$, then $\mathfrak{I}^*_{hk}$, the join of $\mathfrak{I}_h$ and $\mathfrak{I}'_k$, coincides with $\mathfrak{I}'_k$. Then there is a one-to-one correspondence (without exception) between the irreducible subvarieties of U' not in $y_k = 0$ and the irreducible subvarieties of U^* not in $z_{hk} = 0$, corresponding varieties being of the same dimension. Again, since in any case we have $\mathfrak{I}_h \subseteq \mathfrak{I}^*_{hk}$, $\mathfrak{I}'_k \subseteq \mathfrak{I}^*_{hk}$, for all admissible values of h and k, we see that to every variety V^* of U^* there corresponds a unique variety V of U and a unique variety V' of U'. Further, V and V' correspond in Γ, and $Q(V) \subseteq Q(V^*)$, $Q(V') \subseteq Q(V^*)$, while dim $V \leqslant$ dim V^*, dim $V' \leqslant$ dim V^*.

We now seek to determine

(a) the totality of varieties on U' which correspond to a given irreducible variety V on U, and

(b) the totality of varieties on U' which correspond to varieties W contained in a given irreducible variety V on U.

We shall consider the two problems together. In view of what has been said above, it is sufficient to find the varieties of U^* which correspond to V or to subvarieties of V.

Suppose that V is not in $x_h = 0$. We first find the varieties of U^* not in $z_{hk} = 0$ which correspond to subvarieties of V. V is given by a prime ideal $\mathfrak{P}$ of $\mathfrak{J}_h$, and any irreducible variety W^* of U^* not in $z_{hk} = 0$ is given by a prime ideal $\mathfrak{P}^*$ of $\mathfrak{J}^*_{hk}$. The unique variety W of U which corresponds to W^* is given by the prime ideal $\mathfrak{J}_h \wedge \mathfrak{P}^*$ of $\mathfrak{J}_h$, and $W \subseteq V$ if and only if

$$\mathfrak{P} = 0 \ (\mathfrak{J}_h \wedge \mathfrak{P}^*).$$

If $\mathfrak{P} \subseteq \mathfrak{J}_h \wedge \mathfrak{P}^*$,

$$\mathfrak{J}^*_{hk} \cdot \mathfrak{P} \subseteq \mathfrak{J}^*_{hk} \cdot (\mathfrak{J}_h \wedge \mathfrak{P}^*) \subseteq \mathfrak{P}^* \quad \text{[XV, § 4, Th. VI]}.$$

Conversely, if $\mathfrak{J}^*_{hk} \cdot \mathfrak{P} \subseteq \mathfrak{P}^*$,

$$\mathfrak{P} \subseteq \mathfrak{J}_h \wedge (\mathfrak{J}^*_{hk} \cdot \mathfrak{P}) \subseteq \mathfrak{J}_h \wedge \mathfrak{P}^* \quad \text{[XV, § 4, Th. IV]}.$$

Hence W^* corresponds to a variety W contained in V if and only if

$$\mathfrak{J}^*_{hk} \cdot \mathfrak{P} = 0 \ (\mathfrak{P}^*).$$

Now the ideal $\mathfrak{J}^*_{hk} \cdot \mathfrak{P}$ defines an algebraic variety T^*_{hk} on U^* (which may be vacuous, if $\mathfrak{J}^*_{hk} \cdot \mathfrak{P} = \mathfrak{J}^*_{hk}$), and the condition that W^* should correspond to a subvariety of V is that W^* is contained in this variety.

By allowing h to take all values from 0 to m, excluding only those such that V lies in $x_h = 0$, and taking $k = 0, \ldots, n$, it follows that $\dot{\Sigma} T^*_{hk}$ is an algebraic variety of U^* such that any irreducible variety of U^* which corresponds to a subvariety of V is contained in $\dot{\Sigma} T^*_{hk}$, and any irreducible subvariety of this corresponds to a unique subvariety of V. $\dot{\Sigma} T^*_{hk}$ is called the *total transform of V in U^** and is usually denoted by $T^*\{V\}$. Passing from U^* to U' by replacing each component of $T^*\{V\}$ by the corresponding variety on U', we get the total transform $T'\{V\}$ of V on U'.

To find the varieties of U^* not in $z_{hk} = 0$ which correspond to V (and not subvarieties of it), we have to find the ideals $\mathfrak{P}^*$ in $\mathfrak{J}^*_{hk}$ such that $\mathfrak{J}_h \wedge \mathfrak{P}^* = \mathfrak{P}$. Consider any such ideal $\mathfrak{P}^*$. Again we have

$$\mathfrak{J}^*_{hk} \cdot \mathfrak{P} = 0 \ \ (\mathfrak{P}^*);$$

hence any variety of U^* not in $z_{hk} = 0$ which corresponds to V is contained in T^*_{hk}.

Let
$$\mathfrak{J}^*_{hk} \cdot \mathfrak{P} = [\mathfrak{Q}^*_1, \ldots, \mathfrak{Q}^*_l],$$

where $\mathfrak{Q}^*_i$ is a primary ideal whose radical is $\mathfrak{P}^*_i$. Then

$$\mathfrak{P} \subseteq \mathfrak{J}^*_{hk} \cdot \mathfrak{P} \subseteq \mathfrak{Q}^*_i \subseteq \mathfrak{P}^*_i$$

for each value of i, and therefore,

$$\mathfrak{P} = 0 \ \ (\mathfrak{J}_h \wedge \mathfrak{P}^*_i) \quad (i = 1, \ldots, l).$$

Since
$$\mathfrak{J}^*_{hk} \cdot \mathfrak{P} = 0 \ \ (\mathfrak{P}^*),$$
$$\mathfrak{P}^*_i = 0 \ \ (\mathfrak{P}^*)$$

for some value of i. For any such value of i, we have

$$\mathfrak{J}_h \wedge \mathfrak{P}^*_i \subseteq \mathfrak{J}_h \wedge \mathfrak{P}^* = \mathfrak{P}.$$

Hence, for this value of i,

$$\mathfrak{J}_h \wedge \mathfrak{P}^*_i = \mathfrak{P}.$$

Let $\mathfrak{Q}^*_i$ be such that

$$\mathfrak{J}_h \wedge \mathfrak{P}^*_i = \mathfrak{P} \quad (i = 1, \ldots, s) \qquad \text{and} \qquad \mathfrak{J}_h \wedge \mathfrak{P}^*_i \supset \mathfrak{P} \quad (i > s).$$

It follows from what has been proved above that if V^*_i is the variety of U^* defined by $\mathfrak{P}^*_i$, $V^*_1, \ldots, V^*_s$ correspond to V, and any variety of U^* not contained in $z_{hk} = 0$ which corresponds to V is contained in some V_i, for $i \leqslant s$. As when we discussed the total transform of V, we run through all the permissible values of h and k, and obtain an algebraic variety $T^*[V]$, which we call the *transform of* V *on* U^*, which has the properties

(i) every isolated component of $T^*[V]$ corresponds to V, and

(ii) every irreducible variety of U^* which corresponds to V is contained in $T^*[V]$.

Finally, replacing the components of $T^*[V]$ by the corresponding varieties of U' (and omitting any embedded components), we obtain the transform $T'[V]$ of V on U'. Summing up, we have

THEOREM VI. *If V is any irreducible variety of U, there exist two algebraic varieties $T'[V]$ and $T'\{V\}$ of U' corresponding to U with the properties:*

(i) *an irreducible variety W' of U' corresponds to a variety W contained in V if and only if W' is contained in $T'\{V\}$;*

(ii) *every isolated component of $T'[V]$ corresponds to V, and any irreducible variety of U' which corresponds to V is contained in $T'[V]$;*

(iii) $T'[V] \subseteq T'\{V\}$.

Corollary. By the continued application of (iii) *we see that $T'\{V\}$ is the variety formed by zero-dimensional varieties of U' which correspond to zero-dimensional varieties on V.*

The next group of theorems which we prove is established on the assumption that the quotient ring $Q(V)$ of V is integrally closed. The first is the converse of Theorem IV, which was proved without any restriction on $Q(V)$.

THEOREM VII. *Let V be an irreducible subvariety of U whose quotient ring is integrally closed. If V has a unique corresponding variety V' on U', then*

(i) *$Q(V') \subseteq Q(V)$, and the non-units of $Q(V')$ are non-units of $Q(V)$;*

(ii) *there exists an integer k such that $\eta_i/\eta_k \in Q(V)$ $(i = 0, \ldots, n)$, where $(\eta_0, \ldots, \eta_n)$ is a generic point of U' (chosen in Σ).*

(i) Let $\mathfrak{B}$ be any valuation of Σ whose centre on U is V, and let $\mathfrak{R}$ be its valuation ring. Since V' is the only variety of U' which corresponds to V, V' is the centre of $\mathfrak{B}$ on U'. Hence [XVII, § 5, p. 213] $Q(V') \subseteq \mathfrak{R}$, and every non-unit of $Q(V')$ is a non-unit of $\mathfrak{R}$. Since $Q(V)$ is integrally closed, it is the intersection of the valuation rings of all the valuations of Σ whose centre on U is V [XVII, § 5, Th. X]. Since $Q(V')$ is contained in all these valuation rings, $Q(V') \subseteq Q(V)$. Moreover, any non-unit α of $Q(V')$ is a non-unit of $\mathfrak{R}$. Hence α^{-1} is not in $\mathfrak{R}$, and therefore not in $Q(V)$. Therefore α is a non-unit of $Q(V)$. Hence every non-unit of $Q(V')$ is a non-unit of $Q(V)$.

(ii) By (i), $Q(V') \subseteq Q(V)$. Choose k so that $y_k \neq 0$ on V'. Then

$$\eta_i/\eta_k \in Q(V') \subseteq Q(V), \quad \text{for} \quad i = 0, \ldots, n.$$

Corollary. Let η be any element of $K(\xi_1, \ldots, \xi_n)$, and let x' be any point of U which is a generic point of a subvariety V of U whose quotient ring is integrally closed. If, corresponding to the specialisation $\xi \rightarrow x'$, there corresponds a unique finite proper specialisation of η, *then we can write* $\eta = \alpha(\xi)/\beta(\xi)$, *where* $\alpha(\xi)$ *and* $\beta(\xi)$ *are in* $K[\xi_1, \ldots, \xi_n]$, *and* $\beta(x') \neq 0$.

We take U' to be the variety whose generic point is $(\xi_1, \ldots, \xi_n, \eta)$. U' is birationally equivalent to U, and in the natural correspondence between U and U' there is a unique point $(x_1', \ldots, x_n', y')$ corresponding to x'. Hence there is a unique variety V' of U' corresponding to V, and $Q(V') \subseteq Q(V)$. Then $y' \in Q(V') \subseteq Q(V)$; hence $y' = \alpha(\xi)/\beta(\xi)$, where $\alpha(x)$ and $\beta(x)$ belong to $K[x_1, \ldots, x_n]$, and $\beta(x)$ does not vanish on V. Since x' is a generic point of V, this implies that $\beta(x') \neq 0$.

THEOREM VIII. *If* V *is an irreducible variety on* U, *and* $Q(V)$ *is integrally closed, and if there exists a prime section of* U' *which contains no variety of* U' *corresponding to* V, *then there is exactly one variety of* U' *which corresponds to* V.

Without loss of generality, we may assume that the coordinate system in S_n' is chosen so that there is no variety on U' corresponding to V in $y_k = 0$. Hence if $\mathfrak{B}$ is any valuation of Σ whose centre on U is V, its centre on U' is not in $y_k = 0$, and therefore the valuation ring of $\mathfrak{B}$ contains $\mathfrak{J}_k'$. Since $Q(V)$ is the intersection of the valuation rings of valuations of Σ whose centres on U are V, it follows that $\mathfrak{J}_k' \subseteq Q(V)$. Let $\mathfrak{P}$ be the ideal of non-units of $Q(V)$, and let $\mathfrak{P}' = \mathfrak{J}_k' \wedge \mathfrak{P}$. The ideal $\mathfrak{P}'$ of $\mathfrak{J}_k'$ defines an irreducible variety V' on U'. Let ζ be an element of $Q(V')$. Then we can write $\zeta = \alpha/\beta$, where α and β are in $\mathfrak{J}_k'$, and β is not in $\mathfrak{P}'$. Since $\mathfrak{J}_k' \subseteq Q(V)$ and $\mathfrak{P}' = \mathfrak{J}_k' \wedge \mathfrak{P}$, α and β are in $Q(V)$, and β is a unit of $Q(V)$. Therefore $\zeta = \alpha/\beta$ is in $Q(V)$. ζ is a non-unit of $Q(V')$ if and only if α is in $\mathfrak{P}$, and hence in $\mathfrak{P}'$. Hence ζ is a non-unit of $Q(V')$ if and only if it is a non-unit of $Q(V)$. Thus $Q(V') \subseteq Q(V)$, and every non-unit of $Q(V')$ is a non-unit of $Q(V)$. Hence [Th. IV] V' is the only variety on U' which corresponds to V.

Corollary. If $Q(V)$ *is integrally closed and there exist only a finite number of varieties on* U' *corresponding to* V, *there is only one variety on* U' *corresponding to* V.

This follows from the theorem by observing that if there are only a finite number of varieties corresponding to V, we can find a prime section of U' not containing any of them.

The results of Theorems VII and VIII show that more precise results are possible when we consider irreducible varieties V on U whose quotient rings are integrally closed. In particular, we should expect to carry our investigations further in the case in which both U and U' are projectively normal varieties, since in this case the quotient ring of every irreducible subvariety of U or U' is integrally closed. We shall deal with this case systematically in the next section. In the meantime, we introduce an important classification of the irreducible subvarieties of U (or U') whose quotient rings are integrally closed. Let V be an irreducible subvariety of U whose quotient ring is integrally closed. By Theorem II, there is at least one variety V' on U' corresponding to V in Γ.

V is said to be *regular* in Γ if there exists a variety V' of U' corresponding to it such that $Q(V') = Q(V)$. By Theorem IV, V' is the only variety of U' corresponding to V, and $\dim V' = \dim V$. The quotient ring of V', being equal to $Q(V)$, is also integrally closed, and V' is also regular in Γ.

V is said to be *irregular* in Γ if there is a variety V' of U' corresponding to it such that $Q(V') \subset Q(V)$. Then, if $\mathfrak{B}$ is any valuation of Σ whose centre on U is V and whose centre on U' is V', and $\mathfrak{R}$ is the valuation ring of $\mathfrak{B}$, any non-unit of $Q(V')$ is a non-unit of $\mathfrak{R}$, and hence a non-unit of $Q(V)$. Hence [Theorem IV] V' is the only variety of U' which corresponds to V, and $\dim V' \leqslant \dim V$. By Theorem VII it follows that if there is only one variety of U' corresponding to V, V is regular or irregular.

V is said to be *fundamental* in Γ if there is a variety V' corresponding to it on U' such that $Q(V') \nsubseteq Q(V)$. By Theorem VIII we see that there is a variety corresponding to V in every prime section of U'. For otherwise there would be only one variety of U' corresponding to V, and V would therefore be regular or irregular. Hence there is an infinite number of varieties on U' corresponding to a fundamental variety V. Let V'' be any one of these. If

$$Q(V'') \subseteq Q(V),$$

V would be regular or irregular, and V'' would be the only variety of U' corresponding to V. Hence $Q(V'') \nsubseteq Q(V)$.

We also observe that if V is irregular, and if the unique variety V' on U' corresponding to it has an integrally closed quotient ring, then V' is fundamental.

Finally, we define *regular birational correspondences*. Suppose that the correspondence Γ is such that to every irreducible variety V on U there corresponds a variety V' of U' such that

$$Q(V') = Q(V).$$

Then [Theorem IV] V' is the only variety on U' which corresponds to V, and V is the only variety on U which corresponds to V'. (This does not require that $Q(V)$ be integrally closed.) The birational correspondence Γ between U and U' is then said to be *regular*. If Γ is regular, to each irreducible subvariety of U (or U') there corresponds a unique subvariety of U' (or U). While our definition of a regular correspondence does not require that the varieties U and U' be normal, the idea is most useful when they are normal. For many purposes in algebraic geometry it is unnecessary to distinguish between projectively normal varieties which are in regular birational correspondence, on account of the one-to-one correspondence between their irreducible subvarieties.

2. Birational correspondences between normal varieties. Before proceeding to a discussion of birational correspondences between two projectively normal varieties, we consider the birational correspondence between any irreducible variety U in S_m and a derived normal variety $U^{(\delta)} = U'$ in S'_n, where δ is a character of homogeneity of U [Chap. XVI, § 6]. We use the notation of Chapter XVI, § 6.

We may assume that U is not in $x_0 = 0$, and let $(\xi_0, \xi_1, \ldots, \xi_m)$ $(\xi_0 = 1)$ be a normalised generic point of U. Let τ_0 be an indeterminate over the function field $\Sigma = K(\xi_1, \ldots, \xi_m)$ of U, and let $\tau_i = \tau_0 \xi_i$ $(i = 0, \ldots, m)$. Then U' has a generic point $(\omega_0, \ldots, \omega_n)$, where ω_i is in $\Sigma(\tau_0)$, and is integrally dependent on $K[\tau_0, \ldots, \tau_m]$; also, $K[\omega_0, \ldots, \omega_n]$ is integrally closed in its quotient field. We shall assume that $\omega_i = \tau_i^\delta$ $(i = 1, \ldots, m)$. Certainly $\tau_0^\delta, \ldots, \tau_m^\delta$ are linear in $\omega_0, \ldots, \omega_n$. In Chapter XVI we found it convenient to assume that $\omega_0, \ldots, \omega_n$ were linearly independent over K, but this restriction is no longer necessary, and we shall drop it. Any ω_i satisfies an equation

$$\omega_i^{\rho_i} + a_1(\tau)\, \omega_i^{\rho_i - 1} + \ldots + a_{\rho_i}(\tau) = 0,$$

where $a_i(\tau)$ is homogeneous of degree δi in $\tau_0, \ldots, \tau_m$. From our choice of coordinate system in S'_n it follows that the linear space

$$y_i = 0 \quad (i = 0, \ldots, m)$$

has no points in common with U'. Hence if V' is any irreducible variety contained in U', we can find a value of k not exceeding m such that V' does not lie in $y_k = 0$, and hence V' is given by a prime ideal in $\mathfrak{J}'_k$. Hence we have only to consider the rings $\mathfrak{J}'_k$ $(k = 0, \ldots, m)$. Since $K[\omega_0, \ldots, \omega_n]$ is integrally dependent on $K[\tau_0, \ldots, \tau_m]$,

$$\mathfrak{J}'_k = K[\omega_0/\omega_k, \ldots, \omega_n/\omega_k]$$

is integrally dependent on $K[\tau_0/\tau_k, \ldots, \tau_m/\tau_k] = \mathfrak{J}_k$, and since [XVI, § 6, Th. III] $\mathfrak{J}'_k$ is integrally closed, it follows that $\mathfrak{J}'_k$ is the integral closure of $\mathfrak{J}_k$.

Let V' be any irreducible subvariety of U', and choose k $(k \leqslant m)$ so that V' is not in $y_k = 0$. Then, by Theorem V of § 1, there is a unique variety V of U which corresponds to it. Hence each irreducible subvariety of U' corresponds to a unique subvariety of U. Now consider any irreducible subvariety V of U, and suppose that V is not in $x_k = 0$. If $\mathfrak{B}$ is any valuation of Σ whose centre on U is V, and $\mathfrak{R}$ is the valuation ring of $\mathfrak{B}$, $\mathfrak{J}_k \subseteq \mathfrak{R}$. Now $\mathfrak{J}'_k$ is the integral closure of $\mathfrak{J}_k$, and $\mathfrak{R}$ is integrally closed [XVII, § 2, Th. IV]; hence $\mathfrak{J}'_k \subseteq \mathfrak{R}$ and the centre of $\mathfrak{B}$ on U' is not in $y_k = 0$. The varieties on U' which correspond to V are therefore given by the prime ideals of $\mathfrak{J}'_k$ which contract to the prime ideal of V in $\mathfrak{J}_k$. Hence to each variety of U there corresponds a finite number of varieties on U'. Corresponding varieties on U and U' are of the same dimension [XVI, § 5, Th. III].

Now let U'' be another projectively normal variety in S''_n derived from U, corresponding to the character of homogeneity ϵ. We may suppose that the generic point $(\nu_0, \ldots, \nu_n)$ of U'' is such that $\nu_i = \tau_i^\epsilon$ $(i = 0, \ldots, m)$. Then $\mathfrak{J}''_k$ $(k \leqslant m)$ is the integral closure of $\mathfrak{J}_k$, and is therefore equal to $\mathfrak{J}'_k$. It follows that to each variety V' of U' not in $y_k = 0$ there corresponds a unique variety V'' of U'', and to each variety V'' of U'' not in $z_k = 0$ there corresponds a unique variety V' of U'. If V' is given by the prime ideal $\mathfrak{P}'$ of $\mathfrak{J}'_k$, V'' is given by the prime ideal $\mathfrak{P}'' = \mathfrak{P}'$ of $\mathfrak{J}''_k = \mathfrak{J}'_k$, and hence

$$Q(V') = Q(V'').$$

Since this holds for all values of k not exceeding m, it follows that the birational correspondence between U' and U'' is regular. This result also follows from XVI, § 6, Th. II, which implies that to each subvariety of U' (or U'') there corresponds a unique subvariety of U'' (or U'). Thus there are no fundamental varieties on U' or U'', and hence the correspondence is regular.

Now let U and U' be two varieties in birational correspondence, and let $\bar{U}$ and $\bar{U}'$ be projectively normal varieties derived from U and U', respectively. Let V and V' be corresponding subvarieties of U and U'. Then there exists a valuation $\mathfrak{B}$ of the common function field Σ of U, $\bar{U}$, $\bar{U}'$, U' which has centre V on U and V' on U'. Let the centre of $\mathfrak{B}$ on $\bar{U}$ be $\bar{V}$, and let the centre of $\mathfrak{B}$ on $\bar{U}'$ be $\bar{V}'$. Then V and $\bar{V}$ correspond in the birational correspondence between U and $\bar{U}$, $\bar{V}$ and $\bar{V}'$ correspond in the birational correspondence between $\bar{U}$ and $\bar{U}'$, and $\bar{V}'$ and V' correspond in the birational correspondence between $\bar{U}'$ and U'. Conversely, let V, $\bar{V}$, $\bar{V}'$, V' be irreducible subvarieties of U, $\bar{U}$, $\bar{U}'$, U', respectively, such that V and $\bar{V}$ correspond in the birational correspondence between U and $\bar{U}$, $\bar{V}$ and $\bar{V}'$ correspond in the birational correspondence between $\bar{U}$ and $\bar{U}'$, and $\bar{V}'$ and V' correspond in the birational correspondence between $\bar{U}'$ and U'. Let $\mathfrak{B}$ be the valuation of Σ whose centre on $\bar{U}$ is $\bar{V}$, and whose centre on $\bar{U}'$ is $\bar{V}'$. Since V is the only variety on U which corresponds to $\bar{V}$ on $\bar{U}$, V is the centre of $\mathfrak{B}$ on U, and since V' is the only variety on U' which corresponds to $\bar{V}'$ on $\bar{U}'$, V' is the centre of $\mathfrak{B}$ on U'. Since $\mathfrak{B}$ is a valuation of Σ whose centre on U is V and whose centre on U' is V', V and V' correspond in the birational correspondence between U and U'. Hence, given any variety V on U, the corresponding varieties on U' are obtained by selecting a variety $\bar{V}$ on $\bar{U}$ corresponding to V, then a variety $\bar{V}'$ on $\bar{U}'$ corresponding to $\bar{V}$, and finally the unique variety V' of U' which corresponds to $\bar{V}'$. The study of any birational correspondence can thus be split up into a study of the birational correspondence between a variety and a projectively normal variety derived from it, and the study of a birational correspondence between two normal varieties. The remainder of this section is devoted to the latter problem.

Let U and U' now be normal varieties, between which there exists a birational correspondence Γ. U^* will again denote their join. It should be borne in mind that U^* is not necessarily normal. Since U and U' are normal, the quotient ring of every irreducible subvariety V of U (or any subvariety V' of U') is integrally closed. Hence every irreducible subvariety of U (or of U') is either regular, irregular, or fundamental for Γ.

THEOREM I. *If V is regular or irregular for Γ, it is regular for the correspondence between U and U^*, and if V is fundamental for Γ it is fundamental for the correspondence between U and U^*.*

If V is regular or irregular, there is a unique variety V' of U' corresponding to it, and $Q(V')\subseteq Q(V)$ [p. 235]. Let $\mathfrak{B}$ be a valuation of Σ whose centre on U is V, and whose centre on U' is V', and let V^* be the centre of $\mathfrak{B}$ on U^*. We may suppose that V^* is not in $z_{hk} = 0$; then V is not in $x_h = 0$ and V' is not in $y_k = 0$. V^* is given by a prime ideal $\mathfrak{P}^*$ in $\mathfrak{J}^*_{hk}$. Also

$$\mathfrak{J}_h \subseteq Q(V), \quad \mathfrak{J}'_k \subseteq Q(V') \subseteq Q(V).$$

Since $\mathfrak{J}^*_{hk}$ is the join of $\mathfrak{J}_h$ and $\mathfrak{J}'_k$, any element α^* of $\mathfrak{J}^*_{hk}$ is of the form

$$\alpha^* = \alpha_1\alpha'_1 + \ldots + \alpha_r\alpha'_r,$$

where $\alpha_1, \ldots, \alpha_r$ belong to $\mathfrak{J}_h$, and hence to $Q(V)$, and $\alpha'_1, \ldots, \alpha'_r$ belong to $\mathfrak{J}'_k$, and hence to $Q(V)$. Hence α^* belongs to $Q(V)$, that is, $\mathfrak{J}^*_{hk} \subseteq Q(V)$. Now α^* belongs to $\mathfrak{P}^*$ if and only if its value in $\mathfrak{B}$ is positive, and hence, since V is the centre of $\mathfrak{B}$ on U, if and only if α^* is a non-unit of $Q(V)$. It follows that $\mathfrak{J}^*_{\mathfrak{P}} = Q(V^*) \subseteq Q(V)$. But, by Theorem V of § 1, $Q(V) \subseteq Q(V^*)$, since $\mathfrak{J}_h \subseteq \mathfrak{J}^*_{hk}$; hence $Q(V) = Q(V^*)$. Therefore V is regular for the correspondence between U and U^*.

On the other hand, if V is fundamental for Γ, it cannot be regular or irregular for the correspondence between U and U^*. Indeed, if it were, there would be a unique variety V^* of U^* corresponding to it. Since every variety V^* of U^* determines a unique variety V' of U', V' would be the only variety of U' corresponding to V. But, since V is fundamental for Γ, there is an infinite number of varieties on U' corresponding to V in Γ, and we have a contradiction. Hence V must be fundamental for the correspondence between U and U^*.

THEOREM II. *If there are no fundamental varieties on U', and if V is a fundamental variety of U of dimension ρ, then at least one component of $T[V]$ is of dimension greater than ρ, and no component is of dimension less than ρ.*

Since there are no fundamental varieties on U', every irreducible variety V' of U' is regular or irregular, and hence there is only one variety of U corresponding to it, and the dimension of this is not greater than dim V'. If V' is a variety of U' corresponding to V, we must therefore have dim $V' \geqslant \rho$. Since every component of $T[V]$ corresponds to V, every component of $T[V]$ must be of dimension ρ at least. If each component of $T[V]$ were of dimension ρ, these components would be the only varieties on U' corresponding to V. Since $T[V]$ has only a finite number of components, we deduce from the Corollary to Theorem VIII of § 1 that $T[V]$ has only one component, and there is only one variety on U' corresponding to V. Hence V is not fundamental, and the theorem follows.

Zariski has proved the stronger theorem that in the conditions of Theorem II, every component of $T[V]$ is of dimension greater than ρ [see Bibliographical Notes, p. 332].

THEOREM III. *If V is fundamental on U for Γ, every zero-dimensional irreducible subvariety of V is fundamental for Γ.*

Let $\bar{U}^*$ be a normal variety derived from U^*. We consider the birational correspondence between U and $\bar{U}^*$. Let $\bar{W}^*$ be any irreducible subvariety of $\bar{U}^*$. It corresponds to a unique subvariety W^* of U^*, which in turn determines a unique variety W on U. Hence $\bar{W}^*$ is not fundamental on $\bar{U}^*$, and so there are no fundamental varieties on $\bar{U}^*$. Let us now consider the fundamental varieties on U of the correspondence between U and $\bar{U}^*$. If W is fundamental for this correspondence, it follows from Theorem II that corresponding to it there exists a variety $\bar{W}^*$ on $\bar{U}^*$ such that $\dim \bar{W}^* > \dim W$. $\bar{W}^*$ determines a unique variety W^* of U^* which corresponds to U, in the correspondence between U and U^*, and $\dim W^* = \dim \bar{W}^*$. We know that $Q(W^*) \supseteq Q(W)$, and if these rings were equal we should have $\dim W = \dim W^*$. Hence $Q(W^*) \supset Q(W)$, and it follows that W is fundamental for the correspondence between U and U^*. Hence [Theorem I] W is fundamental for Γ. Conversely, if W is fundamental for Γ, it is fundamental for the correspondence between U and U^*, and hence if W^* is any variety of U^* corresponding to W, $Q(W^*) \nsubseteq Q(W)$. Let $\bar{W}^*$ be any variety on $\bar{U}^*$ corresponding to W^*. Then $Q(\bar{W}^*) \supseteq Q(W^*)$, and hence $Q(\bar{W}^*) \nsubseteq Q(W)$. Since $\bar{W}^*$ corresponds to W, it follows that W is fundamental for the correspondence between U and $\bar{U}^*$. Hence the fundamental varieties on U for Γ are just the fundamental varieties on U for the correspondence between U and $\bar{U}^*$. Hence it is sufficient to prove our theorem for the correspondence between U and $\bar{U}^*$; this is equivalent to considering a correspondence Γ for which there are no fundamental varieties on U'. We can then use Theorem II.

Since V is fundamental in Γ, and there are no fundamental varieties on U', it follows from Theorem II that there exists a variety V' on U' which corresponds to V, of dimension greater than $\dim V$. Let ρ be the dimension of V, ρ' that of V' $(\rho' > \rho)$.

Since V and V' correspond in Γ, there exists an irreducible correspondence Δ between V and V' such that any pair of points which correspond in Δ also correspond in Γ. The dimension σ of Δ is not less than ρ'. To a generic point of V there corresponds a variety of dimension $\sigma-\rho$, by the Principle of Counting Constants; hence to any point of V there corresponds a variety of dimension $\sigma-\rho$ at least. Hence to any irreducible variety W of dimension zero contained in V there corresponds a variety W' of dimension $\sigma-\rho>0$, at least, in Δ, and hence in Γ. Since dim $W'>$ dim W, W must be fundamental in Γ.

We next consider the problem of determining which irreducible subvarieties of U are fundamental in the birational correspondence Γ. For this, it is convenient to suppose that the coordinate system in S_n has been chosen so that U' does not lie in $y_k = 0$ for any value of k. Then, if

$$\frac{\eta_i}{\eta_k} = \frac{\phi_i(\xi_0, \ldots, \xi_m)}{\phi_k(\xi_0, \ldots, \xi_m)} \quad (i = 0, \ldots, n)$$

are the equations which arise when we identify the function fields of U and U', none of the homogeneous polynomials $\phi_i(x_0, \ldots, x_m)$ vanish on U. The intersection of U with the primal

$$\phi_i(x_0, \ldots, x_m) = 0$$

is then a pure $(d-1)$-dimensional variety, which we denote by A_i. The varieties $A_0, \ldots, A_n$ may have components in common. We write

$$A_i = B + C_i \quad (i = 0, \ldots, n),$$

where B is common to $A_0, \ldots, A_n$, and no component of any C_i is a component of every other C_j. Let F be the intersection of $C_0, \ldots, C_n$. F is necessarily of dimension less than $d-1$, and may be vacuous. We shall prove that a subvariety V of U is fundamental if and only if V is contained in F.

Let us find an ideal in $\mathfrak{J}_h$ which determines the part of F which does not lie in $x_h = 0$. Since U is supposed to be projectively normal, $\mathfrak{J}_h = K[\xi_0/\xi_h, \ldots, \xi_m/\xi_h]$ is integrally closed. We can therefore apply to it the multiplicative theory of ideals [Chap. XV, § 7]. Let

$$\mathfrak{i}_r = \mathfrak{J}_h \cdot (\phi_r(\xi_0/\xi_h, \ldots, \xi_m/\xi_h)) \quad (i = 0, \ldots, n).$$

The ideal $\mathfrak{i}_r$ determines the part of A_r which does not lie in $x_h = 0$. Let $\mathfrak{M}$ be the highest common quasi-factor of $\mathfrak{i}_0, \ldots, \mathfrak{i}_n$, and let

$$\mathfrak{i}_r \sim \mathfrak{M}\mathfrak{A}_r.$$

The variety on U determined by $\mathfrak{M}$ is pure $(d-1)$-dimensional (unless $\mathfrak{M} = 1$), since all the prime quasi-factors of $\mathfrak{i}_r$ are not quasi-equal to unity, and hence determine irreducible varieties of dimension $d-1$ [Chap. XVI, § 5]. Hence the variety determined by $\mathfrak{M}$ is contained in B. Conversely, if B' is any isolated irreducible component of B (hence a pure $(d-1)$-dimensional variety) not in $x_h = 0$, and if $\mathfrak{P}'$ is the prime ideal determined by B' in $\mathfrak{J}_h$, $\mathfrak{i}_r \geqslant \mathfrak{P}'$, and hence $\mathfrak{P}'$ is a quasi-factor of each $\mathfrak{i}_r$, and is therefore a quasi-factor of $\mathfrak{M}$. Hence $\mathfrak{M}$ defines that part of B which is not in $x_h = 0$, and $\mathfrak{A}_r$ defines that part of C_r not in $x_h = 0$. Hence the part of F not in $x_h = 0$ is given by the ideal

$$\mathfrak{A} = (\mathfrak{A}_0, \ldots, \mathfrak{A}_n).$$

Let V be any irreducible subvariety of U, not in $x_h = 0$, and let $\mathfrak{P}$ be the prime ideal of V in $\mathfrak{J}_h$. Let us suppose first that V is not contained in F. Then V is not contained in C_i for some value of i, say $i = k$; that is, $\mathfrak{A}_k \neq 0 \ (\mathfrak{P})$. Let α be an element of $\mathfrak{A}_k$ not in $\mathfrak{P}$. Then

$$\mathfrak{J}_h \cdot (\alpha) \geqslant \mathfrak{A}_k \quad \text{and} \quad \mathfrak{J}_h \cdot (\eta_i/\eta_k) \sim \mathfrak{A}_i \mathfrak{A}_k^{-1}.$$

Therefore

$$\mathfrak{J}_h \cdot (\alpha\eta_i/\eta_k) \geqslant \mathfrak{A}_i,$$

and hence $\mathfrak{J}_h \cdot (\alpha\eta_i/\eta_k)$ is an integral ideal. It contains $\alpha\eta_i/\eta_k$, and hence we must have $\alpha\eta_i/\eta_k = \beta_i$, where $\beta_i \in \mathfrak{J}_h$. Therefore, since $\alpha \neq 0 \ (\mathfrak{P})$,

$$\frac{\eta_i}{\eta_k} = \frac{\beta_i}{\alpha} \in Q(V) \quad (i = 0, \ldots, n),$$

and it follows from Theorem IV of § 1 that V has a unique transform on U', and therefore it is not fundamental. Thus if V is not contained in F it is not fundamental.

Suppose, on the other hand, that V is not a fundamental variety for Γ. Then [§ 1, Th. VII] there exists an integer k such that η_i/η_k belongs to $Q(V)$, that is, $\eta_i/\eta_k = \beta_{ik}/\gamma_{ik}$, where β_{ik}, γ_{ik} belong to $\mathfrak{J}_h$ and γ_{ik} is not in $\mathfrak{P}$, for $i = 0, \ldots, n$. Since $\mathfrak{P}$ is prime,

$$\alpha_k = \gamma_{0k} \cdots \gamma_{nk}$$

is not in $\mathfrak{P}$, and hence

$$\frac{\eta_i}{\eta_k} = \frac{\alpha_i}{\alpha_k} \quad (i = 0, \ldots, n),$$

where α_k is not in $\mathfrak{P}$, and $\alpha_i = \beta_{ik} \prod_{j \neq i} \gamma_{jk}$. Then

$$\mathfrak{J}_h \cdot (\alpha_i/\alpha_k) \sim \mathfrak{J}_h \cdot (\eta_i/\eta_k) \sim \mathfrak{A}_i \mathfrak{A}_k^{-1},$$

that is $$\mathfrak{A}_k \mathfrak{I}_h . (\alpha_i) \sim \mathfrak{A}_i \mathfrak{I}_h . (\alpha_k) \quad (i = 0, \ldots, n), \tag{1}$$

and $\mathfrak{I}_h . (\alpha_k) \neq 0 \ (\mathfrak{P})$.

If $$\mathfrak{I}_h . (\alpha_k) \sim \mathfrak{P}_1 \ldots \mathfrak{P}_s,$$

where $\mathfrak{P}_1, \ldots, \mathfrak{P}_s$ are minimal prime ideals, the variety on U defined by the principal ideal $\mathfrak{I}_h . (\alpha_k)$ is the sum of the irreducible varieties defined by the $\mathfrak{P}_j$; hence, since this variety does not contain V,

$$\mathfrak{P}_j \neq 0 \ (\mathfrak{P}) \quad (j = 1, \ldots, s). \tag{2}$$

Since $\mathfrak{A}_0, \ldots, \mathfrak{A}_n$ have no common quasi-factor, it follows from (1) that every prime quasi-factor of $\mathfrak{A}_k$ is a quasi-factor of $\mathfrak{I}_h . (\alpha_k)$. From (2), it follows that the pure $(d-1)$-dimensional variety C_k defined by $\mathfrak{A}_k$ does not contain V. Hence V is not contained in F. Thus we have

THEOREM IV. *If there are subvarieties of U which are fundamental for Γ, there is an algebraic locus F on U, of dimension $d-2$ at most, such that an irreducible subvariety V of U is fundamental for Γ if and only if V is contained in F.*

F is called *the fundamental locus of U.*

Let $T'\{F\}$ denote the sum of total transforms of all the isolated components of F. It is clearly an algebraic variety on U'. If V' is any irreducible subvariety of $T'\{F\}$, there corresponds to it a variety V on U which is contained in F. V must therefore be fundamental on U, and hence $Q(V') \nsubseteq Q(V)$. V' must therefore be irregular or fundamental on U'. On the other hand, if V' is a subvariety of U' which is irregular for Γ, there is only one variety V on U corresponding to it, and for this we have $Q(V) \subset Q(V')$. V must therefore be fundamental on U, and hence V' is contained in $T'\{F\}$. Hence we have

THEOREM V. *If F is the fundamental locus on U, and $T'\{F\}$ its total transform on U', any irregular subvariety of U' is contained in $T'\{F\}$, and any irreducible subvariety of $T'\{F\}$ is either irregular or fundamental for Γ.*

From these two theorems, we are able to find all the irregular and fundamental varieties on U and U'.

Immediate deductions from Theorem IV are

THEOREM VI. *Any birational correspondence between projectively normal curves is regular.*

THEOREM VII. *In a birational correspondence between projectively normal surfaces, there is at most only a finite number of fundamental varieties of dimension zero, and there are no fundamental curves on either surface.*

§§ 1 and 2 serve to give an account of the methods used to develop the theory of birational correspondences. In the next section we consider a special class of birational transformations, called *monoidal* transformations, whose importance lies in the fact that they are often used in the course of geometrical investigations when it is necessary to transform a given sublocus V on a variety U into a sublocus on U' whose dimension is greater than that of V. Monoidal transformations are, in fact, the working tools which one uses when applying birational transformations to given varieties.

3. Monoidal transformations. Let U be an irreducible variety of dimension d in S_n, $\xi = (\xi_0, \ldots, \xi_n)$ a generic point of it. Without loss of generality, we may assume that U does not lie in $x_h = 0$, for any value of h, and therefore $\xi_h \neq 0$. In particular, we may assume ξ normalised so that $\xi_0 = 1$. We also assume that U is normal.

Let C be any subvariety of U, of dimension $d-2$ at most. Let ν be any integer such that there exist at least two forms $f(x)$, $g(x)$ of degree ν which vanish on C and are such that $f(\xi)$, $g(\xi)$ are linearly independent over K, and let $\phi_0^{(\nu)}(x), \ldots, \phi_{r_\nu}^{(\nu)}(x)$ be a set of forms of degree ν which vanish on C and have the properties:

(*a*) $\phi_0^{(\nu)}(\xi), \ldots, \phi_{r_\nu}^{(\nu)}(\xi)$ are linearly independent over K;

(*b*) if $f(x)$ is any form of degree ν which vanishes on C, there exist elements $a_0, \ldots, a_{r_\nu}$ of K such that

$$f(\xi) = a_0 \phi_0^{(\nu)}(\xi) + \ldots + a_{r_\nu} \phi_{r_\nu}^{(\nu)}(\xi).$$

Consider the irreducible variety in S_{r_ν} whose generic point is $(\phi_0^{(\nu)}(\xi), \ldots, \phi_{r_\nu}^{(\nu)}(\xi))$. We denote it by $U^{(\nu)}$. It is clear that if $U^{(\nu)}$ is defined, so is $U^{(\nu')}$ for any value of ν' greater than ν. The varieties $U^{(\nu)}$, for different ν, may not be birationally equivalent, or, even if they are birationally equivalent, they may not be in regular correspondence. We can, however, see at once that the function field of $U^{(\nu)}$ is isomorphic with a subfield of the function field Σ of U. Indeed, since by hypothesis $\phi_0^{(\nu)}(\xi) \neq 0$, the function field of $U^{(\nu)}$ is isomorphic to

$$\Sigma^{(\nu)} = K(\phi_1^{(\nu)}(\xi)/\phi_0^{(\nu)}(\xi), \ldots, \phi_{r_\nu}^{(\nu)}(\xi)/\phi_0^{(\nu)}(\xi)) \subseteq K(\xi_1, \ldots, \xi_n).$$

We can, however, show that there exists an integer a such that if ν, ν' are greater than a, $U^{(\nu)}$ and $U^{(\nu')}$ are birationally equivalent to U, and are in regular correspondence with one another. We first show that if $U^{(\nu)}$ exists then $U^{(\nu+1)}$ is birationally equivalent to U. From this, it will follow immediately that $U^{(\nu')}$, for any $\nu' > \nu$, is birationally equivalent to U. Suppose that $U^{(\nu)}$ exists; then $\phi_0^{(\nu)}(\xi)$ exists and is not zero. Then by the property (b) of the forms $\phi_i^{(\nu+1)}(\xi)$,

$$\xi_i \phi_0^{(\nu)}(\xi) = \sum_j a_{ij} \phi_j^{(\nu+1)}(\xi) \quad (a_{ij} \in K),$$

and therefore

$$\xi_i = \frac{\xi_i}{\xi_0} = \frac{\Sigma a_{ij} \phi_j^{(\nu+1)}(\xi)}{\Sigma a_{0j} \phi_j^{(\nu+1)}(\xi)} \subseteq K\left(\frac{\phi_1^{(\nu+1)}(\xi)}{\phi_0^{(\nu+1)}(\xi)}, \ldots, \frac{\phi_{r_{\nu+1}}^{(\nu+1)}(\xi)}{\phi_0^{(\nu+1)}(\xi)}\right) = \Sigma^{(\nu+1)}.$$

Hence
$$\Sigma = K(\xi_1, \ldots, \xi_n) \subseteq \Sigma^{(\nu+1)};$$

and, since we have already proved that $\Sigma^{(\nu+1)} \subseteq \Sigma$, we conclude that $\Sigma = \Sigma^{(\nu+1)}$, that is, that $U^{(\nu+1)}$ is birationally equivalent to U.

We next show that if ν is sufficiently large, $U^{(\nu+1)}$ is the join of U and $U^{(\nu)}$. Let

$$f_i(x) = 0 \quad (i = 1, 2, \ldots, r)$$

be a basis for the equations of C, and let $f_i(x)$ be of degree a_i, and $a = \max[a_1, \ldots, a_r]$. Since C is of dimension $d-2$ at most, at least two of the forms $f_i(\xi)$ are different from zero; hence $U^{(a)}$ is defined. Let $X_j^{(\rho)}(x)$ $(j = 1, 2, \ldots)$ be the various power-products of the x_h of degree ρ. Then, if $\nu \geqslant a$, $X_j^{(\nu-a_i)}(x) f_i(x)$ vanishes on C and is of degree ν; hence

$$X_j^{(\nu-a_i)}(\xi) f_i(\xi) = \Sigma b_{ijk} \phi_k^{(\nu)}(\xi),$$

where $b_{ijk} \in K$. Now

$$\phi_i^{(\nu+1)}(x) = \sum_{j,k} \alpha_{ijk} X_k^{(\nu+1-a_j)} f_j(x) \quad (\alpha_{ijk} \in K),$$

and hence $\quad \phi_i^{(\nu+1)}(\xi) = \Sigma \beta_{ijk} \xi_j \phi_k^{(\nu)}(\xi) \qquad (\beta_{ijk} \in K).$

Also, since $x_j \phi_k^{(\nu)}(x)$ vanishes on C and is of degree $\nu+1$,

$$\xi_j \phi_k^{(\nu)}(\xi) = \Sigma \gamma_{ijk} \phi_i^{(\nu+1)}(\xi) \qquad (\gamma_{ijk} \in K).$$

But the join of U and $U^{(\nu)}$ has the generic point $(\ldots, \xi_i \phi_j^{(\nu)}(\xi), \ldots)$; hence it follows that $U^{(\nu+1)}$ is projectively equivalent to the join of U and $U^{(\nu)}$. By § 1, Th. V, it follows that to each irreducible variety $V^{(\nu+1)}$ on $U^{(\nu+1)}$ there corresponds a unique variety V of U and a unique variety $V^{(\nu)}$ of $U^{(\nu)}$, and

$$Q(V) \subseteq Q(V^{(\nu+1)}), \quad Q(V^{(\nu)}) \subseteq Q(V^{(\nu+1)}).$$

This holds whenever $\nu \geqslant a$. If $\nu > a$, we conclude from this that

$$Q(V) \subseteq Q(V^{(\nu)}).$$

In S_n, S_{r_ν}, $S_{r_{\nu+1}}$ we can choose primes at infinity not containing V, $V^{(\nu)}$, $V^{(\nu+1)}$, and such that, if $\mathfrak{J}$, $\mathfrak{J}^{(\nu)}$, $\mathfrak{J}^{(\nu+1)}$ are the integral domains of U, $U^{(\nu)}$, $U^{(\nu+1)}$ in the derived affine spaces, $\mathfrak{J} \subseteq \mathfrak{J}^{(\nu)} \subseteq \mathfrak{J}^{(\nu+1)}$, and $\mathfrak{J}^{(\nu+1)}$ is the join of $\mathfrak{J}$ and $\mathfrak{J}^{(\nu)}$; that is, $\mathfrak{J}^{(\nu+1)} = \mathfrak{J}^{(\nu)}$. If $\mathfrak{P}^{(\nu+1)}$ is the prime ideal of $V^{(\nu+1)}$ in $\mathfrak{J}^{(\nu+1)}$, the prime ideals $\mathfrak{P}$, $\mathfrak{P}^{(\nu)}$ of V, $V^{(\nu)}$ are

$$\mathfrak{P} = \mathfrak{J}_\wedge \mathfrak{P}^{(\nu+1)}, \quad \mathfrak{P}^{(\nu)} = \mathfrak{J}^{(\nu)}{}_\wedge \mathfrak{P}^{(\nu+1)} = \mathfrak{P}^{(\nu+1)}.$$

Hence $$Q(V^{(\nu)}) = \mathfrak{J}^{(\nu)}_{\mathfrak{P}^{(\nu)}} = \mathfrak{J}^{(\nu+1)}_{\mathfrak{P}^{(\nu+1)}} = Q(V^{(\nu+1)}).$$

Hence to each irreducible variety $V^{(\nu+1)}$ of $U^{(\nu+1)}$ there corresponds a unique variety $V^{(\nu)}$ on $U^{(\nu)}$, and since $Q(V^{(\nu)}) = Q(V^{(\nu+1)})$, $V^{(\nu+1)}$ is the only variety of $U^{(\nu+1)}$ corresponding to $V^{(\nu)}$. Hence the correspondence between $U^{(\nu+1)}$ and $U^{(\nu)}$ is regular.

Thus $U^{(a+1)}, U^{(a+2)}, \ldots$ are in regular correspondence. Any variety $U^{(\nu)}$, where $\nu > a$, is called a *monoidal transform of U with basis C*. We thus have

THEOREM I. *The monoidal transform of a variety U with basis C is uniquely determined to within a regular transformation, and is birationally equivalent to U. To every irreducible variety V' of the monoidal transform U' of U there corresponds a unique variety V of U, and the relation $Q(V) \subseteq Q(V')$ holds.*

We now drop the notation $U^{(\nu)}$, and denote the monoidal transform of U by U'. When we have to refer to ν, the degree of the forms used to define U', we shall call it the *degree* of the monoidal transformation.

It should be pointed out that the monoidal transform U' need not be normal. This results in some minor reservations in using the results of §§ 1 and 2, but no serious difficulties will arise. In practice, it is often convenient to follow the process of constructing a monoidal transform by deriving a normal variety from it, but we shall not follow this practice here.

As an example of a monoidal transformation, let U be a plane, with coordinates (x_0, x_1, x_2), and let C be the variety consisting of the points $(1, 0, 0)$, $(0, 1, 0)$, $(0, 0, 1)$. A basis for the equations of C is

$$x_1 x_2 = 0, \quad x_2 x_0 = 0, \quad x_0 x_1 = 0.$$

Here $a = 2$. $U^{(a)}$ has the generic point $(\xi_1\xi_2, \xi_2\xi_0, \xi_0\xi_1)$, and is a plane. It is birationally equivalent to U, but it is not a monoidal transform, since to the points $(1,0,0)$, $(0,1,0)$, $(0,0,1)$ of $U^{(a)}$ there correspond all the points of the lines $x_0 = 0$, $x_1 = 0$, $x_2 = 0$, respectively. But $U^{(3)} = U'$ is a monoidal transform. It can be represented in S_6 as the variety which has the generic point $(\xi_1^2\xi_2, \xi_1\xi_2^2, \xi_0^2\xi_1, \xi_0\xi_1^2, \xi_2^2\xi_0, \xi_2\xi_0^2, \xi_0\xi_1\xi_2)$ and can be shown to be a surface of order six. This surface contains three lines:

$$(l_1) \quad y_0 = y_1 = y_3 = y_4 = y_6 = 0;$$

$$(l_2) \quad y_1 = y_2 = y_4 = y_5 = y_6 = 0;$$

$$(l_3) \quad y_0 = y_2 = y_3 = y_5 = y_6 = 0.$$

To any point of l_1 there corresponds uniquely the point $(1,0,0)$ of U; similarly, to the points of l_2, l_3 correspond, respectively, $(0,1,0)$ and $(0,0,1)$ of U. Since U' is a monoidal transform of U, every point of U' determines a unique point of U, and the reader may verify that to every point P of U other than $(1,0,0)$, $(0,1,0)$, $(0,0,1)$ there corresponds a unique point P' of U', and $Q(P) = Q(P')$.

Returning now to the general theory of monoidal transformations we prove

THEOREM II. *In a monoidal transformation of U, with basis C, every variety of U is either regular or fundamental, and a variety V is fundamental if and only if $V \subseteq C$.*

The fact that every variety on U is either regular or fundamental follows from Theorem I, which also tells us there are no fundamental varieties on the monoidal transforms. Since the monoidal transforms of U are in regular correspondence, we have only to consider monoidal transforms of U of sufficiently high degree ν. U' is defined by the equations

$$y_i = \phi_i^{(\nu)}(x) \quad (i = 0, \ldots, r_\nu).$$

Hence, by the methods used to establish Theorem IV of § 2, which can be applied because U is normal, the fundamental locus is obtained by removing the components of dimension $d-1$ from the locus on U given by the equations

$$\phi_i^{(\nu)}(x) = 0 \quad (i = 0, \ldots, r_\nu),$$

and hence the fundamental locus is C.

The aggregate of varieties on the monoidal transform U' which correspond to varieties contained in a variety V of U is called the total transform $T\{V\}$ of V. We now prove

THEOREM III. *The total transform $T\{C\}$ of the basis C of a monoidal transformation of U is a pure $(d-1)$-dimensional variety.*

Let U' be a monoidal transform (with basis C) of U, of degree ν. The join U^* of U and U' is projectively equivalent to the monoidal transform of U of degree $\nu+1$, and is therefore in regular correspondence with U'. It is therefore sufficient to prove that the total transform $T^*\{C\}$ on U^* of C is pure $(d-1)$-dimensional. It should be made clear that the device of working on U^* rather than U' is solely to enable us to use a simplified notation.

Without loss of generality, we may assume that x_0 is not zero on any variety V of U which corresponds to any isolated component V' of $T\{C\}$, and that the coordinate system $(y_0, \ldots, y_{r_\nu})$ in S_{r_ν} is chosen so that y_0 is not zero on any component of $T\{C\}$. Then if we denote the coordinate system of the space in which U^* lies by $(z_{00}, \ldots, z_{nr_\nu})$, the generic point of U^* is $(\zeta_{00}, \ldots, \zeta_{nr_\nu})$, where

$$\zeta_{ij} = \xi_i \phi_j^{(\nu)}(\xi),$$

and z_{00} is not zero on any component of $T^*\{C\}$. As usual, we suppose that the coordinates of ξ are normalised so that $\xi_0 = 1$.

Let C^* be the variety on U^* having the equations of U^* and the equations

$$\phi_i^{(\nu)}(z_{0j}, \ldots, z_{nj}) = 0 \quad (i, j = 0, \ldots, r_\nu).$$

We shall first prove that $C^* = T^*\{C\}$.

Suppose that V^* is any isolated component of $T^*\{C\}$, and let V, V' be the corresponding varieties of U and U'. Let $\mathfrak{B}$ be any valuation whose centre on U^* is V^*; then its centre on U is V, and its centre on U' is V'. Since V is not in $x_0 = 0$ and V' is not in $y_0 = 0$, we have

$$v(\xi_i) \geqslant 0, \quad v[\phi_j^{(\nu)}(\xi)/\phi_0^{(\nu)}(\xi)] \geqslant 0,$$

for all admissible values of i and j; and since $V \subseteq C$ (from the definition of $T\{C\}$),

$$v[\phi_j^{(\nu)}(\xi)] > 0 \quad (j = 0, \ldots, r_\nu).$$

We then have

$$\begin{aligned} v[\phi_i^{(\nu)}(\zeta_{0j}, \ldots, \zeta_{nj})/\zeta_{00}^{\nu}] &= v[\phi_i^{(\nu)}(\xi)\{\phi_j^{(\nu)}(\xi)/\phi_0^{(\nu)}(\xi)\}^{\nu}] \\ &= v[\phi_i^{(\nu)}(\xi)] + \nu v[\phi_j^{(\nu)}(\xi)/\phi_0^{(\nu)}(\xi)] \\ &> 0. \end{aligned}$$

Hence $\phi_i^{(\nu)}(z_{0j}, \ldots, z_{nj})$ vanishes on V^*, and therefore $V^* \subseteq C^*$. Therefore $T^*\{C\} \subseteq C^*$.

Next, let V^* be any irreducible variety contained in C^*, and let $\mathfrak{V}$ be a valuation whose centre on U^* is V^*. If V^* is not in $z_{hk} = 0$,

$$v[\phi_i^{(\nu)}(\zeta_{0k}, \ldots, \zeta_{nk})/\zeta_{hk}^\nu] > 0.$$

But
$$\phi_i^{(\nu)}(\zeta_{0k}, \ldots, \zeta_{nk})/\zeta_{hk}^\nu = \phi_i^{(\nu)}(\xi)/\xi_h^\nu;$$

hence, since $v[\phi_i^{(\nu)}(\xi)/\xi_h^\nu] > 0$, the centre V of $\mathfrak{V}$ on U is not in $x_h = 0$ and satisfies the equation

$$\phi_i^{(\nu)}(x) = 0 \quad (i = 0, \ldots, r_\nu).$$

It follows that $V \subseteq C$, and hence $V^* \subseteq T^*\{C\}$. Therefore $C^* \subseteq T^*\{C\}$, and since we have already shown that $T^*\{C\} \subseteq C^*$, we have

$$C^* = T^*\{C\}.$$

We have now to show that C^* is pure $(d-1)$-dimensional. Let

$$\Phi_i(z) = \phi_i^{(\nu)}(z_{0i}, \ldots, z_{ni}) \quad (i = 0, \ldots, r_\nu).$$

Then
$$\begin{aligned}[\phi_i^{(\nu)}(\zeta_{0j}, \ldots, \zeta_{nj})]^{\nu+1} &= [\phi_i^{(\nu)}(\xi)]^{\nu+1}[\phi_j^{(\nu)}(\xi)]^{\nu(\nu+1)} \\ &= \Phi_i(\zeta)[\Phi_j(\zeta)]^\nu.\end{aligned}$$

Hence
$$[\phi_i^{(\nu)}(z_{0j}, \ldots, z_{nj})]^{\nu+1} - \Phi_i(z)[\Phi_j(z)]^\nu$$

vanishes on U^*, and so the locus C^*, which is the intersection of U^* with the locus whose equations are

$$\phi_i^{(\nu)}(z_{0j}, \ldots, z_{nj}) = 0 \quad (i,j = 0, \ldots, r_\nu),$$

is the intersection of U^* with the locus defined by the equation

$$\Phi_i(z_{0i}, \ldots, z_{ni}) = 0 \quad (i = 0, \ldots, r_\nu).$$

By construction, no component of $C^* = T^*\{C\}$ lies in $z_{00} = 0$. Hence C^* is the variety defined by the ideal $\mathfrak{J}^*.[\Phi_0(\zeta), \ldots, \Phi_{r_\nu}(\zeta)]$ in the integral domain $\mathfrak{J}^* = K[\zeta_{00}/\zeta_{00}, \ldots, \zeta_{nr_\nu}/\zeta_{00}]$. But

$$\frac{\Phi_i(\zeta)}{\Phi_0(\zeta)} = \left[\frac{\phi_i^{(\nu)}(\xi)}{\phi_0^{(\nu)}(\xi)}\right]^{\nu+1} = \left(\frac{\zeta_{0i}}{\zeta_{00}}\right)^{\nu+1} \in \mathfrak{J}^*.$$

Hence C^* is defined by the principal ideal $\mathfrak{J}^*.[\Phi_0(\zeta)]$. But the variety on an irreducible variety of dimension d in affine space defined by a principal ideal is a pure $(d-1)$-dimensional variety. Hence $C^* = T^*\{C\}$ is a pure $(d-1)$-dimensional variety.

The next theorem is of use in determining the transform $T[V]$ of an irreducible subvariety V of U, that is, the aggregate of varieties on U' which correspond to V. By § 1, Theorem VI, this is algebraic. What we shall prove is that if U_1 is a monoidal transform of U with basis C_1, where the components of C_1 which contain V are the same as the components of C which contain V, then there is a one-to-one correspondence between the varieties on U' and U_1 which correspond to V, and if V' and V_1 are two such corresponding varieties, $Q(V') = Q(V_1)$. By means of this theorem, if we wish to find the varieties of U' which correspond to V, we can omit the components of C which do not contain V, and consider the monoidal transform of U with this reduced basis. As this new monoidal transformation may be simpler than the original one, the problem of determining $T[V]$ is correspondingly simplified. A short statement of the theorem to be proved is

THEOREM IV. *The varieties in U' which correspond to an irreducible variety V on U depend only on the components of C which contain V.*

Let V be any irreducible variety on U, and let C_1 be the sum of the components of C which contain V. Let ν be an integer so large that monoidal transforms U' and U_1 of U of degree ν for the bases C and C_1 exist. Let $\phi_i^{(\nu)}(x)$ $(i = 0, ..., n_1)$ be a basis for the forms of degree ν, modulo the equations of U, which vanish on C_1, such that $\phi_i^{(\nu)}(x)$ $(i = 0, ..., n' < n_1)$ is a basis for the forms of degree ν, modulo the equations of U, which vanish on C; and let $g(x)$ be any form which vanishes on the components of C which do not pass through V, but which does not vanish on V.

Let V' be any irreducible variety on U' which corresponds to V, and let $\mathfrak{B}$ be any valuation whose centre on U' is V', and whose centre on U is V. Without loss of generality, we may assume that $x_0 \neq 0$ on V, and $y_0 \neq 0$ on V'. Then

$$v(\xi_i) \geqslant 0, \quad v[\phi_j^{(\nu)}(\xi)/\phi_0^{(\nu)}(\xi)] \geqslant 0 \quad (j = 0, ..., n').$$

Now $g(x) \phi_{n'+i}^{(\nu)}(x)$ vanishes on C. Hence we have

$$g(\xi) \phi_{n'+i}^{(\nu)}(\xi) = \sum_{j=0}^{n'} A_{ij}(\xi) \phi_j^{(\nu)}(\xi) \quad (i = 1, ..., n_1 - n'),$$

where $A_{ij}(\xi) \in \mathfrak{J} = K[\xi_1, ..., \xi_n]$. Hence

$$\frac{\phi_{n'+i}^{(\nu)}(\xi)}{\phi_0^{(\nu)}(\xi)} = \sum_{j=0}^{n'} \frac{A_{ij}(\xi)}{g(\xi)} \frac{\phi_j^{(\nu)}(\xi)}{\phi_0^{(\nu)}(\xi)}.$$

Since $g(x)$ does not vanish on V, $A_{ij}(\xi)/g(\xi)$ is contained in $Q(V)$, which is contained in $Q(V')$ [Th. I], and

$$\phi_j^{(\nu)}(\xi)/\phi_0^{(\nu)}(\xi) = \eta_j/\eta_0 \in Q(V');$$

hence
$$\frac{\phi_{n'+i}^{(\nu)}(\xi)}{\phi_0^{(\nu)}(\xi)} \in Q(V').$$

Therefore

$$\mathfrak{J}' = K\left[\frac{\phi_1^{(\nu)}(\xi)}{\phi_0^{(\nu)}(\xi)}, \ldots, \frac{\phi_{n'}^{(\nu)}(\xi)}{\phi_0^{(\nu)}(\xi)}\right] \subseteq \mathfrak{J}_1 = K\left[\frac{\phi_1^{(\nu)}(\xi)}{\phi_0^{(\nu)}(\xi)}, \ldots, \frac{\phi_{n_1}^{(\nu)}(\xi)}{\phi_0^{(\nu)}(\xi)}\right] \subseteq Q(V').$$

Let $\mathfrak{R}$ be the valuation ring of $\mathfrak{B}$, $\mathfrak{p}$ the ideal of non-units of $\mathfrak{R}$. Then the ideal of non-units of $Q(V')$ is $\mathfrak{P} = Q(V')_\wedge \mathfrak{p}$, and the centre V_1 of $\mathfrak{B}$ on U_1 is defined by the prime ideal $\mathfrak{P}_1 = \mathfrak{J}_{1\wedge}\mathfrak{p} = \mathfrak{J}_{1\wedge}\mathfrak{P}$ of $\mathfrak{J}_1$. $\mathfrak{P}$ is determined by V' only, hence $\mathfrak{J}_{1\wedge}\mathfrak{P}$ is determined by V' only, and therefore V_1 is determined by V' uniquely. V' is determined by the ideal $\mathfrak{P}' = \mathfrak{J}'_\wedge \mathfrak{p} = \mathfrak{J}'_\wedge \mathfrak{P}_1$ of $\mathfrak{J}'$. Since

$$\mathfrak{J}' \subseteq \mathfrak{J}_1 \subseteq Q(V'),$$

and
$$\mathfrak{P}' = \mathfrak{J}'_\wedge \mathfrak{P}_1 = \mathfrak{J}'_\wedge \mathfrak{P},$$

we have
$$\mathfrak{J}'_{\mathfrak{P}'} \subseteq \mathfrak{J}_{1\mathfrak{P}_1} \subseteq Q(V')_{\mathfrak{P}}.$$

But
$$\mathfrak{J}'_{\mathfrak{P}'} = Q(V') = Q(V')_{\mathfrak{P}},$$

hence
$$Q(V') = \mathfrak{J}_{1\mathfrak{P}_1} = Q(V_1).$$

On the other hand, let V_1 be any irreducible variety on U_1 which corresponds to V. Let $\mathfrak{B}$ be any valuation whose centres on U and U_1 are, respectively, V and V_1. Without loss of generality, we may assume that the $\phi_i^{(\nu)}(x)$ $(i = 0, \ldots, n')$ are arranged so that

$$v[\phi_i^{(\nu)}(\xi)] \geqslant v[\phi_0^{(\nu)}(\xi)] \quad (i \leqslant n').$$

Then, as above,

$$v[\phi_i^{(\nu)}(\xi)] \geqslant v[\phi_0^{(\nu)}(\xi)] \quad (i = 0, \ldots, n_1),$$

and hence $\mathfrak{J}_1$ is in the valuation ring of $\mathfrak{B}$. Hence V_1, the centre of $\mathfrak{B}$ on U_1, is not in $y_0^{(1)} = 0$, and it is given by an ideal $\mathfrak{P}_1$ in $\mathfrak{J}_1$. The centre V' of $\mathfrak{B}$ on U' is defined by the prime ideal $\mathfrak{J}'_\wedge \mathfrak{P}_1$, and just as above we prove that $Q(V') = Q(V_1)$. Thus there is a one-to-one correspondence between the varieties on U' and U_1 which correspond to V, and if V' and V_1 are related in this correspondence, $Q(V') = Q(V_1)$.

Corollary. The method used to establish Theorem IV enables us to prove a lemma which we shall use later. Let ν be any integer such that a monoidal transform U' of degree ν of U, with basis C, exists, and let $\psi_i(x)$ $(i = 0, \ldots, m)$ be $m+1$ forms of degree ν such that the prime ideals of the components of C in $\mathfrak{J} = K[\xi_1, \ldots, \xi_n]$ are isolated components of the ideal $\mathfrak{i} = \mathfrak{J}.(\psi_0(\xi), \ldots, \psi_m(\xi))$. Let U_2 be the variety whose generic point is $(\psi_0(\xi), \ldots, \psi_m(\xi))$, and suppose that U_2 is birationally equivalent to U and U'. If V is any irreducible variety contained in C, there is a one-to-one correspondence between the varieties on U' and U_2 which correspond to V, and if V' and V_2 are related in this correspondence, $Q(V') = Q(V_2)$, provided that V_2 is not contained in any component of the variety defined by $\mathfrak{i}$ other than those components contained in C.

To prove this, we have only to choose the polynomial $g(x)$ introduced in the proof of Theorem IV so that the section of U by $g(x) = 0$ does not contain V or any component of C, and so that $g(\xi)\,\phi_i^{(\nu)}(\xi)$ belongs to $\mathfrak{i}$. The proof of Theorem IV then applies.

The results so far obtained in this section relate to monoidal transformations of U with any basis C, provided that no component of C is of dimension greater than $d-2$. Monoidal transformations in which C is restricted to be irreducible, and simple on U, have considerable importance in applications of monoidal transformations, and as it is possible to go considerably further in the theory of such monoidal transformations than with general monoidal transformations, we devote the remainder of this section to this case. The results which we shall prove can all be summed up in the following theorem:

THEOREM V. *Let U be an irreducible variety of dimension d, and let C be an irreducible variety of dimension e on U, which is simple for U, where $e \leqslant d-2$. Let U' be the monoidal transform of U with basis C. If V is an irreducible subvariety of C, of dimension f, which is simple both for C and for U, then $T[V]$ is irreducible and simple both for U' and $T[C]$, and is of dimension $d-1-e+f$; and if V' is any variety of $T[V]$ which corresponds to V, V' is simple for $T[V]$, $T[C]$ and U'.*

By taking $V = C$, this implies that $T[C]$ is irreducible and simple on U', and that every variety of U' which corresponds to C is simple for U' and for $T[C]$.

Before proving Theorem V, we shall establish two preliminary results which will simplify the proof. First, we shall show that if the theorem can be proved in the case in which $f = 0$, it is true for any value of f (not exceeding e).

Without loss of generality, we may assume that x_0 is not zero on V, and hence that it is not zero on C or U. Let $\mathfrak{p}$ be the prime ideal of V in $\mathfrak{J} = K[\xi_1, \ldots, \xi_n]$, $\mathfrak{P}$ that of C. Since $\mathfrak{p}$ is of dimension f, we may, again without loss of generality, assume that $\xi_1, \ldots, \xi_f$ are algebraically independent over K modulo $\mathfrak{p}$. Then, since $\mathfrak{P} \subseteq \mathfrak{p}$, $\xi_1, \ldots, \xi_f$ are algebraically independent modulo $\mathfrak{P}$. Let

$$\Omega = K(\xi_1, \ldots, \xi_f),$$

and let $\bar{U}$ be the variety in A_{n-f} over Ω whose generic point is $(\xi_{f+1}, \ldots, \xi_n)$. Since $\xi_1, \ldots, \xi_f$ are independent over K, $\bar{U}$ is of dimension $d-f$. We write $\mathfrak{J}^* = \Omega[\xi_{f+1}, \ldots, \xi_n]$. We consider the ideals $\mathfrak{p}^* = \mathfrak{J}^*.\mathfrak{p}$ and $\mathfrak{P}^* = \mathfrak{J}^*.\mathfrak{P}$ in $\mathfrak{J}^*$. Just as in XVI, § 4, p. 139, we show that these are prime ideals of dimension zero and $e-f$, respectively, and that if $\bar{V}$ and $\bar{C}$ are the irreducible varieties on $\bar{U}$ defined by them, their quotient rings $\bar{Q}(\bar{V})$ and $\bar{Q}(\bar{C})$ are equal, respectively, to $Q(V)$ and to $Q(C)$. Hence, in particular,

$$\bar{Q}(\bar{V}) \subseteq \bar{Q}(\bar{C}),$$

and therefore $\bar{V} \subseteq \bar{C}$. It is also clear that $\mathfrak{p} = \mathfrak{J}_\wedge \mathfrak{p}^*$, and $\mathfrak{P} = \mathfrak{J}_\wedge \mathfrak{P}^*$.

We are given that C is simple on U, and that V is simple on C and on U. These properties may be expressed as follows:

(i) The ideal of non-units of $Q(C)$ has a basis consisting of $d-e = (d-f)-(e-f)$ elements.

(ii) The ideal of non-units of $Q(V)$ has a basis consisting of $d-f$ elements.

(iii) The ideal of non-units of the quotient ring of $\mathfrak{J}/\mathfrak{P}$ with respect to the ideal $\mathfrak{p}/\mathfrak{P}$ has a basis consisting of $e-f$ elements.

Now $Q(C) = \bar{Q}(\bar{C})$, and $Q(V) = \bar{Q}(\bar{V})$. Since $\bar{U}$ is of dimension $d-f$, and $\bar{C}$ is of dimension $e-f$, (i) and (ii) imply that $\bar{C}$ and $\bar{V}$ are simple on $\bar{U}$. Again,

$$\begin{aligned}(\mathfrak{J}/\mathfrak{P})_{\mathfrak{p}/\mathfrak{P}} &= \mathfrak{J}_\mathfrak{p}/\mathfrak{J}_\mathfrak{p}.\mathfrak{P} \quad \text{[XV, § 5, Th. XI]}\\ &= Q(V)/Q(V).\mathfrak{P}.\end{aligned}$$

But $$Q(V) = \bar{Q}(\bar{V}), \quad Q(V).\mathfrak{P} = \bar{Q}(\bar{V}).\mathfrak{P}^*,$$

and hence we have

$$(\mathfrak{J}/\mathfrak{P})_{\mathfrak{p}/\mathfrak{P}} = \bar{Q}(\bar{V})/\bar{Q}(\bar{V}).\mathfrak{P}^* = (\mathfrak{J}^*/\mathfrak{P}^*)_{\mathfrak{p}^*/\mathfrak{P}^*}.$$

Then (iii) tells us that $\bar{V}$ is simple on $\bar{C}$.

We now construct the monoidal transform $\bar{U}'$ of $\bar{U}$ with basis $\bar{C}$. We take the degree ν so large that the monoidal transforms of U

with basis C, and of $\bar{U}$ with basis $\bar{C}$, exist. Let $\phi_i^{(\nu)}(x_0, \ldots, x_n)$ $(i = 0, \ldots, r_\nu)$ be a set of forms serving to define the monoidal transform of order ν of U. Then if $\bar{x}_0, \bar{x}_{f+1}, \ldots, \bar{x}_n$ are the homogeneous coordinates in the space containing $\bar{U}$, we have

$$\phi^{(\nu)}(1, \xi_1, \ldots, \xi_n) \in \mathfrak{P} \subseteq \mathfrak{P}^*,$$

and therefore $\phi^{(\nu)}(\bar{x}_0, \xi_1 \bar{x}_0, \ldots, \xi_f \bar{x}_0, \bar{x}_{f+1}, \ldots, \bar{x}_n)$ vanishes on $\bar{C}$. Now let $\phi(\xi, \bar{x}) = \phi(\xi_1, \ldots, \xi_f, \bar{x}_0, \bar{x}_{f+1}, \ldots, \bar{x}_n)$ be a form of $\Omega[\bar{x}_0, \bar{x}_{f+1}, \ldots, \bar{x}_n]$ of degree ν, with coefficients in $K[\xi_1, \ldots, \xi_f]$, which vanishes on $\bar{C}$. Then if ρ is any integer such that $x_0^\rho \phi(x_1/x_0, \ldots, x_f/x_0, x_0, x_{f+1}, \ldots, x_n)$ is a homogeneous polynomial, this polynomial vanishes on C, since $\phi(\xi_1, \ldots, \xi_f, \xi_{f+1}, \ldots, \xi_n) \in \mathfrak{J}_\wedge \mathfrak{P}^* = \mathfrak{P}$. Therefore, we can deduce that

$$x_0^\rho \phi(x_1/x_0, \ldots, x_f/x_0, x_0, x_{f+1}, \ldots, x_n) = \sum_{i=0}^{r_\nu} a_i(x) \phi_i^{(\nu)}(x)$$

on U, where $a_i(x)$ is a form in $x_0, \ldots, x_n$ over K. From this, we deduce that

$$\bar{x}_0^\rho \phi(\xi, \bar{x}) = \sum_{i=0}^{r_\nu} \alpha_i(\bar{x}) \phi_i^{(\nu)}(\bar{x}_0, \xi_1 \bar{x}_1, \ldots, \xi_f \bar{x}_f, \bar{x}_{f+1}, \ldots, \bar{x}_n)$$

on $\bar{U}$. On account of the presence of the factor $\bar{x}_0^\rho$ on the left-hand side of this equation we are unable to deduce that the monoidal transform $\bar{U}'$ of $\bar{U}$ of degree ν is the variety whose generic point, with respect to Ω, is $(\phi_0^{(\nu)}(\xi), \ldots, \phi_{r_\nu}^{(\nu)}(\xi))$. But we can deduce that $\mathfrak{P}^*$ is an isolated component of $\mathfrak{J}^*.(\phi_0^{(\nu)}(\xi), \ldots, \phi_{r_\nu}^{(\nu)}(\xi))$. Using the Corollary to Theorem IV, we are then able to infer that if $\bar{W}$ is the variety defined over Ω by the generic point $(\phi_0^{(\nu)}(\xi), \ldots, \phi_{r_\nu}^{(\nu)}(\xi))$, $\bar{W}$ is birationally equivalent to $\bar{U}$, and that there is a one-to-one correspondence between the varieties on $\bar{U}'$ and $\bar{W}$ which correspond to any given subvariety of $\bar{C}$, and that the quotient rings of corresponding varieties are equal.

Now let us assume the truth of Theorem V for the case $f = 0$. It follows that, since $\bar{V}$ is of dimension zero, the transform $T[\bar{V}]$ of $\bar{V}$ on $\bar{U}'$ is irreducible and of dimension $(d-f)-1-(e-f) = d-1-e$ and is simple both for $\bar{U}'$ and for $T[\bar{C}]$. Further, if $\bar{V}'$ is any subvariety of $T[\bar{V}]$ which corresponds to $\bar{V}$, $\bar{V}'$ is simple for $T[\bar{V}]$, $T[\bar{C}]$ and $\bar{U}'$. In view of the correspondence between $\bar{U}'$ and $\bar{W}$, we conclude that these results hold when $T[\bar{V}]$ and $T[\bar{C}]$ denote the transforms of $\bar{V}$ and $\bar{C}$ on $\bar{W}$, and $\bar{U}'$ is replaced by $\bar{W}$ whenever it occurs in the statement of the theorem.

We now consider the relationship between $\overline{W}$ and U'. $\overline{W}$ is obtained from U' simply by adding f independent elements of the function field of U' to the ground field. In this case, these f elements are not f of the coordinates of the generic point of U', but it is clear that this makes no essential difference, and that the relation between $\overline{W}$ and U' is, essentially, the same as the relationship between $\overline{U}$ and U. Thus we can pass from the transforms of V and C on U' to their transforms on $\overline{W}$, and from this we deduce immediately the truth of Theorem V. Thus we have only to prove Theorem V when $f = 0$.

From this point, we shall assume that f is zero. Our second simplification of Theorem V is to show that we can replace the monoidal transform U' of U by a birationally equivalent variety which is easier to deal with.

Let
$$\psi_i(x) = 0 \quad (i = 1, \ldots, a)$$
be a basis for the equations of U. Since $C \subset U$, we can obtain a basis for the equations of C by adding a set of equations
$$\psi_i(x) = 0 \quad (i = a+1, \ldots, b)$$
to those of U; and since $V \subseteq C$, we obtain a basis for the equations of V by adding to the equations of C a further set
$$\psi_i(x) = 0 \quad (i = b+1, \ldots, c).$$
Let x' be a generic point of V. Then x' is simple for V, and since V is simple on C and U, x' is simple for C and U. Hence the matrix
$$\left(\frac{\partial \psi_i(x')}{\partial x'_j}\right) \quad (j = 1, \ldots, n;\ i = 1, \ldots, a)$$
is of rank $n-d$; the matrix
$$\left(\frac{\partial \psi_i(x')}{\partial x'_j}\right) \quad (j = 1, \ldots, n;\ i = 1, \ldots, b)$$
is of rank $n-e$; and the matrix
$$\left(\frac{\partial \psi_i(x')}{\partial x'_j}\right) \quad (j = 1, \ldots, n;\ i = 1, \ldots, c)$$
is of rank n. It follows that the matrices
$$\left(\frac{\partial \psi_i(x')}{\partial x'_j}\right) \quad (j = 1, \ldots, n;\ i = b+1, \ldots, c)$$
and
$$\left(\frac{\partial \psi_i(x')}{\partial x'_j}\right) \quad (j = 1, \ldots, n;\ i = a+1, \ldots, c)$$

must be of ranks e and d at least, respectively. Hence, from the set of forms $\psi_i(x)$ $(i = a+1, \ldots, c)$ we can select d, say $\chi_1(x), \ldots, \chi_d(x)$, such that $\chi_1(x), \ldots, \chi_{d-e}(x)$ vanish on C, and the matrix

$$\left(\frac{\partial \chi_i(x')}{\partial x'_j}\right) \quad (j = 1, \ldots, n;\ i = 1, \ldots, d)$$

is of rank d. Let $\chi_i(x)$ be of degree ρ_i, and let ν be an integer such that $\nu \geqslant \max[\rho_1, \ldots, \rho_d]$. We then write

$$\phi_i(x) = x_0^{\nu-\rho_i}\chi_i(x) \quad (i = 1, \ldots, d).$$

Then
$$\left(\frac{\partial \phi_i(x')}{\partial x'_j}\right) \quad (j = 1, \ldots, n;\ i = 1, \ldots, d)$$

and
$$\left(\frac{\partial \phi_i(x')}{\partial x'_j}\right) \quad (j = 1, \ldots, n;\ i = 1, \ldots, d-e)$$

are of rank d and $d-e$ respectively. It follows from XVI, §§ 2 and 3, that if
$$t_i = \phi_i(\xi) \quad (i = 1, \ldots, d),$$
$t_1, \ldots, t_d$ are uniformising parameters on U at V, and $t_1, \ldots, t_{d-e}$ are uniformising parameters at C.

Since $t_1, \ldots, t_{d-e}$ are uniformising parameters at C,

$$Q(C).(t_1, \ldots, t_{d-e})$$

is a prime ideal, and the prime ideal $\mathfrak{P}$ of C in $\mathfrak{J} = K[\xi_1, \ldots, \xi_n]$ is an isolated component of $\mathfrak{J}.(t_1, \ldots, t_{d-e})$. Again, since $Q(V) \subseteq Q(C)$, and $Q(C).(t_1, \ldots, t_{d-e})$ is prime, it follows that $Q(V).(t_1, \ldots, t_{d-e})$ is prime. Hence no component of the variety defined by the ideal $\mathfrak{J}.(t_1, \ldots, t_{d-e})$, other than C, passes through V. We can now apply the corollary to Theorem IV. Let W be the variety whose generic point is

$$\zeta = (\zeta_{01}, \ldots, \zeta_{n\,d-e}) = (\ldots, \xi_i t_j, \ldots) \quad (i = 0, \ldots, n;\ j = 1, \ldots, d-e).$$

The corollary then tells us that there is a one-to-one correspondence between the varieties on W and U' which correspond to V, and that corresponding varieties have the same quotient rings. It follows that in order to establish Theorem V it is sufficient to prove this theorem when the monoidal transform U' is replaced by W. It is considerably simpler to do this. Hence we prove Theorem V when U' is replaced by W, and $T[V]$, $T[C]$ are the transforms of V and C on W.

Since x_0 is not zero on V, it follows that, given any variety V' on W which corresponds to V, we can find a value of h such that V' is not in $z_{0h} = 0$. We first seek to find the varieties not in $z_{0h} = 0$ which correspond to V. The total transform $T\{V\}$ of V consists of all the varieties on W which correspond to varieties contained in V [§ 1, Th. VI, Cor.]. Hence, since V is of dimension zero, $T\{V\}$ is equal to the transform $T[V]$ of V.

Let $\mathfrak{p}$ be the prime ideal of V in $\mathfrak{J} = K[\xi_1, \ldots, \xi_n]$. If $z_{0h} = 0$ is taken as the prime at infinity in the space containing W, the integral domain of W is

$$\mathfrak{J}_h = K[\zeta_{01}/\zeta_{0h}, \ldots, \zeta_{n\,d-e}/\zeta_{0h}] = \mathfrak{J}[t_1/t_h, \ldots, t_{d-e}/t_h] \supseteq \mathfrak{J}.$$

Then the part of $T[V] = T\{V\}$ not in $z_{0h} = 0$ is given by the ideal $\mathfrak{J}_h . \mathfrak{p}$. We prove that this is a prime ideal.

Any element ζ of $\mathfrak{J}_h$ is of the form

$$\zeta = f_\rho(t_1, \ldots, t_{d-e})/t_h^\rho,$$

where $f_\rho(x_1, \ldots, x_{d-e})$ is a homogeneous polynomial in $x_1, \ldots, x_{d-e}$ of degree ρ with coefficients in $\mathfrak{J}$. If the coefficients are in $\mathfrak{p}$, it is clear that $\zeta \in \mathfrak{J}_h . \mathfrak{p}$. Conversely, suppose that ζ is contained in $\mathfrak{J}_h . \mathfrak{p}$. Then it can clearly be written in the form

$$\zeta = \frac{g_\sigma(t_1, \ldots, t_{d-e})}{t_h^\sigma},$$

where the coefficients are in $\mathfrak{p}$. Then

$$t_h^\sigma f_\rho(t_1, \ldots, t_{d-e}) = t_h^\rho g_\sigma(t_1, \ldots, t_{d-e}).$$

We expand both sides of this equation in terms of the local parameters $t_1, \ldots, t_d$ at one of the conjugate points which constitute V. Then the right-hand side has no terms of degree less than $\rho + \sigma + 1$, since the coefficients of g_σ are in $\mathfrak{p}$. If not all the coefficients of f_ρ were in $\mathfrak{p}$, the left-hand side would contain a power-product in $t_1, \ldots, t_d$ of degree $\rho + \sigma$ with non-zero coefficient. Since the expansion of $t_h^\sigma f_\rho = t_h^\rho g_\sigma$ is unique, we obtain a contradiction. Thus the coefficients of f_ρ must be in $\mathfrak{p}$. Thus ζ is in $\mathfrak{J}_h . \mathfrak{p}$ if and only if the coefficients of f_ρ are in $\mathfrak{p}$.

Let
$$\zeta = \frac{f_\rho(t_1, \ldots, t_{d-e})}{t_h^\rho},$$

and
$$\eta = \frac{h_\sigma(t_1, \ldots, t_{d-e})}{t_h^\sigma},$$

be any two elements of $\mathfrak{J}_h$ not in $\mathfrak{J}_h.\mathfrak{p}$. Then $\zeta\eta = f_\rho h_\sigma/t_h^{\rho+\sigma}$. Since the expansions of f_ρ and h_σ in power-series of $t_1, \ldots, t_d$ at one of the points of V have non-zero terms of degree ρ and σ respectively, the expansion of $f_\rho h_\sigma$ has at least one non-zero term of degree $\rho+\sigma$. Hence the coefficients of $f_\rho h_\sigma$ are not in $\mathfrak{p}$. It follows that $\zeta\eta$ is not in $\mathfrak{J}_h.\mathfrak{p}$. Thus $\mathfrak{J}_h.\mathfrak{p}$ is a prime ideal in $\mathfrak{J}_h$. It therefore defines an irreducible variety V^*, which is the only component of $T[V]$ on W not in $z_{0h}=0$.

The elements t_i/t_h $(i = 1, \ldots, h-1, h+1, \ldots, d-e)$ are algebraically independent modulo $\mathfrak{J}_h.\mathfrak{p}$. For otherwise there would exist a non-zero homogeneous polynomial $f_\rho(x_1, \ldots, x_{d-e})$ with coefficients in K such that

$$f_\rho(t_1, \ldots, t_{d-e})/t_h^\rho \in \mathfrak{J}_h.\mathfrak{p}.$$

Hence, as above, the coefficients of f_ρ would lie in $\mathfrak{p}$. Since the coefficients are in K, they must be zero. Thus $f_\rho(x_1, \ldots, x_{d-e})$ is zero, and we have a contradiction. Hence z_{0i} is not zero on V^* $(i = 1, \ldots, d-e)$. If we follow the method described above to find the part of $T[V]$ which is not in $z_{0i} = 0$, for each i, we again get an irreducible variety, and this must coincide with V^*. It follows that $V^* = T[V]$, which is therefore irreducible.

The fact that t_i/t_h $(i = 1, \ldots, h-1, h+1, \ldots, d-e)$ are algebraically independent modulo $\mathfrak{J}_h.\mathfrak{p}$ shows that $\dim V^* \geqslant d-e-1$. Let ζ be any element of $\mathfrak{J}_h$. Then

$$\zeta = \Sigma\alpha_{i_1 \ldots i_{d-e}}(t_1/t_h)^{i_1} \ldots (t_{d-e}/t_h)^{i_{d-e}},$$

where $\alpha_{i_1 \ldots i_{d-e}}$ is in $\mathfrak{J}$. Let z_i be the residue of t_i/t_h modulo $\mathfrak{J}_h.\mathfrak{p}$, and let $a_{i_1 \ldots i_{d-e}}$ be the residue of $\alpha_{i_1 \ldots i_{d-e}}$ modulo $\mathfrak{p}$. Then

$$\Sigma a_{i_1 \ldots i_{d-e}} z_1^{i_1} \ldots z_{d-e}^{i_{d-e}}$$

is the residue of ζ modulo $\mathfrak{J}_h.\mathfrak{p}$. Since $\alpha_{i_1 \ldots i_{d-e}}$ is algebraic over K, this residue is algebraically dependent on $z_1, \ldots, z_{h-1}, z_{h+1}, \ldots, z_{d-e}$ over K. Hence ζ is algebraically dependent on $t_1/t_h, \ldots, t_{h-1}/t_h$, $t_{h+1}/t_h, \ldots, t_{d-e}/t_h$, modulo $\mathfrak{J}_h.\mathfrak{p}$. It follows that $\dim V^* \leqslant d-e-1$. Hence $\dim V^* = d-e-1$.

Since $Q(V) \subseteq Q(V^*)$, we have

$$\begin{aligned} Q(V^*).(\mathfrak{J}_h.\mathfrak{p}) &= Q(V^*).[Q(V).\mathfrak{p}] \\ &= Q(V^*).[Q(V).(t_1, \ldots, t_d)] \\ &= Q(V^*).(t_1, \ldots, t_d). \end{aligned}$$

Moreover, $$t_i/t_h \in \mathfrak{J}_h \subseteq Q(V^*) \quad (i \leqslant d-e);$$

hence $$Q(V^*).(\mathfrak{J}_h.\mathfrak{p}) = Q(V^*).(t_h, t_{d-e+1}, \ldots, t_d).$$

But $Q(V^*).(\mathfrak{J}_h.\mathfrak{p})$ is the ideal of non-units of $Q(V^*)$. Hence the ideal of non-units of $Q(V^*)$ has a basis consisting of $e+1 = d-(d-e-1)$ elements. Since $\dim V^* = d-e-1$, it follows that V^* is simple on W.

Thus we have proved that $T[V]$ is irreducible and simple on W, and that it is of dimension $d-e-1$. By considering the case $V = C$, we deduce that $C^* = T[C]$ is irreducible and simple on W and that it is of dimension $d-1$, and also that the ideal of non-units of $Q(C^*)$ is $Q(C^*).(t_h)$.

Now regard V^* as a subvariety of C^*. Its quotient ring is $Q(V^*)/Q(V^*).\mathfrak{J}_h.\mathfrak{P} = Q(V^*)/Q(V^*).\mathfrak{P}$. Hence the residue classes defined by $t_{d-e+1}, \ldots, t_d$ form a basis for the ideal of non-units of the quotient ring of V^* on C^*. Since $e = (d-1)-(d-e-1)$, and V^* is of dimension $d-e-1$, while C^* is of dimension $d-1$, it follows that V^* is simple on C^*.

To complete the proof of Theorem V, we have to consider a variety V' on W which corresponds to V. Since $V^* = T\{V\}$, $V' \subseteq V^*$. We have to show that V' is simple on V^*, C^* and W. In order to do this, we have only to show that V' contains a zero-dimensional subvariety V'' with these properties. Since $V'' \subseteq V'$, the unique corresponding subvariety of U is contained in V, and therefore, since V is of dimension zero, it must coincide with V. Thus V'' corresponds to V. Therefore, we need only consider the case in which V' is of dimension zero. In what follows, we shall assume that V' is of dimension zero, and does not lie in $z_{0h} = 0$.

Since every variety on W which corresponds to V is irregular for the correspondence between U and W, we have $Q(V) \subset Q(V')$. Let $\mathfrak{p}'$ be the prime ideal of V' in $\mathfrak{J}_h$. Then $\mathfrak{p} = \mathfrak{J}_\wedge \mathfrak{p}'$. Again, since $\mathfrak{J} \subseteq \mathfrak{J}_h$, $\mathfrak{J}_h/\mathfrak{p}'$ contains a subring isomorphic to $\mathfrak{J}/\mathfrak{p}$. Since V is of dimension zero, $\Delta = \mathfrak{J}/\mathfrak{p}$ is a field which is an algebraic extension of K, and since V' is zero-dimensional, $\Delta' = \mathfrak{J}_h/\mathfrak{p}'$ is also an algebraic extension of K. We can regard Δ as a subfield of Δ'. Thus Δ' is an algebraic extension of Δ. Let $\tau_1, \ldots, \tau_{h-1}, \tau_{h+1}, \ldots, \tau_{d-e}$ be the residues of $t_1/t_h, \ldots, t_{h-1}/t_h, t_{h+1}/t_h, \ldots, t_{d-e}/t_h$ in Δ', and let $f_i(z)$ be the characteristic polynomial of τ_i over Δ. The coefficients of $f_i(z)$ are residues of elements of $\mathfrak{J}$ in Δ. We can then construct a polynomial

$F_i(z)$, with coefficients in $\mathfrak{J}$, such that, when we replace these coefficients by their residues, $F_i(z)$ becomes $f_i(z)$. Let

$$u_i = F_i(t_i/t_h) \quad (i = 1, \ldots, h-1, h+1, \ldots, d-e),$$
$$u_i = t_i \qquad (i = h, d-e+1, \ldots, d).$$

$u_h, u_{d-e+1}, \ldots, u_d$ are uniformising parameters at V^* on W, and u_h is the uniformising parameter on W at C^*. If we can show that $u_1, \ldots, u_d$ are uniformising parameters on W at V', it will follow that V' is simple for V^*, C^*, W, by the proof already given to prove that V^* is simple on C^* and on W. Thus in order to complete the proof of Theorem V, we have only to show that $u_1, \ldots, u_d$ are uniformising parameters at V'.

To do this, we have to show that $\mathfrak{p}'$ is an isolated component of $\mathfrak{i} = \mathfrak{J}_h \cdot (u_1, \ldots, u_d)$. We pass from the ring $\mathfrak{J}_h$ to the remainder-class ring defined modulo $\mathfrak{p}^* = \mathfrak{J}_h \cdot \mathfrak{p}$. Since $\mathfrak{p}^* \subseteq \mathfrak{p}'$, $\mathfrak{p}'$ is an isolated component of $\mathfrak{i}$ if and only if $\mathfrak{p}'/\mathfrak{p}^*$ is an isolated component of $\mathfrak{i}/\mathfrak{p}^*$. Since $\mathfrak{p} = \mathfrak{J}_\wedge \mathfrak{p}^*$, the residues of elements of $\mathfrak{J}_h$ which are in $\mathfrak{J}$, modulo $\mathfrak{p}^*$, form a ring isomorphic with $\Delta = \mathfrak{J}/\mathfrak{p}$, and can be identified with the corresponding elements of Δ. Now $t_h, t_{d-e+1}, \ldots, t_d$ belong to $\mathfrak{p}^*$, hence their residues modulo $\mathfrak{p}^*$ are zero. On the other hand, we saw above that the residues $z_1, \ldots, z_{h-1}, z_{h+1}, \ldots, z_{d-e}$ of $t_1/t_h, \ldots, t_{h-1}/t_h, t_{h+1}/t_h, \ldots, t_{d-e}/t_h$ modulo $\mathfrak{p}^*$ are algebraically independent over K, and hence over Δ, which is an algebraic extension of K. Since

$$\mathfrak{J}_h = \mathfrak{J}[t_1/t_h, \ldots, t_{d-e}/t_h],$$

we conclude that

$$\mathfrak{J}_h/\mathfrak{p}^* = \Delta[z, \ldots, z_{h-1}, z_{h+1}, \ldots, z_{d-e}],$$

and $\mathfrak{i}/\mathfrak{p}^*$ is the ideal in this which has the basis

$$f_i(z_i) \quad (i = 1, \ldots, h-1, h+1, \ldots, d-e).$$

On the other hand, any element of $\mathfrak{p}'/\mathfrak{p}^*$ is a polynomial

$$f(z_1, \ldots, z_{h-1}, z_{h+1}, \ldots, z_{d-e})$$

with coefficients in Δ, which vanishes at the point

$$z_i = \tau_i \quad (i = 1, \ldots, h-1, h+1, \ldots, d-e),$$

where $\tau_1, \ldots, \tau_{d-e}$ are the residues of $t_1/t_h, \ldots, t_{d-e}/t_h$, modulo $\mathfrak{p}'$. The prime ideal of this point in the affine space $(z_1, \ldots, z_{h-1}, z_{h+1}, \ldots, z_{d-e})$ over Δ is an isolated component of the ideal in

$$\Delta[z_1, \ldots, z_{h-1}, z_{h+1}, \ldots, z_{d-e}]$$

generated by $f_i(z_i)$ $(i = 1, \ldots, h-1, h+1, \ldots, d-e)$, since each $f_i(z)$ is irreducible over Δ. Hence $\mathfrak{p}'/\mathfrak{p}^*$ is an isolated component of $\mathfrak{i}/\mathfrak{p}^*$.

It follows that $\mathfrak{p}'$ is an isolated component of $\mathfrak{i}$, and hence $u_1, \ldots, u_d$ are uniformising parameters at V'. Thus the proof of Theorem V is complete.

4. The reduction of singularities and the Local Uniformisation Theorem. In the three preceding sections we have studied the theory of birational transformations. For the remainder of this chapter we shall be concerned with applications of that theory to problems involving the multiple points of an algebraic variety. The reader will be aware of the difficulties which arise when an algebraic variety has multiple points. For instance, the theory of intersections given in Chapter XII is much simpler if there are no multiple points. Hence the question arises: can we obtain for any irreducible algebraic variety of dimension d a birational transform which has no multiple points? We have already shown how to obtain such a transform in the case $d = 1$ [XVI, § 6, Th. V], and later we shall prove the existence of a non-singular birational transform of an algebraic surface. Zariski has also successfully dealt with the problem in the case $d = 3$, but the proof is difficult, and we shall not give it. For values of d greater than three our question is as yet unanswered.

However, even though we could prove that there always exists a birational transform without singularities of an irreducible variety, much would be lost if we were to confine our attention to those transforms, that is, to non-singular varieties. For instance, the structure of multiple points is itself a subject of great interest. In order to study a multiple point, it is usually best to perform a birational transformation for which the multiple point is fundamental, and then examine the transform of the point, with a view to deducing the structure of the singularity, and it is important that no subvariety of the transformed variety, or at least no subvariety of the transform of the singular point, should be fundamental. If we could arrange that every point of the transform of the singularity was a simple point of the new variety, we should probably be able to derive all the information which we require, but unfortunately this problem is of the same order of difficulty as that of obtaining a non-singular birational transform of a variety. However, a partial result of this nature, which is of fundamental importance not only for the problem we are discussing but for many other reasons, is provided by

THE LOCAL UNIFORMISATION THEOREM. *Let U be an irreducible algebraic variety with the function field Σ, and let $\mathfrak{B}$ be an s-dimensional valuation of Σ whose centre on U is V. Then there exists a birational transform U' of U on which the centre of $\mathfrak{B}$ is a simple s-dimensional variety V', such that $Q(V) \subseteq Q(V')$.*

Although we do not propose to make a study of the multiple points of an algebraic variety, we may point out briefly how this result can form the basis of the study of an isolated multiple point P of an algebraic variety U, defined over the field of complex numbers. We use non-homogeneous coordinates and take P to be $(0, \ldots, 0)$, and $(\xi_1, \ldots, \xi_n)$ to be the generic point of U. Let $\mathfrak{B}$ be any zero-dimensional valuation of the function field Σ of U whose centre on U is P. Then, by the theorem, there exists a birational transform U' of U on which the centre of $\mathfrak{B}$ is a simple point P', where $Q(P) \subseteq Q(P')$. Let $t_1, \ldots, t_d$ be uniformising parameters at P'. Since $\xi_i \in Q(P) \subseteq Q(P')$, ξ_i can be expanded as a power series in $t_1, \ldots, t_d$. The n series so obtained converge in a finite neighbourhood of P' on U', which we denote by $N(\mathfrak{B})$. We now consider the aggregate of zero-dimensional valuations $\mathfrak{B}$ of Σ whose centre on U is P, and the associated neighbourhoods $N(\mathfrak{B})$. It can be shown that there exists a finite number of them, say $N(\mathfrak{B}_1), \ldots, N(\mathfrak{B}_k)$, such that any zero-dimensional valuation whose centre on U is P has a simple centre in some $N(\mathfrak{B}_i)$. Corresponding to each $\mathfrak{B}_i$ we have a set of expansions for $\xi_1, \ldots, \xi_n$. Each of these determines a partial neighbourhood, or 'wedge', at P, and the wedges cover the whole neighbourhood of P on U. The existence of this finite set of wedges covering the neighbourhood of P forms the basis of the analysis of the singularity at P.

We shall prove the Local Uniformisation Theorem in the following sections, and later use it to obtain certain important results on the removal of the singularities of an algebraic variety. The proof, though direct, is long, and we shall break it up into sections. It is first convenient to notice that it is sufficient to prove the theorem when U is a primal, that is, a variety of d dimensions in a space of $d+1$ dimensions. This can be seen as follows.

Let us suppose that U is given in S_n and refer it to non-homogeneous coordinates with regard to which the centre V of $\mathfrak{B}$ is at a finite distance. Let $(\xi_1, \ldots, \xi_n)$ be a generic point of U. In the ring $\mathfrak{J} = K[\xi_1, \ldots, \xi_n]$ we can find $d+1$ elements $\eta_1, \ldots, \eta_{d+1}$ such that

(a) $\eta_1, \ldots, \eta_d$ are algebraically independent over K,

(b) η_{d+1} is a primitive element of Σ (which is the quotient field of $\mathfrak{J}$) over $K(\eta_1, \ldots, \eta_d)$, and

(c) any element of $\mathfrak{J}$ is integrally dependent on $K[\eta_1, \ldots, \eta_d]$.

Let $f(y_1, \ldots, y_{d+1}) = 0$ be the unique irreducible equation satisfied by $\eta_1, \ldots, \eta_{d+1}$. In the affine space A'_{d+1} the equation

$$f(y_1, \ldots, y_{d+1}) = 0$$

defines an irreducible variety U' whose generic point is $(\eta_1, \ldots, \eta_{d+1})$, and whose integral domain is $\mathfrak{J}' = K[\eta_1, \ldots, \eta_{d+1}]$. By construction, the quotient field of U' is Σ, that is, U' is birationally equivalent to U. The equations of the birational correspondence include equations

$$y_i = \phi_i(x_1, \ldots, x_n) \quad (i = 1, \ldots, d+1),$$

where

$$\eta_i = \phi_i(\xi_1, \ldots, \xi_n) \quad (i = 1, \ldots, d+1),$$

$\phi_i(x)$ being a polynomial. By construction, the integral domain $\mathfrak{J} = K[\xi_1, \ldots, \xi_n]$ is integrally dependent on the integral domain $\mathfrak{J}' = K[\eta_1, \ldots, \eta_{d+1}]$.

Let V' be the centre of $\mathfrak{B}$ on U'. If the Local Uniformisation Theorem is known to be true when U' is a primal, there exists a variety U'' in birational correspondence with U and U' on which the centre V'' of $\mathfrak{B}$ is simple, and of dimension s. Further,

$$Q(V') \subseteq Q(V'').$$

Since V'' is simple, $Q(V'')$ is integrally closed, and is the intersection of the valuation rings which contain it. Since $Q(V') \subseteq Q(V'')$, all these valuation rings contain $Q(V')$, and hence $\mathfrak{J}'$. But $\mathfrak{J}$ is integrally dependent on $\mathfrak{J}'$, and it follows from the fact that a valuation ring is integrally closed that $\mathfrak{J}$ is in the valuation ring of every valuation whose ring contains $Q(V'')$. Hence $\mathfrak{J} \subseteq Q(V'')$. Every element of $\mathfrak{J}$ which is not in the prime ideal $\mathfrak{p}$ of V has value zero in $\mathfrak{B}$; hence it is a unit of $Q(V'')$. Hence $\mathfrak{J}_{\mathfrak{p}} = Q(V)$ is contained in $Q(V'')$. Thus from the assumption that the Local Uniformisation Theorem is true for the primal U' we deduce that it is true for any variety U.

The first stage in the proof of the Local Uniformisation Theorem for the primal U is to prove it when $\mathfrak{B}$ is a zero-dimensional valuation of rank one. This is the most difficult part of the whole proof.

We shall actually prove a stronger result in this case than is required by the general theorem, for we shall prove that we can take the variety U' to be a primal too, and that the equations of the correspondence between U and U' (in non-homogeneous coordinates) are of the form

$$\left.\begin{aligned} x_i &= P_i(y_1, \ldots, y_{d+1}) \quad (i = 1, \ldots, d+1), \\ y_i &= R_i(x_1, \ldots, x_{d+1}) \quad (i = 1, \ldots, d+1), \end{aligned}\right\} \tag{1}$$

where the P_i are polynomials and the R_i rational functions; moreover, we shall show that these equations define a *Cremona transformation*, that is, a birational correspondence between the $(d+1)$-spaces containing U and U', and not merely a birational correspondence between U and U'. This stronger form of the theorem is of interest in itself, but we give it because it is essential at a later stage of the proof of the general theorem. It should also be noticed that the fact that the P_i are polynomials enables us to deduce at once that $Q(V) \subseteq Q(V')$. In fact, if $(\eta_1, \ldots, \eta_{d+1})$ is a generic point of U', we have, from the equations of the correspondence,

$$\mathfrak{J} = K[\xi_1, \ldots, \xi_n] \subseteq K[\eta_1, \ldots, \eta_{d+1}] = \mathfrak{J}'.$$

If $\mathfrak{P}'$ is the prime ideal of V' in $\mathfrak{J}'$, the prime ideal of V is $\mathfrak{P} = \mathfrak{J} \wedge \mathfrak{P}'$, and therefore $\mathfrak{J}_{\mathfrak{P}} \subseteq \mathfrak{J}_{\mathfrak{P}'}$, that is, $Q(V) \subseteq Q(V')$. Hence the first stage of our proof of the Local Uniformisation Theorem is to show that if U is a primal in A_{d+1} and $\mathfrak{B}$ is a zero-dimensional valuation of rank one of its function field, then there exists a Cremona transformation of the form (1) of A_{d+1} into A'_{d+1} which transforms U into a primal U' on which the centre of $\mathfrak{B}$ is simple.

Once this result is proved, the general Local Uniformisation Theorem is proved by an induction argument. We show that if it is true for a zero-dimensional valuation of rank k, then it is true for an s-dimensional valuation of rank k, and that if it is true for any valuation of rank k it is true for a zero-dimensional valuation of rank $k+1$.

Turning now to the problem of the elimination of the singularities of an algebraic variety with function-field Σ, what we have to do is to find a *projective model* of Σ; that is, an algebraic variety whose function field is Σ, on which every zero-dimensional valuation of Σ has as centre a simple subvariety. For every zero-dimensional subvariety of a projective model of Σ is the centre of some zero-dimensional valuation of Σ, and if the centres of all such valuations Σ are simple, every zero-dimensional subvariety of the model, and hence every point of the model, is simple. Unfortunately, we cannot as yet prove the existence of such a projective model, but we can prove that if Σ is the function field of an algebraic variety there exists *a finite resolving system*, that is, a finite number of projective models of Σ with the property that any zero-dimensional valuation of Σ has a centre on at least one of the models which is a simple subvariety of it. Clearly, the elimination of the singularities of an algebraic variety is equivalent to showing that its function field has a resolving system consisting of a single projective model. The natural way to try to show this is to take a resolving system consisting of k models, and to try to reduce it to one consisting of $k-1$ models. We shall show how this can be done in the case of algebraic surfaces, in which the reduction from k to $k-1$ models is relatively easy. For varieties of dimension three, the reduction has been achieved by Zariski, but it is difficult owing to the existence of fundamental curves in the correspondences between the various models. A reduction in the case of varieties of dimension greater than three is still awaited.

Note. The next three sections are devoted to the proof of the Local Uniformisation Theorem. Since the details of this proof are somewhat complicated, the reader may prefer to accept the theorem at first, and pass directly to § 8.

5. Some Cremona transformations. The proof of the Local Uniformisation Theorem for zero-dimensional valuations of rank one is based on the construction of a series of Cremona transformations, and as a preliminary we consider the properties of certain Cremona transformations which will be used later.

Let U be a d-dimensional irreducible variety in the affine space A_{d+1} with the coordinate system $(x_1, \ldots, x_{d+1})$, and let $(\xi_1, \ldots, \xi_{d+1})$ be a generic point of U. The transformations which we consider are all of the type

$$(C)\begin{cases} x_i = P_i(y_1, \ldots, y_{d+1}) & (i = 1, \ldots, d+1), \\ y_i = R_i(x_1, \ldots, x_{d+1}) & (i = 1, \ldots, d+1), \end{cases}$$

where P_i is a polynomial in $y_1, \ldots, y_{d+1}$ with coefficients in K, and R_i is a rational function of $x_1, \ldots, x_{d+1}$ over K. It will save interrupting our argument with trivial definitions if we agree to denote the transform of $(\xi_1, \ldots, \xi_{d+1})$ by any transformation such as (C) by the Greek letters corresponding to the Latin letters used to denote the coordinate system in the transformed space.

Let $\mathfrak{B}$ be a zero-dimensional valuation of rank one of the field $K(\xi_1, \ldots, \xi_{d+1})$. Its centre is a zero-dimensional variety on U, that is, a set of conjugate points $(\alpha_1^{(i)}, \ldots, \alpha_{d+1}^{(i)})$ $(i = 1, \ldots, g)$. We shall be much concerned with the field R obtained by adjoining $\alpha_j^{(i)}$ $(j = 1, \ldots, d+1; i = 1, \ldots, g)$ to K. It should be noted that if we perform the transformation (C), the centre of $\mathfrak{B}$ in the y-space is a set of conjugate points $(\beta_1^{(i)}, \ldots, \beta_{d+1}^{(i)})$ $(i = 1, \ldots, h)$. The field $R' = K(\beta_1^{(1)}, \ldots, \beta_1^{(h)}, \ldots, \beta_{d+1}^{(1)}, \ldots, \beta_{d+1}^{(h)})$ is not necessarily the same as R. But given i, it is easily seen that there exists an integer j $(1 \leqslant j \leqslant h)$ such that

$$\alpha_k^{(i)} = P_k(\beta_1^{(j)}, \ldots, \beta_{d+1}^{(j)}) \quad (k = 1, \ldots, d+1);$$

hence $R \subseteq R'$; it will be useful to bear in mind this property of the transformations (C).

R is an algebraic extension of K of order g; by making, if necessary, a linear transformation on $x_1, \ldots, x_{d+1}$, we can arrange that $\alpha_j^{(1)}, \ldots, \alpha_j^{(g)}$ are all distinct for each value of j. Then $R = K(\alpha_j^{(1)}, \ldots, \alpha_j^{(g)})$ $(j = 1, \ldots, d+1)$. It will be convenient to suppose that the coordinate system has been chosen initially so that $\alpha_j^{(1)}, \ldots, \alpha_j^{(g)}$ are distinct. When we construct the series of transformations which we shall ultimately use, it is not usually permissible to introduce a linear transformation to ensure that the jth coordinates of the set of conjugate points which constitute the centre of $\mathfrak{B}$ in the space considered are all distinct, but as a rule this property will be true as a consequence of the properties of the transformations used. The property is essential at certain stages of our work.

We shall often have to extend the ground field K to some algebraic extension K^*. Usually, we shall take $K^* = R$. When we extend K to K^*, the variety U may become reducible. $(\xi_1, \ldots, \xi_{d+1})$ is then a generic point over K^* of one of the components, say U^*, into which U breaks up, and there is a valuation $\mathfrak{B}^*$ of $K^*(\xi_1, \ldots, \xi_{d+1})$ which is an extension of $\mathfrak{B}$. If $f(x_1, \ldots, x_{d+1})$ is any polynomial over K which does not vanish on the centre of $\mathfrak{B}$, $v[f(\xi)] = 0$, hence $v^*[f(\xi)] = 0$, so that $f(x_1, \ldots, x_{d+1})$ does not vanish at the centre of $\mathfrak{B}^*$. Hence the centre of $\mathfrak{B}^*$ is contained in the centre of $\mathfrak{B}$; and, in particular, if $K^* \supseteq R$, the centre of $\mathfrak{B}^*$ is at one of the points $(\alpha_1^{(i)}, \ldots, \alpha_{d+1}^{(i)})$. We shall assume that, whenever we extend K, we at the same time select a corresponding extension $\mathfrak{B}^*$ of $\mathfrak{B}$, and when there is no risk of confusion we shall, for simplicity of notation, continue to denote it by $\mathfrak{B}$.

When we take R as the ground field we may suppose that the centre of $\mathfrak{B}$ is $(\alpha_1, \ldots, \alpha_{d+1}) = (\alpha_1^{(1)}, \ldots, \alpha_{d+1}^{(1)})$. Then $\tau_i = v(\xi_i - \alpha_i) > 0$. The transformations with which we are concerned in this section are of two types: (i) they are transformations of type (C) on only m variables, say $x_1, \ldots, x_m$, in the case in which $\tau_1, \ldots, \tau_m$ are rationally independent real numbers; (ii) they are transformations of type (C) on $m+1$ variables, say $x_1, \ldots, x_{m+1}$, where $\tau_1, \ldots, \tau_m$ are rationally independent, and τ_{m+1} is rationally dependent on $\tau_1, \ldots, \tau_m$. We begin with an algebraic lemma on rational approximations to a set of m rationally independent real numbers.

Lemma I. (Perron's algorithm.) *If $\tau_1, \ldots, \tau_m$ are m rationally independent positive real numbers, there exist m series of non-negative integers $A_i^{(0)}, A_i^{(1)}, A_i^{(2)}, \ldots$ $(i = 1, \ldots, m)$ such that*

$$\text{(i)} \qquad \begin{vmatrix} A_1^{(h)} & A_1^{(h+1)} & \cdots & A_1^{(h+m-1)} \\ \cdot & \cdot & \cdots & \cdot \\ A_m^{(h)} & A_m^{(h+1)} & \cdots & A_m^{(h+m-1)} \end{vmatrix} = \pm 1,$$

$$\text{(ii)} \qquad \lim_{h\to\infty} \frac{A_i^{(h)}}{A_1^{(h)}} = \frac{\tau_i}{\tau_1}.$$

If ξ is any positive real number, we denote the integral part of it by $[\xi]$. Let $a_j^{(0)} = [\tau_j/\tau_1]$, and write

$$\tau_1 = \tau_m^{(1)}, \quad \tau_i = \tau_{i-1}^{(1)} + a_i^{(0)}\tau_m^{(1)} \quad (i = 2, \ldots, m).$$

The numbers $\tau_1^{(1)}, \ldots, \tau_m^{(1)}$ are rationally independent. For if $\lambda_1, \ldots, \lambda_m$ are integers such that

$$\lambda_1\tau_1^{(1)} + \ldots + \lambda_m\tau_m^{(1)} = 0,$$

we have $$\left(\lambda_m - \sum_1^{m-1} \lambda_i a_{i+1}^{(0)}\right)\tau_1 + \lambda_1\tau_2 + \ldots + \lambda_{m-1}\tau_m = 0,$$

and hence, from the rational independence of $\tau_1, \ldots, \tau_m$ we deduce that $\lambda_1 = \lambda_2 = \ldots = \lambda_m = 0$. Also

$$\frac{\tau_i^{(1)}}{\tau_m^{(1)}} = \frac{\tau_{i+1}}{\tau_1} - \left[\frac{\tau_{i+1}}{\tau_1}\right] < 1;$$

hence $\tau_i^{(1)} < \tau_m^{(1)}$. We define

$$A_1^{(k)} = 0 \quad (k = 1, \ldots, m-1), \qquad A_1^{(m)} = 1,$$

$$A_2^{(1)} = 1, \qquad A_2^{(k)} = 0 \quad (k = 2, \ldots, m-1), \qquad A_2^{(m)} = a_2^{(0)}, \ldots$$

$$A_j^{(i-1)} = \delta_j^i \quad (i \leqslant m), \qquad A_j^{(m)} = a_j^{(0)} \quad (j = 3, \ldots, m).$$

Then $$\tau_i = \sum_{j=1}^{m} A_i^{(j)} \tau_j^{(1)} \quad (i = 1, \ldots, m).$$

The construction of the numbers $a_i^{(0)}$, $A_j^{(i)}$ $(i, j \leqslant m)$, $\tau_1^{(1)}, \ldots, \tau_m^{(1)}$ will be referred to as *the first stage of Perron's algorithm.*

We now start with the rationally independent numbers $\tau_1^{(1)}, \ldots, \tau_m^{(1)}$, and repeat the process, obtaining m rationally independent numbers $\tau_1^{(2)}, \ldots, \tau_m^{(2)}$. We define $a_j^{(1)} = [\tau_j^{(1)}/\tau_1^{(1)}]$ and write

$$\tau_1^{(1)} = \tau_m^{(2)}, \quad \tau_i^{(1)} = \tau_{i-1}^{(2)} + a_i^{(1)} \tau_m^{(2)} \quad (i = 2, \ldots, m),$$

where $\tau_i^{(2)} < \tau_m^{(2)}$. In this case, however, we have $a_m^{(1)} \geqslant 1$, since $\tau_1^{(1)} < \tau_m^{(1)}$ and $a_m^{(1)} \geqslant a_i^{(1)} \geqslant 0$.

Proceeding in this way, we produce a series of sets of rationally independent numbers $\tau_1^{(h)}, \ldots, \tau_m^{(h)}$ $(h = 1, 2, 3, \ldots)$, connected by the relations

$$\tau_1^{(h-1)} = \tau_m^{(h)}, \quad \tau_i^{(h-1)} = \tau_{i-1}^{(h)} + a_i^{(h-1)} \tau_m^{(h)} \quad (i = 2, \ldots, m),$$

$$a_i^{(h-1)} = [\tau_i^{(h-1)}/\tau_1^{(h-1)}] \geqslant 0,$$

where $$a_m^{(h-1)} \geqslant 1, \quad a_m^{(h-1)} \geqslant a_i^{(h-1)} \quad (h \geqslant 2).$$

Combining the various equations, we obtain

$$\tau_i = \sum_{j=1}^{m} p_{ij}^{(h)} \tau_j^{(h)},$$

where the $p_{ij}^{(h)}$ are defined by the recurrence relations

$$\begin{pmatrix} p_{11}^{(h)} & \cdots & p_{1m}^{(h)} \\ \cdot & \cdots & \cdot \\ \cdot & \cdots & \cdot \\ p_{m1}^{(h)} & \cdots & p_{mm}^{(h)} \end{pmatrix} = \begin{pmatrix} p_{11}^{(h-1)} & \cdots & p_{1m}^{(h-1)} \\ \cdot & \cdots & \cdot \\ \cdot & \cdots & \cdot \\ p_{m1}^{(h-1)} & \cdots & p_{mm}^{(h-1)} \end{pmatrix} \begin{pmatrix} 0 & 0 & \cdots & 0 & 1 \\ 1 & 0 & \cdots & 0 & a_2^{(h-1)} \\ \cdot & \cdot & \cdots & \cdot & \cdot \\ 0 & \cdot & \cdots & 1 & a_m^{(h-1)} \end{pmatrix}.$$

From these relations we deduce at once that we can write $p_{ij}^{(h)} = A_i^{(h+j-1)}$, and that

$$\begin{vmatrix} A_1^{(h)} & \cdots & A_1^{(h+m-1)} \\ \cdot & \cdots & \cdot \\ \cdot & \cdots & \cdot \\ A_m^{(h)} & \cdots & A_m^{h+m-1} \end{vmatrix} = (-1)^{h(m-1)}.$$

$A_i^{(h)}$ is a non-negative integer. The matrix equation written above, with $h+1$ written for h, gives the equation for $p_{mi}^{(h+1)}$ in the form

$$A_i^{(h+m)} = A_i^{(h)}a_1^{(h)} + \dots + A_i^{(h+m-1)}a_m^{(h)} \quad (a_1^{(h)} = 1). \tag{1}$$

The process of passing from $\tau_1^{(h-1)}, \dots, \tau_m^{(h-1)}$ to $\tau_1^{(h)}, \dots, \tau_m^{(h)}$, and defining $a_i^{(h-1)}$, etc., will be called *the h-th stage of Perron's algorithm.*

Since $a_i^{(h)} \geqslant 0$, $a_m^{(h)} \geqslant 1$, it follows from (1) that $A_i^{(h+m)} \geqslant A_i^{(h+m-1)}$, and since $A_i^{(h)}, \dots, A_i^{(h+m-1)}$ cannot all be zero, it follows that there exists a finite integer k such that $A_j^{(i)} > 0$ if $i > k$. In what follows we shall assume that h exceeds this value k. For a sufficiently large h, we have, from (1),

$$\frac{A_i^{(h+m)}}{A_1^{(h+m)}} = \sum_{j=0}^{m-1} \lambda_j^{(h)} \frac{A_i^{(h+j)}}{A_1^{(h+j)}}, \tag{2}$$

where
$$\lambda_j^{(h)} = a_{j+1}^{(h)} \frac{A_1^{(h+j)}}{A_1^{(h+m)}} \geqslant 0,$$

and
$$\sum_{j=0}^{m-1} \lambda_j^{(h)} = \frac{\sum_0^{m-1} a_{j+1}^{(h)} A_1^{(h+j)}}{A_1^{(h+m)}} = 1. \tag{3}$$

Now
$$\frac{\lambda_j^{(h)}}{\lambda_{m-1}^{(h)}} = \frac{a_{j+1}^{(h)}}{a_m^{(h)}} \frac{A_1^{(h+j)}}{A_1^{(h+m-1)}} \leqslant 1 \quad (j < m),$$

since, as we have seen, $a_{j+1}^{(h)} \leqslant a_m^{(h)}$, $A_1^{(h+j)} \leqslant A_1^{(h+m-1)}$. From (3) we conclude that $\lambda_{m-1}^{(h)} \geqslant 1/m$.

Let
$$g_i^{(h)} = \max\left[\frac{A_i^{(h)}}{A_1^{(h)}}, \dots, \frac{A_i^{(h+m-1)}}{A_1^{(h+m-1)}}\right],$$

and
$$k_i^{(h)} = \min\left[\frac{A_i^{(h)}}{A_1^{(h)}}, \dots, \frac{A_i^{(h+m-1)}}{A_1^{(h+m-1)}}\right].$$

Equation (2) tells us that $g_i^{(h)} \geqslant g_i^{(h+1)}$, $k_i^{(h+1)} \geqslant k_i^{(h)}$. Hence, generally, we have

$$g_i^{(h)} \geqslant g_i^{(h+1)} \geqslant g_i^{(h+2)} \geqslant \dots \geqslant k_i^{(h+2)} \geqslant k_i^{(h+1)} \geqslant k_i^{(h)}.$$

It follows that $G_i = \lim_{h\to\infty} g_i^{(h)}$ and $K_i = \lim_{h\to\infty} k_i^{(h)}$ both exist, and that $G_i \geqslant K_i$. We shall show that $G_i = K_i$.

Given any $\epsilon > 0$, we can choose h_0 so that

$$0 \leqslant g_i^{(h)} - G_i < \epsilon, \quad 0 \leqslant K_i - k_i^{(h)} < \epsilon,$$

whenever $h > h_0$. Then

$$\frac{A_i^{(h+j)}}{A_1^{(h+j)}} > K_i - \epsilon, \quad \text{if} \quad h > h_0.$$

But
$$\frac{A_i^{(h+m)}}{A_1^{(h+m)}} = \sum_{j=0}^{m-2} \lambda_j^{(h)} \frac{A_i^{(h+j)}}{A_1^{(h+j)}} + \lambda_{m-1}^{(h)} \frac{A_i^{(h+m-1)}}{A_1^{(h+m-1)}}$$

$$> \sum_{j=0}^{m-2} \lambda_j^{(h)}(K_i - \epsilon) + \lambda_{m-1}^{(h)} \frac{A_i^{(h+m-1)}}{A_1^{(h+m-1)}}$$

$$= K_i - \epsilon + \lambda_{m-1}^{(h)}\left(\frac{A_i^{(h+m-1)}}{A_1^{(h+m-1)}} - K_i + \epsilon\right)$$

$$\geqslant K_i - \epsilon + m^{-1}\left(\frac{A_i^{(h+m-1)}}{A_1^{(h+m-1)}} - K_i + \epsilon\right),$$

if $h > h_0$. Hence, if $h > h_0$,

$$\frac{A_i^{(h+m)}}{A_1^{(h+m)}} - K_i > m^{-1}\left(\frac{A_i^{(h+m-1)}}{A_1^{(h+m-1)}} - K_i\right) - \epsilon\left(1 - \frac{1}{m}\right).$$

Repeating this process $s = m - r + 1$ times, we have

$$\frac{A_i^{(h+m)}}{A_1^{(h+m)}} - K_i > m^{-s}\left(\frac{A_i^{(h+r-1)}}{A_1^{(h+r-1)}} - K_i\right) - \epsilon\left(1 - \frac{1}{m^s}\right),$$

provided h is sufficiently large. Since the $k_i^{(h)}$ increase with h, it is clearly possible, in view of (2), to choose $h+m$ so that

$$A_i^{(h+m)}/A_1^{(h+m)} = k_i^{(h+1)},$$

and then choose r $(r \leqslant m)$ so that $A_i^{(h+r-1)}/A_1^{(h+r-1)} = g_i^{(h+1)}$. When we do this, we obtain the inequality

$$g_i^{(h+1)} - K_i < \epsilon(m^s - 1) + m^s(k_i^{(h+1)} - K_i)$$
$$\leqslant \epsilon(m^s - 1),$$

and hence
$$G_i - K_i \leqslant \epsilon(m^s - 1).$$

It follows that $G_i = K_i$.

We then see that $\lim_{h \to \infty} A_i^{(h)}/A_1^{(h)} = G_i = K_i$. Choose h_1 so that $|A_i^{(h)}/A_1^{(h)} - G_i| < \epsilon$, whenever $h > h_1$. We have

$$\tau_i = \sum_{j=1}^{m} A_i^{(h+j-1)} \tau_j^{(h)},$$

and hence

$$\left|\frac{\tau_i}{\tau_1} - G_i\right| = \left|\frac{\sum_{j=1}^{m} (A_i^{(h+j-1)} - G_i A_1^{(h+j-1)})\,\tau_j^{(h)}}{\sum_{j=1}^{m} A_1^{(h+j-1)}\tau_j^{(h)}}\right|$$

$$\leqslant \left|\frac{\epsilon \sum A_1^{(h+j-1)}\tau_j^{(h)}}{\sum A_1^{(h+j-1)}\tau_j^{(h)}}\right| = \epsilon,$$

since $A_1^{(h+j-1)} > 0$, $\tau_j^{(h)} > 0$. We conclude that

$$\lim_{h\to\infty} \frac{A_i^{(h)}}{A_1^{(h)}} = \frac{\tau_i}{\tau_1}.$$

We require also a second preliminary lemma dealing with a set of conjugate points $(\alpha_1^{(i)}, \ldots, \alpha_m^{(i)})$ $(i = 1, \ldots, g)$ in the affine space A_m, defined over the ground field K. Here R denotes the field

$$K(\alpha_1^{(1)}, \ldots, \alpha_1^{(g)}, \ldots, \alpha_m^{(1)}, \ldots, \alpha_m^{(g)}).$$

Lemma II. Let $X_i(x_1, \ldots, x_m)$ *and* $Y_i(x_1, \ldots, x_m)$ $(i = 1, \ldots, g)$ *be two sets of polynomials in* $\mathfrak{J} = R[x_1, \ldots, x_m]$ *which are conjugate over* K, *and suppose that* $X_i(\alpha_1^{(i)}, \ldots, \alpha_m^{(i)})$ *is not zero. Then there exists a polynomial* $H(x_1, \ldots, x_m)$ *in* $K[x_1, \ldots, x_m]$ *such that if* $X_i H + Y_i$ *is written as a polynomial in* $x_1 - \alpha_1^{(i)}, \ldots, x_m - \alpha_m^{(i)}$, *it has no terms of degree less than* ρ_0, ρ_0 *being any assigned integer.*

Let $\mathfrak{p}_i$ be the ideal with basis $(x_1 - \alpha_1^{(i)}, \ldots, x_m - \alpha_m^{(i)})$ in the integral domain $\mathfrak{J} = R[x_1, \ldots, x_m]$. Since the points $(\alpha_1^{(i)}, \ldots, \alpha_m^{(i)})$ $(i = 1, \ldots, g)$ are all distinct, the g ideals $\mathfrak{p}_1, \ldots, \mathfrak{p}_g$ are distinct maximal ideals, and hence $(\mathfrak{p}_i^{\rho_0}, \mathfrak{p}_j^{\rho_0}) = \mathfrak{J}$ $(i \neq j)$. Hence we can find g conjugate elements $\zeta_1, \ldots, \zeta_g$ in $\mathfrak{J}$ such that

$$\zeta_i = \delta_{ij} \quad (\mathfrak{p}_j^{\rho_0}).$$

Since $X_i \neq 0$ $(\mathfrak{p}_i)$, by hypothesis,

$$\mathfrak{J}\cdot(X_i, \mathfrak{p}_i^{\rho_0}) = \mathfrak{J};$$

hence we can find g conjugate elements $h_1, \ldots, h_g$ in $\mathfrak{J}$, and a set of conjugate elements $\eta_1, \ldots, \eta_g$, where η_i is in $\mathfrak{p}_i^{\rho_0}$, such that

$$h_i X_i + \eta_i = -Y_i \quad (i = 1, \ldots, g).$$

We take $$H = h_1\zeta_1 + \ldots + h_g\zeta_g.$$

Then $$X_i H + Y_i = h_i X_i + Y_i = -\eta_i = 0 \quad (\mathfrak{p}_i^{\rho_0}),$$

and $H = H(x_1, \ldots, x_m)$ is a polynomial over R, self-conjugate over K, and therefore in $K[x_1, \ldots, x_m]$, which satisfies the required conditions.

We now proceed to build up a series of Cremona transformations over K on $x_1, \ldots, x_m$ corresponding to the various stages of Perron's algorithm performed on the values $\tau_1, \ldots, \tau_m$ of $\xi_1 - \alpha_1, \ldots, \xi_m - \alpha_m$ in the valuation $\mathfrak{B}$, which we assume to be rationally independent. For convenience in future applications we assume, as we may, that if $R = K(\alpha_1^{(1)}, \ldots, \alpha_{d+1}^{(g)})$ is an extension of K of order g, the g conjugates $\alpha_j^{(1)}, \ldots, \alpha_j^{(g)}$ are distinct, for each value of j. It is worth noting, however, that we only use the fact that $\alpha_1^{(1)}, \ldots, \alpha_1^{(g)}$ are distinct. Since the transformation only involves $x_1, \ldots, x_m$, we shall omit $x_{m+1}, \ldots, x_{d+1}$ from our equations.

Let $\psi(x) = \prod_{i=1}^{g} (x - \alpha_1^{(i)})$ be the characteristic polynomial of $\alpha_1^{(1)} = \alpha_1$ over K. Defining the integers $a_i^{(0)}$ as in Lemma I, we consider a transformation

$$T_1 \begin{cases} x_1 = x'_m, \\ x_i = x'_{i-1}[\psi(x'_m)]^{a_i^{(0)}} + \phi_i(x'_m) \quad (i = 2, \ldots, m), \end{cases}$$

where $\phi_i(x)$ is a polynomial over K to be determined. We notice that the equations of T_1 can be solved to give

$$x'_i = [x_{i+1} - \phi_{i+1}(x_1)]/[\psi(x_1)]^{a_i^{(0)}} \quad (i = 1, \ldots, m-1),$$
$$x'_m = x_1;$$

hence T_1 is a Cremona transformation of the type (C).

Let $\alpha'_1, \ldots, \alpha'_{m-1}$ be any elements of R such that the g conjugates of α'_j $(j = 1, \ldots, m-1)$ over K are distinct, and let $\alpha'_m = \alpha_1$. If we write $\psi(x) = (x - \alpha_1)\psi^*(x)$, the equations of T_1 can be written (over R) in the form

$$x_1 - \alpha_1 = x'_m - \alpha'_m,$$
$$x_i - \alpha_i = (x'_m - \alpha'_m)^{a_i^{(0)}} [\psi^*(x'_m)]^{a_i^{(0)}} (x'_{i-1} - \alpha'_{i-1})$$
$$+ \alpha'_{i-1}[\psi(x'_m)]^{a_i^{(0)}} - \alpha_i + \phi_i(x'_m) \quad (i = 2, \ldots, m).$$

In Lemma II, take $m = 1$, $\rho_0 = \rho + a_i^{(0)}$, where ρ is an arbitrarily large integer, and $X_1 = 1$, $Y_1 = \alpha'_{i-1}[\psi(x)]^{a_i^{(0)}} - \alpha_i$. We put $\phi_i(x)$ equal to the polynomial $H(x)$ constructed in the lemma. It follows that

$$x_i - \alpha_i = (x'_m - \alpha'_m)^{a_i^{(0)}} \Delta_{i-1}(x'),$$

where

$$\Delta_i(x') = [\psi^*(x'_m)]^{a_i^{( )}} (x'_i - \alpha'_i) + (x'_m - \alpha'_m)^{\rho} G_i(x'_m) \quad (i = 1, \ldots, m-1).$$

We now consider the values of $\xi_i' - \alpha_i'$ in the valuation $\mathfrak{B}$. From the equations of T_1,

$$\tau_m' = v(\xi_m' - \alpha_m') = v(\xi_1 - \alpha_1) = \tau_1 > 0.$$

Also, since $\psi^*(\alpha_1) = \psi^*(\alpha_m') \neq 0$, $v[\psi^*(\xi_m')] = 0$. If ρ is sufficiently large, $\rho\tau_1 > v(\xi_i' - \alpha_i')$, since the value group is Archimedean-ordered. Hence, provided that σ is sufficiently large, we have, for $i < m$,

$$v[\Delta_i(\xi')] = \min\left[v\{[\psi^*(\xi_m')]^{a_i^{(0)}}(\xi_i' - \alpha_i')\}, \quad \rho\tau_1 + v\{G_i(\xi_m')\}\right]$$
$$= v(\xi_i' - \alpha_i') = \tau_i' \quad \text{(say)}.$$

Hence
$$\tau_i = v(\xi_i - \alpha_i) = a_i^{(0)} v(\xi_m' - \alpha_m') + \tau_{i-1}'$$
$$= a_i^{(0)} \tau_m' + \tau_{i-1}'.$$

By comparison with the equations obtained in the first stage of Perron's algorithm, we see that

$$\tau_i' = \tau_i^{(1)} \quad (i = 1, \ldots, m).$$

We sum up the essential properties of the transformation T_1 corresponding to the first stage of Perron's algorithm as follows (writing $\Delta_m(x')$ for $x_m' - \alpha_m'$):

(i) T_1 is a transformation of type (C) which, over R, can be written in the form

$$x_1 - \alpha_1 = \Delta_m(x'), \quad x_i - \alpha_i = \Delta_{i-1}(x')\Delta_m^{a_i^{(0)}}(x') \quad (i = 2, \ldots, m),$$

that is, equivalently,

$$x_i - \alpha_i = \Delta_1^{A_i^{(1)}}(x') \ldots \Delta_m^{A_i^{(m)}}(x') \quad (i = 1, \ldots, m);$$

(ii) $\Delta_i(x') = A_i(x_1', \ldots, x_m')(x_i' - \alpha_i') + (x_m' - \alpha_m')^\rho B_i(x_1', \ldots, x_m'),$

where $\quad A_i(\alpha_1', \ldots, \alpha_m') \neq 0 \quad (i \neq m), \qquad \Delta_m(x') = x_m' - \alpha_m';$

(iii) the centre of $\mathfrak{B}$ in the transformed space is

$$(\alpha_1', \ldots, \alpha_m', \alpha_{m+1}, \ldots, \alpha_{d+1}),$$

and its coordinates lie in R; the conjugates of α_i' over K are all distinct;

(iv) $v(\xi_i' - \alpha_i') = v[\Delta_i(\xi')] = \tau_i^{(1)};$

(v) ρ is a large integer.

Corresponding to the second stage of Perron's algorithm, we can construct a transformation $(x_1', \ldots, x_m')$ to $(x_1'', \ldots, x_m'')$ similar to T_1;

the transformation from $(x_1, \ldots, x_m)$ to $(x''_1, \ldots, x''_m)$ will be denoted by T_2. Proceeding in this way, we obtain a transformation T_h corresponding to the h-th stage of the algorithm. This is clearly a transformation of the type (C):

$$x_i = P_i(y_1, \ldots, y_m) \quad (i = 1, \ldots, m),$$

where P_i is a polynomial over K. We shall prove that T_h has the following properties:

(i) over R, T_h can be written in the form

$$x_i - \alpha_i = \Delta_1^{A_i^{(h)}}(y) \ldots \Delta_m^{A_i^{(h+m-1)}}(y)\, L_i(y_1, \ldots, y_m) \quad (i = 1, \ldots, m),$$

where
$$L_i(\beta_1, \ldots, \beta_m) \neq 0;$$

(ii)
$$\Delta_i(y) = A_i(y_1, \ldots, y_m)(y_i - \beta_i) + (y_m - \beta_m)^\rho B_i(y_1, \ldots, y_m) \quad (i < m),$$
$$\Delta_m(y) = y_m - \beta_m,$$

where
$$A_i(\beta_1, \ldots, \beta_m) \neq 0;$$

(iii) the centre $\mathfrak{B}$ in the transformed space is

$$(\beta_1, \ldots, \beta_m, \alpha_{m+1}, \ldots, \alpha_{d+1}),$$

and its coordinates lie in R; the conjugates of β_i over K are all distinct;

(iv)
$$v(\eta_i - \beta_i) = v[\Delta_i(\eta)] = \tau_i^{(h)};$$

(v) ρ is a large integer.

We have seen that these conditions are satisfied in the case $h = 1$. We therefore assume that T_h has the above properties, and prove that T_{h+1} also has these properties.

We pass from T_h to T_{h+1} by subjecting $y_1, \ldots, y_m$ to a transformation $T_1^{(h)}$ over K similar to T_1 which has, over R, the form

$$y_1 - \beta_1 = y'_m - \beta'_m,$$
$$y_i - \beta_i = (y'_m - \beta'_m)^{a_i^{(h)}} [C_i(y'_m)(y'_{i-1} - \beta'_{i-1}) + (y'_m - \beta'_m)^{\rho'} D_i(y'_m)] \quad (i = 2, \ldots, m),$$

where $(\beta'_1, \ldots, \beta'_m, \alpha_{m+1}, \ldots, \alpha_{d+1})$ is the centre of $\mathfrak{B}$ in the y'-space, and has coordinates in R; also the conjugates of β'_i over K are distinct [Property (iii)]. $C_i(\beta'_m) \neq 0$, and ρ' is a large integer.

From the properties of T_1 we also see that, if ρ' is sufficiently large,

$$v[C_i(\eta'_m)(\eta'_{i-1}-\beta'_{i-1})+(\eta'_m-\beta'_m)^{\rho'}D_i(\eta'_m)] = v(\eta'_{i-1}-\beta'_{i-1}) = \tau'_{i-1}, \text{ say} \quad (i = 2, \ldots, m),$$

and
$$\tau'_m = v(\eta'_m-\beta'_m) = v(\eta_1-\beta_1) = \tau_1^{(h)}.$$

Hence
$$\tau_i^{(h)} = v(\eta_i-\beta_i) = a_i^{(h)}\tau_1^{(h)}+\tau'_{i-1} \quad (i = 2, \ldots, m).$$

Comparing these equations with the equations of Perron's algorithm, we find that $\tau'_i = \tau_i^{(h+1)}$ $(i = 1, \ldots, m)$.

We now consider $\Delta_1(y), \ldots, \Delta_m(y)$.

$$\begin{aligned}\Delta_1(y) &= A_1(y)(y_1-\beta_1)+(y_m-\beta_m)^{\rho}B_1(y)\\ &= (y'_m-\beta'_m)[A_0(y')+(y'_m-\beta'_m)^{\rho a_m^{(h)}-1}B_0(y')],\end{aligned}$$

where $A_0(y')$ and $B_0(y')$ are polynomials in $y'_1, \ldots, y'_m$, and

$$A_1(y) = A_0(y').$$

Since $v[A_1(\eta)] = 0$, $v[A_0(\eta')] = 0$; that is $A_0(\beta') \neq 0$. Now $a_m^{(h)} > 0$; hence, if ρ is sufficiently large,

$$v[A_0(\eta')+(\eta'_m-\beta'_m)^{\rho a_m^{(h)}-1}B_0(\eta')] = v[A_0(\eta')] = 0.$$

Hence
$$\Delta_1(y) = (y'_m-\beta'_m)M(y'_1, \ldots, y'_m),$$

where $M(\beta'_1, \ldots, \beta'_m) \neq 0$. If $1 < i < m$,

$$\begin{aligned}\Delta_i(y) &= A_i(y)(y_i-\beta_i)+(y_m-\beta_m)^{\rho}B_i(y)\\ &= A'_i(y')(y'_m-\beta'_m)^{a_i^{(h)}}\{C_i(y'_m)(y'_{i-1}-\beta'_{i-1})+(y'_m-\beta'_m)^{\rho'}D_i(y'_m)\}\\ &\quad +(y'_m-\beta'_m)^{\rho a_m^{(h)}}\{C_m(y'_m)(y'_{m-1}-\beta'_{m-1})+(y'_m-\beta'_m)^{\rho'}D_m(y'_m)\}^{\rho}\\ &\qquad \times B'_i(y'),\end{aligned}$$

where
$$A_i(y) = A'_i(y'), \quad B_i(y) = B'_i(y').$$

Hence if ρ, ρ' are sufficiently large, we can write

$$\Delta_i(y) = (y'_m-\beta'_m)^{a_i^{(h)}}[A^*_{i-1}(y')(y'_{i-1}-\beta'_{i-1})+(y'_m-\beta'_m)^{\sigma}B^*_{i-1}(y')],$$

where σ is an assigned large integer, and $B^*_{i-1}(y')$ is a polynomial, while

$$A^*_{i-1}(y') = A'_i(y')C_i(y'_m).$$

Since
$$v[A'_i(\eta')] = v[C_i(\eta')] = 0, \quad A^*_{i-1}(\beta') \neq 0.$$

We write, for $1 < i < m$,

$$\Delta^*_{i-1}(y') = A^*_{i-1}(y')(y'_{i-1}-\beta'_{i-1})+(y'_m-\beta'_m)^{\sigma}B^*_{i-1}(y').$$

Then, if σ is sufficiently large,

$$v[\Delta^*_{i-1}(\eta')] = v(\eta'_{i-1}-\beta'_{i-1}) = \tau^{(h+1)}_{i-1}.$$

Finally, we have

$$\begin{aligned}\Delta_m(y) &= y_m - \beta_m \\ &= (y'_m-\beta'_m)^{a^{(h)}_m}\Delta^*_{m-1}(y'),\end{aligned}$$

where

$$\begin{aligned}\Delta^*_{m-1}(y') &= C_m(y'_m)(y'_{m-1}-\beta'_{m-1}) + (y'_m-\beta'_m)^{\rho'} D_m(y'_m) \\ &= A^*_{m-1}(y')(y'_{m-1}-\beta'_{m-1}) + (y'_m-\beta'_m)^{\sigma} B^*_{m-1}(y'),\end{aligned}$$

if $\rho' \geqslant \sigma$. Again, if σ is sufficiently large, we have

$$v[\Delta^*_{m-1}(\eta')] = v(\eta'_{m-1}-\beta'_{m-1}) = \tau^{(h+1)}_{m-1}.$$

If we write $\Delta^*_m(y') = y'_m - \beta'_m$, we see that $\Delta^*_1(y'), \ldots, \Delta^*_m(y')$ satisfy the condition (ii); also that (iv) is satisfied with h replaced by $h+1$. The integer σ satisfies the condition (v) for ρ.

We have still to consider condition (i). We have

$$x_i - \alpha_i = \Delta_1^{A_i^{(h)}}(y)\ldots\Delta_m^{A_i^{(h+m-1)}}(y)\,L_i(y_1,\ldots,y_m),$$

and

$$\Delta_1(y) = \Delta^*_m(y')\,M(y'_1,\ldots,y'_m),$$

$$\Delta_i(y) = \Delta^*_{i-1}(y')\,[\Delta^*_m(y')]^{a^{(h)}_i} \quad (i = 2,\ldots,m).$$

Hence

$$\begin{aligned}x_i-\alpha_i = [\Delta^*_1(y')]^{A_i^{(h+1)}}\ldots[\Delta^*_{m-1}(y')]^{A_i^{(h+m-1)}}&[\Delta^*_m(y')]^{k_i} \\ &\times L_i(y_1,\ldots,y_m)\,[M(y')]^{A_1^{(h)}},\end{aligned}$$

where

$$k_i = \sum_{j=1}^{m} A_i^{(h+j-1)}a_j^{(h)} = A_i^{(h+m)} \quad (a_1^{(h)}=1).$$

Also if $L^*_i(y'_1,\ldots,y'_m) = L_i(y_1,\ldots,y_m)\,[M(y'_1,\ldots,y'_m)]^{A_1^{(h)}}$,

$$L^*_i(\beta'_1,\ldots,\beta'_m) = L_i(\beta_1,\ldots,\beta_m)\,[M(\beta'_1,\ldots,\beta'_m)]^{A_1^{(h)}} \neq 0.$$

Hence, replacing $\Delta_1,\ldots,\Delta_m, L_i$ by $\Delta^*_1,\ldots,\Delta^*_m, L^*_i$ in (i) and h by $h+1$, we see that T_{h+1} satisfies property (i), and therefore all the required properties.

The transformations T_h perform a fundamental role in the proof of the Local Uniformisation Theorem.

We now establish an important property of them. Let

$$F(x_1,\ldots,x_{d+1})$$

be a polynomial with coefficients in R. We write

$$F(x) = \Sigma a_i(x_{m+1},\ldots,x_{d+1})(x_1-\alpha_1)^{\lambda_{i1}}\ldots(x_m-\alpha_m)^{\lambda_{im}}.$$

If $i \neq j$, we cannot have

$$\lambda_{i1}\tau_1 + \ldots + \lambda_{im}\tau_m = \lambda_{j1}\tau_1 + \ldots + \lambda_{jm}\tau_m,$$

since the fact that $\tau_1, \ldots, \tau_m$ are rationally independent would imply that $\lambda_{ik} = \lambda_{jk}$, for $k = 1, \ldots, m$. We may suppose that the terms are arranged so that

$$\lambda_{11}\tau_1 + \ldots + \lambda_{1m}\tau_m < \lambda_{j1}\tau_1 + \ldots + \lambda_{jm}\tau_m \quad (j > 1).$$

We say that $F(x)$ is *monovalent* in $x_1, \ldots, x_m$ if

$$v[a_1(\xi_{m+1}, \ldots, \xi_{d+1})] = 0.$$

If $F(x)$ is monovalent, and $j > 1$,

$$v[a_j(\xi)(\xi_1 - \alpha_1)^{\lambda_{j1}} \ldots (\xi_m - \alpha_m)^{\lambda_{jm}}] = v[a_j(\xi)] + \sum_k \lambda_{jk}\tau_k$$
$$> v[a_1(\xi)] + \sum_k \lambda_{1k}\tau_k;$$

hence $$v[F(\xi)] = \lambda_{11}\tau_1 + \ldots + \lambda_{1m}\tau_m.$$

THEOREM I. *If $F(x_1, \ldots, x_{d+1})$ is monovalent in $x_1, \ldots, x_m$, then, if h is sufficiently large, T_h transforms $F(x)$ into a polynomial of the form*

$$\Delta_1^{\lambda_1}(y) \ldots \Delta_m^{\lambda_m}(y)\, G(y_1, \ldots, y_{d+1}),$$

where $$v[G(\eta)] = 0 \quad \textit{and} \quad \lambda_i = \sum_{j=1}^{m} \lambda_{1j} A_j^{(h+i-1)}.$$

Since T_h transforms

$$(x_j - \alpha_j) \quad \text{into} \quad \Delta_1^{A_j^{(h)}}(y) \ldots \Delta_m^{A_j^{(h+m-1)}}(y)\, L_j(y_1, \ldots, y_m),$$

where $v[L_j(\eta)] = 0$, it transforms

$$\prod_{j=1}^{m} (x_j - \alpha_j)^{\lambda_{ij}} \quad \text{into} \quad \prod_{j=1}^{m} \Delta_j^{\mu_{ij}}(y)\, M_i(y_1, \ldots, y_m),$$

where $$v[M_i(\eta)] = 0 \quad \text{and} \quad \mu_{ij} = \sum_{k=1}^{m} \lambda_{ik} A_k^{(h+j-1)}.$$

Now, by hypothesis,

$$\sum_k (\lambda_{ik} - \lambda_{1k})\tau_k > 0 \quad (i > 1).$$

But $A_k^{(h+j-1)}/A_1^{(h+j-1)}$ tends to τ_k/τ_1, and if h is sufficiently large, $A_1^{(h+j-1)}$, τ_1, τ_k are positive. Hence, if h is sufficiently large,

$$\sum_k (\lambda_{ik} - \lambda_{1k})\, A_k^{(h+j-1)} > 0.$$

We choose h so large that this holds of all values of i greater than one, and for $j = 1, \ldots, m$. Then $\mu_{ij} > \mu_{1j} = \lambda_j$, say. Then

$$a_1 \prod_{j=1} (x_j - \alpha_j)^{\lambda_{1j}} = \Delta_1^{\lambda_1}(y) \ldots \Delta_m^{\lambda_m}(y) \, N(y_1, \ldots, y_m),$$

where $v[N(\eta)] = 0$; and, if $k > 1$,

$$a_k \prod_{j=1}^{m} (x_j - \alpha_j)^{\lambda_{kj}} = \Delta_1^{\lambda_1+1}(y) \ldots \Delta_m^{\lambda_m+1}(y) \, H_k(y_1, \ldots, y_m).$$

Hence

$$F(x) = \Delta_1^{\lambda_1}(y) \ldots \Delta_m^{\lambda_m}(y) \, [N(y_1, \ldots, y_m) + \Delta_1(y) \ldots \Delta_m(y) \sum_{k \geq 2} H_k(y)].$$

Since

$$v[\Delta_i(\eta)] > 0,$$

$$\begin{aligned} v[G(\eta)] &= v[N(\eta) + \Delta_1(\eta) \ldots \Delta_m(\eta) \, \Sigma H_k(\eta)] \\ &= v[N(\eta)] = 0. \end{aligned}$$

The theorem is therefore proved.

Corollary. If $F(x)$ and $G(x)$ are monovalent in $x_1, \ldots, x_m$, and $v[F(\xi)] = v[G(\xi)]$, then, if h is sufficiently large, T_h transforms $F(x)$ and $G(x)$ into polynomials in $y_1, \ldots, y_m$ which differ only by factors $F_1(y)$ and $G_1(y)$ such that $v[F_1(\eta)] = v[G_1(\eta)] = 0$.

We now consider some Cremona transformations on $x_1, \ldots, x_{m+1}$, where the value τ_{m+1} of $\xi_{m+1} - \alpha_{m+1}$ is rationally dependent on $\tau_1, \ldots, \tau_m$, and $\tau_1, \ldots, \tau_m$ are rationally independent.

A set of forms $\Delta_1(x), \ldots, \Delta_m(x)$ in $x_1, \ldots, x_m$, with coefficients in R, such that

$$\Delta_i(x) = A_i(x_1, \ldots, x_m) \, (x_i - \alpha_i) + (x_m - \alpha_m)^\rho \, B_i(x_1, \ldots, x_m),$$

where ρ is a large integer, and $v[A_i(\xi)] = 0$, will be called a *normal m-set* in $x_1, \ldots, x_m$. We note that

$$\Delta_m(x) = (x_m - \alpha_m) \, C(x_1, \ldots, x_m),$$

where $v[C(\xi_1, \ldots, \xi_m)] = 0$; hence we can, in fact, always take $\Delta_m(x)$ to be $x_m - \alpha_m$. In the case in which τ_{m+1} is rationally dependent on $\tau_1, \ldots, \tau_m$, we have also to consider $(m+1)$-sets in $x_1, \ldots, x_{m+1}$, such as $\delta_1(x), \ldots, \delta_{m+1}(x)$, where

$$\delta_i(x) = a_i(x_1, \ldots, x_{m+1}) \, (x_i - \alpha_i) + (x_m - \alpha_m)^\rho \, b_i(x_1, \ldots, x_{m+1}),$$

where $v[a_i(\xi)] = 0$, and ρ is a large integer.

We seek a Cremona transformation of type (C) which will reduce an $(m+1)$-set to a special form. We prove

THEOREM II. *Given an $(m+1)$-set $\delta_1(x), \ldots, \delta_{m+1}(x)$, we can construct a Cremona transformation*

$$x_i = P_i(y_1, \ldots, y_{m+1}) \quad (i = 1, \ldots, m+1)$$

with the following properties:

(i) *if $\mathfrak{B}$ is regarded as a valuation over R, the centre of $\mathfrak{B}$ in the y-space is a zero-dimensional variety with generic point*

$$(\beta_1, \ldots, \beta_{m+1}, \alpha_{m+2}, \ldots, \alpha_{d+1}),$$

where β_i $(i \leqslant m)$ is in R, and the conjugates of β_i $(i \leqslant m)$ over K are distinct;

(ii) $$\alpha_i = P_i(\beta_1, \ldots, \beta_m, y_{m+1}) \quad (i = 1, \ldots, m+1);$$

(iii) $$\delta_i(x) = \Delta_1^{b_{i1}}(y) \ldots \Delta_{m+1}^{b_{i\,m+1}}(y)\, G_i(y_1, \ldots, y_{m+1}),$$

where $b_{ij} \geqslant 0$, $\det(b_{ij}) = \pm 1$, $\Delta_1(y), \ldots, \Delta_m(y)$ is a normal m-set in $y_1, \ldots, y_m$, and

$$\Delta_{m+1}(y) = A_{m+1}(y_1, \ldots, y_{m+1})\, y_{m+1} + (y_m - \beta_m)^\sigma B_{m+1}(y_1, \ldots, y_{m+1}),$$

and $$v[A_{m+1}(\eta)] = v[G_i(\eta)] = 0,$$

σ being a large integer;

(iv) *$A_{m+1}(\beta_1, \ldots, \beta_m, y_{m+1})$ and $G_i(\beta_1, \ldots, \beta_m, y_{m+1})$ are non-zero constants in R;*

(v) $v(\eta_{m+1}) = 0$.

In the proof of this theorem it is essential to suppose that α_j $(j = 1, \ldots, m)$ and its conjugates over K form a K-basis for R; we have seen that by a suitable choice of coordinate systems it is possible to arrange for this in any application of the theorem.

Since τ_{m+1} is rationally dependent on $\tau_1, \ldots, \tau_m$, we have a relation

$$\lambda \tau_{m+1} = \lambda_1 \tau_1 + \ldots + \lambda_m \tau_m,$$

where $\lambda, \lambda_1, \ldots, \lambda_m$ are integers without a common factor, and we may assume that λ is positive. Since $\tau_1, \ldots, \tau_m$ are rationally independent, this relation is unique. For simplicity, we shall break up the construction of the transformation into three stages. Stage A will consist of transformations of type T_h on $x_1, \ldots, x_m$, designed to lead us to a situation in which $\lambda_1, \ldots, \lambda_m$ are all positive and λ_m is greater than λ, and is not divisible by λ, if $\lambda > 1$. Stage B will enable us to reduce the case $\lambda > 1$ to $\lambda = 1$, and Stage C will deal with the case $\lambda = 1$, $\lambda_i \geqslant 1$. In special cases Stages A or B may be absent, but Stage C is always present. This is important in order to establish properties (ii) and (iv). It will help the reader to appreciate the cumulative properties of the transformations used at different stages if we verify at each stage that the transformations have, essentially, property (iii).

Stage A. We perform a transformation of type T_h on $x_1, \ldots, x_m$. Since $\delta_i(x)$ $(i \leqslant m)$ is clearly monovalent in $x_1, \ldots, x_m$, we have, by Theorem I, for $i \leqslant m$,

$$\delta_i(x) = \Delta_1^{A_i^{(h)}}(y) \ldots \Delta_m^{A_i^{(h+m-1)}}(y)\, L_i(y_1, \ldots, y_{m+1}), \quad v[L_i(\eta)] = 0.$$

$\Delta_1(y), \ldots, \Delta_m(y)$ is a normal m-set, so if $\Delta_{m+1}(y) = \delta_{m+1}(x)$, $\Delta_1(y), \ldots, \Delta_{m+1}(y)$ is an $(m+1)$-set, and since $\det(A_i^{(h+j-1)}) = \pm 1$, $A_i^{(h+j-1)} \geqslant 0$, we see that our transformation has the required property (iii).

We have
$$\tau_i = \sum_{j=1}^{m} A_i^{(h+j-1)} \tau_j^{(h)};$$
hence
$$\lambda \tau_{m+1} = \lambda_1^{(h)} \tau_1^{(h)} + \ldots + \lambda_m^{(h)} \tau_m^{(h)},$$
where
$$\lambda_i^{(h)} = \lambda_1 A_1^{(h+i-1)} + \ldots + \lambda_m A_m^{(h+i-1)}.$$

Since $\lambda_i^{(h)}/A_1^{(h+i-1)} \to \left(\sum_{i=1}^{m} \lambda_i \tau_i\right)\Big/\tau_1 = \lambda\tau_{m+1}/\tau_1$ as h tends to infinity, and $A_1^{(h+i-1)}$, λ, τ_{m+1} and τ_1 are positive, $\lambda_i^{(h)}$ is positive if h is sufficiently large. Also, since $A_i^{(h)} \to \infty$, we can arrange that $\lambda_i^{(h)} > \lambda$. Also, since (if $\lambda > 1$) $\lambda_1, \ldots, \lambda_m$ are not all divisible by λ, and $\det(A_i^{(h+j-1)}) = \pm 1$, $\lambda_1^{(h)}, \ldots, \lambda_m^{(h)}$ are not all divisible by λ. Since $\lambda_i^{(h+1)} = \lambda_{i+1}^{(h)}$ $(i < m)$, we can clearly choose h so that $\lambda_m^{(h)}$ is not divisible by λ. Thus, we can find T_h so that $\lambda_i^{(h)} > 0$, $\lambda_m^{(h)} > \lambda$ and $\lambda_m^{(h)}$ is not divisible by λ. Further, by the properties of T_h discussed earlier, we know that the coordinates of the centre of $\mathfrak{B}$ in the y-space are in R, and that the conjugates of each of the coordinates over K are all distinct.

Thus, by suitably choosing T_h, we have reduced our problem to that of proving Theorem II when $\lambda_1, \ldots, \lambda_m$ are positive, $\lambda_m > \lambda$, and λ_m is divisible by λ only if $\lambda = 1$. If $\lambda = 1$, we pass straight to Stage C. If $\lambda > 1$, we come to Stage B.

Stage B. In considering Stage B, we recall that one of the properties of T_h is that $\tau_m^{(h)} > \tau_1^{(h)}$. We therefore assume that

$$\lambda \tau_{m+1} = \lambda_1 \tau_1 + \ldots + \lambda_m \tau_m,$$

where $\lambda_i > 0$, $\lambda_m > \lambda > 1$, $\tau_m > \tau_1$, and λ_m is not divisible by λ. Then we can write
$$\lambda_m = g\lambda + \lambda',$$

where $\lambda > \lambda' > 0$ and $g \geqslant 1$. In the manner in which we constructed T_1, we can construct a Cremona transformation over K on x_m, x_{m+1} only, of the form

$$x_m = x'_m,$$
$$x_{m+1} = x'_{m+1}P(x'_m) + Q(x'_m),$$

which, over R, reduces to the form

$$x_m - \alpha_m = x'_m - \alpha'_m,$$
$$x_{m+1} - \alpha_{m+1} = (x'_m - \alpha'_m)^g [C(x'_m)(x'_{m+1} - \alpha'_{m+1}) + (x'_m - \alpha'_m)^\sigma D(x'_m)].$$

By the properties of the transformation T_h used in Stage A, the conjugates of $\alpha'_i = \alpha_i$ $(i \leqslant m)$ over K are all distinct, and we can choose α'_{m+1} so that its conjugates over K are all distinct.

Again, $$\delta_{m+1}(x) = (x'_m - \alpha'_m)^g \delta'_{m+1}(x'),$$
where

$$\delta'_{m+1}(x') = a_{m+1}(x) C(x'_m)(x'_{m+1} - \alpha'_{m+1}) + (x'_m - \alpha'_m)^{\rho - g} b_{m+1}(x) + (x'_m - \alpha'_m)^\sigma D(x'_m).$$

If we write $\delta'_i(x) = \delta_i(x)$ $(i = 1, \ldots, m)$, $\delta'_1(x'), \ldots, \delta'_{m+1}(x')$ is an $(m+1)$-set and

$$\delta_i(x) = \delta'_i(x') \quad (i = 1, \ldots, m), \qquad \delta_{m+1}(x) = [\delta'_m(x')]^g \delta'_{m+1}(x'),$$

so property (iii) is fulfilled. Moreover, we have

$$\tau'_{m+1} = v(\xi'_{m+1} - \alpha'_{m+1}) = v[\delta'_{m+1}(\xi')] = \tau_{m+1} - g\tau_m,$$

and so we have

$$\lambda\tau'_{m+1} = \lambda_1\tau_1 + \ldots + \lambda_{m-1}\tau_{m-1} + \lambda'\tau_m,$$

where $0 < \lambda' < \lambda$, $\lambda_i > 0$.

It should be noted at this stage that since the relation connecting $\tau_1, \ldots, \tau_m, \tau'_{m+1}$ is unique, and each τ_i is present, as well as τ'_{m+1}, any m of the numbers $\tau_1, \ldots, \tau_m$, τ'_{m+1} must be rationally independent. We now construct another Cremona transformation similar to T_1, this time on x'_1, x'_m, x'_{m+1} (or on x'_1, x'_2 if $m = 1$). If $m = 1$, the transformation is of the form

$$x'_1 = y_2 R(y_1) + S(y_1),$$
$$x'_2 = y_1,$$

and reduces over the field R to the form

$$x'_1 - \alpha'_1 = (y_1 - \beta_1)[E(y_1)(y_2 - \beta_2) + (y_1 - \beta_1)^\tau F(y_1)],$$
$$x'_2 - \alpha'_2 = y_1 - \beta_1;$$

and if $m > 1$, it is of the form

$$\begin{aligned} x'_1 &= y_m, \\ x'_m &= y_{m+1} R(y_m) + S(y_m), \\ x'_{m+1} &= y_1, \end{aligned}$$

and over the field R it reduces to the form

$$\begin{aligned} x'_1 - \alpha'_1 &= y_m - \beta_m, \\ x'_m - \alpha'_m &= (y_m - \beta_m)\,\Delta_{m+1}(y), \\ x'_{m+1} - \alpha'_{m+1} &= y_1 - \beta_1. \end{aligned}$$

$(\beta_1, \ldots, \beta_{m+1}, \alpha_2, \ldots, \alpha_{d-1})$ is the centre of $\mathfrak{B}$ in the y-space, and the conjugates of β_i over K are all distinct. $\Delta_{m+1}(y)$ is of the form

$$E(y_m)\,(y_{m+1} - \beta_{m+1}) + (y_m - \beta_m)^\tau F(y_m),$$

where $v[E(\eta_m)] = 0$, and τ is a large integer.

It is sufficient to consider the case $m > 1$, the differences in the case $m = 1$ being trivial. We have

$$\begin{aligned} \delta'_1(x') &= a'_1(x')\,(x'_1 - \alpha'_1) + (x'_m - \beta'_m)^\sigma b'_1(x') \\ &= (y_m - \beta_m)\,\phi(y_1, \ldots, y_{m+1}), \end{aligned}$$

where $$\phi(y) = a'_1(x) + (y_m - \beta_m)^{\sigma-1} \Delta^\sigma_{m+1}(y)\, b'_1(x'),$$

and hence $\phi(\beta_1, \ldots, \beta_{m+1}) \neq 0$, that is, $v[\phi(\eta)] = 0$.

We write $$\delta'_i(x') = \Delta_i(y) \quad (i = 2, \ldots, m-1).$$

If we put $$\Delta_m(y) = y_m - \beta_m,$$

then we have $$\delta'_m(x') = \Delta_m(y)\,\Delta_{m+1}(y).$$

Finally

$$\begin{aligned} \delta'_{m+1}(x') &= a'_{m+1}(x')\,(x'_{m+1} - \alpha'_{m+1}) + (x'_m - \alpha'_m)^\sigma b'_{m+1}(x') \\ &= a'_{m+1}(x')\,(y_1 - \beta_1) + (y_m - \beta_m)^\sigma \Delta^\sigma_{m+1}(y)\, b'_{m+1}(x') \\ &= A_{m+1}(y)\,(y_1 - \beta_1) + (y_m - \beta_m)^\sigma B_{m+1}(y) \\ &= \Delta_1(y). \end{aligned}$$

Then $\Delta_1(y), \ldots, \Delta_{m+1}(y)$ is an $(m+1)$-set, and

$$\begin{aligned} \delta'_1(x') &= \Delta_m(y)\,\phi(y), \\ \delta'_i(x') &= \Delta_i(y) \quad (i = 2, \ldots, m-1), \\ \delta'_m(x') &= \Delta_m(y)\,\Delta_{m+1}(y), \\ \delta'_{m+1}(x') &= \Delta_1(y); \end{aligned}$$

so property (iii) is satisfied.

We have, also,

$$\begin{aligned}\bar{\tau}_{m+1} = v[\Delta_{m+1}(\eta)] &= v(\eta_{m+1}-\beta_{m+1}) \\ &= v(\xi'_m-\alpha'_m)-v(\eta_m-\beta_m) \\ &= \tau_m-\tau_1;\end{aligned}$$

$$\bar{\tau}_1 = v(\eta_1-\beta_1) = \tau'_{m+1};$$

$$\bar{\tau}_m = v(\eta_m-\beta_m) = \tau_1;$$

$$\bar{\tau}_i = \tau_i \quad (i \neq 1, m, m+1).$$

Hence $\lambda'\bar{\tau}_{m+1} = \lambda\bar{\tau}_1-\lambda_2\bar{\tau}_2-\dots-\lambda_{m-1}\bar{\tau}_{m-1}-(\lambda_1+\lambda')\bar{\tau}_m.$

By our construction, $\bar{\tau}_1, \dots, \bar{\tau}_m$ are rationally independent, and the conjugates of β_i over K are all distinct. Thus we have returned to the position in which we started, but with the integer λ replaced by the smaller integer λ'. By repeating Stages A and B alternately, we can eventually reduce our consideration of Theorem II to the consideration of the case in which

$$\tau_{m+1} = \lambda_1\tau_1+\dots+\lambda_m\tau_m,$$

where $\lambda_i > 0$. This is the case considered at Stage C.

Stage C. As a preliminary to the consideration of this stage of our construction, we shall require another lemma.

Lemma III. Given any positive integer ρ_0, which we suppose sufficiently large, there exists a polynomial $\phi(x_1, \dots, x_m)$, with coefficients in K, and m sets $\epsilon_i^{(1)}, \dots, \epsilon_i^{(g)}$ $(i = 1, \dots, m)$, of polynomials over R which are conjugate over K, such that

(a) $\phi(x)-\alpha_{m+1}^{(j)} = \epsilon_1^{(j)} \dots \epsilon_m^{(j)}$;

(b) $\epsilon_i^{(j)} = A_i^{(j)}(x_1, \dots, x_i)\,(x_i-\alpha_i^{(j)})^{\lambda_i}$ + *terms of degree not less than* ρ_0 *in* $x_1-\alpha_1^{(j)}, \dots, x_i-\alpha_i^{(j)}$, *and*

$$A_i^{(j)}(\alpha_1^{(j)}, \dots, \alpha_i^{(j)}) \neq 0, \quad \epsilon_i^{(j)}(\alpha_1^{(k)}, \dots, \alpha_i^{(k)}) \neq 0 \quad (k \neq j).$$

We recall that $(\alpha_1^{(j)}, \dots, \alpha_{d+1}^{(j)})$ $(j = 1, \dots, g)$ are the conjugates of the centre of $\mathfrak{B}$ in the x-space, and that $\alpha_i^{(j)}$ $(j = 1, \dots, g)$ are assumed all distinct. This is, indeed, one of the points in our argument at which the assumption is essential, at least as regards values of i not exceeding m.

We prove this result by an inductive argument on m, first considering the case $m = 1$. By elementary algebra, we can construct a polynomial $\phi(x)$, with coefficients in K, such that

$$\phi(\alpha_1^{(j)}) = \alpha_{m+1}^{(j)},$$

$$\phi^{(\nu)}(\alpha_i^{(j)}) = 0 \quad (\nu = 1, \dots, \lambda-1,\ \lambda+1, \dots, \rho_0-1),$$

$$\phi^{(\lambda)}(\alpha_1^{(j)}) = 1,$$

where $j = 1, \ldots, g$, and $\phi^{(\nu)}(x)$ is the νth derivative of $\phi(x)$. Then the polynomials

$$\phi(x_1) \quad \text{and} \quad \epsilon_1^{(j)} = \phi(x_1) - \alpha_{m+1}^{(j)}$$

satisfy the requirements of the lemma.

We now suppose that we have constructed functions $\phi'(x)$, $\epsilon_i^{(j)}$ $(i = 1, \ldots, m)$ for a given m and given $\lambda_1, \ldots, \lambda_m$. If $\lambda_m = 0$, we can put $\epsilon_m^{(j)} = 1$, and we are in the case $m-1$. We now prove the theorem for $m, \lambda_1, \ldots, \lambda_{m-1}, \lambda_m + 1$. We write

$$\epsilon_i = \prod_{j=1}^{g} \epsilon_i^{(j)}, \quad \epsilon_i^{*(j)} = \epsilon_i / \epsilon_i^{(j)}.$$

ϵ_i is in $K[x_1, \ldots, x_i]$. We show that the function

$$\phi(x) = \phi'(x) + [cx_m + H(x_1, \ldots, x_{m-1})]\,\epsilon_1 \ldots \epsilon_m$$

satisfies the requirements of the lemma, with λ_m replaced by $\lambda_m + 1$, provided that c is a suitable element of K, and $H(x_1, \ldots, x_{m-1})$ is suitably chosen in $K[x_1, \ldots, x_{m-1}]$. We have

$$\phi(x) - \alpha_{m+1}^{(j)} = \epsilon_1^{(j)} \ldots \epsilon_m^{(j)} \bar{\epsilon}^{(j)},$$

where
$$\bar{\epsilon}^{(j)} = 1 + [cx_m + H(x_1, \ldots, x_{m-1})]\,\epsilon_1^{*(j)} \ldots \epsilon_m^{*(j)}.$$

Now $\epsilon_1^{*(j)} \ldots \epsilon_m^{*(j)} = a^{(j)} + \Sigma b_i^{(j)}(x_i - \alpha_i^{(j)}) +$ terms of higher degree,

and in virtue of the properties of $\epsilon_i^{(j)}$, $a^{(j)}$ is not zero. If

$$H(\alpha_1^{(j)}, \ldots, \alpha_{m-1}^{(j)}) = h_0,$$

$$\bar{\epsilon}^{(j)} = 1 + (c\alpha_m^{(j)} + h_0)\,a^{(j)} + [(c\alpha_m^{(j)} + h_0)\,b_m^{(j)} + ca^{(j)}]\,(x_m - \alpha_m^{(j)}) + \ldots.$$

Eventually, we shall choose $H(x)$ so that this expansion has no constant term. Then the coefficient of $(x_m - \alpha_m^{(j)})$ is $ca^{(j)} - b_m^{(j)}/a^{(j)}$. We take c to be any element of K which does not make this vanish. Now let X_j be the polynomial obtained from $\epsilon_1^{*(j)} \ldots \epsilon_m^{*(j)}$ by putting $x_m = \alpha_m^{(j)}$, and let $Y_j = 1 + c\alpha_m^{(j)} X_j$, and then apply Lemma II (p. 271) to obtain $H(x_1, \ldots, x_{m-1})$. It follows that when we put $x_m = \alpha_m^{(j)}$ in $\bar{\epsilon}^{(j)}$ we obtain a polynomial $B^{(j)}(x_1, \ldots, x_{m-1})$, which, when expanded in terms of $x_1 - \alpha_1^{(j)}, \ldots, x_{m-1} - \alpha_{m-1}^{(j)}$, has no terms of degree less than ρ_0. Hence

$$\bar{\epsilon}^{(j)} = A^{(j)}(x_1, \ldots, x_m)\,(x_m - \alpha_m^{(j)}) + B^{(j)}(x_1, \ldots, x_{m-1}),$$

and since the coefficient of $x_m - \alpha_m^{(j)}$ in the expansion of $\bar{\epsilon}^{(j)}$ is not zero, we have

$$A^{(j)}(\alpha_1^{(j)}, \ldots, \alpha_m^{(j)}) \neq 0.$$

Also $$\bar{\epsilon}^{(j)}(\alpha_1^{(k)}, \ldots, \alpha_m^{(k)}) = 1, \quad \text{if} \quad j \neq k.$$

We put $\bar{\bar{\epsilon}}_m^{(j)} = \epsilon_m^{(j)}\bar{\epsilon}^{(j)}$. Then it is clear that $\phi(x), \epsilon_1^{(j)} \ldots \epsilon_{m-1}^{(j)}\bar{\bar{\epsilon}}_m^{(j)}$ satisfy the lemma when λ_m is replaced by $\lambda_m + 1$.

We return to the problem of constructing a Cremona transformation to satisfy Theorem II.

We construct $\phi(x)$, $\epsilon_i^{(j)}$ for the numbers $\lambda_1, \ldots, \lambda_m$ which appear in the relation
$$\tau_{m+1} = \lambda_1\tau_1 + \ldots + \lambda_m\tau_m,$$
where, we recall, $\lambda_i > 0$. Let
$$y_{m+1} = \frac{x_{m+1} - \phi(x_1, \ldots, x_m)}{\epsilon_1(x) \ldots \epsilon_m(x)} + S(x_1, \ldots, x_m),$$
where $S(x_1, \ldots, x_m)$ is a polynomial over K to be determined. We have
$$y_{m+1} = \frac{x_{m+1} - \alpha_{m+1}^{(1)}}{\epsilon_1(x) \ldots \epsilon_m(x)} - \frac{1}{\epsilon_1^{*(1)}(x) \ldots \epsilon_m^{*(1)}(x)} + S(x).$$
We use Lemma II to determine $S(x)$ so that
$$S(x)\,\epsilon_1^{*(j)}(x) \ldots \epsilon_m^{*(j)}(x) - 1 = C^{(j)}(x_1, \ldots, x_m),$$
when expanded in terms of $x_i - \alpha_i^{(j)}$ $(i = 1, \ldots, m)$, begins with terms of degree σ_0 at least, σ_0 being a large integer. We consider the Cremona transformation
$$\mathscr{T}\begin{cases} x_{m+1} = \phi(y_1, \ldots, y_m) + [y_{m+1} - S(y_1, \ldots, y_m)]\,\epsilon_1(y) \ldots \epsilon_m(y), \\ x_i = y_i \quad (i \neq m+1). \end{cases}$$
We first consider the centre of $\mathfrak{B}$ in the y-space (the ground field being taken to be R). If $i \neq m+1$, we write $\beta_i = \alpha_i$; we then have
$$v(\eta_i - \beta_i) = v(\xi_i - \alpha_i) = \tau_i > 0.$$
We also have
$$\begin{aligned} x_{m+1} - \alpha_{m+1} &= x_{m+1} - \alpha_{m+1}^{(1)} \\ &= \epsilon_1^{(1)}(y) \ldots \epsilon_m^{(1)}(y)\,[1 + (y_{m+1} - S(y))\,\epsilon_1^{*(1)}(y) \ldots \epsilon_m^{*(1)}(y)] \\ &= \epsilon_1^{(1)}(y) \ldots \epsilon_m^{(1)}(y)\,D^{(1)}(y_1, \ldots, y_{m+1}), \end{aligned}$$
where
$$D^{(1)}(y_1, \ldots, y_{m+1}) = \epsilon_1^{*(1)}(y) \ldots \epsilon_m^{*(1)}(y)\,y_{m+1} - C^{(1)}(y_1, \ldots, y_m).$$
Now if the ρ_0 of Lemma III is sufficiently large,
$$v[\epsilon_i^{(1)}(\eta)] = \lambda_i v(\eta_i - \alpha_i) = \lambda_i\tau_i,$$
and, since $$\epsilon_i^{*(1)}(\alpha) \neq 0, \quad v[\epsilon_i^{*(1)}(\eta)] = 0.$$

Therefore $$\sum_{i=1}^{m} \lambda_i \tau_i = \tau_{m+1} = v[\epsilon_1^{(1)}(\eta) \ldots \epsilon_m^{(1)}(\eta) D^{(1)}(\eta)]$$
$$= \sum_{i=1}^{m} \lambda_i \tau_i + v[D^{(1)}(\eta)],$$

that is, $$0 = v[D^{(1)}(\eta)] = v[\epsilon_1^{*(1)}(\eta) \ldots \epsilon_m^{*(1)}(\eta) \eta_{m+1}],$$

if σ_0 is sufficiently large. Hence we conclude that $v(\eta_{m+1}) = 0$.

We therefore see that $v(\eta_i) \geqslant 0$ $(i = 1, \ldots, d+1)$, and hence that the centre of $\mathfrak{B}$ in the y-space is at a finite distance. It lies in the primes $y_i = \beta_i = \alpha_i$ $(i = 1, \ldots, m, m+2, \ldots, d+1)$, defined over R, and it does not lie in $y_{m+1} = 0$. It is a zero-dimensional variety over R, but need not be a point. Its generic point is of the form $(\beta_1, \ldots, \beta_{d+1})$, where β_{m+1} is algebraic over R. The g conjugates of β_i $(i \leqslant m)$ over K are all distinct.

We now point out an important property of the transformation $\mathscr{T}$. Let $f(x_1, \ldots, x_{m+1})$ be any polynomial in $x_1, \ldots, x_{m+1}$, and let $g(y_1, \ldots, y_{m+1})$ be the polynomial derived from it by $\mathscr{T}$. Then

$$g(y_1, \ldots, y_{m+1})$$
$$= f(y_1, \ldots, y_m, \phi(y_1, \ldots, y_m) + [y_{m+1} - S(y)]\epsilon_1(y) \ldots \epsilon_m(y)).$$

Hence, since $$\epsilon_i(\beta) = 0,$$

$$g(\beta_1, \ldots, \beta_m, y_{m+1}) = f(\alpha_1, \ldots, \alpha_m, \phi(\alpha_1, \ldots, \alpha_m)),$$

that is, it is independent of y_{m+1}.

The final transformation is a transformation of type T_h on $y_1, \ldots, y_m$, and hence it does not destroy the property of $\mathscr{T}$ just described. If T_h transforms $y_1, \ldots, y_{m+1}$ into $z_1, \ldots, z_{m+1}$, and $(\beta_1, \ldots, \beta_{d+1})$ into $(\gamma_1, \ldots, \gamma_{d+1})$, $\gamma_i \in R$ $(i \neq m)$, and we obtain

$$\delta_i(x) = \Delta_1^{A_i^{(h)}}(z) \ldots \Delta_m^{A_i^{(h+m-1)}}(z) L_i(z) \quad (i = 1, \ldots, m),$$

where $\Delta_1(z), \ldots, \Delta_m(z)$ is a normal m-set and $v[L_i(\zeta)] = 0$. Also

$$\delta_{m+1}(x) = a_{m+1}(x)(x_{m+1} - \alpha_{m+1}) + (x_m - \alpha_m)^\rho b_{m+1}(x)$$
$$= a'(y)\epsilon_1^{(1)}(y) \ldots \epsilon_m^{(1)}(y) D^{(1)}(y_1, \ldots, y_{m+1}) + (y_m - \beta_m)^\rho b'(y),$$

where $$a_{m+1}(x) = a'(y), \quad b_{m+1}(x) = b'(y).$$

We note that $a'(\beta_1, \ldots, \beta_m, y_{m+1})$ is independent of y_{m+1} by the special property of $\mathscr{T}$.

We now pass from the y-space to the z-space. Since

$$\epsilon_i^{(1)}(y), \ldots, \epsilon_m^{(1)}(y),\; y_m - \beta_m$$

depend only on $y_1, \ldots, y_m$, an immediate application of Theorem I tells us that

$$\epsilon_i^{(1)}(y) = \Delta_1^{\lambda_i A_i^{(h)}}(z) \ldots \Delta_m^{\lambda_i A_i^{(h+m-1)}}(z)\, N_i(z) \quad (i = 1, \ldots, m),$$

$$y_m - \beta_m = \Delta_1^{A_m^{(h)}}(z) \ldots \Delta_m^{A_m^{(h+m-1)}}(z)\, N(z),$$

where $N_i(z)$, $N(z)$ depend only on $z_1, \ldots, z_m$, and

$$v[N_i(\zeta)] = v[N(\zeta)] = 0.$$

Let $\qquad a'(y) = \alpha(z) \quad \text{and} \quad b'(y) = \beta(z).$

These are polynomials in $z_1, \ldots, z_{m+1}$, but, by a remark made above,

$$\alpha(\gamma_1, \ldots, \gamma_m, z_{m+1}) = a'(\beta_1, \ldots, \beta_m, y_{m+1})$$

is a non-zero constant.

Further, we have

$$\begin{aligned} D^{(1)}(y) &= \epsilon_1^{*(1)}(y) \ldots \epsilon_m^{*(1)}(y)\, y_{m+1} - C^{(1)}(y_1, \ldots, y_m) \\ &= P(z_1, \ldots, z_m)\, z_{m+1} + \Delta_1^{\sigma_1}(z) \ldots \Delta_m^{\sigma_m}(z)\, Q'(z); \end{aligned}$$

since $C^{(1)}(y)$ has no terms of degree less than σ_0, which can be made arbitrarily large, we may assume that σ_m is as large as we please. Let $\sigma_m \geqslant \rho'$, where ρ' is a large integer. We then have

$$D^{(1)}(y) = P(z)\, z_{m+1} + (z_m - \gamma_m)^{\rho'} R(z),$$

where $\qquad v[P(\zeta)] = v[\epsilon_1^{*(1)}(\eta), \ldots, \epsilon_m^{*(1)}(\eta)] = 0,$

and $\qquad v(\zeta) = 0.$

We therefore have

$$\begin{aligned} \delta_{m+1}(x) &= a'(y)\, \epsilon_1^{(1)}(y) \ldots \epsilon_m^{(1)}(y)\, D^{(1)}(y) + (y_m - \beta_m)^{\rho}\, b'(y) \\ &= \alpha(z) \prod_1^m \Delta_i^{\mu_i}(z) \prod_1^m N_i(z)\, [P(z)\, z_{m+1} + (z_m - \gamma_m)^{\rho'} R(z)] \\ &\qquad + \prod_1^m \Delta_i^{\rho A_m^{(h+i-1)}}(z)\, N^{\rho}(z)\, \beta(z), \end{aligned}$$

where $\qquad \mu_i = \lambda_1 A_1^{(h+i-1)} + \ldots + \lambda_m A_m^{(h+i-1)}.$

We choose ρ so that

$$\rho A_m^{h+i-1} \geqslant \mu_i \quad (i = 1, \ldots, m-1),$$

and

$$\rho A_m^{(h+m-1)} \geqslant \mu_m + \rho'.$$

Then if we write $$A_{m+1}(z) = \alpha(z) \prod_1^m N_i(z) P(z),$$

$A_{m+1}(\gamma_1, \ldots, \gamma_m, z_{m+1})$ is a non-zero constant, and if we write

$$B_{m+1}(z) = \alpha(z) \prod_1^m N_i(z) R(z) + \prod_1^{m-1} \Delta_i^{\rho A_m^{(h+i-1)} - \mu_1}(z) \Delta_m^{\rho A_m^{(h+m-1)} - \mu_m - \rho'}(z) N^\rho(z) \beta(z),$$

we have $$\delta_{m+1}(x) = \Delta_1^{\mu_1}(z) \ldots \Delta_m^{\mu_m}(z) \Delta_{m+1}(z),$$

where $$\Delta_{m+1}(z) = A_{m+1}(z) z_{m+1} + (z_m - \gamma_m)^{\rho'} B_{m+1}(z).$$

This virtually completes the proof of Theorem II. When we combine all the transformations into one, we clearly obtain a Cremona transformation of the form

$$x_i = P_i(y_1, \ldots, y_{m+1}) \quad (i = 1, \ldots, m+1),$$

where $P_i(y)$ is a polynomial over K. We have shown that the resulting transformation has the properties (iv) and (v); and since at each stage we have seen that (iii) holds for the constituent transformations it necessarily holds for the resultant transformation. We have also shown that (i) holds, so it only remains to show that (ii) is true.

Let the result of combining the transformations of Stages A and B be

$$x_i = p_i(X_1, \ldots, X_{m+1}) \quad (i = 1, \ldots, m+1).$$

The transformation

$$X_i = q_i(y_1, \ldots, y_{m+1}) \quad (i = 1, \ldots, m+1)$$

of Stage C transforms any polynomial $f(X)$ into a polynomial $g(y)$ such that $g(\beta_1, \ldots, \beta_m, y_{m+1})$ is a constant. Taking $f(X)$ to be $\phi_i(X_1, \ldots, X_{m+1})$, we see that, in the combined total transformation

$$x_i = P_i(y_1, \ldots, y_{m+1}) \quad (i = 1, \ldots, m+1),$$

$P_i(\beta_1, \ldots, \beta_m, y_{m+1})$ is independent of y_{m+1}. But when $y_{m+1} = \beta_{m+1}$, we have

$$\alpha_i = P_i(\beta_1, \ldots, \beta_{m+1}) \quad (i = 1, \ldots, m+1).$$

Hence it follows that

$$\alpha_i = P_i(\beta_1, \ldots, \beta_m, y_{m+1}) \quad (i = 1, \ldots, m+1),$$

as required by condition (ii).

Thus we satisfy ourselves that all the requirements of Theorem II are fulfilled.

Our next problem should be to generalise Theorem I so as to remove the condition that $F(x)$ be monovalent in $x_1, \ldots, x_m$. We shall consider this generalisation in the next section. At present we confine ourselves to a special result which will be useful. Let $\delta_1(x), \ldots, \delta_{m+1}(x)$ be an $(m+1)$-set as defined above, and consider any number of products

$$\pi_i(x) = \prod_{j=1}^{m+1} \delta_j^{\lambda_{ij}}(x).$$

The value of $\pi_i(\xi)$ is $\sum_{j=1}^{m+1} \lambda_{ij}\tau_j$, and since $\tau_1, \ldots, \tau_{m+1}$ are rationally dependent, there may be several $\pi_i(\xi)$ with the same values. Suppose that

$$v[\pi_\alpha(\xi)] = v[\pi_\beta(\xi)] = \ldots = v[\pi_\delta(\xi)] < v[\pi_i(\xi)] \quad (i \neq \alpha, \beta, \ldots, \delta).$$

We apply the transformation of Theorem II. Then

$$\pi_i(x) = \prod_{j=1}^{m+1} \Delta_j^{\mu_{ij}}(y)\, G_i(y),$$

where $v[G_i(\eta)] = 0$ and $\Delta_1(y), \ldots, \Delta_{m+1}(y)$ are as described in the theorem. Also

$$\mu_{ij} = \sum_{k=1}^{m+1} \lambda_{ik} b_{kj}.$$

We have
$$v[\pi_i(\xi)] = \sum_{j=1}^{m} \mu_{ij} v[\Delta_j(\eta)];$$

and since $v[\Delta_1(\eta)], \ldots, v[\Delta_m(\eta)]$ are rationally independent, we conclude that

$$\mu_{\alpha j} = \mu_{\beta j} = \ldots = \mu_{\delta j} \leqslant \mu_{ij} \quad (j = 1, \ldots, m),$$

and that $\mu_{\delta j} < \mu_{ij}$ for at least one j, if $i \neq \alpha, \beta, \ldots, \delta$. (By performing another transformation of type T_h, if necessary, it can easily be arranged that $\mu_{\delta j} < \mu_{ij}$ for all j.)

We consider two products $\pi_\alpha(\xi)$, $\pi_\beta(\xi)$ with the same value. Then

$$\sum_{k=1}^{m+1} (\lambda_{\alpha k} - \lambda_{\beta k})\, b_{kj} = 0 \quad (j = 1, \ldots, m),$$

$$\sum_{k=1}^{m+1} (\lambda_{\alpha k} - \lambda_{\beta k})\, b_{k\,m+1} = \mu_{\alpha\, m+1} - \mu_{\beta\, m+1}.$$

By Theorem II, we know that $\det(b_{ij}) = \epsilon = \pm 1$. Let ϵd be the determinant of the m-rowed matrix (b_{ij}) $(i, j \leqslant m)$. Then if we solve the above equations, we have

$$\lambda_{\alpha\, m+1} - \lambda_{\beta\, m+1} = d(\mu_{\alpha\, m+1} - \mu_{\beta\, m+1}).$$

If $d = 0$, $\lambda_{\alpha\, m+1} = \lambda_{\beta\, m+1}$. In this case, since $v[\delta_1(\xi)], \ldots, v[\delta_m(\xi)]$ are rationally independent, the condition $v[\pi_\alpha(\xi)] = v[\pi_\beta(\xi)]$ would imply that $\lambda_{\alpha j} = \lambda_{\beta j}$ $(j = 1, \ldots, m+1)$, and hence $\alpha = \beta$. Since, however, we have seen that we can find two distinct products $\pi_\alpha(\xi)$ and $\pi_\beta(\xi)$ with the same values, we conclude that $d \neq 0$. The important point to note is that d is a non-zero integer depending only on the transformation used in Theorem II, and not on products $\pi_i(x)$ considered. The particular result which we shall quote later can be stated as

THEOREM III. *If $\delta_1(x), \ldots, \delta_{m+1}(x)$ is an $(m+1)$-set, and if*

$$\pi_\alpha(x) = \prod_{i=1}^{m+1} \delta_i^{\alpha_i}(x), \quad \pi_\beta(x) = \prod_{i=1}^{m+1} \delta_i^{\beta_i}(x),$$

where

$$v[\pi_\alpha(\xi)] = v[\pi_\beta(\xi)],$$

then the transformation of Theorem II transforms the products into

$$\prod_{i=1}^{m+1} \Delta_i^{\alpha'_i}(y)\, G_\alpha(y) \quad \textit{and} \quad \prod_{i=1}^{m+1} \Delta_i^{\beta'_i}(y)\, G_\beta(y),$$

where (i) $v[G_\alpha(\eta)] = v[G_\beta(\eta)] = 0$;

(ii) $\alpha'_i = \beta'_i \quad (i = 1, \ldots, m)$;

(iii) $\beta'_{m+1} - \alpha'_{m+1} = (\beta_{m+1} - \alpha_{m+1})/d$,

where d is a non-zero integer depending only on the transformation.

6. The Local Uniformisation Theorem: the main case. In this section we prove what may be called the main case of the Local Uniformisation Theorem, namely, the case of a zero-dimensional valuation of rank one. Explicitly, we state the theorem to be proved as

THEOREM I. *Let U be a primal in the affine space $A_{d+1}(x_1, \ldots, x_{d+1})$, and let $\mathfrak{B}$ be a zero-dimensional valuation of its function field of rank one, whose centre on U is at a finite distance in A_{d+1}. Then there exists a Cremona transformation of A_{d+1} into A'_{d+1} $(y_1, \ldots, y_{d+1})$ of the form*

$$x_i = P_i(y_1, \ldots, y_{d+1}) \quad (i = 1, \ldots, d+1),$$
$$y_i = R_i(x_1, \ldots, x_{d+1}) \quad (i = 1, \ldots, d+1),$$

where P_i is a polynomial over K and R_i is a rational function with coefficients in K, which transforms U into a birationally equivalent variety U', on which the centre of $\mathfrak{B}$ is a simple zero-dimensional variety at a finite distance.

The proof of this theorem is very closely linked with the proof of a theorem which is the most general form of a result of which Theorem I of § 5 is a special case. The extension of this theorem which we shall prove is

THEOREM II. *Let $\mathfrak{B}$ be a zero-dimensional valuation of rank one of the function field $K(\xi_1, \ldots, \xi_{d+1})$ of the irreducible primal U in $(x_1, \ldots, x_{d+1})$-space, whose centre is at a finite distance. Let R' be the least normal extension of K over which the centre of $\mathfrak{B}$ on U reduces to a set of points, and let $\mathfrak{B}'$ be a valuation of $R'(\xi_1, \ldots, \xi_{d+1})$ which is an extension of $\mathfrak{B}$. If $F(x_1, \ldots, x_r)$ $(r \leqslant d+1)$ is any polynomial in $x_1, \ldots, x_r$ with coefficients in R', such that $F(\xi_1, \ldots, \xi_r) \neq 0$, there exists a Cremona transformation of $(x_1, \ldots, x_{d+1})$ into $(y_1, \ldots, y_{d+1})$ of the form*

$$x_i = P_i(y_1, \ldots, y_r) \quad (i = 1, \ldots, r), \qquad x_i = y_i \quad (i > r),$$

where P_i is a polynomial in $y_1, \ldots, y_r$ with coefficients in K, such that

$$F(x_1, \ldots, x_r) = \Delta_1^{\mu_1}(y) \ldots \Delta_m^{\mu_m}(y)\, G(y_1, \ldots, y_r)$$

over the field R^* which is the least normal extension of R' over which the centre of $\mathfrak{B}'$ in the y-space reduces to a set of points.

Let $\mathfrak{B}^*$ be a valuation of $R^*(\xi_1, \ldots, \xi_{d+1})$ which is an extension of $\mathfrak{B}'$, and let $(\beta_1, \ldots, \beta_{d+1})$ be the centre of $\mathfrak{B}^*$ in the y-space, $(\eta_1, \ldots, \eta_{d+1})$ the transform of $(\xi_1, \ldots, \xi_{d+1})$. Then,

(i) *R^* is the field obtained by adjoining to K the element β_j and its conjugates over K, for any j not exceeding m;*

(ii) $v^*[\Delta_i(\eta)] = v^*(\eta_i - \beta_i) = \tau_i^*$ $(i = 1, \ldots, m)$, *where* $\tau_1^*, \ldots, \tau_m^*$ *are rationally independent, and* $\Delta_1(y), \ldots, \Delta_m(y)$ *is a normal m-set in* $y_1, \ldots, y_m$;

(iii) $v^*[G(\eta_1, \ldots, \eta_r)] = 0$.

Some remarks on this enunciation will help towards a fuller understanding of the proof. It may be pointed out that the property (i) of Theorem II has little intrinsic interest, but must be included in the statement of the theorem for the sake of the induction argument. We begin the proof of Theorem II for a polynomial $G(x_1, \ldots, x_{r+1})$ by applying, as hypothesis of induction, the theorem to a derived polynomial $F(x_1, \ldots, x_r)$. The next stage is to apply Theorem II of § 5 to $\Delta_1(y), \ldots, \Delta_m(y), y_{r+1} - \beta_{r+1}$, and in order to do this the property of β_j $(j \leqslant m)$ given in Theorem II (i) is essential. This will explain why at one stage of our argument a diversion has to be made to ensure that (i) is satisfied.

It should be noted that in the statement of Theorem II, the field R' which contains the coefficients of $F(x_1, \ldots, x_r)$ is determined completely by the variety which is the centre of the valuation $\mathfrak{B}$ on U. In our inductive proof, we shall make the truth of Theorem II for $F(x_1, \ldots, x_r)$ depend on the truth of the theorem in an earlier case applied to a polynomial $G(z_1, \ldots, z_s)$ in a transformed space. The coefficients of $G(z_1, \ldots, z_s)$ may not be in R', but as they will lie in the least normal extension of K over which the centre of $\mathfrak{B}$ in the z-space reduces to a set of points, the hypothesis of induction will be valid.

Since it is important to distinguish between valuations $\mathfrak{B}$ of $K(\xi_1, \ldots, \xi_r)$ and extended valuations $\mathfrak{B}'$, $\mathfrak{B}^*$, etc., of $R'(\xi_1, \ldots, \xi_r)$, $R^*(\xi_1, \ldots, \xi_r)$, etc., we shall find it more convenient in this section to denote extended and induced valuations by different symbols. $\mathfrak{B}$ will always denote the basic valuation of $K(\xi_1, \ldots, \xi_r)$, and the valuation for extended ground fields will bear a distinguishing mark bearing an obvious reference to the corresponding ground field.

The valuation $\mathfrak{B}^*$ defined in the theorem has its centre at the point $(\beta_1, \ldots, \beta_{d+1})$. The condition (iii), $v^*[G(\eta)] = 0$, is equivalent to saying that the equation

$$G(\beta_1, \ldots, \beta_{r-1}, y) = 0$$

does not have $y = \beta_r$ as a root. Let us consider the effect of removing the restriction $F(\xi_1, \ldots, \xi_r) \neq 0$. The argument used to prove Theorem II will lead in this case to an equation

$$F(x) = \Delta_1^{\mu_1}(y) \ldots \Delta_m^{\mu_m}(y)\, G(y_1, \ldots, y_r),$$

where conditions (i) and (ii) of Theorem II are satisfied. But if $F(\xi) = 0$, we deduce that $G(\eta) = 0$. Hence the equation

$$G(\beta_1, \ldots, \beta_{r-1}, y) = 0$$

must have $y = \beta_r$ as a root. We can, however, state certain results on the multiplicity of this root. We consider the particular case in which $F(x)$ is an irreducible polynomial with coefficients in K. Then $F(x) = 0$ is the equation of a variety U over K. In this case, we can show that

$$G(\beta_1, \ldots, \beta_{r-1}, y) = 0$$

has $y = \beta_r$ as a simple root, provided that the Cremona transformation is suitably chosen. If $A(y)$ is the polynomial of least degree over K which contains $\Delta_1^{\mu_1}(y) \ldots \Delta_m^{\mu_m}(y)$ as a factor, $A(y) \neq 0$, since $A(y)$ is a polynomial in $y_1, \ldots, y_m$ only, and

$$v^*(\eta_i - \beta_i) \quad (i = 1, \ldots, m)$$

are rationally independent. Hence

$$F(x) = A(y) \Pi G_i^{\rho_i}(y),$$

where the $G_i(y)$ are essentially distinct polynomials in $y_1, \ldots, y_r$ with coefficients in K, and where, for one value of i, $G_i(\eta) = 0$. $G_i(y) = 0$ is then the equation of the transform U' of U by the Cremona transformation. Suppose that the factors are arranged so that $G_1(\eta) = 0$. Since $G_1(\beta) = 0$, and

$$\Pi G_i^{\rho_i}(\beta_1, \ldots, \beta_{r-1}, y) = 0$$

has $y = \beta_r$ as a simple root, it follows that

$$G_1(\beta_1, \ldots, \beta_{r-1}, y) = 0$$

has $y = \beta_r$ as a simple root, and that $\rho_1 = 1$. Moreover,

$$\frac{\partial G_1}{\partial y_r}(\beta_1, \ldots, \beta_r) \neq 0;$$

hence $(\beta_1, \ldots, \beta_{d+1})$ is a simple point of U'. But $(\beta_1, \ldots, \beta_{d+1})$ is one of the conjugate points which make up the centre of $\mathfrak{B}$ on U', and therefore the centre of $\mathfrak{B}$ on U' is simple. This result is simply Theorem I.

Thus the proof of Theorems I and II are closely related; indeed, we find it convenient to prove the theorems together. The procedure, in fact, is to prove Theorem II by induction on r, and then to deduce Theorem I from it. If the equation $F(x) = 0$ of U over K involves only the coordinates $x_1, \ldots, x_{r+1}$, we use Theorem II as applied to polynomials in $x_1, \ldots, x_r$. But the method of proof differs very little from the method of deducing Theorem II for $r+1$ variables from Theorem II for r variables; hence it is convenient to carry out the two proofs simultaneously. It should also be mentioned that in view of our initial choice of the coordinate system in the x-space, the equation of U involves the $d+1$ coordinates $x_1, \ldots, x_{d+1}$; nevertheless, for formal reasons, we assume the truth of Theorem II for r variables and then consider a polynomial $F(x_1, \ldots, x_{r+1})$, distinguishing the cases $F(\xi) = 0$ [Theorem I] and $F(\xi) \neq 0$ [Theorem II].

We must first begin our induction by considering a form $F(x_1)$ in x_1 only. In the case of Theorem I, $F(x_1) = 0$ is taken as the irreducible equation of U, which is therefore a set of conjugate primes

$$x_1 = \alpha_1^{(i)} \quad (i = 1, \ldots, g).$$

Since the centre of $\mathfrak{B}$ is at a finite distance, it lies on just one of these primes, and hence it is already simple, so there is nothing to prove. In the case of Theorem II, we can apply Theorem I of § 5 at once to obtain the required result. We now assume the truth of Theorem II for polynomials in r variables, and consider a polynomial $F(x_1, \ldots, x_{r+1})$ in $r+1$ variables, with coefficients in R. In the case of Theorem I, $F(x_1, \ldots, x_{r+1}) = 0$ is the irreducible equation of U over K; hence no polynomial in $x_1, \ldots, x_r$ over K vanishes when we replace x by ξ; thus in dealing with auxiliary polynomials in r variables we are only concerned with Theorem II.

Let $(\alpha_1, \ldots, \alpha_{d+1})$ be the centre of $\mathfrak{B}'$ in the x-space. In the case of Theorem I we may make a preliminary linear transformation on $x_1, \ldots, x_{r+1}$ to ensure that $F(\alpha_1, \ldots, \alpha_r, x_{r+1})$ is not identically zero, and let us suppose that $x_{r+1} = \alpha_{r+1}$ is an s-fold root of this polynomial. Since $(\alpha_1, \ldots, \alpha_{d+1})$ is on U, $s > 0$. If $s = 1$, it is a simple point of U, and therefore the centre of $\mathfrak{B}$ is simple on U. We therefore suppose that Theorem I is true if $s < s_0$, and consider the case in which $s = s_0$. Our object is then to obtain a Cremona transformation of the required type which reduces U to a primal for which $s < s_0$. In the case of Theorem II we can, by a preliminary transformation, reduce $F(x)$ to a form in which $F(\alpha_1, \ldots, \alpha_r, x_{r+1})$ is not identically zero, but our intermediary transformations may cause us to depart from this condition. However, it is clear that by an initial choice of coordinate system we can arrange that

$$F(x) = A(x_1, \ldots, x_r)\, F_1(x_1, \ldots, x_{r+1}) \ldots F_t(x_1, \ldots, x_{r+1}),$$

where $F_i(x_1, \ldots, x_{r+1})$ is irreducible over R, and $x_{r+1} = \alpha_{r+1}$ is an s_i-fold root of

$$F_i(\alpha_1, \ldots, \alpha_r, x_{r+1}) = 0,$$

where s_i is finite ($i = 1, \ldots, t$). The intermediary transformations will not destroy this condition. If $s = \max[s_1, \ldots, s_t]$, then $s = 0$ implies that $s_i = 0$ ($i = 1, \ldots, t$) and hence

$$F_i(\alpha_1, \ldots, \alpha_{r+1}) \neq 0,$$

and therefore $v'[F_i(\xi)] = 0$. If we then apply Theorem II, for r variables, to $A(x_1, \ldots, x_r)$, we immediately reduce $F(x)$ to the form required by Theorem II, which is thus established for $F(x)$. We therefore assume that Theorem II is true for polynomials $F(x)$ for which $s < s_0$, and consider a polynomial $F(x)$ for which $s = s_0$. We then seek a transformation which reduces $F(x)$ to a polynomial for which $s < s_0$. If we can show that a transformation of the required type can always be found with this property, Theorem II will follow by an induction argument. Thus in the case of Theorem I or of Theorem II, we have only to show that a transformation of the type described in the statement of Theorem II exists which will reduce the problem to one in which $s < s_0$.

In the case of Theorem I we write

$$F(x) = \sum_i \pi_i(x), \quad \text{where} \quad \pi_i(x) = a_i(x_1, \ldots, x_r)\,(x_{r+1} - \alpha_{r+1})^i,$$

and in the case of Theorem II we write

$$F_h(x) = \sum_i \pi_i^{(h)}(x), \quad \text{where} \quad \pi_i^{(h)}(x) = \alpha_i^{(h)}(x_1, \ldots, x_r)\,(x_{r+1} - \alpha_{r+1})^i,$$

the summation being over the non-zero terms only. By construction,

$$a_{s_0}(\alpha_1, \ldots, \alpha_r) \neq 0, \quad a_{s_h}^{(h)}(\alpha_1, \ldots, \alpha_r) \neq 0.$$

We apply Theorem II for the r variables $x_1, \ldots, x_r$ to $\prod_i a_i(x_1, \ldots, x_r)$ [Theorem I], or to $A(x_1, \ldots, x_r) \prod_{h,i} a_i^{(h)}(x_1, \ldots, x_r)$ [Theorem II]. We then have

$$\pi_i(x) = b_i(y_1, \ldots, y_r)\, \Delta_1^{\lambda_{i1}}(y) \ldots \Delta_m^{\lambda_{im}}(y)\,(y_{r+1} - \beta_{r+1})^i \quad \text{[Theorem I]};$$

$$\pi_i^{(h)}(x) = b_i^{(h)}(y_1, \ldots, y_r)\, \Delta_1^{\lambda_{i1}^{(h)}}(y) \ldots \Delta_m^{\lambda_{im}^{(h)}}(y)\,(y_{r+1} - \beta_{r+1})^i,$$

and

$$A(x) = a(y_1, \ldots, y_r)\, \Delta_1^{\lambda_1}(y) \ldots \Delta_m^{\lambda_m}(y) \quad \text{[Theorem II]},$$

where the coefficients are in a field R_1 which contains the centre $(\beta_1, \ldots, \beta_{r+1})$ of an extended valuation $\mathfrak{B}_1$ of $\mathfrak{B}$ in the y-space, and where

$$v_1[b_i(\eta)] = v_1[b_i^{(h)}(\eta)] = v_1[a(\eta)] = 0,$$

and $\Delta_1(y), \ldots, \Delta_m(y)$ is a normal m-set on $y_1, \ldots, y_m$, and the conjugates of β_i $(i \leqslant m)$ over K are all distinct.

We have $v_1(\eta_{r+1} - \beta_{r+1}) > 0$. Let us consider first the case in which

$$v_1[\Delta_1(\eta)], \ldots, v_1[\Delta_m(\eta)], v_1(\eta_{r+1} - \beta_{r+1})$$

are rationally independent. Then we are only concerned with Theorem II. For if we were dealing with Theorem I, the relation

$$\Sigma \pi_i(\xi) = 0$$

implies that there exist two distinct suffices i and j such that $v_1[\pi_i(\xi)] = v_1[\pi_j(\xi)]$. Since

$$v_1[\Delta_1(\eta)], \ldots, v_1[\Delta_m(\eta)], v_1(\eta_{r+1} - \beta_{r+1})$$

are rationally independent, this implies that

$$\lambda_{ik} = \lambda_{jk} \quad (k = 1, \ldots, m) \quad \text{and} \quad i = j,$$

and so we have a contradiction. In the case of Theorem II,

$$\Delta_1(y), \ldots, \Delta_m(y), y_{r+1} - \beta_{r+1}$$

form a normal $(m+1)$-set in $y_1, \ldots, y_m, y_{r+1}$. We apply to $y_1, \ldots, y_m, y_{r+1}$ a transformation of type T_h, so that in the new coordinate system $(z_1, \ldots, z_{d+1})$ we have

$$\pi_i^{(h)}(x) = \bar{\Delta}_1^{\mu_{i1}^{(h)}}(z) \ldots \bar{\Delta}_{m+1}^{\mu_{im+1}^{(h)}}(z)\, \bar{b}_i^{(h)}(z_1, \ldots, z_{r+1}),$$
$$A(x) = \bar{\Delta}_1^{\mu_1}(z) \ldots \bar{\Delta}_{m+1}^{\mu_{m+1}}(z)\, \bar{a}(z_1, \ldots, z_{r+1}),$$

where

$$v_1[\bar{b}_i^{(h)}(\zeta)] = v_1[\bar{a}(\zeta)] = 0,$$

and $\bar{\Delta}_1(z), \ldots, \bar{\Delta}_{m+1}(z)$ is a normal $(m+1)$-set in $z_1, \ldots, z_m, z_{r+1}$. Also we may arrange that if $(\gamma_1, \ldots, \gamma_{d+1})$ is the centre of $\mathfrak{B}_1$ in the z-space, the field obtained by adjoining γ_i and its conjugates over K to K is R_1 (all i), and that if

$$v_1[\pi_i^{(h)}(\xi)] < v_1[\pi_j^{(h)}(\xi)], \quad \mu_{ik}^{(h)} < \mu_{jk}^{(h)} \quad (k = 1, \ldots, m+1).$$

If

$$\mu_k^{(h)} = \min[\mu_{ik}^{(h)}, \mu_{jk}^{(h)}, \ldots] \quad (k = 1, \ldots, m+1),$$

and

$$\nu_k = \mu_k + \sum_h \mu_k^{(h)},$$

then $$F(x) = \bar{\Delta}_1^{\nu_1}(z) \ldots \bar{\Delta}_{m+1}^{\nu_{m+1}}(z)\, G(z_1, \ldots, z_{r+1}),$$

where $v_1[G(\zeta)] = 0$, and Theorem II as applied to $F(x)$ is then satisfied.

We therefore have only to consider the case in which $v_1(\eta_{r+1} - \beta_{r+1})$ is rationally dependent on $v_1[\Delta_1(\eta)], \ldots, v_1[\Delta_m(\eta)]$. Since the conjugates of β_i over K are all distinct $(i \leqslant m)$ [Theorem II (i) for a polynomial in m variables], we can apply Theorem II of § 5 to the $(m+1)$-set $\Delta_1(y), \ldots, \Delta_m(y), y_{r+1} - \beta_{r+1}$ in $y_1, \ldots, y_m, y_{r+1}$. By this theorem, we can construct a transformation of the required type which transforms $\pi_i(x)$ into

$$\pi_i(x) = c_i(z_1, \ldots, z_{r+1})\, \delta_1^{\mu_{i1}}(z) \ldots \delta_m^{\mu_{im}}(z)\, \delta^{\nu_i}(z),$$

where (a) $\delta_1(x), \ldots, \delta_m(z)$ is a normal m-set in $z_1, \ldots, z_m$;

(b) $$\delta(z) = P(z_1, \ldots, z_m, z_{r+1})\, z_{r+1} + (z_m - \gamma_m)^\rho\, Q(z_1, \ldots, z_m, z_{r+1}),$$

and $P(\gamma_1, \ldots, \gamma_m, z_{r+1})$ is a non-zero constant $\bar{p}$;

(c) the centre of $\mathfrak{B}_1$ in the z-space has generic point $(\gamma_1, \ldots, \gamma_{r+1})$, where $\gamma_i\ (i \neq r+1)$ is in R_1, and γ_{r+1} is algebraic over R_1;

(d) $c_i(z_1, \ldots, z_{r+1})$ has the property that $c_i(\gamma_1, \ldots, \gamma_r, z_{r+1})$ is a non-zero constant $\bar{c}_i$;

(e) $v_1(\zeta_{r+1}) = 0$.

Property (d) follows from the fact that $c_i(z_1, \ldots, z_{r+1})$ is the product of $b_i(y_1, \ldots, y_r)$ and a product of powers of the polynomials G_i of Theorem II of § 5. Now, by a result established in § 5,

$$b_i(y_1, \ldots, y_r) = d_i(z_1, \ldots, z_{r+1}),$$

where $d_i(\gamma_1, \ldots, \gamma_r, z_{r+1})$ is a constant, necessarily different from zero because $v[b_i(\eta)] = 0$, and since $G_i(\gamma_1, \ldots, \gamma_r, z_{r+1})$ is a non-zero constant, (d) follows.

In the case of Theorem II, we have similar formulae for $\pi_i^{(h)}(x)$, $A(x)$.

We now consider the polynomials $\pi_i(\xi)$ which have least value in $\mathfrak{B}'$, or $\mathfrak{B}_1$ [Theorem I], or the $\pi_i^{(h)}(\xi)$ which have least value for a given h [Theorem II]. Suppose that

$$v_1[\pi_\alpha(\xi)] = v_1[\pi_\beta(\xi)] = \ldots = v_1[\pi_\delta(\xi)] < v_1[\pi_i(\xi)] \quad (i \neq \alpha, \beta, \ldots, \delta),$$

or

$$v_1[\pi_\alpha^{(h)}(\xi)] = v_1[\pi_\beta^{(h)}(\xi)] = \ldots = v_1[\pi_\delta^{(h)}(\xi)] < v_1[\pi_i^{(h)}(\xi)] \quad (i \neq \alpha, \beta, \ldots, \delta),$$

where $\alpha < \beta < \ldots < \delta$. Then if we write

$$\mu_j = \mu_{\alpha j} = \mu_{\beta j} = \ldots = \mu_{\delta j} < \mu_{ij},$$

or
$$\mu_j^{(h)} = \mu_{\alpha j}^{(h)} = \mu_{\beta j}^{(h)} = \ldots = \mu_{\delta j}^{(h)} < \mu_{ij}^{(h)},$$

we have

$$\begin{aligned} F(x) &= \delta_1^{\mu_1}(z) \ldots \delta_m^{\mu_m}(z)\,[c_\alpha(z)\,\delta^{\nu_\alpha}(z) + \ldots \\ &\qquad + c_\delta(z)\,\delta^{\nu_\delta}(z) + \delta_1(z) \ldots \delta_m(z)\,\phi(z)] \\ &= \delta_1^{\mu_1}(z) \ldots \delta_m^{\mu_m}(z)\,H(z), \end{aligned}$$

and similarly $\qquad F_h(x) = \delta_1^{\mu_1^{(h)}}(z) \ldots \delta_m^{\mu_m^{(h)}}(z)\,H_h(z).$

Also, by Theorem III of § 5,

$$\nu_i - \nu_j = (i-j)/d, \quad \nu_i^{(h)} - \nu_j^{(h)} = (i-j)/d,$$

where d is a non-zero integer, depending only on the transformation. We use this to determine a relation between δ and s_0 or s_h. We have

$$v_1[\pi_i(\xi)] = v_1[a_i(\xi)] + iv_1(\xi_{r+1} - \alpha_{r+1}).$$

But $\qquad v_1[a_i(\xi)] \geqslant 0, \quad v_1[a_{s_0}(\xi)] = 0, \quad v_1(\xi_{r+1} - \alpha_{r+1}) > 0;$

hence $\qquad v_1[\pi_i(\xi)] > v_1[\pi_{s_0}(\xi)] \quad (i > s_0).$

Then $\delta \leqslant s_0$. Similarly, in the case of $F_h(x)$, $\delta \leqslant s_h$. Hence

$$|\nu_\delta - \nu_\alpha| = (\delta - \alpha)/|d| \leqslant s_0 \quad (\text{or } s_h).$$

Let us first suppose that we are dealing with Theorem I. Since $v[\delta_i(\zeta)]$ is finite, and $F(\xi) = 0$, $H(\zeta) = 0$, and we conclude that the irreducible polynomial $\bar{F}(z)$ which vanishes on the transform of U in the z-space is a factor of $H(z)$. We consider the value of s for $\bar{F}(z)$.

By the properties of the transformation used in Theorem II, § 5, $v_1(\zeta_i - \gamma_i) > 0$ $(i \neq r+1)$, $v_1(\zeta_{r+1}) = 0$. Hence the centre of $\mathfrak{B}_1$ in the z-space is at a finite distance, but is not in $z_{r+1} = 0$. Therefore γ_{r+1} is a non-zero root of the equation

$$H(\gamma_1, \ldots, \gamma_r, t) = 0,$$

and from the form of $H(z)$ we see that this equation reduces to

$$\bar{c}_\alpha \bar{p}^{\nu_\alpha} t^{\nu_\alpha} + \bar{c}_\beta \bar{p}^{\nu_\beta} t^{\nu_\beta} + \ldots + \bar{c}_\delta \bar{p}^{\nu_\delta} t^{\nu_\delta} = 0,$$

where $\bar{p}$ is defined, as above, to be the value of the coefficient of z_{m+1} in $\delta(z)$ at the centre of the valuation. The multiplicity of $z = \gamma_{r+1}$ as a root of $\bar{F}(\gamma_1, \ldots, \gamma_r, z) = 0$ therefore cannot exceed

$|\nu_\delta - \nu_\alpha| \leqslant s_0$. Thus, in general, the number s for the transform of U in the z-space is less than s_0, and may therefore be reduced to one, by the hypothesis of our induction. In the exceptional case when there is no reduction in s, we must have $|\nu_\delta - \nu_\alpha| = s_0$, and hence $\alpha = 0$, $\delta = s_0$. Moreover, the equation of order s_0 for γ_{r+1} has all its roots equal, and since $\bar{c}_\alpha, \bar{c}_\beta, \ldots, \bar{c}_\delta, \bar{p}$ are in R_1, so is γ_{r+1}. Further,

$$\frac{\pi_i(x)}{\pi_s(x)} = \frac{c_i(z)}{c_s(z)}[\delta(z)]^{\nu_i-\nu_s} \quad (i \leqslant s),$$

and hence the residue of $\pi_i(\xi)/\pi_s(\xi)$ in $\mathfrak{B}_1$ is

$$\rho_i = \frac{\bar{c}_i}{\bar{c}_s}\bar{p}^{\nu_i-\nu_s}\gamma_{r+1}^{\nu_i-\nu_s},$$

and the equation

$$\rho_0 + \rho_1 t + \ldots + \rho_{s-1}t^{s-1} + t^s = 0$$

has all its roots equal, and since the unique root is not zero, $\rho_i \neq 0$.

Again, since $(\alpha, \beta, \ldots, \delta) = (0, 1, \ldots, s_0)$,

$$\begin{aligned} v_1[\pi_0(\xi)] &= v_1[\pi_2(\xi)] = \ldots = v_1[\pi_{s_0}(\xi)] \\ &= v_1[a_0(\xi)] = v_1[a_1(\xi)] + v_1(\xi_{r+1} - \alpha_{r+1}) = \ldots \\ &\qquad = v_1[a_{s_0}(\xi)] + s_0 v_1(\xi_{r+1} - \alpha_{r+1}), \end{aligned}$$

we have $$v_1(\xi_{r+1} - \alpha_{r+1}) = v_1[a_{s_0-1}(\xi)];$$

that is, *the value of $\xi_{r+1} - \alpha_{r+1}$ is equal to the value of a polynomial in $\xi_1, \ldots, \xi_r$ with coefficients in R'.*

We next consider the case of Theorem II. We have

$$F_h(x) = \delta_1^{\mu_1^{(h)}}(z) \ldots \delta_m^{\mu_m^{(h)}}(z)\, H_h(z).$$

If $s_h = 0$, we clearly have $\mu_i^{(h)} = 0$ and $v[H_h(\zeta)] = 0$. Hence the number s'_h of $H_h(z)$ is zero. If $s_h > 0$, the reasoning used above for the case of Theorem I shows that $s'_h \leqslant s_h$, and that if $s'_h = s_h$, for any h, γ_{r+1} is in R_1, the equation

$$\rho_0 + \rho_1 t + \ldots + \rho_{s_h-1}\, t^{s_h-1} + t^{s_h} = 0,$$

where ρ_i is the non-zero residue of $\pi_i^{(h)}(\xi)/\pi_s^{(h)}(\xi)$ in $\mathfrak{B}_1$, has all its roots equal, and the value of $\xi_{r+1} - \alpha_{r+1}$ is equal to the value of an element of the ring $R'[\xi_1, \ldots, \xi_r]$.

We have assumed that $s_0 = \max[s_1, \ldots, s_t]$ is greater than zero; otherwise there is nothing to prove. Let us first consider the case

$s_0 = 1$, and suppose that for each value of h for which $s_h = 1$, $s'_h = 0$. Then clearly we have an equation

$$F(x) = B(z)\,\delta_1^{\lambda_1}(z)\ldots\delta_m^{\lambda_m}(z),$$

where $v[B(\zeta)] = 0$. Then we have achieved the transformation of $F(x)$ required by Theorem II, except that we may not have satisfied condition (ii), since γ_{r+1} may be algebraic over R_1. Let R^* be the least normal extension of R_1 which contains γ_{r+1}, and let $\mathfrak{B}^*$ be an extension of $\mathfrak{B}_1$ which is a valuation of $R^*(\xi_1, \ldots, \xi_{r+1})$. The centre of $\mathfrak{B}^*$ in the z-space is $(\gamma_1, \ldots, \gamma_{r+1})$. If R^* is a true extension of R_1, the field obtained by adjoining γ_i $(i \leqslant m)$ and its g conjugates over K to K is not R^*, but only R_1.

To overcome this difficulty, we construct a polynomial $f(x)$ with coefficients in K having the properties:

(a) the h conjugates of $f(\gamma_{r+1})$ over R_1 are distinct;

(b) the first $\rho_0 - 1$ derivatives of $f(x)$ vanish when $x = \gamma_{r+1}$.

For each value of i $(1 \leqslant i \leqslant m)$ the gh numbers obtained from $\gamma_i + a_i f(\gamma_{r+1})$ by replacing γ_i by its g conjugates and γ_{r+1} by its h conjugates over R_1 are all distinct, where a_i is any element of K (with the possible exception of a finite number of values of a_i). If we put $\bar{\gamma}_i = \gamma_i + a_i f(\gamma_{r+1})$, *the field obtained by adjoining $\bar{\gamma}_i$ and its conjugates over K to K is R^*.* Now make the Cremona transformation

$$z_i = \bar{z}_i - a_i f(\bar{z}_{r+1}) \quad (i = 1, \ldots, m),$$

$$z_i = \bar{z}_i \quad (i > m).$$

We have
$$z_i - \gamma_i = \bar{z}_i - \bar{\gamma}_i + a_i[f(\gamma_{r+1}) - f(z_{r+1})] \quad (i \leqslant m)$$
$$= \bar{z}_i - \bar{\gamma}_i + (z_{r+1} - \gamma_{r+1})^{\rho_0} b_i(z).$$

The centre of $\mathfrak{B}^*$ in the $\bar{z}$-space is $(\bar{\gamma}_1, \ldots, \bar{\gamma}_m, \gamma_{m+1}, \ldots, \gamma_{r+1})$, and if ρ_0 is sufficiently large, $v^*(z_i - \gamma_i) = v^*(\bar{z}_i - \bar{\gamma}_i)$, and hence the values $v^*(\bar{z}_i - \bar{\gamma}_i)$ $(i = 1, \ldots, m)$ are rationally independent. Let

$$\bar{\delta}_i(\bar{z}) = \delta_i(z) \quad (i = 1, \ldots, m).$$

It is clear that if $a_1, \ldots, a_m$ are chosen to avoid at most a finite number of possible values in K, $\bar{\delta}_i(z)$ is monovalent in $\bar{z}_1, \ldots, \bar{z}_m$. Then, by Theorem I of § 5, a transformation of the type T_h on $\bar{z}_1, \ldots, \bar{z}_m$ enables us to express $F(x)$ in a form which satisfies all the requirements of Theorem II.

Suppose next that $s_0 = \max[s_1, \ldots, s_t] > 1$, and that $s'_h < s_h$ for every h for which $s_h > 0$. We then have

$$F(x) = \delta_1^{\lambda_1}(z) \ldots \delta_m^{\lambda_m}(z)\, a(z) \prod_{h=1}^{t} H_h(z),$$

where $v_1[a(\zeta)] = 0$. Now perform on $z_1, \ldots, z_{r+1}$ a general linear transformation (with coefficients in K). In the new coordinate system we have $F(x) = G(u)$ expressed as a product of

$$\lambda_1 + \ldots + \lambda_m + 1 + t$$

factors, and for these the numbers s_i can be determined. The factor arising from $\delta_i(z)$ has $s_i = 1$; that from $a(z)$ has $s_i = 0$, and the factor corresponding to $H_i(z)$ has $s_i = s'_h$. Thus since $s_0 > 1$, the maximum of the s_i for the $G(u)$ is at most equal to $\max[1, s'_1, \ldots, s'_t]$; it may be less if $H_h(z)$ is reducible, giving factors with multiplicity s''_h, s'''_h where $s''_h + s'''_h = s'_h$. Hence for $G(u)$ the number s is less than s_0, and we are in a position to apply our induction hypothesis.

We have still to consider the excepted case in which there is no reduction in s_0. In the case of Theorem II, this implies that there is a value of h for which $s'_h = s_h$. To treat the two cases together, we define $f(x)$ and s to be $F(x)$ and s_0 in the case of Theorem I, and to be $F_h(x)$ and s_h in the case of Theorem II. Then we write

$$f(x) = \pi_0(x) + \ldots + \pi_s(x) + \sum_{i>s} \pi_i(x),$$

where $$\pi_i(x) = a_i(x_1, \ldots, x_r)\,(x_{r+1} - \alpha_{r+1})^i.$$

We know that

$$v'[a_s(\xi)] = 0, \quad v'(\xi_{r+1} - \alpha_{r+1}) = v'[a_{s-1}(\xi)].$$

Also, $$\begin{aligned} v'[a_i(\xi)] &= v'[a_s(\xi)] + (s-i)\, v'(\xi_{r+1} - \alpha_{r+1}) \\ &= v'[a_s(\xi)] + (s-i)\, v'[a_{s-1}(\xi)]. \end{aligned}$$

Hence $$v'\left[\frac{\xi_{r+1} - \alpha_{r+1}}{a_{s-1}(\xi)}\right] = 0 = v'\left[\frac{a_i(\xi)}{a_s(\xi)\, a_{s-1}^{s-i}(\xi)}\right],$$

and therefore when we pass to the residue field of the valuation $\mathfrak{B}'$, the residues c and d_i of $\dfrac{\xi_{r+1} - \alpha_{r+1}}{a_{s-1}(\xi)}$ and $\dfrac{a_i(\xi)}{a_s(\xi)\, a_{s-1}^{s-i}(\xi)}$ are not zero.

Now $$\frac{\pi_i(\xi)}{\pi_s(\xi)} = \frac{a_i(\xi)}{a_s(\xi)\, a_{s-1}^{s-i}(\xi)} \left[\frac{a_{s-1}(\xi)}{\xi_{r+1} - \alpha_{r+1}}\right]^{s-i},$$

and passing to the residue field we have

$$d_i = \rho_i c^{s-i}.$$

We recall that the residues $\rho_0, \ldots, \rho_s = 1$ of $\pi_0(\xi)/\pi_s(\xi), \ldots, \pi_s(\xi)/\pi_s(\xi)$ have the property that the equation

$$\rho_0 + \rho_1 t + \ldots + \rho_{s-1} t^{s-1} + t^s = 0$$

has all its roots equal [p. 299]. From this it follows that the equation

$$d_0 + d_1 t + \ldots + d_{s-1} t^{s-1} + t^s = 0$$

has all its roots equal. The unique root is $-\frac{1}{s} d_{s-1}$. Since d_{s-1} is the residue of $a_s^{-1}(\xi)$, where $a_s(\xi)$ is in $R'[\xi_1, \ldots, \xi_r]$, and the residues of the ξ_i are in R', it follows that the root is in R'.

We next show that the root of this equation is in fact c. We have

$$\frac{f(\xi)}{a_s(\xi)\, a_{s-1}^s(\xi)} = \sum_i \frac{a_i(\xi)}{a_s(\xi)\, a_{s-1}^{s-i}(\xi)} \left[\frac{\xi_{r+1} - \alpha_{r+1}}{a_{s-1}(\xi)}\right]^i;$$

hence, passing to the residue field, we have

$$d_0 + d_1 c + \ldots + d_{s-1} c^{s-1} + c^s$$

equal to the residue of $X = f(\xi)/a_s(\xi)\, a_{s-1}^s(\xi)$. In the case of Theorem I, $f(\xi) = 0$; hence the residue of X is zero. In the case of Theorem II,

$$f(x) = F_h(x) = \delta_1^{\mu_1^{(h)}}(z) \ldots \delta_m^{\mu_m^{(h)}}(z)\, H_h(z),$$

where $\quad v_1[\delta_1^{\mu_1^{(h)}}(\zeta) \ldots \delta_m^{\mu_m^{(h)}}(\zeta)] = v_1[\pi_0^{(h)}(\xi)] = \ldots = v_1[\pi_s^{(h)}(\xi)],$

and, since $s_h' > 0$, $v_1[H_h(\zeta)] > 0$. Therefore,

$$0 < v'\left[\frac{f(\xi)}{\pi_s(\xi)}\right] = v'\left[\frac{f(\xi)}{a_s(\xi)\, a_{s-1}^s(\xi)}\right] = v'[X].$$

Therefore in either case the residue of X is zero, and, we have

$$d_0 + d_1 c + \ldots + d_{s-1} c^{s-1} + c^s = 0,$$

which was what we wished to prove. We can therefore conclude that there exists a constant c in R' such that

$$v'\left[\frac{\xi_{r+1} - \alpha_{r+1}}{a_{s-1}(\xi)} - c\right] > 0.$$

The position which we have now reached can be stated as follows. Whether we are dealing with Theorem I or with Theorem II, in the excepted case the value of $\xi_{r+1}-\alpha_{r+1}$ is equal to the value of a polynomial in $\xi_1, \ldots, \xi_r$ with coefficients in R', which we shall now denote by $H_0(\xi)$, and there exists an element c of R', different from zero, such that $v[(\xi_{r+1}-\alpha_{r+1})/H_0(\xi)-c]>0$.

We now use Lemma II of § 5 to construct a polynomial $Q_0(x_1, \ldots, x_r)$ with coefficients in K such that $Q_0(x)-cH_0(x)$, when expanded in powers of $x_1-\alpha_1, \ldots, x_r-\alpha_r$, begins with terms of degree greater than ρ_0, where ρ_0 is a suitably large integer. We then construct a transformation

$$x_i = x_i^{(1)} \quad (i=1, \ldots, r),$$
$$x_{r+1} = x_{r+1}^{(1)} + Q_0(x_1^{(1)}, \ldots, x_r^{(r)}).$$

This is clearly a Cremona transformation of the required type. We then have

$$\begin{aligned} v'(\xi_{r+1}^{(1)}-\alpha_{r+1}) &= v'[\xi_{r+1}-\alpha_{r+1}-Q_0(\xi_1, \ldots, \xi_r)] \\ &\geqslant \min\{v'[\xi_{r+1}-\alpha_{r+1}-cH_0(\xi)], v'[Q_0(\xi)-cH_0(\xi)]\} \\ &= v'[\xi_{r+1}-\alpha_{r+1}-cH_0(\xi)] \quad (\text{if } \rho_0 \text{ is sufficiently large}) \\ &= v'[H_0(\xi)]+v'[(\xi_{r+1}-\alpha_{r+1})/H_0(\xi)-c] \\ &> v'[H_0(\xi)] = v'(\xi_{r+1}-\alpha_{r+1}). \end{aligned}$$

The function $F(x)$ transforms into

$$F^{(1)}(x^{(1)}) = F(x_1^{(1)}, \ldots, x_r^{(1)}, x_{r+1}^{(1)}+Q_0(x^{(1)})).$$

It is clear that $F^{(1)}(x^{(1)})$ has the same numbers $s_0, s_1, \ldots, s_t$ as $F(x)$. We then try to obtain a reduction of these numbers by the methods already used. If we succeed, we have obtained the required reduction for $F(x)$, and Theorems I and II follow. If no reduction can be obtained for $F^{(1)}(x^{(1)})$, we have $v'(\xi_{r+1}^{(1)}-\alpha_{r+1})$ equal to the value of a polynomial in $(\xi_1^{(1)}, \ldots, \xi_r^{(1)}) = (\xi_1, \ldots, \xi_r)$, say $H_1(\xi)$, and as above we can construct a Cremona transformation

$$x_i^{(1)} = x_i^{(2)} \quad (i=1, \ldots, r),$$
$$x_{r+1}^{(1)} = x_{r+1}^{(2)} + Q_1(x_1^{(2)}, \ldots, x_r^{(2)}),$$

such that
$$v'(\xi_{r+1}^{(2)}-\alpha_{r+1}) > v'(\xi_{r+1}^{(1)}-\alpha_{r+1}).$$

We define $F^{(2)}(x^{(2)})$ by the equation

$$F^{(2)}(x^{(2)}) = F^{(1)}(x_1^{(2)}, \ldots, x_r^{(2)}, x_{r+1}^{(2)}+Q_1(x_1^{(2)}, \ldots, x_r^{(2)})),$$

and repeat the reasoning. We continue in this way. If we can show that after a finite number of stages we necessarily reach a polynomial $F^{(t)}(x^{(t)})$ for which a reduction in s_0 by our previous methods is possible, then our induction proof of Theorems I and II is complete. Hence we have only to show that we cannot carry out our construction for $F^{(1)}(x^{(1)}), F^{(2)}(x^{(2)}), \ldots$ indefinitely.

Suppose that we can carry out the construction indefinitely. We have an infinite sequence of polynomials in $x_1, \ldots, x_r$ over R', $H_0(x), H_1(x), H_2(x), \ldots$ such that

$$v'(\xi^{(t)}_{r+1} - \alpha_{r+1}) = v'[H_t(\xi)],$$

and since

$$v'(\xi^{(t+1)}_{r+1} - \alpha_{r+1}) > v'(\xi^{(t)}_{r+1} - \alpha_{r+1}),$$

we have

$$v'[H_0(\xi)] < v'[H_1(\xi)] < v'[H_2(\xi)] < \ldots.$$

We define an element $\Phi(x)$ of $R'[x_1, \ldots, x_{r+1}]$ and an integer σ as follows: in the case of Theorem I, $\Phi(x) = \dfrac{\partial F}{\partial x_{r+1}}$, $\sigma = s_0 - 1$, and in the case of Theorem II, $\Phi(x) = F_h(x)$, $\sigma = s_h$. If $\sigma = 0$, Theorem I or II is already established, so we may assume that $\sigma > 0$. We have

$$\Phi^{(t)}(x^{(t)}) = \Phi^{(t-1)}(x^{(t)}_1, \ldots, x^{(t)}_r, x^{(t)}_{r+1} + Q_{t-1}(x^{(t)}_1, \ldots, x^{(t)}_r)).$$

Then

$$\Phi(\xi) = \Phi^{(1)}(\xi^{(1)}) = \Phi^{(2)}(\xi^{(2)}) = \ldots \neq 0,$$

and hence $v'[\Phi^{(t)}(\xi^{(t)})] = M$ is independent of t and finite. Moreover, from the definition of σ,

$$M = v'[\Phi^{(t)}(\xi^{(t)})] \geqslant \sigma v'(\xi^{(t)}_{r+1} - \alpha_{r+1}) \geqslant v'(\xi^{(t)}_{r+1} - \alpha_{r+1}) = v'[H_t(\xi)].$$

We denote by Ω the algebraic closure of K, and let $\overline{\mathfrak{B}}$ be a valuation of $\Omega(\xi_1, \ldots, \xi_r)$ which is an extension of the valuation $\mathfrak{B}'$ of $R'(\xi_1, \ldots, \xi_r)$. Let $\tau = \min[\bar{v}(\xi_1 - \alpha_1), \ldots, \bar{v}(\xi_r - \alpha_r)]$, and let m be the least integer such that $m\tau > M$. Let $\alpha_1(\xi), \ldots, \alpha_a(\xi)$ be an Ω-basis for the polynomials in $\xi_1 - \alpha_1, \ldots, \xi_r - \alpha_r$ of degree less than m. We may suppose that the α_i are arranged so that $\bar{v}[\alpha_1(\xi)] \leqslant \bar{v}[\alpha_j(\xi)]$ $(j > 1)$. Then, since Ω is algebraically closed and $\overline{\mathfrak{B}}$ is of dimension zero, there exists an element c_j of Ω such that

$$\bar{v}[\alpha_j(\xi) - c_j\alpha_1(\xi)] > \bar{v}[\alpha_1(\xi)] \quad (j > 1).$$

Without loss of generality we may assume the α_j arranged so that

$$\bar{v}[\alpha_j(\xi) - c_j\alpha_1(\xi)] \geqslant \bar{v}[\alpha_2(\xi) - c_2\alpha_1(\xi)].$$

We then find d_j in Ω so that

$$\bar{v}[\alpha_j(\xi)-c_j\alpha_1(\xi)-d_j\alpha_2(\xi)+d_jc_2\alpha_1(\xi)]>\bar{v}[\alpha_2(\xi)-c_2\alpha_1(\xi)].$$

If we take $\beta_1(\xi)=\alpha_1(\xi),\quad \beta_2(\xi)=\alpha_2(\xi)-c_2\alpha_1(\xi),$

$$\beta_3(\xi)=\alpha_3(\xi)-c_3\alpha_1(\xi)-d_3\alpha_2(\xi)+d_3c_2\alpha_1(\xi),\ \ldots,$$

$\beta_1(\xi),\beta_2(\xi),\ldots$ is an Ω-basis for the polynomials in $\xi_1-\alpha_1,\ldots,\xi_r-\alpha_r$ of degree less than m, and

$$\bar{v}[\beta_1(\xi)]<\bar{v}[\beta_2(\xi)]<\bar{v}[\beta_3(\xi)]<\ldots.$$

If $P(x)$ is any polynomial in $\Omega[x_1,\ldots,x_r]$,

$$P(\xi)=\sum_{i=1}^{a}b_i\beta_i(\xi)+Q(\xi),$$

where $Q(\xi)$ is a polynomial in $\xi_1-\alpha_1,\ldots,\xi_r-\alpha_r$ with no terms of degree less than m, and $b_1,\ldots,b_a$ are in Ω. Then $\bar{v}[Q(\xi)]\geqslant m\tau>M$. Hence if $\bar{v}[P(\xi)]\leqslant M$,

$$\begin{aligned}\bar{v}[P(\xi)]&=\bar{v}[\sum_i b_i\beta_i(\xi)]\\&=\bar{v}[\beta_\rho(\xi)],\end{aligned}$$

where ρ is the least value of i for which $b_i\neq 0$.

Now we have seen that

$$v'[H_t(\xi)]=\bar{v}[H_t(\xi)]\leqslant M.$$

Hence the only possible values of $v'[H_t(\xi)]$ are

$$\bar{v}[\beta_1(\xi)],\quad \ldots,\quad \bar{v}[\beta_a(\xi)].$$

Since, however, $\quad v'[H_0(\xi)]<v'[H_1(\xi)]<\ldots$

the sequence $H_0,H_1,\ldots$ cannot contain more than a terms, which contradicts our assumption that the sequence is infinite.

We therefore conclude that after a finite number of stages we arrive at a polynomial $F^{(t)}(x^{(t)})$ for which it is possible to reduce s by our earlier methods. This completes our induction argument, and allows us to conclude the truth of Theorems I and II.

7. Valuations of dimension *s* and rank *k*. The rather lengthy series of constructions carried out in §§ 5 and 6 have enabled us to establish the Local Uniformisation Theorem, as stated at the beginning of § 4, in the case of zero-dimensional valuations of rank one. In actual fact, we have in this case gone further, for we have shown that if U is a primal in $A_{d+1}(x_1, \ldots, x_{d+1})$ and $\mathfrak{B}$ is a zero-dimensional valuation of the function field of U, then there exists a Cremona transformation of A_{d+1} of the form

$$x_i = P_i(y_1, \ldots, y_{d+1}) \quad (i = 1, \ldots, d+1),$$

in which $P_i(y_1, \ldots, y_{d+1})$ lies in $K[y_1, \ldots, y_{d+1}]$, which transforms U into a variety U' on which the centre of $\mathfrak{B}$ is simple. It will appear later that this additional information plays an important part in the reasoning of this section. Our object here is to deduce from the special case of the Local Uniformisation Theorem (L.U.T. for short) already proved that the theorem is true for a valuation of dimension s and rank k.

We shall do this in two stages. First, we shall show that if the L.U.T. is true for a valuation of rank k and of dimension zero, then it is true for a valuation of rank k and of any admissible dimension s; and secondly, we shall show that if the L.U.T. is true for a valuation of rank k and any admissible dimension s, then it is true for a valuation of rank $k+1$. These two results, together with Theorem I of § 6, clearly establish the L.U.T. quite generally.

Let U be an irreducible algebraic variety of dimension d in the affine space A_n, and let $(\xi_1, \ldots, \xi_n)$ be a generic point of U. Suppose that we can assume that the L.U.T. is true for any valuation of the function field $\Sigma = K(\xi_1, \ldots, \xi_n)$ of U of rank k and of dimension zero. We now consider a valuation $\mathfrak{B}$ of Σ of rank k and of dimension s. Without loss of generality, we may assume that the centre V of $\mathfrak{B}$ on U is at a finite distance. Let $\mathfrak{R}$ be the valuation ring of $\mathfrak{B}$, and let $\mathfrak{p}$ be the ideal of non-units of $\mathfrak{R}$. Since V is at a finite distance,

$$\mathfrak{J} = K[\xi_1, \ldots, \xi_n] \subseteq \mathfrak{R};$$

and V is given by the prime ideal $\mathfrak{P} = \mathfrak{J} \wedge \mathfrak{p}$ of $\mathfrak{J}$.

Let V be of dimension s'. Then we know that $s' \leqslant s$. If $s' < s$, we proceed as follows. There are sets of s' of the ξ_i which are algebraically independent modulo $\mathfrak{P}$; and we may assume that the coordinates are so arranged that $\xi_1, \ldots, \xi_{s'}$ form such a set. $\xi_1, \ldots, \xi_{s'}$ are then also algebraically independent modulo $\mathfrak{p}$. Since $\mathfrak{B}$ is of dimension s, we can find $s-s'$ further elements $\eta_{s'+1}, \ldots, \eta_s$ of $\mathfrak{R}$ such that $\xi_1, \ldots, \xi_{s'}, \eta_{s'+1}, \ldots, \eta_s$ are algebraically independent modulo $\mathfrak{p}$. We now consider the variety U' in affine space of dimension $n+s-s'$

having the generic point $(\xi_1, \ldots, \xi_n, \eta_{s'+1}, \ldots, \eta_s)$. U' is clearly birationally equivalent to U, since $K(\xi_1, \ldots, \xi_n, \eta_{s'+1}, \ldots, \eta_s) = \Sigma$.

Let $\mathfrak{J}'$ be the integral domain $K[\xi_1, \ldots, \xi_n, \eta_{s'+1}, \ldots, \eta_s]$. Then

$$\mathfrak{J} \subseteq \mathfrak{J}' \subseteq \mathfrak{R},$$

and the centre V' of $\mathfrak{B}$ on U' is given by the prime ideal $\mathfrak{P}' = \mathfrak{J}' \wedge \mathfrak{p}$ of $\mathfrak{J}'$. Since $\xi_1, \ldots, \xi_{s'}, \eta_{s'+1}, \ldots, \eta_s$ are algebraically independent modulo $\mathfrak{p}$ and lie in $\mathfrak{J}'$, they are algebraically independent modulo $\mathfrak{P}'$; hence V' is of dimension s at least, and since the dimension of the centre of an s-dimensional valuation cannot exceed s, the dimension of V' must be exactly s. Again, we have

$$\mathfrak{P} = \mathfrak{J} \wedge \mathfrak{p} = \mathfrak{J} \wedge \mathfrak{J}' \wedge \mathfrak{p} = \mathfrak{J} \wedge \mathfrak{P}';$$

and from the relations

$$\mathfrak{J} \subseteq \mathfrak{J}', \quad \mathfrak{P} = \mathfrak{J} \wedge \mathfrak{P}',$$

we deduce
$$Q(V) = \mathfrak{J}_{\mathfrak{p}} \subseteq \mathfrak{J}'_{\mathfrak{p}'} \subseteq Q(V').$$

Hence it is sufficient to prove the L.U.T. for the variety U' and the valuation $\mathfrak{B}$.

Thus we may assume that the centre V of $\mathfrak{B}$ on U is of dimension s, and that $\xi_1, \ldots, \xi_s$ are algebraically independent modulo $\mathfrak{P}$. Let Ω be the field $K(\xi_1, \ldots, \xi_s)$. Any non-zero polynomial in $\xi_1, \ldots, \xi_s$ with coefficients in K is a unit of $\mathfrak{R}$; hence the valuation of Ω induced by the valuation $\mathfrak{B}$ of Σ is trivial. Taking Ω as a new ground field, we consider the variety $\bar{U}$ defined over it by the generic point $(\xi_1, \ldots, \xi_n)$. It is at once seen to be of dimension $d-s$. Since

$$\Omega = K(\xi_1, \ldots, \xi_s),$$

$$\Sigma = K(\xi_1, \ldots, \xi_n) = \Omega(\xi_1, \ldots, \xi_n)$$

is the function field $\bar{\Sigma}$ of $\bar{U}$, and hence $\mathfrak{B}$ is a valuation of $\bar{\Sigma}$. As a valuation of $\bar{\Sigma}$ it is of rank k; but since $\mathfrak{B}$ is of dimension s and $\xi_1, \ldots, \xi_s$ are algebraically independent over K, modulo the ideal $\mathfrak{p}$ of non-units of $\mathfrak{R}$, any element η of $\mathfrak{R}$ is algebraically dependent on $\xi_1, \ldots, \xi_s$ over K, modulo $\mathfrak{p}$, and hence any element of η is algebraic over Ω modulo $\mathfrak{p}$. Thus $\mathfrak{B}$ is a zero-dimensional valuation of $\bar{\Sigma}$. Let $\bar{V}$ be its centre on $\bar{U}$. It is defined by the ideal $\Omega . (\mathfrak{J} \wedge \mathfrak{p}) = \Omega . \mathfrak{P}$ of $\Omega . \mathfrak{J}$. Further, since every element of Ω, other than zero, is a unit of $Q(V)$, we see that the quotient ring $\bar{Q}(\bar{V})$ of $\bar{V}$ is equal to $Q(V)$.

By our hypothesis of induction, there exists a variety $\bar{U}'$ defined over Ω, which is birationally equivalent to $\bar{U}$, and on which the centre $\bar{V}'$ of $\mathfrak{B}$ is a simple zero-dimensional variety, where $\bar{Q}(\bar{V}) \subseteq \bar{Q}(\bar{V}')$. $\bar{U}'$ may be regarded as a variety in affine space on which $\bar{V}'$ is at a finite distance. Let $(\eta_1, \ldots, \eta_m)$ be a generic point of $\bar{U}'$. Let $\bar{U}^*$ be the variety defined over Ω with generic point $(\xi_1, \ldots, \xi_n, \eta_1, \ldots, \eta_m)$ (it is in regular correspondence with the join of $\bar{U}$ and $\bar{U}'$), and let U^* be the variety defined over K by the generic point $(\xi_1, \ldots, \xi_n, \eta_1, \ldots, \eta_m)$.

Since $\eta_i \in \Omega(\xi_1, \ldots, \xi_n) = \bar{\Sigma} = \Sigma$, it is clear that U^* is birationally equivalent to U. Let V^* be the centre of $\mathfrak{B}$ on U^*. We prove that V^* is simple on U^* and of dimension s, and that $Q(V) \subseteq Q(V^*)$. Since V and $\bar{V}'$ are each at a finite distance in their respective spaces, we have

$$K[\xi_1, \ldots, \xi_n] \subseteq \mathfrak{R}, \quad \Omega[\eta_1, \ldots, \eta_m] \subseteq \mathfrak{R};$$

hence $$\mathfrak{J}^* = K[\xi_1, \ldots, \xi_n, \eta_1, \ldots, \eta_m] \subseteq \mathfrak{R},$$

that is, V^* is at a finite distance, and it is given by the ideal $\mathfrak{P}^* = \mathfrak{J}^* {}_\wedge \mathfrak{p}$ of $\mathfrak{J}^*$. Since $\xi_1, \ldots, \xi_s$ are algebraically independent elements of $\mathfrak{J}^*$ modulo $\mathfrak{P}^*$, V^* is of dimension s. Again, since $\bar{Q}(\bar{V}) \subseteq \bar{Q}(\bar{V}')$, the unique variety $\bar{V}^*$ on the join $\bar{U}^*$ of $\bar{U}$ and $\bar{U}'$ which corresponds to $\bar{V}$ and $\bar{V}'$ (that is, the centre of $\mathfrak{B}$ on $\bar{U}^*$) fulfils the relation $\bar{Q}(\bar{V}^*) = \bar{Q}(\bar{V}')$. Hence we have

$$Q(V) = \bar{Q}(\bar{V}) \subseteq \bar{Q}(\bar{V}') = \bar{Q}(\bar{V}^*);$$

and just as we saw that $Q(V) = \bar{Q}(\bar{V})$, so we now see that

$$\bar{Q}(\bar{V}^*) = Q(V^*).$$

Hence $$Q(V) \subseteq Q(V^*).$$

We recall that $\bar{V}'$ is a simple zero-dimensional variety on $\bar{U}'$, which is of dimension $d-s$. The ideal of non-units of the quotient ring $Q(V^*) = \bar{Q}(\bar{V}^*) = \bar{Q}(\bar{V}')$ therefore has a basis consisting of $d-s$ elements. But U^* is of dimension d, and V^* is of dimension s. This therefore implies that V^* is simple on U^*. U^* is therefore a birational transform of U on which the centre of $\mathfrak{B}$ is a simple s-dimensional variety V^*, where $Q(V) \subseteq Q(V^*)$. Hence on the assumption that the L.U.T. is true for zero-dimensional valuations of rank k, we can conclude that it is true for all valuations of rank k, whatever their dimension.

To complete the proof of the theorem, we have to show that if the L.U.T. is true for valuations of rank k and any admissible dimension, it is also true for zero-dimensional valuations of rank $k+1$. We consider a d-dimensional variety U and a zero-dimensional valuation $\mathfrak{B}$ of rank $k+1$ of its function field, and we may suppose that non-homogeneous coordinates are chosen so that the centre V of $\mathfrak{B}$ is at a finite distance. We denote the generic point of U by $(\xi_1, \ldots, \xi_n)$.

Since $\mathfrak{B}$ is of rank $k+1$, it can be compounded from a k-dimensional valuation $\mathfrak{B}'$ of $\Sigma = K(\xi_1, \ldots, \xi_n)$ and a valuation $\mathfrak{B}_1$ of the residue field Σ_1 of $\mathfrak{B}'$ which is of rank one. Since $\mathfrak{B}$ is of dimension zero, $\mathfrak{B}_1$ must also be of dimension zero. On the other hand, the dimension s of $\mathfrak{B}'$ is greater than zero, since $\mathfrak{B}_1$ is not trivial, and hence Σ_1 is of positive dimension over K. Let W be the centre of $\mathfrak{B}'$ on U; if W is of dimension s', $0 \leqslant s' \leqslant s$. Since $V \subseteq W$, and V is at a finite distance, W must also be at a finite distance.

By the hypothesis of our induction, there exists a birational transform U' of U on which the centre W' of $\mathfrak{B}'$ is a simple s-dimensional variety such that $Q(W) \subseteq Q(W')$. Since $\mathfrak{B}$ is compounded from $\mathfrak{B}'$, the centre V' of $\mathfrak{B}$ on U' is contained in W'. We choose non-homogeneous coordinates so that V', and hence W', is at a finite distance, and denote the generic point of U' by $(\eta_1, \ldots, \eta_m)$. Finally, we denote by U^* the variety in $(n+m)$-space whose generic point is $(\xi_1, \ldots, \xi_n, \eta_1, \ldots, \eta_m)$.

Let $\mathfrak{R}$ be the valuation ring of $\mathfrak{B}$, and $\mathfrak{p}$ the ideal of non-units of $\mathfrak{R}$. Then, since V and V' are at a finite distance,

$$\mathfrak{J} = K[\xi_1, \ldots, \xi_n] \subseteq \mathfrak{R}, \quad \mathfrak{J}' = K[\eta_1, \ldots, \eta_m] \subseteq \mathfrak{R},$$

and therefore $$\mathfrak{J}^* = K[\xi_1, \ldots, \xi_n, \eta_1, \ldots, \eta_m] \subseteq \mathfrak{R}.$$

The centre V^* of $\mathfrak{B}$ on U^* is therefore at a finite distance, and is given by the ideal $\mathfrak{P}^* = \mathfrak{J}^* \wedge \mathfrak{p}$ of $\mathfrak{J}^*$. If W^* is the centre of $\mathfrak{B}'$ on U^*, $V^* \subseteq W^*$, and hence W^* is also at a finite distance. Then, as we have already seen, the relation $Q(W) \subseteq Q(W')$, together with the fact that $\mathfrak{J}^*$ is the join of $\mathfrak{J}$ and $\mathfrak{J}'$, implies that $Q(W') = Q(W^*)$. W^* is therefore a simple s-dimensional subvariety of U^*. Again, since V and V' are given by the ideals $\mathfrak{P} = \mathfrak{J} \wedge \mathfrak{p}$ and $\mathfrak{P}' = \mathfrak{J}' \wedge \mathfrak{p}$ of $\mathfrak{J}$ and $\mathfrak{J}'$ respectively, we have

$$\mathfrak{P} = \mathfrak{J} \wedge \mathfrak{p} = \mathfrak{J} \wedge \mathfrak{J}^* \wedge \mathfrak{p} = \mathfrak{J} \wedge \mathfrak{P}^*, \quad \mathfrak{J} \subseteq \mathfrak{J}^*,$$

and from this we deduce that $\mathfrak{J}_{\mathfrak{p}} \subseteq \mathfrak{J}^*_{\mathfrak{p}}$, that is, $Q(V) \subseteq Q(V^*)$.

It follows that if we can prove our theorem for the variety U^* we can prove it for U. In other words, we need only prove the theorem on the assumption that the centre W of $\mathfrak{B}'$ on U is an s-dimensional subvariety of U which is simple on U. We next propose to show that we can further reduce the problem to the case in which, in addition to the restrictions already stated, U is a primal.

We suppose that W is simple and of dimension s on U, and let it be given by the prime ideal $\mathfrak{P}_s$ of $\mathfrak{J} = K[\xi_1, \ldots, \xi_n]$. Without any essential lack of generality, we may assume that the coordinate system has been chosen to satisfy the following conditions.

(i) The intersection of U with the space Π_{n-s},

$$x_1 = \ldots = x_s = 0,$$

is a $(d-s)$-dimensional variety on which each component of the intersection of Π_{n-s} and W, which consists of a set of zero-dimensional varieties $F_1, \ldots, F_t$, is simple for W and U. Since F_i is simple on U, we can find in $\mathfrak{J}$ uniformising parameters $\eta_{s+1}, \ldots, \eta_d$ of U at W such that $\xi_1, \ldots, \xi_s, \eta_{s+1}, \ldots, \eta_d$ are uniformising parameters at F_i. Moreover [cf. XVI, § 3], we can choose $\eta_{s+1}, \ldots, \eta_d$, and an element ζ in $\mathfrak{J}$ so that the following properties hold:

(ii) $\xi_1, \ldots, \xi_s, \eta_{s+1}, \ldots, \eta_d$ are algebraically independent over K;

(iii) $\Sigma = K(\xi_1, \ldots, \xi_n) = K(\xi_1, \ldots, \xi_s, \eta_{s+1}, \ldots, \eta_d, \zeta)$;

(iv) $\mathfrak{J}$ is integrally dependent on $K[\xi_1, \ldots, \xi_s, \eta_{s+1}, \ldots, \eta_d]$;

(v) if $f(x_1, \ldots, x_s, y_{s+1}, \ldots, y_d, z)$ is the irreducible polynomial over K such that

$$f(\xi_1, \ldots, \xi_s, \eta_{s+1}, \ldots, \eta_d, \zeta) = 0,$$

then the equation $\quad f(0, \ldots, 0, 0, \ldots, 0, z) = 0$

has all its roots simple.

Let U' be the variety in A_{d+1} whose generic point is $(\xi_1, \ldots, \xi_s, \eta_{s+1}, \ldots, \eta_d, \zeta)$, and let $\mathfrak{J}' = K[\xi_1, \ldots, \xi_s, \eta_{s+1}, \ldots, \eta_d, \zeta]$. Since $\mathfrak{J}' \subseteq \mathfrak{J}$, the centre of $\mathfrak{B}'$ on U' is the variety W' defined by the ideal $\mathfrak{P}'_s = \mathfrak{J}' \wedge \mathfrak{P}_s$ of $\mathfrak{J}'$. We prove first that W' is of dimension s. Any element of $\mathfrak{J}$ is integrally dependent on $\xi_1, \ldots, \xi_s, \eta_{s+1}, \ldots, \eta_d$, and hence on $\xi_1, \ldots, \xi_s$ modulo $\mathfrak{P}_s$. Since W is of dimension s, $\xi_1, \ldots, \xi_s$ are therefore algebraically independent modulo $\mathfrak{P}_s$ and hence, *a fortiori*, modulo $\mathfrak{P}'_s$. W' is therefore of dimension s at least, and since it is the centre of an s-dimensional valuation we conclude that it is of dimension s exactly.

We next show that W' is simple on U'. Condition (v), above, implies that

$$\frac{\partial f(\xi, \eta, \zeta)}{\partial \zeta} \neq 0 \quad (\mathfrak{P}_s).$$

Since

$$\frac{\partial f(\xi, \eta, \zeta)}{\partial \zeta} \in \mathfrak{J}',$$

this implies that

$$\frac{\partial f(\xi, \eta, \zeta)}{\partial \zeta} \neq 0 \quad (\mathfrak{P}'_s),$$

which tells us that W' is simple for U'.

Finally, since, by construction, $\mathfrak{J}$ is integrally dependent on $\mathfrak{J}'$, the argument used in § 4, p. 263, tells us that $Q(V) \subseteq Q(V')$. Hence if we can prove the L.U.T. for the variety U' and the valuation $\mathfrak{B}$, we can deduce it for U. Now U' is the primal in A_{d+1} with the equation

$$f(x_1, \ldots, x_s, y_{s+1}, \ldots, y_d, z) = 0.$$

Thus we have only to prove our theorem in the case in which U is a primal and W, the centre of $\mathfrak{B}'$ on U, is simple and s-dimensional. It will be convenient to state explicitly what we shall now assume about U and $\mathfrak{B}$. U is a primal in A_{d+1} having the equation

$$f(x_1, \ldots, x_{d+1}) = 0,$$

and generic point $(\xi_1, \ldots, \xi_{d+1})$. $\mathfrak{B}$ is a zero-dimensional valuation of $\Sigma = K(\xi_1, \ldots, \xi_{d+1})$ of rank $k+1$, which is compounded of an s-dimensional valuation $\mathfrak{B}'$ of Σ of rank k and a zero-dimensional valuation $\mathfrak{B}_1$ of the residue field of $\mathfrak{B}'$. The centres V and W of $\mathfrak{B}$ and $\mathfrak{B}'$ on U are at a finite distance, and W is of dimension s and simple on U. The elements $\xi_{s+1}, \ldots, \xi_d$ are uniformising parameters on U at W. If $\mathfrak{J} = K[\xi_1, \ldots, \xi_{d+1}]$, and $\mathfrak{P}_s$ is the prime ideal of W in $\mathfrak{J}$,

$$\frac{\partial f}{\partial \xi_{d+1}} \neq 0 \quad (\mathfrak{P}_s).$$

$\xi_1, \ldots, \xi_s$ are algebraically independent modulo $\mathfrak{P}_s$.

We pass from $\mathfrak{J}$ to the residue class ring $\mathfrak{J}/\mathfrak{P}_s$. The element of this corresponding to ξ_i will be denoted by $\bar{\xi}_i$. Then

$$\bar{\xi}_{s+1} = \ldots = \bar{\xi}_d = 0.$$

On the other hand, $\bar{\xi}_1, \ldots, \bar{\xi}_s$ are algebraically independent over K. Since

$$f(\bar{\xi}_1, \ldots, \bar{\xi}_s, 0, \ldots, 0, \bar{\xi}_{d+1}) = 0,$$

$\bar{\xi}_{d+1}$ is algebraically dependent on $\bar{\xi}_1, \ldots, \bar{\xi}_s$, and hence [III, § 3, Th. VI] there is an essentially unique irreducible relation connecting $\bar{\xi}_1, \ldots, \bar{\xi}_s, \bar{\xi}_{d+1}$, say

$$g(\bar{\xi}_1, \ldots, \bar{\xi}_s, \bar{\xi}_{d+1}) = 0,$$

or, equivalently, $\quad g(\xi_1, \ldots, \xi_s, \xi_{d+1}) = 0 \ (\mathfrak{P}_s).$
Then

$$f(x_1, \ldots, x_s, 0, \ldots, 0, x_{d+1}) = g(x_1, \ldots, x_s, x_{d+1})\, \phi(x_1, \ldots, x_s, x_{d+1}),$$

where $\phi(x_1, \ldots, x_s, x_{d+1})$ is a polynomial over K. Hence

$$f(x_1, \ldots, x_{d+1}) = g(x_1, \ldots, x_s, x_{d+1})\, \phi(x_1, \ldots, x_s, x_{d+1}) + \sum_{i=s+1}^{d} A_i(x_1, \ldots, x_{d+1})\, x_i.$$

From the equation

$$\frac{\partial f(x)}{\partial x_{d+1}} = g(x) \frac{\partial \phi(x)}{\partial x_{d+1}} + \frac{\partial g(x)}{\partial x_{d+1}} \phi(x) + \sum_{i=s+1}^{d} \frac{\partial}{\partial x_{d+1}} A_i(x_1, \ldots, x_{d+1})\, x_i,$$

we deduce that $\quad \dfrac{\partial f(\xi)}{\partial \xi_{d+1}} = \dfrac{\partial g(\xi)}{\partial \xi_{d+1}} \phi(\xi) \ (\mathfrak{P}_s).$

Since, by hypothesis $\quad \dfrac{\partial f(\xi)}{\partial \xi_{d+1}} \neq 0 \ (\mathfrak{P}_s),$

we have $\phi(\xi) \neq 0 \ (\mathfrak{P}_s)$. Equivalently, $v'[\phi(\xi)] = 0$.

We return to the mapping of $\mathfrak{J}$ on $\mathfrak{J}/\mathfrak{P}_s$. Since $\mathfrak{P}_s$ is the ideal in $\mathfrak{J}$ of the centre of $\mathfrak{V}'$, $\mathfrak{J}/\mathfrak{P}_s$ is isomorphic with an integral domain contained in the residue field of $\mathfrak{V}'$. $\mathfrak{J}/\mathfrak{P}_s$, is, however, the integral domain of W. Hence the function field $\bar{\Sigma}$ of W is isomorphic with a subfield of the residue field of $\mathfrak{V}'$, and may be identified with it. The valuation $\mathfrak{V}_1$, which, compounded with $\mathfrak{V}'$, gives $\mathfrak{V}$, induces a valuation of $\bar{\Sigma}$, which we denote by $\bar{\mathfrak{V}}$.

W is the variety in the A_{s+1} given by $x_{s+1} = \ldots = x_d = 0$ with the generic point $(\bar{\xi}_1, \ldots, \bar{\xi}_d, \bar{\xi}_{d+1})$, and $\bar{\mathfrak{V}}$ is a valuation of its function field, of dimension zero and rank one. Hence, by Theorem I of § 6, there exists a Cremona transformation

$$x_i = P_i(y_1, \ldots, y_s, y_{d+1}) \quad (i = 1, \ldots, s, d+1),$$

where P_i is a polynomial over K, which transforms W into a variety W' on which the centre of $\bar{\mathfrak{V}}$ is simple. Let

$$g(x) = G(y_1, \ldots, y_s, y_{d+1})\, \psi(y_1, \ldots, y_s, y_{d+1}),$$

where $G(y) = 0$ is the irreducible equation which is satisfied by W'. Without loss of generality, we shall suppose that $\partial G/\partial y_{d+1}$ does not vanish at the centre of $\overline{\mathfrak{B}}$, which is simple on W'. We have to show that $\psi(y)$ does not vanish on W'.

We consider the unique s-dimensional valuation $\mathfrak{B}^*$ of the $(s+1)$-dimensional field $K(y_1, \ldots, y_s, y_{d+1})$ whose centre in the $(y_1, \ldots, y_s, y_{d+1})$-space is W' [XVII, § 5, Th. VII]. It is a discrete valuation, and $v^*[G(y)] > 0$. Since $G(y)$ is irreducible, we deduce that $v^*[G(y)] = 1$, if the valuation group is taken to be the additive group of integers. In the birational correspondence between the spaces $(y_1, \ldots, y_s, y_{d+1})$ and $(x_1, \ldots, x_s, x_{d+1})$, W' is not fundamental, since the dimension of any fundamental variety is less than s [§ 2, Th. IV]. There is therefore a unique variety W^* in $(x_1, \ldots, x_s, x_{d+1})$ corresponding to W', and W^* is the centre of $\mathfrak{B}^*$ in the x-space. Now if $(\overline{\eta}_1, \ldots, \overline{\eta}_s, \overline{\eta}_{d+1})$ is a generic point of W', and

$$\overline{\xi}_i = P_i(\overline{\eta}_1, \ldots, \overline{\eta}_s, \overline{\eta}_{d+1}) \quad (i = 1, \ldots, s, d+1),$$

$(\overline{\xi}_1, \ldots, \overline{\xi}_s, \overline{\xi}_{d+1})$ is a generic point of W^*. But by the properties of the Cremona transformation, $(\overline{\xi}_1, \ldots, \overline{\xi}_s, \overline{\xi}_{d+1})$ is a generic point of W. Hence $W = W^*$, which is therefore of dimension s. Then, since $g(x)$ is irreducible and vanishes on W, we see, as above, that $v^*[g(x)] = 1$. Hence, since

$$v^*[g(x)] = v^*[G(y)] + v^*[\psi(y)]$$

and
$$v^*[g(x)] = 1 = v^*[G(y)],$$

we have
$$v^*[\psi(y)] = 0,$$

that is
$$\psi(\overline{\eta}) \neq 0.$$

We define $\eta_1, \ldots, \eta_s, \eta_{d+1}$ by means of the equations

$$\xi_i = P_i(\eta_1, \ldots, \eta_s, \eta_{d+1}) \quad (i = 1, \ldots, s, d+1).$$

Then $\overline{\eta}_i$ can be taken to be the residue of η_i in $\mathfrak{B}'$. Since $v'(\eta_i) \geqslant 0$ and $\psi(\overline{\eta}) \neq 0$, we must have

$$v'[\psi(\eta)] = 0.$$

The centre of $\overline{\mathfrak{B}}$ on W' is the variety whose generic point is $(\eta_1^*, \ldots, \eta_s^*, \eta_{d+1}^*)$, where η_i^* is the residue of $\overline{\eta}_i$ in $\overline{\mathfrak{B}}$, that is, the residue of η_i in $\mathfrak{B}$. The fact that $\partial G/\partial y_{d+1}$ does not vanish at the centre of $\overline{\mathfrak{B}}$ can now be expressed by the inequality

$$\frac{\partial G}{\partial \eta_{d+1}^*} \neq 0.$$

We now return to the variety U in the $(d+1)$-dimensional affine space $(x_1, \ldots, x_{d+1})$. We perform the transformation

$$x_i = P_i(y_1, \ldots, y_{d+1}) \quad (i = 1, \ldots, d+1),$$

where $P_i(y_1, \ldots, y_{d+1}) = P_i(y_1, \ldots, y_s, y_{d+1}) \quad (i = 1, \ldots, s, d+1),$

and $P_i(y_1, \ldots, y_{d+1}) = y_i \psi(y) \chi(y) \quad (i = s+1, \ldots, d).$

Here $\chi(y) = \phi(x_1, \ldots, x_s, x_{d+1}) = \phi[P_1(y), \ldots, P_s(y), P_{d+1}(y)].$

The transformation is clearly a Cremona transformation. We denote by $(\eta_1, \ldots, \eta_{d+1})$ the transform of $(\xi_1, \ldots, \xi_{d+1})$, and observe that $\eta_1, \ldots, \eta_s, \eta_{d+1}$ are the elements of Σ considered above in our discussion of the $(s+1)$-space $(y_1, \ldots, y_s, y_{d+1})$. Then since $\phi(x)$ does not vanish on W, $v'[\phi(\xi)] = v'[\chi(\eta)] = 0$. Since

$$\xi_i = \eta_i \psi(\eta) \chi(\eta) \quad (i = s+1, \ldots, d),$$

and $$v'(\xi_i) > 0 \quad (i = s+1, \ldots, d),$$

we have $$\begin{aligned} 0 < v'(\xi_i) &= v'(\eta_i) + v'[\psi(\eta)] + v'[\chi(\eta)] \\ &= v'(\eta_i). \end{aligned}$$

We now consider the transform U' of U by the Cremona transformation, and the centre of $\mathfrak{B}$ on U'. We have

$$\begin{aligned} f(x) &= g(x)\phi(x) + \sum_{i=s+1}^{d} A_i(x_1, \ldots, x_{d+1})\, x_i \\ &= G(y)\psi(y)\chi(y) + \sum_{i=s+1}^{d} B_i(y)\, y_i \psi(y)\chi(y), \end{aligned}$$

where $B_i(y) = A_i(x)$. We write

$$H(y) = G(y) + \sum_{i=s+1}^{d} B_i(y)\, y_i.$$

Since $\psi(\eta)$ and $\chi(\eta)$ are not zero, the equation $f(\xi) = 0$ implies $H(\eta) = 0$; hence

$$H(y_1, \ldots, y_{d+1}) = 0$$

is satisfied by U'. Since $G(y)$ is irreducible and is independent of $y_{s+1}, \ldots, y_d$, $H(y)$ can only be reducible if it contains a factor of the form

$$T(y) = 1 + \sum_{i=s+1}^{d+1} C_i(y)\, y_i.$$

$T(y)$ cannot vanish on U'. For the centre of $\mathfrak{B}'$ on U' is at a finite distance, since $v'(\eta_i) \geq 0$ $(i = 1, \ldots, d+1)$, and has generic point $(\bar{\eta}_1, \ldots, \bar{\eta}_{d+1})$, where $\bar{\eta}_i$ is the residue of η_i in $\mathfrak{B}'$. Since

$$v'(\eta_i) > 0 \quad (i = s+1, \ldots, d),$$

we have

$$\bar{\eta}_i = 0 \quad (i = s+1, \ldots, d).$$

Clearly this cannot satisfy $T(y) = 0$, and since $\bar{\eta}$ lies on U', it follows that $T(y)$ does not vanish on U'. It follows easily that the irreducible equation of U' is of the form

$$H'(y) = G(y) + \sum_{i=s+1}^{d} B'_i(y)\, y_i.$$

To find the centre of $\mathfrak{B}$ on the variety U', we construct the residues η_i^* of η_i in $\mathfrak{B}$ $(i = 1, \ldots, d+1)$. η^* is then a generic point of the centre of $\mathfrak{B}$. Since $\mathfrak{B}$ is compounded from $\mathfrak{B}'$, and $v'(\eta_i) > 0$ $(i = s+1, \ldots, d)$, we have $\eta_i^* = 0$ $(i = s+1, \ldots, d)$. The remaining η_i^* have already been defined in connection with the centre of $\overline{\mathfrak{B}}$ over W'. Now

$$\frac{\partial H'}{\partial y_{d+1}} = \frac{\partial G}{\partial y_{d+1}} + \sum_{i=s+1}^{d} \frac{\partial B'_i(y)}{\partial y_{d+1}} y_i.$$

Hence

$$\frac{\partial H'}{\partial \eta^*_{d+1}} = \frac{\partial G}{\partial \eta^*_{d+1}} \neq 0,$$

that is, $\partial H/\partial y_{d+1}$ does not vanish at the centre of $\mathfrak{B}$, which is therefore simple on U'.

Thus we have obtained a transform of U on which the centre V' of $\mathfrak{B}$ is simple. Since $K[\xi_1, \ldots, \xi_{d+1}] \subseteq K[\eta_1, \ldots, \eta_{d+1}]$, we deduce, as usual, that $Q(V) \subseteq Q(V')$. This is the last case of the L.U.T. to be considered. We now conclude that *the Local Uniformisation Theorem is true in all cases.*

Finally, using XVI, § 5, Th. III and XVI, § 6, Th. III, we observe that, if U^* is a projectively normal variety derived from U', there is a unique variety V^* on U^* which is of dimension s and corresponds to V', and $Q(V') = Q(V^*)$. Hence V^* is simple on U^*, and the centre of $\mathfrak{B}$ on U^* is V^*. Thus in the statement of the Local Uniformisation Theorem, we can require that the birational transform of U on which the centre of $\mathfrak{B}$ is to be simple be projectively normal.

8. Resolving systems. The Local Uniformisation Theorem is very valuable in the study of a single valuation of the function field Σ of an algebraic variety U. But in many problems of algebraic geometry we have to consider not a single valuation, but an aggregate. For instance, we may be concerned with the aggregate of all the valuations of Σ whose centres on U lie on a given subvariety V of U; such considerations arise, for instance, when we wish to examine the structure of an irreducible variety V which is singular for U. It would obviously be advantageous if we could find a birational transform of U on which the centre of every valuation of the aggregate is simple. Ideally, we should like to obtain a birational transform U' of U on which the centre of every valuation of Σ is simple. Since we can find a valuation of Σ whose centre is any given irreducible subvariety of U', this is equivalent to obtaining a birational transform of U which has no singular points. As we have already pointed out, it has as yet only been proved that a non-singular birational transform of U exists in the cases in which the dimension of U does not exceed three. However, a partial result of some importance which can be proved for varieties U of any dimension is contained in the following theorem.

THEOREM I. *Given an irreducible algebraic variety U with the function field Σ, there exists a finite set of varieties $U_1, \ldots, U_k$, each of which is birationally equivalent to U, such that if $\mathfrak{B}$ is any valuation of Σ, the centre of $\mathfrak{B}$ on at least one of the varieties U_i is simple.*

A set of varieties $U_1, \ldots, U_k$ with the properties described in this theorem is called a *finite resolving system* for Σ. Any aggregate $\{U_\alpha\}$, whether finite or not, of varieties which are birationally equivalent to U and which has the property that, given any valuation of Σ, there exists a variety of the system $\{U_\alpha\}$ on which the centre of the valuation is simple, is called a *resolving system* for Σ. The L.U.T. tells us that, given any algebraic variety U, there always exists a resolving system for its function field; in particular, the aggregate of all birational transforms of U is a resolving system. In order to prove Theorem I, we have only to show that the existence of a resolving system implies the existence of a *finite* resolving system.

An immediate simplification of our problem suggests itself. In fact, we need only prove the existence of a finite resolving system for the zero-dimensional valuations of Σ; that is, the existence of a set of varieties $U_1, \ldots, U_k$ each having Σ as function field with the property that if $\mathfrak{B}$ is any zero-dimensional valuation of Σ, there exists an i $(1 \leqslant i \leqslant k)$ such that the centre of $\mathfrak{B}$ on U_i is simple. In fact, suppose that $U_1, \ldots, U_k$ form a resolving system for the zero-dimensional valuations of Σ, and let $\mathfrak{B}$ be any valuation of Σ of dimension greater than zero. Then [XVII, § 5, Ths. VIII and IX] we can compound $\mathfrak{B}$ with a valuation of its residue field to give a valuation $\mathfrak{B}_0$ of Σ of dimension zero. Let U_i be a member of the resolving system on which the centre V_i of $\mathfrak{B}_0$ is simple, and let W_i be the centre of $\mathfrak{B}$ on U_i. Since $\mathfrak{B}_0$ is compounded from $\mathfrak{B}$, $V_i \subseteq W_i$; and since V_i is simple on U_i, not all points of W_i are singular for U_i; that is, W_i is simple on U_i. It follows that the system $U_1, \ldots, U_k$ is a resolving system for *all* the valuations of Σ. For this reason we may confine our attention to zero-dimensional valuations, and for the remainder of this section every valuation considered will be zero-dimensional.

As a first step towards the proof of Theorem I, we prove a lemma.

Lemma I. Let U be an irreducible algebraic variety, and let $\{V_\alpha\}$ be any aggregate of subvarieties of U. Then if $\{V_\alpha\}$ has the property that every finite subset $V_{\alpha_1}, \ldots, V_{\alpha_k}$ has a non-empty intersection, there exists at least one point common to every variety of the set.

Let us assume that the lemma is not true. Consider any variety V_{α_1} of $\{V_\alpha\}$. It cannot be empty, otherwise the intersection of any finite subset of $\{V_\alpha\}$ which includes V_{α_1} would be empty, contrary to hypothesis. Not all varieties of $\{V_\alpha\}$ contains V_{α_1} as a subvariety, otherwise any point of V_{α_1} would be common to all V_α, whereas we have assumed that there is no point common to all these varieties. There therefore exists a variety V_{α_2} of $\{V_\alpha\}$ which does not contain V_{α_1}; hence

$$V_{\alpha_1} \supset V_{\alpha_1} \wedge V_{\alpha_2}.$$

Repeating the reasoning, we show that $V_{\alpha_1} \wedge V_{\alpha_2}$ is not empty, and that not all the varieties of $\{V_\alpha\}$ contain $V_{\alpha_1} \wedge V_{\alpha_2}$. There therefore exists a variety V_{α_3} of $\{V_\alpha\}$ such that

$$V_{\alpha_1} \supset V_{\alpha_1} \wedge V_{\alpha_2} \supset V_{\alpha_1} \wedge V_{\alpha_2} \wedge V_{\alpha_3}.$$

Proceeding in this way, our assumption that there is no point common to all the varieties V_α leads to the construction of an infinite strictly decreasing sequence of subvarieties of U. But [Chap. X, § 1] any strictly decreasing system of varieties is finite, and we have a contradiction. Our initial assumption is therefore false; that is, the lemma is true.

In what follows it will be convenient to use the symbol $\prod_\lambda M_\lambda$ for the intersection of a set $\{M_\lambda\}$ of subsets of a given set M, and $\prod_{i=1}^{k} M_{\lambda_i}$ for the intersection of the finite set of subsets $M_{\lambda_1}, \ldots, M_{\lambda_k}$. Thus the lemma proved can be stated as follows: If $\prod_{i=1}^{k} V_{\alpha_i} \neq 0$ for every finite set $(\alpha_1, \ldots, \alpha_k)$, then $\prod_\alpha V_\alpha \neq 0$.

Now consider any algebraic variety U and its function field Σ, and let $\{U_\alpha\}$ be the aggregate of all algebraic varieties birationally equivalent to U. The varieties U_α are usually called *projective models* of Σ. The suffixes α range over some indexing system, and for later purposes we shall assume that they give a well-ordering of the projective models. Let M denote the set of all zero-dimensional valuations of Σ. A subset M' of M is said to be an *algebraic subset* of M if and only if there exists a projective model U of Σ and a subvariety V of U such that the valuation $\mathfrak{B}$ belongs to M' if and only if its centre on U is contained in V.

Lemma II. The intersection of any two (and hence of any finite number of) algebraic subsets of M is an algebraic subset of M, or else is vacuous.

Let M', M'' be algebraic subsets, and let U', U'' be two projective models of Σ such that M' is the set of valuations of Σ whose centres on U' are contained in the subvariety V' of U', and M'' is the set of valuations of Σ whose centres on U'' are contained in the subvariety V'' of U''. Let U be the join of U' and U'', and let V_1 be the total transform of V' on U, and V_2 the total transform of V'' on U. By definition, V_1 is the locus on U of the centres of all zero-dimensional valuations of Σ whose centres on U' lie on V', and is algebraic. Hence the centre on U of any valuation of M' lies on V_1. On the other hand, since U is the join of U' and U'', to each subvariety of U there corresponds a unique subvariety of U'. Consider any zero-dimensional valuation $\mathfrak{B}$ of Σ whose centre W is contained in V_1. Since $W \subseteq V_1$, there must exist some valuation $\mathfrak{B}'$ of Σ contained in M' whose centre is at W. The centre W' of $\mathfrak{B}'$ on U' is contained in V'; hence there is a variety W' contained in V' which corresponds to W. W' is therefore the unique variety of U' which corresponds to W. Hence the centre of $\mathfrak{B}$ on U' is W', contained in V', and therefore $\mathfrak{B}$ is contained in M'. Thus M' may be defined as the set of zero-dimensional valuations whose centres on U lie in V_1.

Similarly, M'' is the set of valuations whose centres on U lie in V_2, which is algebraic. The intersection of M' and M'' is therefore the set of valuations of Σ whose centres on U lie on $V_1 \wedge V_2$; hence it is an algebraic subset of M, or is vacuous.

Lemma III. If $\{M_\lambda\}$ *is any aggregate of algebraic subsets of* M *such that* $\prod_{i=1}^{k} M_{\lambda_i} \neq 0$, *for all finite sets* $(\lambda_1, \ldots, \lambda_k)$, *then* $\prod_\lambda M_\lambda \neq 0$.

Let U_α be any projective model of Σ, and let V_α^λ be the locus of the centres of the valuations of M_λ on V_α. M_λ is the set of valuations of Σ whose centres lie on a subvariety V_β of some projective model U_β. The locus V_α^λ of centres of the valuations of M_λ on V_α therefore forms the total transform of V_β on U_α, and is therefore algebraic. We should note, however, that there may exist valuations of Σ whose centres on U_α are contained in V_α^λ and which do not belong to M_λ; for if W is a zero-dimensional variety of V_α^λ which is fundamental for the birational correspondence between U_α and U_β, there may exist a zero-dimensional variety W' of U_β corresponding to W which is not contained in V_β. A zero-dimensional valuation of Σ whose centre on U_α is W and whose centre on U_β is W' has centre contained in V_α^λ and yet is not contained in M_λ.

We first prove that $\prod_\lambda V_\alpha^\lambda \neq 0$. Let $\lambda_1, \ldots, \lambda_k$ be any finite set of indices. By hypothesis, $\prod_{i=1}^{k} M_{\lambda_i} \neq 0$; hence there is a valuation $\mathfrak{B}$ common to $M_{\lambda_1}, \ldots, M_{\lambda_k}$. Its centre on U_α is common to $V_\alpha^{\lambda_1}, \ldots, V_\alpha^{\lambda_k}$; that is, $\prod_{i=1}^{k} V_\alpha^{\lambda_i} \neq 0$. From Lemma I we conclude that $\prod_\lambda V_\alpha^\lambda \neq 0$.

We have supposed that the set of projective models $\{U_\alpha\}$ is well-ordered. Let W_1 be an irreducible zero-dimensional variety contained in $\prod_\lambda V_1^\lambda$. Commencing with W_1, we construct by transfinite induction an irreducible zero-dimensional variety W_α on each U_α with the properties:

(i) W_α is contained in $\prod_\lambda V_\alpha^\lambda$;

(ii) given any finite set $\alpha_1, \ldots, \alpha_k$ of α's, and any finite set $\lambda_1, \ldots, \lambda_r$ of λ's, there exists a valuation $\mathfrak{B}$ contained in $M_{\lambda_1}, \ldots, M_{\lambda_r}$ whose centre on U_{α_i} is W_{α_i} $(i = 1, \ldots, k)$.

Suppose that W_α has been constructed for all α satisfying $\alpha < \beta$; we show how to construct W_β. Let $\pi_{\alpha_1 \ldots \alpha_k}^\lambda$ denote the algebraic locus on V_β formed by the centres of the valuations of M_λ whose centres on V_{α_i} are at W_{α_i} $(i = 1, \ldots, k)$. Clearly, $\pi^{\lambda}_{\alpha_1\ldots\alpha_k} \subseteq V^{\lambda}_{\beta}$. Now $\pi^{\lambda}_{\alpha_1\ldots\alpha_k}$ is defined for any set $\alpha_1, \ldots, \alpha_k$ of α's subject to the condition $\alpha_i < \beta$; and by the hypothesis of induction not only $\pi^{\lambda}_{\alpha_1\ldots\alpha_k}$, but also $\prod_{i=1}^{r} \pi^{\lambda_i}_{\alpha_1\ldots\alpha_k}$, is not empty, for any finite set $\lambda_1, \ldots, \lambda_r$. Hence, by Lemma I, $\pi_{\alpha_1\ldots\alpha_k} = \prod_{\lambda} \pi^{\lambda}_{\alpha_1\ldots\alpha_k} \neq 0$, and $\pi_{\alpha_1\ldots\alpha_k} \subseteq \prod_{\lambda} V^{\lambda}_{\beta}$. Now consider any finite set of varieties $\pi_{\alpha_1\ldots\alpha_k}$ (subject to the condition that the α_i which appear in the suffixes of any of the varieties precede β), and let $\gamma_1, \ldots, \gamma_t$ be the aggregate of suffixes involved. Then, clearly, $\pi_{\gamma_1\ldots\gamma_t} \subseteq \pi_{\alpha_1\ldots\alpha_k}$, and therefore the intersection of the set of $\pi_{\alpha_1\ldots\alpha_k}$ is not zero. Hence, again using Lemma I, the intersection $\prod_{\alpha_1\ldots\alpha_k} \pi_{\alpha_1\ldots\alpha_k}$, where $\alpha_1, \ldots, \alpha_k$ ranges over all finite sets of α which precede β, is not empty. If W_{β} is any zero-dimensional variety contained in $\Pi\pi_{\alpha_1\ldots\alpha_k}$, it follows that the varieties W_{α} $(\alpha \leqslant \beta)$ satisfy the requirements (i) and (ii) and hence that we can find W_{α} for every α to fulfil those conditions.

We now consider the quotient rings of the varieties W_{α}. These are all contained in the function field Σ. Let $\mathfrak{R}$ be the join of these rings, that is, the set of elements of Σ which can be written as finite sums $\xi_{\alpha_1} + \ldots + \xi_{\alpha_k}$, where $\xi_{\alpha_i} \in Q(W_{\alpha_i})$. Let $\xi = \xi_{\alpha_1} + \ldots + \xi_{\alpha_k}$, where $\xi_{\alpha_i} \in Q(W_{\alpha_i})$, and $\eta = \eta_{\beta_1} + \ldots + \eta_{\beta_s}$, where $\eta_{\beta_j} \in Q(W_{\beta_j})$, be two elements of $\mathfrak{R}$. Then, clearly, $\xi + \eta \in \mathfrak{R}$. We show that $\xi\eta$ also belongs to $\mathfrak{R}$. Let $U_{\gamma_{ij}}$ be the join of U_{α_i} and U_{β_j}. Since there exists a valuation of Σ whose centre on $U_{\gamma_{ij}}$ is $W_{\gamma_{ij}}$ and whose centre on U_{α_i} is W_{α_i}, W_{α_i} and $W_{\gamma_{ij}}$ correspond in the natural birational correspondence between U_{α_i} and $U_{\gamma_{ij}}$; and since $U_{\gamma_{ij}}$ is the join of U_{α_i} and U_{β}, there is only one variety on U_{α_i} corresponding to $W_{\gamma_{ij}}$. It follows that $Q(W_{\alpha_i}) \subseteq Q(W_{\gamma_{ij}})$ [§ 1, p. 227]. Similarly $Q(W_{\beta_i}) \subseteq Q(W_{\gamma_{ij}})$. Hence $\xi_{\alpha_i}\eta_{\beta_j} \subseteq Q(W_{\gamma_{ij}})$. Hence $\xi\eta = \sum_{i=1}^{k} \sum_{j=1}^{s} \xi_{\alpha_i}\eta_{\beta_j}$ belongs to $\mathfrak{R}$, which is therefore a ring. It is clear that $\mathfrak{R}$ is an integral domain whose quotient field is Σ. We prove that *if ω_{α} is any non-unit of $Q(W_{\alpha})$, ω_{α} is a non-unit of $\mathfrak{R}$.*

Suppose that ω_{α} is a unit of $\mathfrak{R}$. Then

$$\omega_{\alpha}^{-1} = \xi_{\alpha_1} + \ldots + \xi_{\alpha_k},$$

where ξ_{α_i} is an element of $Q(W_{\alpha_i})$. Let U_{β} be the join of the varieties $U_{\alpha}, U_{\alpha_1}, \ldots, U_{\alpha_k}$. Then, by the fundamental property of the varieties W_{α}, there exists a valuation $\mathfrak{B}$ whose centres on U_{α}, U_{α_i} $(i = 1, \ldots, k)$,

U_β are, respectively, W_α, W_{α_i} ($i = 1, \ldots, k$), W_β. Since U_β is the join of U_α and $U_{\alpha_i}, \ldots, U_{\alpha_k}$, it follows that $W_\alpha, W_{\alpha_1}, \ldots, W_{\alpha_k}$ are the unique varieties corresponding to W_β, and

$$Q(W_\alpha) \subseteq Q(W_\beta), \quad Q(W_{\alpha_i}) \subseteq Q(W_\beta).$$

Hence ω_α is in $Q(W_\beta)$ and $\omega_\alpha^{-1} \subseteq Q(W_\beta)$. Therefore ω_α is a unit of $Q(W_\beta)$. Its value in $\mathfrak{B}$ is therefore zero. The centre of $\mathfrak{B}$ on U_α is W_α. But the properties $\omega_\alpha \in Q(W_\alpha)$, $v(\omega_\alpha) = 0$ imply that ω_α is a unit of $Q(W_\alpha)$, and we have a contradiction. Hence ω_α must be a non-unit of $\mathfrak{R}$.

A particular consequence of this result is that $\mathfrak{R}$ is a proper subring of Σ. Let ζ be any non-zero element of Σ, and let U_α be any normal projective model of Σ. We choose non-homogeneous coordinates in the space containing U_α so that W_α is not in $x_0 = 0$. Then, if $(1, \xi_1, \ldots, \xi_n)$ is a generic point of U_α,

$$K[\xi_1, \ldots, \xi_n] \subseteq Q(W_\alpha).$$

Let U_β be the projective model of Σ having the generic point $(1, \xi_1, \ldots, \xi_n, \zeta)$, and let $\mathfrak{B}$ be a valuation whose centre on U_α is W_α, and whose centre on U_β is W_β. Then $v(\xi_i) \geqslant 0$. If $v(\zeta) \geqslant 0$, we have $K[\xi_1, \ldots, \xi_n, \zeta] \subseteq Q(W_\beta)$, and hence $\zeta \in Q(W_\beta)$. If $v(\zeta) < 0$, we have $K[\zeta^{-1}, \zeta^{-1}\xi_1, \ldots, \zeta^{-1}\xi_n] \subseteq Q(W_\beta)$ and hence $\zeta^{-1} \in Q(W_\beta)$. Thus if ζ is any non-zero element of Σ, either ζ or ζ^{-1} is in $\mathfrak{R}$. *R is therefore a valuation ring* [XVII, § 2, Th. III].

Let $\mathfrak{B}$ be the valuation of Σ whose valuation ring is $\mathfrak{R}$. It is a non-trivial valuation, since its valuation ring does not coincide with Σ. Let us suppose that its dimension s is positive. Then there exists a projective model U_α of Σ on which the centre V of $\mathfrak{B}$ is of dimension s. Since $Q(W_\alpha) \subseteq \mathfrak{R}$, we have $W_\alpha \subseteq V$ [XVII, § 5, Th. IV]. W_α is of dimension zero, and V is of dimension greater than zero; hence there are non-units of $Q(W_\alpha)$ which are units of $Q(V)$. But any unit of $Q(V)$ is a unit of $\mathfrak{R}$ [XVII, § 5, Th. IV]; hence there are non-units of $Q(W_\alpha)$ which are units of $\mathfrak{R}$. This, however, contradicts an earlier property of $\mathfrak{R}$; hence our assumption that $\mathfrak{B}$ is of dimension greater than zero is false.

Finally, we prove that $\mathfrak{B}$ belongs to each set M_λ of valuations. Let U_α be a projective model of Σ on which there exists a variety V_α with the property that M_λ is the set of all zero-dimensional valuations of Σ whose centres on U_α lie on V_α. U_α exists because M_λ is an algebraic subset of M. Then $W_\alpha \subseteq V_\alpha$. Since $Q(W_\alpha) \subseteq \mathfrak{R}$, W_α is contained in the centre of $\mathfrak{B}$ on U_α, and therefore, since $\mathfrak{B}$ is zero-dimensional, must coincide with it. The centre of $\mathfrak{B}$ on U_α is contained in V_α; hence $\mathfrak{B}$ is contained in M_λ. Since this result holds for all λ, we have $\prod_\lambda M_\lambda \neq 0$, and the proof of Lemma III is complete.

The proof of Theorem I follows quickly from Lemma III. Let $\{U_\alpha\}$ again denote the set of all projective models of Σ, and now define M_α to be the set of all zero-dimensional valuations of Σ whose centres are at singular loci on U_α. We prove that $\prod_\alpha M_\alpha = 0$. If this is not so, there is a valuation $\mathfrak{B}$ contained in each M_α. Its centre on each projective model is therefore singular. This, however, contradicts the Local Uniformisation Theorem, which tells us that there exists a projective model of Σ on which the centre of $\mathfrak{B}$ is simple. Hence we must have $\prod_\alpha M_\alpha = 0$.

From Lemma III, it follows that there exists a finite set $\alpha_1, \ldots, \alpha_k$ of suffixes α such that $\prod_{i=1}^{k} M_{\alpha_i} = 0$. We consider the corresponding projective models $U_{\alpha_1}, \ldots, U_{\alpha_k}$. Let $\mathfrak{B}$ be any zero-dimensional valuation of Σ. Its centre on U_{α_i} must be simple, for some value of i, for otherwise $\mathfrak{B} \subseteq M_{\alpha_i}$ $(i = 1, \ldots, k)$, and hence $\prod_{i=1}^{k} M_{\alpha_i} \neq 0$. Thus $U_{\alpha_1}, \ldots, U_{\alpha_k}$ is a finite resolving system of Σ, and Theorem I is proved.

9. The reduction of the singularities of an algebraic variety. We have already pointed out that the problem of finding a birational transform of an algebraic variety U which is free from singular points is the same as that of finding a resolving system for the function field of U consisting of only one member. When the dimension of U is one, the existence of such a resolving system has already been established; any derived normal variety of U is a projective model of its function field without singularities [XVI, § 6, Th. V]. When dim $U > 1$, the projectively normal varieties birationally equivalent to U may have singular points, and the existence of a non-singular projective model has to be proved by other means. The obvious line of approach is to consider a finite resolving system $U_1, \ldots, U_k$ for the function field of U, and, if $k > 1$, to try to replace this by a resolving system with $k-1$ members, repeating the operation until we get a resolving system of one model. In this section we shall show how this can be done when U is of dimension two. The reader will observe that we have to use the condition $\dim U = 2$ on several occasions during the argument; and it is a matter of great difficulty to extend the argument to varieties of dimension greater than two. We may (and shall) assume that the varieties of the resolving system $U_1, \ldots, U_k$ are projectively normal [§ 7, p. 315]. Now any subvariety of dimension $d-1$ on a normal variety of dimension d is simple [XVI, § 6, Th. IV]; hence in our present case, $d = 2$, the singular varieties on $U_1, \ldots, U_k$ are all zero-dimensional.

Our object is to find a variety U^*_{k-1} such that $U_1, \ldots, U_{k-2}, U^*_{k-1}$ form a resolving system for Σ. The considerations which enable us to prove the existence of a variety U^*_{k-1} with this property deal only with U_{k-1} and U_k. Let N be the set of zero-dimensional valuations of Σ whose centres are simple on one at least of U_{k-1} and U_k. U_{k-1} and U_k form a resolving system for N. If we construct a variety U^*_{k-1} such that any valuation of N has simple centre on U^*_{k-1}, then, $U_1, \ldots, U_{k-2}, U^*_{k-1}$ is a resolving system for Σ; for if $\mathfrak{B}$ is any zero-dimensional valuation of Σ either (i) $\mathfrak{B}$ belongs to N, and hence has simple centre on U^*_{k-1}, or (ii) $\mathfrak{B}$ does not belong to N and so has centre at a singular variety both on U_{k-1} and U_k, in which case it must have simple centre on one of the varieties $U_1, \ldots, U_{k-2}$. Our problem is thus reduced to showing that if U, U' are two two-dimensional varieties forming a resolving system for a set N of zero-dimensional valuations, then there exists a resolving system for N consisting of a single variety.

In the course of establishing this result we shall have to perform a series of birational transformations on U and U'. A large number of these will consist of a monoidal transformation with zero-dimensional base, followed by the operation of forming a derived normal variety. For brevity we shall describe this combined operation as a *quadratic transformation*. We begin with a series of lemmas on quadratic transformations.

Let $V_1, V_2, V_3, \ldots$ be an infinite sequence of birationally equivalent varieties of dimension two, and suppose that V_{i+1} is obtained from V_i by a quadratic transformation whose base on V_i is W_i, and that W_{i+1} corresponds to W_i in the birational correspondence between V_{i+1} and V_i. The sequence $(V_1, W_1), (V_2, W_2), \ldots$ defined in this way is called a *normal sequence*. Our lemmas concern the normal sequence $(V_1, W_1), (V_2, W_2), \ldots$.

We recall that, in the correspondence between V_i and V_{i+1}, W_i is fundamental, while there are no fundamental varieties on V_{i+1}. Hence, since W_i and W_{i+1} correspond, we must have

$$Q(W_i) \subset Q(W_{i+1}),$$

and so
$$Q(W_1) \subset Q(W_2) \subset \ldots.$$

We shall denote the ideal of non-units of $Q(W_i)$ by $\mathfrak{P}_i$.

The rings $Q(W_i)$ have all the same quotient field, namely, the common function field Σ of $V_1, V_2, \ldots$. Let α and β be two elements of Σ such that $\alpha \in Q(W_i)$ for some integer i, and $\beta \in Q(W_j)$ for some integer j. If $k = \max[i,j]$, α and β are contained in $Q(W_k)$, and hence $\alpha+\beta$ and $\alpha\beta$ are contained in $Q(W_k)$. It follows that the elements of Σ which are contained in $Q(W_i)$ for some value of i form a ring Ω. We shall usually write $\Omega = \lim_{i\to\infty} Q(W_i)$. It is clear that Σ is the quotient field of Ω, and that any non-zero element of the ground field K is a unit of Ω.

Lemma I. Let $\mathfrak{B}$ be a one-dimensional valuation of Σ whose valuation ring $\mathfrak{R}$ contains Ω. Then there exists an integer k such that, if $i \geqslant k$, then the centre of $\mathfrak{B}$ on V_i is one-dimensional.

We first observe that if we can find k such that the centre of $\mathfrak{B}$ on V_k is one-dimensional, the rest follows. Indeed, if the centre W of $\mathfrak{B}$ on V_i is one-dimensional, W is not fundamental for the transformation from V_i to V_{i+1}, and hence it has a unique transform W'. Since there are no fundamental varieties on V_{i+1}, it follows that W' is one-dimensional. But W' is the centre of $\mathfrak{B}$ on V_{i+1}, and the result follows by induction.

On the other hand, if the centre of $\mathfrak{B}$ on V_i is zero-dimensional it is necessarily W_i. For $Q(W_i) \subseteq \Omega \subseteq \mathfrak{R}$, hence W_i is contained in the centre of $\mathfrak{B}$ on V_i. Since this centre is of dimension zero and irreducible, it must coincide with W_i.

If the centre of $\mathfrak{B}$ on V_1 is one-dimensional, there is nothing to prove. Let us suppose that it is zero-dimensional. Since $\mathfrak{B}$ is one-dimensional, there exists an element α of the valuation ring $\mathfrak{R}$ whose residue in $\mathfrak{B}$ is transcendental over K. We can write $\alpha = \xi_1/\eta_1$, where ξ_1 and η_1 are in $Q(W_1)$. Now since W_1 is the centre of $\mathfrak{B}$, the residue of any element of $Q(W_1)$ is obtained by replacing the coordinates of the generic point of V_1 by the coordinates of the generic point of W_1, and hence the residue of any element of $Q(W_1)$ is algebraic over K. Since the residue of α is transcendental, we must have $\eta_1 = 0\ (\mathfrak{P}_1)$. Since the residue of α is not zero, η_1/ξ_1 is likewise an element of $\mathfrak{R}$ whose residue is transcendental, and we obtain $\xi_1 = 0\ (\mathfrak{P}_1)$.

Now consider the quadratic transformation between V_1 and V_2. We recall [§ 3] that we introduce a set of forms $\psi_0^{(1)}(x), \ldots, \psi_r^{(1)}(x)$ of like degree such that $\psi_0^{(1)}(\xi), \ldots, \psi_r^{(1)}(\xi)$ form a basis for the ideal $\mathfrak{P}_1$. We then consider the variety V_1' whose generic point is $(\zeta_{00}, \ldots, \zeta_{rn})$, where $\zeta_{ij} = \psi_i^{(1)}\xi_j$. Since V_2 is obtained from V_1' by forming the derived normal variety, there corresponds to W_2 a unique variety W_1' on V_1'. We may suppose that z_{00} is not zero on W_1'. We then have

$$Q(W_1) \subset Q(W_1') \subseteq Q(W_2),$$

$$\frac{\psi_i^{(1)}}{\psi_0^{(1)}} \in Q(W_1') \subseteq Q(W_2) \quad (i = 1, \ldots, r).$$

If ζ_1 is any element of $\mathfrak{P}_1$, we have

$$\begin{aligned} \zeta_1 &= \lambda_0 \psi_0^{(1)} + \ldots + \lambda_r \psi_r^{(1)} \quad (\lambda_i \in Q(W_1)) \\ &= \psi_0^{(1)} \sum_{i=0}^{r} \lambda_i \psi_i^{(1)}/\psi_0^{(1)} \\ &= \psi_0^{(1)} \zeta_2, \end{aligned}$$

where ζ_2 is in $Q(W_2)$. Let $\mathfrak{B}'$ be any valuation whose centre on V_1 is W_1 and whose centre on V_2 is W_2. Since $\psi_0^{(1)} = 0\ (\mathfrak{P}_1)$, $v'(\psi_0^{(1)}) > 0$, and hence $\psi_0^{(1)} = 0\ (\mathfrak{P}_2)$.

We now consider the centre of $\mathfrak{B}$ on V_2. If this is one-dimensional, we have found the required value of k. If not, the centre of $\mathfrak{B}$ is W_2, and hence, since $\psi_0^{(1)} = 0\ (\mathfrak{P}_2)$, $v(\psi_0^{(1)}) > 0$. We then have, by the remark just made,

$$\xi_1 = \psi_0^{(1)}\xi_2, \quad \eta_1 = \psi_0^{(1)}\eta_2,$$

and

$$\alpha = \xi_2/\eta_2,$$

where

$$\xi_2 = 0\ (\mathfrak{P}_2), \quad \eta_2 = 0\ (\mathfrak{P}_2),$$

since the residue of α in $\mathfrak{B}$ is transcendental over K. Now since $\mathfrak{B}$ is one-dimensional and Σ is of dimension two, $\mathfrak{B}$ is a discrete valuation. Taking its value group to be the group of integers, the fact that $v(\psi_0^{(1)}) > 0$ implies that

$$v(\eta_1) \geqslant v(\eta_2) + 1,$$

and since $\eta_2 = 0\ (\mathfrak{P}_2)$, we have $v(\eta_2) > 0$.

We repeat the argument until we reach a V_k on which the centre of $\mathfrak{B}$ is one-dimensional. If at the ith stage we have not reached this situation, we have $\alpha = \xi_i/\eta_i$,

$$v(\eta_1) \geqslant v(\eta_i) + i - 1, \quad v(\eta_i) > 0.$$

But $v(\eta_1)$ is a fixed finite integer. Hence we cannot carry out the process indefinitely; that is, we reach a V_k on which the centre of $\mathfrak{B}$ is one-dimensional. Lemma I is thus proved.

Lemma II. If $\mathfrak{B}_1$ is a one-dimensional valuation whose valuation ring $\mathfrak{R}_1$ contains Ω, there exists a zero-dimensional valuation ring $\mathfrak{B}_0$, whose valuation ring $\mathfrak{R}_0$ satisfies the relation $\mathfrak{R}_1 \supseteq \mathfrak{R}_0 \supseteq \Omega$.

Since $\mathfrak{R}_1$ contains Ω, it follows from Lemma I that there exists an integer k such that, if $i \geqslant k$, the centre C_i of $\mathfrak{B}_1$ on V_i is one-dimensional. For values of i not less than k we define the ideal $\mathfrak{p}_i$ to be the intersection of $Q(W_i)$ with the ideal of non-units of $\mathfrak{R}_1$. (If we pass from $Q(W_i)$ to the integral domain $\mathfrak{J}_i$ of V_i, $\mathfrak{J}_i \wedge \mathfrak{p}_i$ is the prime ideal of the centre of $\mathfrak{B}_1$ on V_i.) $Q(W_i)$ is integrally closed, and the Basis Theorem holds for it; hence [XV, § 8, Th. III] we can apply the multiplicative theory of ideals to ideals in $Q(W_i)$; and since $\mathfrak{p}_i$ is one-dimensional it is a minimal prime ideal, and hence equal to its own reduced ideal [XV, § 7, Th. VI].

Let α be any element of Σ whose value in $\mathfrak{B}_1$ is zero. We can write $\alpha = \xi/\eta$, where ξ and η belong to $Q(W_i)$, and from the multiplicative theory of ideals we can write $\xi = \xi_1\zeta^a$, $\eta = \eta_1\zeta^b$, where $\zeta = 0 \ (\mathfrak{p}_i)$, $\xi_1 \neq 0 \ (\mathfrak{p}_i)$, $\eta_1 \neq 0 \ (\mathfrak{p}_i)$. Since $v(\alpha) = 0$, $v(\zeta) > 0$, $a = b$, and hence $\alpha = \xi_1/\eta_1$ [cf. XVII, § 5, Th. VII]. The residue of α in $\mathfrak{B}_1$ is therefore equal to the ratio of the images of ξ_1 and η_1 in $Q(W_i)/\mathfrak{p}_i$. The residue field Σ^* of $\mathfrak{B}_1$ is therefore equal to the quotient field of $Q(W_i)/\mathfrak{p}_i$, which is the function field of the centre C_i of $\mathfrak{B}_1$ on V_i. Σ^* can be written as the quotient field of a finite integral domain $\mathfrak{J}^*$, that is, an integral domain obtained by adjoining a finite number of elements to K, and $Q(W_i)/\mathfrak{p}_i$ is a quotient ring in this. In fact, it is the quotient ring of W_i, regarded as a subvariety of C_i. There is always at least one, and at most a finite number, of valuations of Σ^* whose centre on C_i is W_i, that is, whose valuation rings contain $Q(W_i)/\mathfrak{p}_i$. Since

$$Q(W_i) \subset Q(W_{i+1}), \quad \mathfrak{p}_i = Q(W_i) \wedge \mathfrak{p}_{i+1},$$

$Q(W_i)/\mathfrak{p}_i$ can be regarded as a subring of $Q(W_{i+1})/\mathfrak{p}_{i+1}$. Any valuation of Σ^* whose valuation ring contains $Q(W_{i+1})/\mathfrak{p}_{i+1}$ contains $Q(W_i)/\mathfrak{p}_i$, and since a valuation ring containing $Q(W_j)/\mathfrak{p}_j$ exists for any $j \geq k$, we conclude that there is a valuation $\mathfrak{V}^*$ of Σ^* whose valuation ring contains each $Q(W_i)/\mathfrak{p}_i$ (all i). Let $\mathfrak{V}_0$ be the valuation of Σ obtained by compounding $\mathfrak{V}_1$ and $\mathfrak{V}^*$. It is zero-dimensional, and its valuation ring $\mathfrak{R}_0$ is contained in $\mathfrak{R}_1$. Further, since $Q(W_i)/\mathfrak{p}_i$ is contained in the valuation ring of $\mathfrak{V}^*$, $Q(W_i) \subseteq \mathfrak{R}_0$ for all i. Hence $\Omega \subseteq \mathfrak{R}_0$, and the valuation $\mathfrak{V}_0$ satisfies all the requirements of the lemma.

Lemma III. $\Omega = \lim_{i \to \infty} Q(W_i)$ *is a zero-dimensional valuation ring.*

Since each V_i is normal, each $Q(W_i)$ is integrally closed. Hence Ω is integrally closed. It is therefore the intersection of the valuation rings of Σ which contain it [XVII, § 2, Th. VI]; by Lemma II we can go further and say that Ω is the intersection of the valuation rings of zero-dimensional valuations which contain it. To prove the lemma above, it is sufficient to show that there cannot exist two distinct zero-dimensional valuations $\mathfrak{V}'$ and $\mathfrak{V}''$ whose valuation rings contain Ω.

Suppose there were two such valuations $\mathfrak{V}'$ and $\mathfrak{V}''$. Since they are distinct their valuation rings are distinct, and we can find an element ζ which is a unit of the one and a non-unit of the other. If $(1, \xi_1, \ldots, \xi_n)$ is a generic point of V_1, the variety V^* in S_{n+1} whose generic point is $(1, \xi_1, \ldots, \xi_n, \zeta)$ is birationally equivalent to $V_1, V_2, \ldots$ and on it $\mathfrak{V}'$ and $\mathfrak{V}''$ have different centres, say C' and C'', since $v'(\zeta) = 0$, $v''(\zeta) > 0$, or $v'(\zeta) > 0$, $v''(\zeta) = 0$.

Since $\mathfrak{V}'$ is zero-dimensional, and its valuation ring contains $Q(W_i)$, W_i is the centre of $\mathfrak{V}'$ on V_i. Similarly, it is the centre of $\mathfrak{V}''$ on V_i. Hence, in the birational correspondence between V_i and V^*, both C' and C'' correspond to W_i, which is therefore fundamental for the correspondence. Hence, corresponding to W_i there is a curve Γ_i on V^*. Let Γ_{i+1}^j be any irreducible component of Γ_{i+1}. There is a one-dimensional valuation $\mathfrak{V}_1$ of Σ whose centre on V^* is Γ_{i+1}^j, and whose centre on V_{i+1} is W_{i+1}. Since $Q(W_i) \subset Q(W_{i+1})$, W_i is the only variety on V_i which corresponds to W_{i+1}, and hence the centre of $\mathfrak{V}_1$ on V_i is W_i. Thus $\Gamma_{i+1} \subseteq \Gamma_i$, and since Γ_i exists for all i, there exists an irreducible variety Γ of dimension one on V^* which is the centre of a valuation $\mathfrak{V}$ of Σ on V^*, and whose centre on V_i is W_i. Let $\mathfrak{R}$ be the valuation ring of $\mathfrak{V}$. Then $Q(W_i) \subseteq \mathfrak{R}$ $(i = 1, 2, \ldots)$; therefore $\Omega \subseteq \mathfrak{R}$. Hence, by Lemma I, there is a value of k such that the centre of $\mathfrak{V}$ on V_k is one-dimensional. But the centre of $\mathfrak{V}$ is

W_k, and we have a contradiction. Our assumption that $\mathfrak{B}' \neq \mathfrak{B}''$ is therefore false, and the lemma follows.

Lemma IV. If $\mathfrak{B}$ is the unique valuation of Σ whose valuation ring is Ω, and if W' is the centre of $\mathfrak{B}$ on any projective model V' of Σ, then there exists an integer k such that $Q(W') \subseteq Q(W_i)$ $(i \geqslant k)$.

Since $Q(W_i) \subset Q(W_{i+1})$, and every non-unit of $Q(W_i)$ has positive values in $\mathfrak{B}$, and is therefore in $\mathfrak{P}_{i+1}$, $\mathfrak{P}_i \subseteq \mathfrak{P}_{i+1}$, and hence

$$Q(W')_\wedge \mathfrak{P}_i \subseteq Q(W')_\wedge \mathfrak{P}_{i+1}.$$

Let ζ be any element of Σ which lies in $Q(W')_\wedge \mathfrak{P}_i$, for some value of i. Since ζ is in $\mathfrak{P}_i$, its value in $\mathfrak{B}$ is positive. It also lies on $Q(W')$, and since its value in $\mathfrak{B}$ is positive, it must therefore lie in $\mathfrak{P}'$, the ideal of the non-units of $Q(W')$. On the other hand, if ζ is any element of $\mathfrak{P}'$, $v(\zeta) > 0$. Hence ζ is a non-unit of Ω. It therefore lies in $Q(W_i)$ for some value of i, and is a non-unit of this ring. Thus it lies in $\mathfrak{P}_i$. Since it is contained in $\mathfrak{P}'$, it is contained in $Q(W')$, and hence in $Q(W')_\wedge \mathfrak{P}_i$ for some value of i. From these two results we have

$$\lim_{i\to\infty} Q(W')_\wedge \mathfrak{P}_i = \mathfrak{P}'.$$

Now since $Q(W')$ is the quotient ring of an algebraic variety on V', the Basis Theorem holds in it. But in $Q(W')$,

$$Q(W')_\wedge \mathfrak{P}_1 \subseteq Q(W')_\wedge \mathfrak{P}_2 \subseteq Q(W')_\wedge \mathfrak{P}_3 \subseteq \dots$$

is an increasing sequence of ideals. Hence there exists an integer k such that

$$Q(W')_\wedge \mathfrak{P}_k = Q(W')_\wedge \mathfrak{P}_{k+1} = Q(W')_\wedge \mathfrak{P}_{k+2} = \dots,$$

and hence $$\lim_{i\to\infty} Q(W')_\wedge \mathfrak{P}_i = Q(W')_\wedge \mathfrak{P}_k.$$

Hence $$Q(W')_\wedge \mathfrak{P}_k = \mathfrak{P}',$$

and therefore $$\mathfrak{P}' \subseteq \mathfrak{P}_k.$$

Now consider any zero-dimensional valuation $\mathfrak{B}_0$ of Σ whose centre on V_k is W_k. Since $\mathfrak{P}' \subseteq \mathfrak{P}_k$, its centre on V' is necessarily W'. Hence W' must be the only variety on V' which corresponds to W_k on V_k. Hence W_k is not fundamental in the birational correspondence between V_k and V'; therefore $Q(W') \subseteq Q(W_k)$.

With the aid of these lemmas we can now prove

THEOREM I. *If U, U' is a resolving system for a set N of zero-dimensional valuations of Σ, which is of dimension two, there exists* *a projective model U^* of Σ on which the centre of every valuation of N is simple.*

U and U' may be taken as normal surfaces; the only singular subvarieties of these are necessarily zero-dimensional. They are in birational correspondence, and in this correspondence there may be fundamental elements on either surface, or on both, but these fundamental elements are necessarily zero-dimensional and hence finite in number. Suppose that there are fundamental elements on U. Taking one of these, W say, as basis, we perform a quadratic transformation which transforms U into U_1. Since the quadratic transformation does not alter the quotient ring of any subvariety of U other than W, the fundamental varieties on U_1 for the birational correspondence between U_1 and U' are the varieties on U_1 arising from the fundamental varieties on U which are different from W, together, possibly, with zero-dimensional subvarieties of the transform on U_1 of W. If there are fundamental varieties on U_1, we repeat the process, and go on until we obtain a variety $\bar{U}$ such that in the birational correspondence between $\bar{U}$ and U' there are no fundamental varieties on $\bar{U}$. It is necessary to prove that we can reach the variety $\bar{U}$ after a finite number of steps. It is clear that if this is not the case we can find a fundamental variety W on U such that (U, W) is the beginning of a normal sequence

$$(U, W), \quad (U_1, W_1), \quad (U_2, W_2), \quad \ldots,$$

where W_i is fundamental in the correspondence between U_i and U'. Let $\Omega = \lim_{i \to \infty} Q(W_i)$, and let $\mathfrak{B}$ be the zero-dimensional valuation of Σ whose valuation ring is Ω [Lemma III]. If W' is the centre of $\mathfrak{B}$ on U', we see from Lemma IV that there exists an integer k such that $Q(W') \subseteq Q(W_k)$. From this it follows that W_k is not fundamental for the birational correspondence between U_k and U', and we thus have a contradiction. Hence we reach $\bar{U}$ after a finite number of steps.

A quadratic transformation with a zero-dimensional basis transforms any simple variety, whether fundamental or not, into a variety of which every point is simple. Hence if $\mathfrak{B}$ is any valuation whose centre on U is simple, the centre of $\mathfrak{B}$ on $\bar{U}$ is also simple. It follows that $\bar{U}$ and U' constitute a resolving system for N.

We next operate on U' as we operated on U, but confining ourselves to varieties on U' which are fundamental for the correspondence between U' and $\bar{U}$, *and are also simple for* U'. By a finite number of quadratic transformations, we can transform U' into a variety $\bar{U}'$ on which there are no *simple* fundamental varieties for the correspondence between $\bar{U}'$ and $\bar{U}$. It should be noted, however, that for this correspondence there may be fundamental varieties on $\bar{U}$. Finally, we construct the join U^* of $\bar{U}$ and $\bar{U}'$.

We now prove that if $\mathfrak{B}$ is any valuation of N, its centre on U^* is simple. Let its centres on $\bar{U}$, U', $\bar{U}'$, U^* be, respectively, $\bar{W}$, W', $\bar{W}'$, W^*. We have to consider three possible cases.

(i) Suppose that W' is singular on U'. Since $\bar{U}$, U' is a resolving system for N, the centre of $\mathfrak{B}$ on $\bar{U}$, that is, $\bar{W}$, must be simple on $\bar{U}$. Since there are no fundamental varieties on $\bar{U}$ for the correspondence between $\bar{U}$ and U',

$$Q(W') \subseteq Q(\bar{W}).$$

Since W' is singular on U', it is not affected by the transformation from U' to $\bar{U}'$; hence

$$Q(\bar{W}') = Q(W') \subseteq Q(\bar{W}),$$

and from the properties of the join we have

$$Q(W^*) = Q(\bar{W}).$$

Since $\bar{W}$ is simple, the ideal of non-units of $Q(\bar{W})$ (and hence of $Q(W^*)$) has a basis consisting of two elements. Hence W^* is simple on U^*.

(ii) Suppose that W' is simple, but not fundamental for the correspondence between U' and $\bar{U}$. Then, since $\bar{W}$ is also not fundamental, $Q(\bar{W}) = Q(W')$.

Since W' is not fundamental it is not a basis for any of the quadratic transformations used to pass from U' to $\bar{U}'$. Hence

$$Q(\bar{W}) = Q(W') = Q(\bar{W}').$$

Passing to the join U^*, we have, as before,

$$Q(W^*) = Q(\bar{W}') = Q(W'),$$

and since W' is simple, so is W^*.

(iii) Suppose that W' is simple, and that it is fundamental for the correspondence between U' and $\bar{U}$. Then, since W' is simple, and any variety on $\bar{U}'$ which corresponds to a simple variety on U' is simple (by the properties of quadratic transformations), $\bar{W}'$ is simple. Again, $\overline{W}'$ is not fundamental for the correspondence between $\overline{U}'$ and $\overline{U}$; hence

$$Q(\overline{W}) \subseteq Q(\overline{W}').$$

Passing to the join, we have

$$Q(W^*) = Q(\overline{W}'),$$

and hence W^* is simple.

Since we have exhausted all possibilities for W', we conclude that the centre on U^* of any valuation of N is simple. Hence Theorem I is proved.

We have already seen how Theorem I can be used to replace a finite resolving system for a two-dimensional field Σ, consisting of k projective models, by a resolving system consisting of $k-1$ models, and by repeated application of this result we can establish the existence of a resolving system formed by a single projective model. This then enables us to state the theorem:

THEOREM II. *Given any irreducible algebraic surface U, there exists a birational transform of it on which there are no singular points.*

In conclusion, we state, without proof, Zariski's result:

THEOREM III. *Given any irreducible algebraic variety of dimension three, there exists a birational transform of it on which there are no singular points.*

BIBLIOGRAPHICAL NOTES

BOOK V

Chapter XV. The material of this chapter is to be found in many of the standard works on Modern Algebra, such as Albert (**1**), Krull (**2**) and van der Waerden (**7**). The fullest account, that of Krull, contains references to the original sources of the main theorems. The presentation given here is designed to provide a sufficient introduction to ideal theory for the purposes of this volume, and makes no claim to originality or completeness.

Chapter XVI. The developments of this chapter follow most closely the work of Zariski, but are confined to the case of ground fields without characteristic. Many of the ideas are treated, in a somewhat different manner, and without restriction on the ground field, by Weil (**9**). The main results of the chapter will be found in Zariski (**10, 11**) and Muhly (**5**). In another paper (**12**) Zariski has examined in detail the notion of a simple point on a variety over any ground field.

Chapter XVII. The account of general valuation theory given here is designed for the purposes of geometry, and no reference is made to other branches of the subject, such as the applications to number theory. It is based mainly on Krull (**3**), though many of the earlier results are to be found in Krull (**2**), in which references to original sources are to be found. The section on the centre of a valuation is based mainly on Zariski (**13**). In this paper will be found the proof of Theorem XI of § 5. The properties of zero-dimensional valuations of rank one referred to on p. 207 are to be found in MacLane and Schilling (**4**).

Chapter XVIII. The first three sections of this chapter are based on Zariski (**13**). The theorem quoted on p. 240 is the 'Main Theorem' of this paper. Zariski has given (**14**) a simplified proof of the theorem, and a third proof, not using valuation theory and using the results of Weil (**9**), has been given by D. G. Northcott (**6**). The Local Uniformisation Theorem, which occupies §§ 5–7, follows closely the lines of Zariski's proof (**15**), the only existing proof of the theorem. The proof of the existence of a finite resolving system is based on Zariski (**16**). Zariski's proof is based on topological considerations, but it is here given in purely algebraic terms, the topological concept of bicompactness being replaced by the Basis Theorem of Hilbert. The proof of the reduction of the singularities of an algebraic surface is, essentially, Zariski's 'Simplified Proof' (**17**), but for completeness we give certain lemmas from his original arithmetic proof (**18**). The history of this theorem is long and involved, and is described in the first chapter of Zariski (**19**). The only other proof of the theorem (valid when the ground field is the field of complex numbers) which has withstood all criticism is that of Walker (**8**). The proof of the reduction of singularities of a V_3 is to be found in Zariski (**20**).

BIBLIOGRAPHY

(1) Albert, A. A. *Modern Higher Algebra* (Cambridge, 1938).
(2) Krull, W. *Idealtheorie* (Ergebnisse der Mathematik, iv, 3) (Berlin, 1935).
(3) Krull, W. *Journal für die reine und angewandte Mathematik*, **167** (1932), 160.
(4) MacLane, S. and Schilling, O. F. G. *Annals of Mathematics*, **40** (1939), 507.
(5) Muhly, H. T. *Annals of Mathematics*, **42** (1941), 921.
(6) Northcott, D. G. *Proceedings of London Mathematical Society*, (3), **1** (1951), 129.
(7) van der Waerden, B. L. *Moderne Algebra* (Berlin, 1930).
(8) Walker, R. J. *Annals of Mathematics*, **36** (1935), 336.
(9) Weil, A. *Foundations of Algebraic Geometry* (New York, 1946).
(10) Zariski, O. *American Journal of Mathematics*, **61** (1939), 249.
(11) Zariski, O. *American Journal of Mathematics*, **62** (1940), 187.
(12) Zariski, O. *Transactions of American Mathematical Society*, **62** (1947), 1.
(13) Zariski, O. *Transactions of American Mathematical Society*, **53** (1943), 490.
(14) Zariski, O. *Proceedings of National Academy of Sciences, U.S.A.*, **34** (1948), 62.
(15) Zariski, O. *Annals of Mathematics*, **41** (1940), 852.
(16) Zariski, O. *Bulletin of American Mathematical Society*, **50** (1944), 683.
(17) Zariski, O. *Annals of Mathematics*, **43** (1942), 583.
(18) Zariski, O. *Annals of Mathematics*, **40** (1939), 639.
(19) Zariski, O. *Algebraic Surfaces* (Ergebnisse der Mathematik, iii, 5) (Berlin, 1935).
(20) Zariski, O. *Annals of Mathematics*, **45** (1944), 472.

INDEX

(The numbers refer to the page where the term occurs for the first time or is defined)

Printed in the United States
By Bookmasters